Electric and Magnetic Fields

From Numerical Models to Industrial Applications

Electric and Magnetic Fields

From Numerical Models to Industrial Applications

Edited by

André Nicolet

University of Liège
Liège, Belgium

and

R. Belmans

Katholieke Universiteit Leuven
Leuven, Belgium

SPRINGER SCIENCE+BUSINESS MEDIA, LLC

Library of Congress Cataloging-in-Publication Data

Electric and magnetic fields : from numerical models to industrial applications / edited by André Nicolet and R. Belmans.
p. cm.
"Proceedings of the Second International Workshop on Electric and Magnetic Fields: From Numerical Models to Industrial Applications, held May 17-20, 1994, in Leuven, Belgium"--T.p. verso.
Includes bibliographical references and index.
ISBN 978-1-306-44991-8 ISBN 978-1-4615-1961-4 (eBook)
DOI 10.1007/978-1-4615-1961-4
1. Electric machinery--Congresses. 2. Electromagnetic fields--Mathematical models--Congresses. I. Nicolet, André. II. Belmans, R., 1956- . III. International Workshop on Electric and Magnetic Fields: From Numerical Models to Industrial Applications (2nd : 1994 : Louvain, Belgium)
TK2000.E37 1995
621.31'042--dc20 95-13990
CIP

Proceedings of the Second International Workshop on Electric and Magnetic Fields: From Numerical Models to Industrial Applications, held May 17–20, 1994, in Leuven, Belgium

ISBN 978-0-306-44991-8

Originally published by Plenum Press in 1995

10 9 8 7 6 5 4 3 2 1

PREFACE

This book contains the edited versions of the papers presented at the Second International Workshop on Electric and Magnetic Fields held at the Katholieke Universiteit van Leuven (Belgium) in May 1994.

This Workshop deals with numerical solutions of electromagnetic problems in real life applications. The topics include coupled problems (thermal, mechanical, electric circuits), CAD & CAM applications, 3D eddy current and high frequency problems, optimisation and application oriented numerical problems.

This workshop was organised jointly by the AIM (Association of Engineers graduated from de Montefiore Electrical Institute) together with the Departments of Electrical Engineering of the Katholieke Universiteit van Leuven (Prof. R. Belmans), the University of Gent (Prof. J. Melkebbek) and the University of Liège (Prof. W. Legros). These laboratories are working together in the framework of the *Pôle d'Attraction Interuniversitaire* - Inter-University-Attractie-Pole 51 - on electromagnetic systems led by the University of Liège and the research work they perform covers most of the topics of the Workshop.

One of the principal aims of this Workshop was to provide a bridge between the electromagnetic device designers, mainly industrialists, and the electromagnetic field computation developers. Therefore, this book contains a continuous spectrum of papers from application of electromagnetic models in industrial design to presentation of new theoretical developments.

In order to keep the publication delay reasonable and in agreement with the lively character of the Workshop, heavy reviewing procedures are avoided. The papers have been selected by the International Scientific Committee on the basis of a short version. The full paper is under the entire responsibility of their authors.

Special thanks are due to the Members of the Organizing Committee and International Scientific Committee for helping in promoting the meeting and ensuring the overall success of the event.

We hope that you will find in this book abundant, valuable and up-to-date information and that it will make you feel like attending the Third International Workshop on Electric and Magnetic Fields to be held in Liège, Belgium, in May 1996.

The Editors

CONTENTS

FIELD PROBLEMS IN ELECTROHEAT SYSTEMS

COUPLED PROBLEMS

NUMERICAL PROBLEMS

NONDESTRUCTIVE TESTING

HIGH FREQUENCY PROBLEMS

SYSTEM OPTIMISATION AND DESIGN

APPLICATIONS

A CHALLENGE FOR MAGNETIC SCALAR POTENTIAL FORMULATIONS OF 3-D EDDY CURRENT PROBLEMS: MULTIPLY CONNECTED CUTS IN MULTIPLY CONNECTED REGIONS WHICH NECESSARILY LEAVE THE CUT COMPLEMENT MULTIPLY CONNECTED

P.W. Gross and P.R. Kotiuga

Boston University
Department of Electrical, Computer, and Systems Engineering
44 Cummington St.
Boston, MA 02215

Abstract – Formulations for making cuts for the magnetic scalar potential in 3-dimensional finite element meshes often assume a priori that cuts should render the nonconducting region simply-connected in order to have a single-valued scalar potential. Starting with the Biot-Savart integral and its interpretation through the Gauss linking number, this paper develops a (co)homology-based definition of cuts which make the scalar potential single-valued but do not guarantee simple connectivity. The variational problem resulting from an algorithm to compute the cuts is then discussed and used to reinforce the central theme of linking. The assumption of simple connectivity is examined in light of the relationship between homotopy and homology to show that it is over-restrictive and moreover, not related to Ampère's Law.

1 INTRODUCTION

Suppose it is wished to calculate the magnetic field exterior to a current-carrying mass of wire, tangled and knotted like a ball of wool, in free space. In order to reduce computation time, many formulations of three-dimensional magnetoquasistatics problems in finite element meshes have sought to take advantage of the irrotational nature of the magnetic field in current-free regions by using a magnetic scalar potential [23, 24]. However, the magnetic scalar potential is multivalued because the current-free region contains closed paths which link nonzero current. In order to make the magnetic scalar potential single-valued, cuts, analogous to branch cuts in complex analysis, must be made in the current-free region such that closed paths link zero current. In three-dimensional eddy current problems, even relatively simple configurations may require quite complicated cuts [13] (Figure 1), let alone cuts for "current-carrying balls of

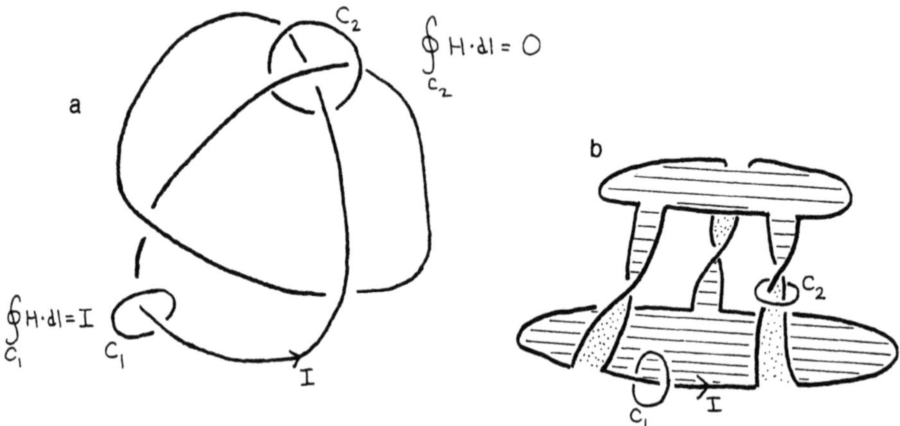

Figure 1. (a) Current-carrying "trefoil knot" and (b) the cut needed to make the scalar potential single-valued. Curve c_1 links current and intersects the cut while c_2 links zero current and does not intersect the cut. A slight deformation of the knot enables the cut to be illustrated as two disks joined by three twisted ribbons. The boundary of the surface is a curve knotted in the same way as that shown in (a).

wool". Hence, for 3-d finite element meshes one would like a general computational algorithm to perform the difficult task of topological reasoning and making cuts.

With few exceptions the cuts literature has centered on algorithmic solutions to making cuts in finite element meshes while largely ignoring the fundamental question of the definition of a cut. On one hand, some formulations are based on the assumption that a set of cuts $\{S_i\}$ must make the (multiply connected) current-free region V simply connected. That is, once the cuts are introduced, any closed path in $V - \bigcup_i S_i$ can be contracted to a point without leaving the region. On the other hand, requiring only that a closed path link zero current after the cuts are introduced is less restrictive but sufficient, though they may not render the region simply connected as illustrated in Figure 1. This has led to some controversy which we presently hope to clarify, but first the history of cuts is summarized in order to gain some perspective on the disagreement. An extensive review of the history of cuts for two-dimensional problems can be found in [1].

Over a century ago the problem of making cuts for the magnetic scalar potential in multiply connected regions was identified in Maxwell's *Treatise on Electricity and Magnetism* [28]. Maxwell gives some general intuition on the nature of cuts and scalar potentials and even points to classical forms of topological duality theorems required for a precise definition of cuts. However he also assumes that the cuts render the current-free region simply connected by saying that all paths should be "reconcilable" on $V - \bigcup_i S_i$. The same assumption is implied in Stratton [39] and made by Lamb [27] in the context of cuts for the velocity potential for irrotational fluid flow in multiply connected regions.

While Maxwell, Stratton, and others speak of multiply connected regions in terms of reconcilable and irreconcilable paths, Brown [7], in the first attempt to treat the three-dimensional cuts problem rigorously, correctly identifies the problem as one of linking zero current and appeals to the notion of homology. Assuming that the problem domain is "contained in a sufficiently large sphere," Brown succeeds in finding the boundaries of cuts and goes on to find spanning cut surfaces which are not embedded manifolds. Brown's assumption amounts to an overly restrictive topological constraint which can be avoided by systematic use of duality theorems discussed in Section 3.

The rules proposed by Vourdas and Binns [40] and the algorithm of Harrold and Simkin [17] are guided by the assumption that cuts should render multiply connected regions simply connected. This is acceptable in the context of planar problems, however in three-dimensional problems such intuition is not equivalent to that of linking zero current. The difference is expressed formally in terms of the relationship between homotopy and homology as discussed in Section 5. To recognize that this relationship is quite subtle one need only look as far as Poincaré who confused these two notions in his work on multivalued functions until realizing that the first homology group is actually an abelianized version of the first homotopy group [32]. The present controversy surrounding cuts [5, 26] goes back to the same problem which plagued Poincaré.

In a series of papers [20, 21, 23, 22], Kotiuga gives a precise definition of cuts in terms of (co)homology and develops an algorithm for their construction. Kotiuga's approach shows that for any three-dimensional finite element mesh, the absolute and relative (co)homology groups computed with integer coefficients are torsion-free and are therefore equivalent to the required groups. It is also shown that the cuts can be realized as orientable, embedded manifolds and that these cuts ensure that there is a single-valued scalar function in the current-free region where the cuts have been introduced. Moreover the cuts are related to a harmonic map from the current-free region to the circle whereby an algorithm for the computation of the cuts is given in [22] and its implementation is given in Murphy [29]. Murphy's implementation was used in [16] to demonstrate multiply connected cuts which leave the current-free region multiply connected while making the scalar potential single-valued. This paper discusses the conceptual background of [16].

Unfortunately, the degree of mathematical formalism used in several key papers [20, 21, 23, 22, 24], has made them somewhat unaccessible. On the other hand, such formalism naturally extends from physical considerations and can be made intuitive through the Biot-Savart law and its interpretation via linking numbers. As shown in Section 3 such an approach leads to homology, cohomology, and the duality theorems which form the basis of Kotiuga's algorithm.

The remainder of this paper is organized as follows. Section 2 reviews the Biot-Savart law and its relation to Gauss' linking number [14] which plays a central role in the cuts formalism. In Section 3 the mathematical formalism is developed to arrive at a definition for cuts which proceeds from a development of linking. Ampère's law is used to argue that cuts should prevent closed paths of integration from linking nonzero current. In Section 4 the variational aspects of Kotiuga's algorithm [22] are discussed in light of the definition described in Section 3. Throughout these discussions, the use of differential forms is avoided and cohomology is discussed in terms of vector fields. Only in the Appendix are differential forms used to discuss a formal view of the linking number. Finally, Section 5 employs the relationship between homotopy and homology to discuss the problems encountered if one insists on cuts that make $V - \bigcup_i S_i$ simply connected.

2 THE BIOT-SAVART LAW AND LINKING NUMBERS REVISITED

In order to simplify the following discussion on linking numbers, we will think of current flowing on a set of thin wires or curves. Such systems have an infinite amount of energy [31], however we can regard wires as tubular neighborhoods of curves. Thinking in these terms does not limit the generality of the arguments which hold for surface and volume distributions of current [21], and for general constitutive laws.

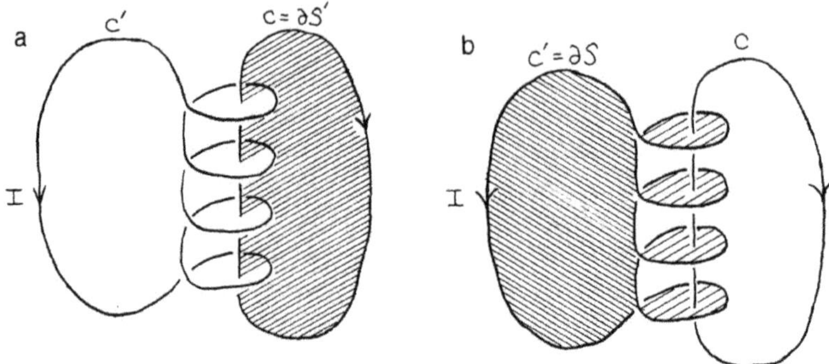

Figure 2. (a) Current on curve c' tranversely crossing a surface S'. Note that $Link(c, c') = 4$. (b) Configuration of (a) showing surface S. S (and S') can be deformed into a disk.

Consider two nonintersecting curves c and c' as shown in Figure 2. Curve c is the boundary of a surface S' ($c = \partial S'$), and a current I flows on curve c' which transversely crosses S'. For magnetoquasistatics, displacement current is assumed to be negligible such that Ampère's Law is

$$\oint_c \mathbf{H} \cdot d\mathbf{r} = \int_{S'} \mathbf{J} \cdot \hat{n}\, ds = I\, Link(c, c') \tag{1}$$

where $Link(c, c')$ is the oriented linking number of curves c and c', $\mathbf{H}$ is the "magnetic field intensity", and $\mathbf{J}$ is the conduction current density. Note that

$$\int_{S'} \mathbf{J} \cdot \hat{n}\, ds = I \quad \text{if } Link(c, c') = 1.$$

In a region where $\mathbf{J} = 0$ the vector field $\mathbf{H}$ is irrotational (curl $\mathbf{H} = 0$) and $\mathbf{H}$ may be expressed as the gradient of a magnetic scalar potential,

$$\mathbf{H}(\mathbf{r}) = -\nabla\psi \tag{2}$$

so that

$$\psi(p) - \psi(p_0) = -\int_{p_0}^{p} \mathbf{H} \cdot d\mathbf{r}. \tag{3}$$

The region in question is often multiply connected so that an integration path in the form of a closed loop c may link a current I. Hence the scalar potential is multivalued, picking up integral multiples of I depending on c. If c links the current n times, the value of the scalar potential at a point will have an added factor nI after traversing c (Figure 2).

In general the magnetic flux $\mathbf{B}$ can be expressed in terms of a vector potential $\mathbf{A}$, so that in linear, isotropic, homogeneous media we have

$$\nabla \times \mathbf{H} = \nabla \times (\frac{1}{\mu}\nabla \times \mathbf{A}) = \mathbf{J}. \tag{4}$$

Note, however, that the choice of constitutive law is not important since we are primarily concerned with topological issues. When the Coulomb gauge (div $\mathbf{A}$=0) is applied in

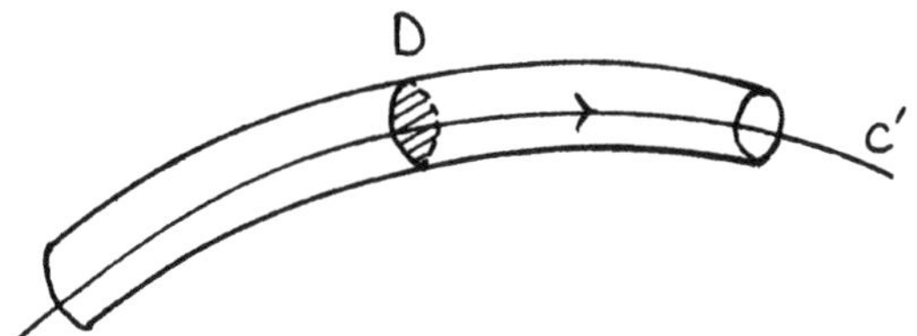

Figure 3. Tubular neighborhood of c'.

the magnetoquasistatic case, and $\mathbf{A}$ is expressed in terms of Cartesian coordinates, the components of $\mathbf{A}$ satisfy

$$\nabla^2 A_i = -\mu J_i \qquad (i = 1, 2, 3). \tag{5}$$

Then, for a vanishing vector potential as $\mathbf{r} \to \infty$ we have the Green's function solution

$$\mathbf{A}(\mathbf{r}) = \frac{\mu}{4\pi} \int\limits_{V'} \frac{\mathbf{J}(\mathbf{r}')}{|\mathbf{r} - \mathbf{r}'|} dV' \tag{6}$$

where $\mathbf{r}'$ is a source point and the integral is over the conducting region. Noting that

$$\nabla_{\mathbf{r}} \times \frac{\mathbf{J}(\mathbf{r}')}{|\mathbf{r} - \mathbf{r}'|} = \frac{\mathbf{J}(\mathbf{r}') \times (\mathbf{r} - \mathbf{r}')}{|\mathbf{r} - \mathbf{r}'|^3} \tag{7}$$

where the $\mathbf{r}$ in $\nabla_{\mathbf{r}}$ refers to differentiation with respect to unprimed variables, equations (4), (6), and (7) give:

$$\mathbf{H}(\mathbf{r}) = \frac{1}{4\pi} \int\limits_{V'} \frac{\mathbf{J}(\mathbf{r}') \times (\mathbf{r} - \mathbf{r}')}{|\mathbf{r} - \mathbf{r}'|^3} dV'. \tag{8}$$

Now consider c' contained in a tubular neighborhood as shown in Figure 3 such that $V' = c' \times D$ where D is a disc transverse to c'. Assume, for a point $\mathbf{r}'$ on c', that $\mathbf{r} - \mathbf{r}'$ does not vary significantly on D, then (8) can be evaluated over D to give the total current I times an integral on c', or

$$\mathbf{H}(\mathbf{r}) = \frac{I}{4\pi} \oint\limits_{c'} \frac{d\mathbf{r}' \times (\mathbf{r} - \mathbf{r}')}{|\mathbf{r} - \mathbf{r}'|^3} \tag{9}$$

which is the Biot-Savart law. Putting (9) into (3), we get

$$\psi(p) - \psi(p_0) = -\frac{I}{4\pi} \int\limits_{p_0}^{p} \oint\limits_{c'} \frac{[(\mathbf{r} - \mathbf{r}') \times d\mathbf{r}'] \cdot d\mathbf{r}}{|\mathbf{r} - \mathbf{r}'|^3}. \tag{10}$$

Note also, the related expression for the linking number obtained by putting equation (9) into (1) and cancelling I on each side of the resulting equation:

$$Link(c, c') = \frac{1}{4\pi} \oint\limits_{c} \oint\limits_{c'} \frac{[d\mathbf{r}' \times (\mathbf{r} - \mathbf{r}')] \cdot d\mathbf{r}}{|\mathbf{r} - \mathbf{r}'|^3} \tag{11}$$

which is due to Gauss [14]. Equation (10) is an exact formula if we started with a current-carrying knot and (11) always yields an integer if c' and c do not intersect (see appendix).

Gauss [14] approached equation (11) through the notion of solid angle which we discuss here in order to develop a geometric understanding of the linking number. The

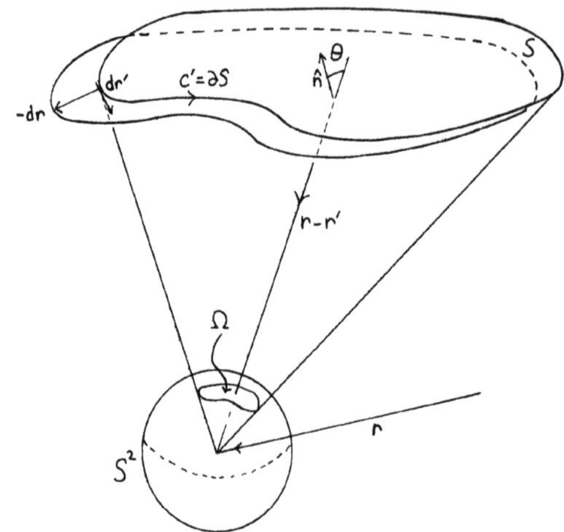

Figure 4. Change in Ω by moving observation point is same as that by moving circuit.

solid angle Ω is defined as the area on the sphere S^2 bounded by the intersection of S^2 and a conical surface with vertex at the center of S^2, as shown in Figure 4. If c' bounds a surface s, then the solid angle at an observation point $\mathbf{r}$ is easily shown to be [8]

$$\Omega = \int_s \frac{\cos\theta}{|\mathbf{r}-\mathbf{r}'|^2}\, ds \tag{12}$$

where θ is the angle between $\mathbf{r}-\mathbf{r}'$ and the normal to s. Thus equation (12) can be written as

$$\Omega = \int_s \frac{(\mathbf{r}-\mathbf{r}')\cdot d\mathbf{s}}{|\mathbf{r}-\mathbf{r}'|^3}. \tag{13}$$

Suppose the observation point $\mathbf{r}$ is moved by an amount $d\mathbf{r}$. This is the same as moving the circuit by $-d\mathbf{r}$, whereby the shifting circuit sweeps out an area $|d\mathbf{s}| = |d\mathbf{r} \times d\mathbf{r}'|$ where $d\mathbf{r}'$ is on c'. So from (13), the change in solid angle is [31]

$$\begin{aligned} d\Omega &= \oint_{c'} \frac{(\mathbf{r}-\mathbf{r}')\cdot(d\mathbf{r}\times d\mathbf{r}')}{|\mathbf{r}-\mathbf{r}'|^3} \\ &= \oint_{c'} \frac{[(\mathbf{r}-\mathbf{r}')\times d\mathbf{r}']\cdot d\mathbf{r}}{|\mathbf{r}-\mathbf{r}'|^3}. \end{aligned}$$

If the observation point is moved through a closed path c, the total change in Ω is the expression for the linking number given by equation (11) up to a factor of $1/4\pi$. Note that (11) is symmetric so that, up to a sign, integration on either c' or c gives the same result. A further development of the linking number can be found in Spivak [36].

Figure 5a shows that the solid angle provides beautiful geometric insight into the linking number via multivalued scalar potentials. The figure shows that Ω picks up a factor of 4π each time the paths link. Figure 5b shows that there exist situations in which two unseparable curves have $Link = 0$.

Formal interpretations of the linking number, as discussed in the following section, lead to an understanding of cuts and the intuition behind the cuts algorithm. The appendix uses the formalism of differential forms to verify the integrality of the linking number. Other uses of the linking number in magnetics can be found in [25].

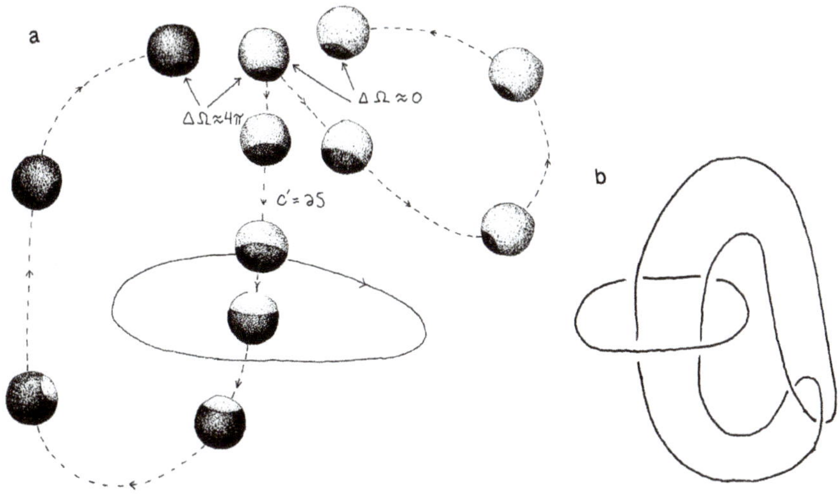

Figure 5. (a) Solid angle on the unit sphere S^2, showing $\Delta\Omega$ on the path c'. (b) *Link* = 0 for two unseparable curves.

3 LINKING, HOMOLOGY, AND DUALITY – A FORMAL FRAMEWORK FOR DEFINING CUTS

Prior to discussing some formal reasons for the integer nature of the linking number, we will add to the geometric intuition gained from Figures 4 and 5. As before, consider two closed, nonintersecting, oriented curves c and c' in $\mathbf{R}^3$ such as those in Figure 2. One of the curves, say c, can be expressed as $c = \partial S'$ where S' is a two-sided (orientable) surface. Then the linking number can be found by taking the sum of oriented intersections of S' and c',

$$Int\,(S', c') = \sum_{c' \cap S'} \pm 1 = Link(c, c')$$

where $Int\,(\cdot,\cdot)$ denotes the intersection number [23].

The intersection number $Int\,(S', c') = Int\,(S, c)$ where $\partial S = c'$. In the case of Figure 2b, S can be deformed so that it is simply a disk. Figure 1 showed S for a current-carrying trefoil knot. For a knot the surface S exists though it is not always intuitive. In any case an algorithm for the construction of the surface results from any constructive proof of the fact that such a surface is realizable as a compact, orientable manifold [22]. It will turn out that S is the cut needed to make the magnetic scalar potential single-valued.

Prior to introducing duality theorems we briefly summarize the notion of the pth absolute homology group of a region V denoted by $H_p(V)$, and the pth relative homology group of a region V "modulo" its boundary ∂V, denoted by $H_p(V, \partial V)$. Excellent introductions to these concepts are to be found in [9, 18] and set in the context of electrical circuit theory in [4].

Consider a triangulated region V (think of a tetrahedral finite element mesh) where linear combinations of oriented *0*-simplexes (vertices), *1*-simplexes (edges), *2*-simplexes (faces), *3*-simplexes (tetrahedra), or generally, p-simplexes, are called *0*-, *1*-, *2*-, *3*-, p-chains, respectively. If we take formal linear combinations of p-simplexes with real

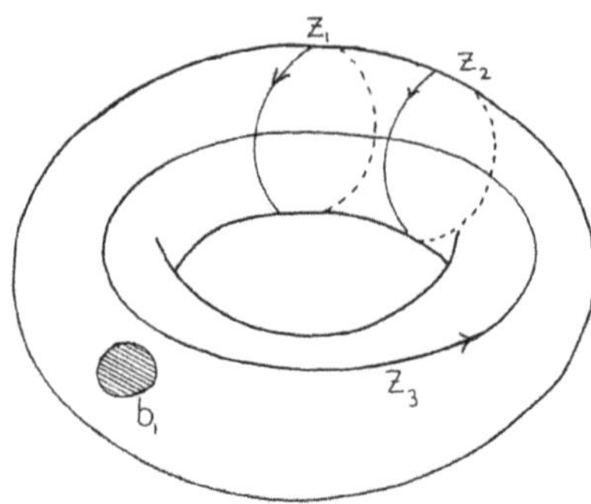

Figure 6. Hollow torus for example 1. Cycles z_1 and z_2 are not boundaries, however together they bound a cylindrical surface, hence they are homologous. Another *1*-cycle is z_3. b_1 is a *1*-boundary and is said to be homologous to zero.

coefficients, then we have vector spaces C_p, of p-chains on V, where an element $c_p \in C_p$ is represented by the vector $(a_1, a_2, \ldots)^t$, $a_i \in \mathbf{R}$, whose components are indexed by a p-simplex in V. It can be shown however, that it is only necessary to consider integer coefficients [20] in which case the C_p are commutative groups, but it is helpful to keep vector spaces in mind.

The intuitive idea of a boundary can now be formally introduced as a linear operator:

$$\partial_p : C_p(V) \to C_{p-1}(V), \quad p \geq 1$$

where we define $\partial_0(c_0) = 0$ for $c_0 \in C_0$. Hence boundaries of *3*-chains are linear combinations of *2*-chains, boundaries of *2*-chains are linear combinations of *1*-chains, and boundaries of *1*-chains are linear combinations of *0*-chains. In electrical circuit theory the boundary operator is realized by an incidence matrix.

A "p-cycle" z_p is a p-chain with zero boundary, $\partial_p(z_p) = 0$. A p-boundary b_p is a p-chain such that $b_p = \partial_{p+1}(c_{p+1})$ where c_{p+1} is a $p+1$-chain. In general $\partial_p(\partial_{p+1}(c_{p+1})) = 0$, that is, the "boundary of a boundary is zero." The "group" B_p of p-boundaries is a subgroup of the group Z_p of p-cycles.In network theory this manifests itself in the orthogonality of mesh-loop and edge-node incidence matrices [3].

We would like to differentiate between between bounding and nonbounding cycles. Boundaries are swept under the rug by building a quotient group $H_p(V)$ which is the set of equivalence classes of p-cycles modulo p-boundaries such that two p-cycles are equivalent, or homologous, if they differ by a p-boundary. Homologous cycles are denoted by $z_p^1 \sim z_p^2$ and are related by $z_p^2 - z_p^1 = \partial(c_{p+1})$ for some $c_{p+1} \in C_{p+1}$.

Example 1. Consider a toroidal surface T as shown in Figure 6. Closed curves z_1, z_2, and z_3 are nonbounding *1*-cycles while b_1 is a *1*-boundary because it bounds a circular patch on the torus. Cycles z_1 and z_2 are homologous since $z_1 - z_2$ is the boundary of a cylinder between them. Any *1*-cycle on T is homologous to $nz_1 + mz_3$, $n, m \in \mathbf{Z}$, where z_1 and z_3 are generators of $H_1(T)$. Therefore, $H_1(T) \cong \mathbf{Z} \oplus \mathbf{Z}$. That is, the first homology group is the direct sum of two copies of the group of integers. In terms of $H_2(T)$ it is apparent that there is only one *2*-cycle, the torus itself, so that $H_2(T) \cong \mathbf{Z}$. All *0*-chains are homologous such that $H_0(T) \cong \mathbf{Z}$.

Now consider a punctured torus T' where the disk bounded by b_1 in Figure 6 has been removed. In this case, there are no *2*-cycles since T' has boundary b_1. Thus $H_2(T') \cong 0$. However the generators for $H_1(T')$ are the same as those of $H_1(T)$, so puncturing the torus doesn't change H_1. The same is true of $H_0(T')$.

The number of copies of $\mathbf{Z}$ in $H_p(V)$ is known as the pth Betti number of V and is denoted by β_p. Intuitively, β_0 gives the number of disjoint pieces of the region in question, while β_1 is the number of holes in the region. Formally, β_p is the rank of the free part of $H_p(V)$ or, when real coefficients are used, β_p is the dimension of the vector space $H_p(V)$. □

For our purposes, a "relative p-cycle" is a p-chain on V with boundary in ∂V. A "relative p-boundary" is a relative p-cycle which can be augmented by a p-chain on ∂V to form the boundary of a $p+1$-chain on V. The quotient group $H_p(V, \partial V)$ is the set of equivalence classes of relative p-cycles modulo the relative boundaries. Two relative p-cycles are equivalent (homologous) if they differ by a relative p-boundary. Since we are interested in $H_2(V, \partial V)$, we can think of it geometrically as the set of equivalence classes of surfaces with boundaries in ∂V which, when augmented by a surface in ∂V, cannot form the boundary of a volume in V. Within a class, surfaces are related by the equivalence relation noted above and these will turn out to be the cuts we seek.

Cohomology groups are "dual" to homology groups. We restrict ourselves to the first cohomology group, $H^1(V)$ which can be regarded as the set of equivalence classes of curl-free vector fields (cocycles) where vector fields F_1 and F_2 are in the same equivalence class (cohomologous) if

$$F_2 - F_1 = \operatorname{grad} \phi$$

for some ϕ where $\operatorname{grad} \phi$ is called a coboundary. These represent classes of irrotational magnetic vector fields. By duality we will see that they correspond to equivalence classes of cuts. Relative cohomology is used to treat boundary conditions and does not concern us in the cuts problem.

A final comment on (co)homology. Formal linear combinations of chains with integer coefficients arise naturally in homology theory because of its simplicial roots while real coefficients are natural for vector fields and cohomology. However as shown in [20], since there is no torsion in the (co)homology groups of interest, there is no loss of generality in expressing both in terms of integers when considering 3-dimensional regions. This is of great practical advantage because it allows computation to be carried out purely in terms of integers. Thus rounding errors and ill-conditioning are never a problem regardless of mesh size.

Example 2. Consider a nonconducting surface S whose boundary ∂S is an interface to conducting regions. For example, S may be in the transverse plane of a transmission line which carries a set of n currents as shown in Figure 7. Curves c_i', $1 \leq i \leq \beta_1(S, \partial S)$, are relative *1*-cycles and are generators of $H_1(S, \partial S)$. They are used to find the $\beta_1(S, \partial S)$ independent potential differences in the problem by evaluating

$$\int_{c_i'} \mathbf{E} \cdot dl = V_i.$$

The c_i' are also cuts for the magnetic scalar potential. This is evident by noting that every closed path c_i which links current intersects a c_i'. This enables one to define a jump in the scalar potential across c_i', making the potential single-valued. The matrix of intersections of c_i' and c_i (which are generators of $H_1(S)$) is guaranteed to be nonsingular as a consequence of the Lefschetz duality theorem discussed below. Lefschetz duality generalizes beyond this planar problem to n dimensions. □

The duality theorems of algebraic topology contain the key to understanding cuts as well as the bridge between lumped parameter and field descriptions of electromagnetic problems [19]. If $\mathbf{R}^3$ is separated into a nonconducting region V and a conducting

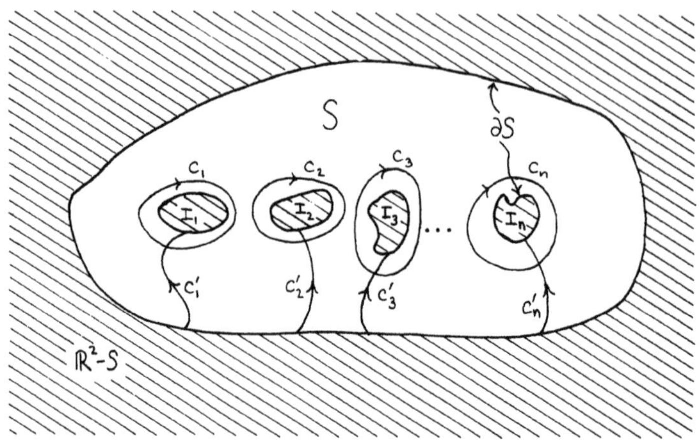

Figure 7. Cross-section of a transmission line with n conductors. In the nonconducting region S, the c_i' comprise a complete set of $\beta_1(S) = n$ relative *1*-cycles which generate $H_1(S, \partial S)$. Note that the boundaries (endpoints) of the c_i' are in ∂S. The set $\{c_j\}$, $1 \leq j \leq \beta_1(S)$ of curves which link current is a basis for $H_1(S)$.

region $\mathbf{R}^3 - V$, then the ranks β_k of the kth homology groups in each region are related thus:

$$\begin{aligned} \beta_1(V) &= \beta_1(\mathbf{R}^3 - V) \\ 1 + \beta_p(V) &= \beta_{2-p}(\mathbf{R}^3 - V) \text{ for } p \neq 1. \end{aligned}$$

This classical form of Alexander duality, known to Maxwell [28], is brought about by the fact that p-cycles in V are linked with *2-p*-cycles in $\mathbf{R}^3 - V$ [35] and represents a corollary of a more general form which states:

$$H^p(V) = H_{3-p}(\mathbf{R}^3, \mathbf{R}^3 - V).$$

The classical version results when one applies the general statement to the long exact homology sequence as shown in [19] and [15].

We are now in a position to understand the meaning of cuts. Consider a set $\{c_i\}$, $1 \leq i \leq \beta_1(V)$, of *1*-cycles in V illustrated in Figure 8 for $\beta_1(V) = 2$. These are generators of $H_1(V; \mathbf{Z})$ and comprise the set of interesting curves used to evaluate the left-hand side of Ampère's law (1) since they link current. They are boundaries $c_i = \partial S_i'$, of surfaces S_i' in $\mathbf{R}^3 - V$ used to measure current flux.

Now consider the set $\{c_j'\}$, $1 \leq j \leq \beta_1(\mathbf{R}^3 - V)$ which forms a basis for $H_1(\mathbf{R}^3 - V)$, and $Link(c_i, c_j')$, the intersection number of S_i' with c_j'. Alexander duality guarantees that the $\beta_1 \times \beta_1$ intersection matrix which has $Link(c_i, c_j')$ for its ijth entry is nonsingular.

The symmetry of the linking number suggests that $\{c_j'\}$ are also boundaries, $c_j' = \partial S_j$, where S_j is a surface in V. The *1*-cycles which link current are generators of $H_1(V; \mathbf{Z})$ and intersect the S_j which are generators of equivalence classes in $H_2(\mathbf{R}^3, \mathbf{R}^3 - V; \mathbf{Z})$. Alexander duality guarantees that the S_j are dual to $H^1(V; \mathbf{Z})$ insuring that $\mathbf{H}$ can be expressed as the gradient of a single-valued scalar potential. Hence if we let the scalar potential have a discontinuous jump I_j across S_j, then S_j is the cut surface and furthermore S_j can also be used as a surface for calculating magnetic flux. The set $\{S_j\}$ of cuts in V allow ψ to be single-valued on $V - (\bigcup S_j)$. Closed curves in $V - (\bigcup_j S_j)$ link zero current as illustrated in Figure 1. Note, however, that there is no promise

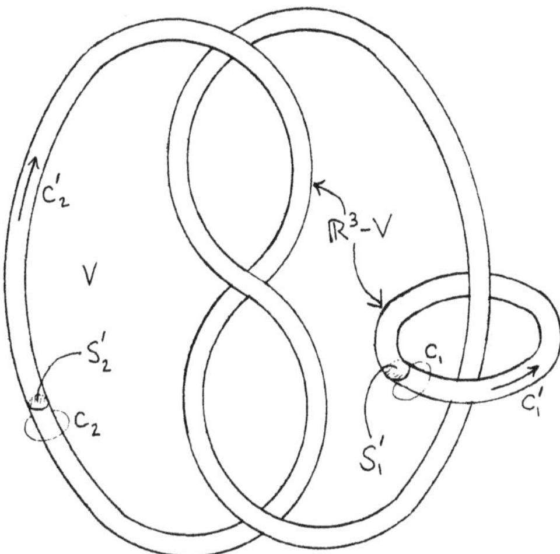

Figure 8. *1*-cycles c_1 and c_2 are generators of $H_1(V;\mathbf{Z})$ while c_1' and c_2' are generators of $H_1(\mathbf{R}^3 - V;\mathbf{Z})$. So $\beta_1(V) = 2$. Cut for the knot is same as that in Figure 1. Cut for loop is an annulus.

that $V - (\bigcup_j S_j)$ is simply-connected! In [20] it was shown that these cuts are compact, orientable, embedded surfaces, and in [21] intersections between cuts are discussed at length.

While Alexander duality provides an intuitive way of defining cuts, it is phrased in terms of the current-carrying region, making it useless for a finite elements algorithm which must be expressed entirely in terms of V and ∂V, the nonconducting region and its boundary. To express cuts in terms of a mesh which represents the current-free region, a limiting process [15] takes Alexander duality to Lefschetz duality:

$$H^p(V) \cong H_{n-p}(V, \partial V).$$

Now we have a duality theorem in terms of the region and its boundary. A constructive proof that generators of $H_2(V, \partial V)$ are realizable as compact, orientable, embedded manifolds [20] then gives rise to an algorithm for finding cuts [22]. This algorithm to find cuts which do not necessarily leave the cut complement simply connected is the topic of the next section.

4 VARIATIONAL ASPECTS OF A CUTS ALGORITHM

This section begins by summarizing the formulation underlying the cuts algorithm described in [22] and implemented in [29]. The remainder of the section investigates the Euler-Lagrange equation of (19) and discusses the exact solutions of this variational problem while hinting at its topological flavor. Though not necessary for continuity to Section 5, we employ the framework developed above to find an explicit solution to the nonconvex variational problem and verify that the resulting function is single-valued.

The cuts algorithm [22] consists of finding a solution to the variational problem of minimizing:

$$F(f) = \int_V \nabla f^* \cdot \nabla f \, dV \tag{14}$$

subject to

$$f^* f = 1 \text{ in } V \tag{15}$$

and for $1 \leq j \leq \beta_1(V)$, the jth cut requires

$$\frac{1}{2\pi i} \oint_{c_k} \mathrm{grad}\,(\ln f) \cdot dl = \delta_{jk} \tag{16}$$

for $1 \leq k \leq \beta_1(V)$. Here f is a map from V to $\mathbf{C}$, and c_l, $1 \leq l \leq \beta_1(V)$ are curves representing a basis for $H_1(V;\mathbf{Z})$. Equation (15) shows that the solution to the above problem defines a map to the unit circle in the complex plane

$$f : V \longrightarrow S^1. \tag{17}$$

Taking the inverse image of a regular value on S^1 (i.e., a point p on S^1 such that the gradient of f is nonzero at every point in the preimage), we end up with a surface whose boundary lies in ∂V. This "cut surface" represents a relative homology class in $H_2(V, \partial V;\mathbf{Z})$ which is dual to the constraint represented by (16). This comes about because S^1 is an Eilenberg-MacLane space $K(\mathbf{Z}, 1)$ [20, 6].

Our immediate objective is to make a distinction between how the variational problem (14)–(16) is handled numerically and analytically. First we note that one can handle the constraint (15) by setting

$$f = e^{2\pi i \psi} \tag{18}$$

where ψ is a real differentiable function which, by (16), must be multivalued. Substituting (18) into (14) gives

$$F(e^{2\pi i \psi}) = 4\pi^2 \int_V \mathrm{grad}\,\psi \cdot \mathrm{grad}\,\psi \;\, dV. \tag{19}$$

The starting point for the algorithm described by Kotiuga [22] and implemented in Murphy [29] is the observation that the Euler-Lagrange equation corresponding to (19) is just Laplace's equation. Hence, in principle, an algorithm to find cuts is easily implemented once one can modify existing finite element code for solving Laplace's equation. Two subtleties which must be addressed are, first, interelement continuity conditions must be modified in order to respect (16) and second, from equation (17), the inverse image of f can be obtained by considering the equipotentials of ψ modulo integers. Addressing these two subtleties, an algorithm to find cuts in any region can be implemented provided a "reasonable" finite element mesh exists, that is, a mesh on which Laplace's equation can be solved. Storage requirement for the algorithm is $\mathcal{O}(n)$ where n is the number of elements in the mesh. Time requirement is $\mathcal{O}(n^2)$ [16].

For a deeper understanding of situations where a complete set of cuts $\{S_i\}$, $1 \leq i \leq \beta_1(V)$, enable one to use a single-valued scalar potential in

$$\widetilde{V} = V - \{\bigcup_{i=1}^{\beta_1(V)} S_i\},$$

but $\widetilde{V}$ is not simply connected, we need a better understanding of the solution of the variational problem (14)-(16). To this end, we will now handle the constraint (15) by a Lagrange multiplier which can be eliminated to obtain a "harmonic map equation" for f. When confronted with this nonlinear partial differential equation we will use what we have established regarding magnetic scalar potentials, the Biot-Savart law, and linking numbers, to produce an explicit solution. The remainder of the paper can then deal with the difference between algorithms approached via homology versus those approached via homotopy notions.

If we append to the functional (14) a Lagrange multiplier term corresponding to the constraint (15), we end up with a variational problem for the functional

$$\tilde{F}(f,\lambda) = \int_V \nabla f^* \cdot \nabla f + \lambda(f^* f - 1)\, dV \tag{20}$$

subject to the constraint (16). When the first variation of this functional with respect to f is set to zero we obtain the weak Galerkin form

$$0 = \int_V [(\nabla \delta f) \cdot (\nabla f^*) + (\nabla \delta f^*) \cdot (\nabla f) + \lambda(f\, \delta f^* + f^*\, \delta f)]\, dV \tag{21}$$

If we eliminate the derivatives of the variation of f through integration by parts, we obtain

$$\begin{aligned} 0 &= \int_V \left\{ \delta f[-\nabla^2 f^* + f^*] + \delta f^*[-\nabla^2 f + \lambda f] \right\} dV \\ &\quad + \int_{\partial V} (\delta f \frac{\partial f^*}{\partial \hat{n}} + \delta f^* \frac{\partial f}{\partial \hat{n}})\, dS. \end{aligned}$$

Writing $f = f_r + i f_i$ where f_r and f_i are real functions which can be varied independently, one finds that the vanishing of the above expression for all admissible δf implies

$$\left. \begin{array}{ll} \nabla^2 f = \lambda f & \text{in } V \\ \frac{\partial f}{\partial n} = 0 & \text{on } \partial V. \end{array} \right\} \tag{22}$$

When the variation of the functional (20) with respect to λ is set to zero, we recover the constraint (15). We begin to eliminate the Lagrange multiplier from equation (22) by first taking the Laplacian of the constraint (15) to obtain

$$0 = \nabla^2(f^* f) = (\nabla^2 f^*) f + 2\nabla f^* \cdot \nabla f + f^* \nabla^2 f$$

or

$$\Re(f^* \nabla^2 f) = -|\nabla f|^2 \tag{23}$$

where $\Re(\cdot)$ denotes the real part of $(\cdot)$. Multiplying equation (22) by f^* and using equations (15) and (23) we can solve for λ:

$$-|\nabla f|^2 = \Re(f^* \nabla^2 f) = \lambda f^* f = \lambda$$

and rephrase (22) as

$$\left. \begin{array}{ll} \nabla^2 f = -|\nabla f|^2 f & \text{in } V \\ \frac{\partial f}{\partial n} = 0 & \text{on } \partial V. \end{array} \right\} \tag{24}$$

Equations (24) and (16) provide a set of equations for the single-valued function defined in the discussion leading to equation (17). At first sight, the solution to these equations in the exterior of current carrying wires is not obvious. If we perform the substitution given by equation (18) then (24) reduces to a boundary value problem involving Laplace's equation and a multivalued function. In the algorithm for computing cuts it was necessary to deal with this multivaluedness in the context of interelement constraints – appealing to physical intuition would have lead to a circular argument where the magnetic scalar potential would be needed to define the cuts for the magnetic scalar potential! In the present case we want to develop our intuition about cuts and seek explicit expressions for the cuts. Hence we are free to use the equipotentials of the multivalued scalar potential tied to any easy-to-use constitutive law as equivalent cuts.

In order to find the function f satisfying (15) and (24) we first find a set of integer-valued currents in the exterior of V which insure that the corresponding scalar potential satisfies (16). Suppose this is accomplished by imposing $\beta_1(V)$ integer valued currents $\{n_i\}$ on $\beta_1(V)$ closed curves $\{c_i'\}$ in the exterior of V. The scalar potential is given by the line integral of the magnetic field found through the use of the Biot-Savart Law:

$$\psi(p) - \psi(p_0) = \frac{1}{4\pi} \int_{p_0}^{p} \sum_{i=1}^{\beta_1(V)} n_i \oint_{c_i'} \frac{[(\mathbf{r} - \mathbf{r}') \times d\mathbf{r}] \cdot d\mathbf{r}'}{|\mathbf{r} - \mathbf{r}'|^3}. \tag{25}$$

Note that we can modify this ψ so that it satisfies a Neumann boundary condition on ∂V by adding a single valued harmonic function which vanishes at p_0. From equation (18), the desired f is easily seen to be:

$$f(\mathbf{r}) = f_0 \exp\left(\frac{i}{2} \int_{p_0}^{p} \sum_{l=1}^{\beta_1(V)} n_l \oint_{c_l'} \frac{[(\mathbf{r} - \mathbf{r}') \times d\mathbf{r}] \cdot d\mathbf{r}'}{|\mathbf{r} - \mathbf{r}'|^3}\right) \tag{26}$$

where the multiplicative factor

$$f_0 = e^{2\pi i\, \psi(p_0)} \tag{27}$$

is an arbitrary complex number of unit length. Note that if we go from p_0 to p_0 along a closed loop c the expression for the linking number (11) shows us that the expression for f changes by

$$\exp[2\pi i \sum_{l=1}^{\beta_1(V)} n_l \, Link(c, c_l')]$$

which is equal to 1. This shows that a concrete understanding of the linking number makes the single-valuedness of f manifest and finally, up to boundary conditions, equation (26) is indeed the solution to the variational problem defined by (14)–(16) and the boundary value problem defined by (24) and (16). Lefschetz duality and the theory underlying the algorithm ensure that cuts take care of current linkage, but nowhere has anything been done to make $V - (\bigcup S_i)$ simply connected.

5 HOMOTOPY-BASED ALGORITHMS – THE LURE OF π_1 AND THE PIT OF FIRE

"By some divine justice the homotopy groups of a finite polyhedron or a manifold seem as difficult to compute as they are easy to define." [6]

RAOUL BOTT

Authors in various fields [28, 39, 27, 40, 2, 17], have assumed that the scalar potential is single-valued when the cut makes the (non-conducting) region simply-connected. The pursuit of simple connectivity has led to formulations based on homotopy where, for two-dimensional problems, the assumption leads to the right cuts. The relation between homotopy and homology reveals that in three-dimensional problems such an assumption is not equivalent to the intuition gained from Ampère's law that the problem of making cuts is one of linking zero current rather than one of making the current-free region simply connected.

The first homotopy group π_1 of a region V embedded in $\mathbf{R}^3$ is an algebraic classification of all closed loops in V which are topologically different. In order to illustrate homotopy and make the above description more precise, consider a closed, oriented curve such as the current-carrying trefoil knot shown in Figure 9, where V denotes the region complementary to the knot and $\mathbf{R}^3 - V$ is the knot itself. An arbitrary point p in V is chosen as a base for drawing oriented, closed paths $a, b, c, \ldots$ in V. The set

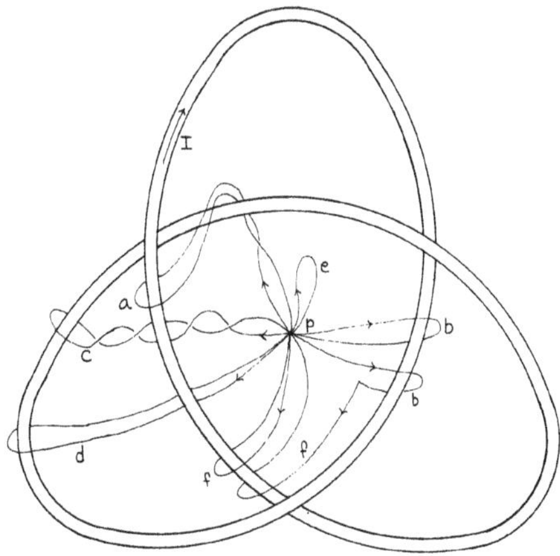

Figure 9. Closed loops based at p in the region complementary to the current-carrying knot. Loops a and b are homotopic. So are c and d. Loop e is trivial since it can be contracted to p. Product bf is shown.

of all closed curves based at p can be partitioned into equivalence classes called homotopy classes where two closed paths are homotopic if either curve can be continuously deformed into the other. The path a represents or generates a class $[a]$ which contains all paths homotopic to a. It can be shown that composition of paths induces a product law for homotopy classes:

$$[a][b] = [ab]$$

where we think of ab as traversing a followed by b as shown in Figure 9. The loop a^{-1} is a with opposite orientation and generates its own homotopy class $[a^{-1}]$ while the constant, or identity, loop e is a path which can be contracted to the basepoint without bumping into the knot. These notions are discussed and extensively illustrated in [30].

It can also be shown that homotopy classes satisfy the following properties:

$$\begin{aligned} &([a][b])[c] = [a]([b][c]) && \text{(associativity)},\\ &[a][e] = [a] = [e][a] && \text{(identity)},\\ &[a][a^{-1}] = [e] = [a^{-1}][a] && \text{(inverse)}. \end{aligned}$$

Hence, the set of all homotopy classes in V is said to form a group, called π_1, where the group law is written as a multiplication. Note that a region is said to be simply-connected if all loops in the region are homotopic to the identity, in which case π_1 is trivial. Formally, a multiply connected region is one which is not simply-connected in this sense.

Of particular significance to π_1 are the homotopy classes of the form $[xyx^{-1}y^{-1}]$, called commutators. If π_1 is commutative, commutators are equal to the identity, that is xy is deformable to yx so that both represent the same homotopy class, $[x][y] = [y][x]$. In general π_1 is not commutative; commutators are not equal to the identity and π_1 has a commutator subgroup $[\pi_1, \pi_1]$ generated by all possible commutator products.

Example 1 revisited. Briefly returning to Example 1 we note that on the torus π_1 is generated by z_1 and z_3 (see Figure 6). In this case the commutator subgroup is trivial so that π_1 is commutative. On the other hand, π_1 of the punctured torus has a nontrivial commutator subgroup since the commutator is homotopic to the boundary of the hole. Note that in homology the effect of puncturing the torus was felt only in the second homology group. □

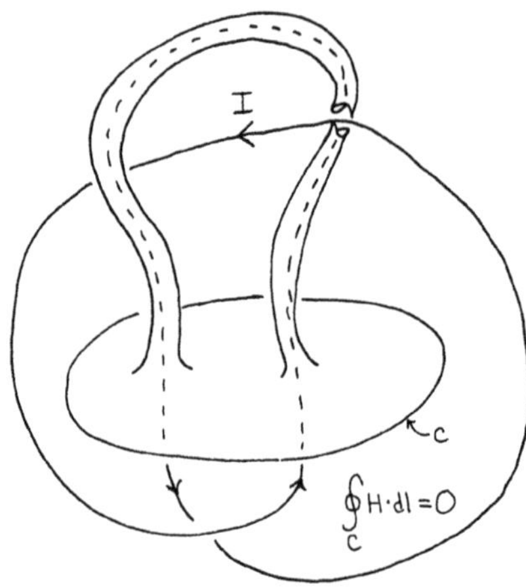

Figure 10. Commutator c links zero current and bounds a surface (a disk with a handle) in V.

The formal relationship between the first homotopy and homology groups is a homomorphism, $\pi_1(V) \to H_1(V)$. The Poincaré isomorphism theorem states that the kernel of the homomorphism is $[\pi_1, \pi_1]$ so that there is an isomorphism

$$\pi_1/[\pi_1, \pi_1] \simeq H_1$$

for any region V. When $[\pi_1, \pi_1]$ is nontrivial, commutators are nontrivial closed loops which, by virtue of the Poincaré isomorphism, are zero-homologous. It can be shown that this is the case [38, 15], thus commutators are boundaries of surfaces which lie entirely in V; that is, the homology classes of *2*-chains bounded by zero-homologous paths can be represented by orientable manifolds in V [38]. Hence no current can be linked by a commutator and surfaces bounded by commutators are unrelated to surfaces used in Ampère's law. A proof of the π_1-H_1 relation can be found in [15] and is discussed in the context of Riemann surfaces and complex analysis in [37]. A discussion of the consequences of the π_1-H_1 relation for computation can be found in [9].

Figure 10 illustrates a commutator element and the surface bounded by the commutator for the current-carrying trefoil knot where c denotes a class in $[\pi_1, \pi_1]$. Since $c \in [\pi_1, \pi_1]$ and $[\pi_1, \pi_1]$ is the kernel of the Poincaré map, c is zero-homologous meaning that a surface S such that $c = \partial S$ exists. By Ampère's Law we then have

$$\begin{aligned} \int_c \mathbf{H} \cdot d\mathbf{r} &= \int_S \mathbf{J} \cdot \hat{n} ds \\ &= 0 \end{aligned}$$

because S lies in V and $\mathbf{J} = 0$ in V. It follows that $Link(c, c') = 0$ for $c \in [\pi_1, \pi_1]$ and $c' \in H_1(\mathbf{R}^3 - V)$.

To make the nonconducting region simply-connected requires one to eliminate all nontrivial closed paths in the region. In problems where $[\pi_1, \pi_1]$ is nontrivial, one is forced to confront commutator loops which are zero-homologous. Since commutators link zero current, they are unimportant in light of Ampère's law; and they are irrelevant to finding cuts which make the scalar potential single-valued.

The opening quote suggests that there exist other fundamental problems with a homotopy approach to cuts. While π_1 is appealing because it "accurately" describes holes in a multiply-connected region, practical aspects of its computation continue to be open problems in mathematics [11, 38]. While π_1 is computable for the torus, it is difficult in general to compute nonabelian (i.e., noncommutative) groups. In homology, groups are abelian and can be expressed as matrices with integer entries. The matrices are sparse with $\mathcal{O}(n)$ nonzero entries and can be computed with graph-theoretical techniques. This is made clear in [4] and [34] for electrical circuits and is the idea behind [22] and [29]. Furthermore, no implementation of a homotopy-based algorithm for making cuts in three-dimensional problems where $[\pi_1, \pi_1]$ is nontrivial has been reported in the literature.

6 CONCLUSION

Linking numbers are essential for understanding of Ampère's Law and the Biot-Savart integral. They naturally make way for the formalism of (co)homology used to define cuts which make the magnetic scalar potential single-valued in multiply-connected regions. It must be stressed that the cuts may not leave the region simply-connected.

Any constructive proof that cuts are realizable as orientable embedded manifolds leads to an algorithm for making cuts. The algorithm pursued by the authors [22, 29, 16] is expressed in its variational form to show how a proper set of cuts can render the current-free region multiply-connected.

Finally, the Poincaré isomorphism theorem which relates homology and homotopy groups, shows that the assumption of simple connectivity and intuition gained from Ampère's Law are fundamentally different. Attempting to make the region simply-connected leads to algorithms based on homotopy concepts and requires computation of nonabelian groups when the commutator subgroup of π_1 is nontrivial.

ACKNOWLEDGEMENTS

The senior author was supported by an NSF research initiation award. The junior author recognizes support of a U.S. Department of Education GRATE fellowship and is indebted to his M.S. thesis committee, S.R. Eisenberg, R. Giles, P.R. Kotiuga, and T. Smith for valuable comments.

APPENDIX: A FORMAL VIEW OF THE INTEGRAL NATURE OF THE LINKING NUMBER

Before laying linking numbers to rest, we consider the degree of a map which explains why $Link(c, c')$ is integer-valued. It also explains why the linking integral is the same as the intersection number definition employed in Section III. While Section III gives an informal and intuitive description of cohomology, here we must use differential forms to represent cohomology classes [12, 10]. This discussion is based on ideas developed in [6] and [33].

Consider a map between compact n-dimensional manifolds $f : M^n \to N^n$ such that f^{-1} maps compact sets to compact sets. Let α be a differential form representing a generator of the cohomology class $H^n(N^n)$ where α has total integral one and support near a regular value q (i.e., a point q on N^n such that the Jacobian matrix has full rank) of the map f. The "pullback" operation on a differential form takes p-forms on a space N to p-forms on M and is a generalization of the Jacobian rule for transformation of variables in multiple integrals. It has a surprising connection to linking. Formally, the pullback $f^*\alpha$ is an n-form (volume form) with support near points $f^{-1}(q)$ and carries α to a multiple of α. The multiple is an integer and is called the degree of f such that

$$\begin{aligned} \deg(f) &= \int_{\mathbf{R}^n} f^*\alpha \quad \text{by definition, and} \\ &= \sum_{f^{-1}(q)} \pm 1 \end{aligned}$$

since f preserves the integral up to a sign. The preimage of a regular value of the map $f : M^n \to N^n$ consists of a number of points counted with multiplicity ± 1, called the degree of f which is the same for any regular value. See [6] for a proof and a picture.

From this point of view, the Gauss integral is the pullback of the volume form on the sphere [12, 6]. Consider two nonintersecting curves c and c', letting $c, c' : S^1 \to \mathbf{R}^3$ and identify $S^3 - \{p\}$ with $\mathbf{R}^3$, where p is not a point of c or c'. Here we think of a sphere with unit volume instead of unit radius although we use the usual notation S^2. A map from the torus to the sphere in $\mathbf{R}^3$, $f : S^1 \times S^1 \to S^2$ can be defined by

$$f(\mathbf{x}, \mathbf{y}) = \frac{\mathbf{x} - \mathbf{y}}{|\mathbf{x} - \mathbf{y}|}$$

where $|\mathbf{x} - \mathbf{y}|$ is the Euclidean length in $\mathbf{R}^3$. Give the "usual" orientations to the torus and S^2. Then

$$Link(c, c') = \deg(f).$$

One can also think of $\deg(f)$ as the winding number of f around the sphere which is evidently an integer because the curves are closed.

REFERENCES

[1] T. Archibald, *Connectivity and smoke-rings: Green's second identity in its first fifty years*, Mathematics Magazine, 62 (1989), pp. 219–232.

[2] H. Ashley and M. Landahl, *Aerodynamics of Wings and Bodies*, Addison-Wesley, Reading, MA, 1965. Section 2-7.

[3] N. Balabanian and T. A. Bickart, *Electrical Network Theory*, John Wiley and Sons, New York, 1969. p. 80.

[4] P. Bamberg and S. Sternberg, *A course in mathematics for students of physics: 2*, Cambridge University Press, New York, 1990. Chapter 12.

[5] A. Bossavit *on* A. Vourdas and K. J. Binns, *Magnetostatics with scalar potentials in multiply connected regions*, IEE Proc. A, 136 (1989), pp. 260–261.

[6] R. Bott and L. W. Tu, *Differential Forms in Algebraic Topology*, Springer-Verlag, New York, 1982. pp. 40–42, 234, 258, 240.

[7] M. L. Brown, *Scalar potentials in multiply connected regions*, Int. J. Numer. Meth. Eng., 20 (1984), pp. 665–680.

[8] R. Courant, *Differential and Integral Calculus (1936)*, vol. 2, Interscience, New York, 1953. pp. 408–411.

[9] F. H. Croom, *Basic Concepts of Algebraic Topology*, Springer-Verlag, New York, 1978. Chaps. 2, 7.3, 4.5.

[10] G. A. DESCHAMPS, *Electromagnetics and differential forms*, IEEE Proc., 69 (1981), pp. 676–696.

[11] E. DOMÍNGUEZ AND J. RUBIO, *Computers in algebraic topology*, in The Mathematical Heritage of C.F. Gauss, G. M. Rassias, ed., World Scientific Publ. Co., Singapore, 1991, pp. 179–194.

[12] H. FLANDERS, *Differential Forms with Applications to the Physical Sciences*, Dover, New York, 1989. pp. 79–81.

[13] G. K. FRANCIS AND B. COLLINS, *On knot-spanning surfaces: An illustrated essay on topological art*, in The Visual Mind, M. Emmer, ed., MIT Press, Cambridge, MA, 1993, ch. 11.

[14] C. F. GAUSS, *Zur mathematischen Theorie der electodynamischen Wirkungen*, in Werke V, Teubner, 1877, p. 605.

[15] M. J. GREENBERG AND J. R. HARPER, *Algebraic Topology*, Benjamin/Cummings, Reading, MA, 1981. P. 235, 63–66.

[16] P. W. GROSS, *The commutator subgroup of the first homotopy group and cuts for scalar potentials in multiply connected regions*, Master's thesis, Dept. of Biomedical Engineering, Boston University, Boston, MA, September 1993.

[17] C. S. HARROLD AND J. SIMKIN, *Cutting multiply connected domains*, IEEE Trans. Mag., MAG-21 (1985), pp. 2495–2498.

[18] R. C. HWA AND V. L. TEPLITZ, *Homology and Feynman Integrals*, W.A. Benjamin, Inc., New York, 1966. Chap. 2.

[19] P. R. KOTIUGA, *Hodge Decompositions and Computational Electromagnetics*, PhD thesis, McGill University, Montreal, 1984. p. 123.

[20] ——, *On making cuts for magnetic scalar potentials in multiply connected regions*, J. Appl. Phys., 61 (1987), pp. 3916–3918.

[21] ——, *Toward an algorithm to make cuts for magnetic scalar potentials in finite element meshes*, J. Appl. Phys., 64 (1988), pp. 3357–3359. Erratum: 64, (8), 4257, (1988).

[22] ——, *An algorithm to make cuts for magnetic scalar potentials in tetrahedral meshes based on the finite element method*, IEEE Trans. Mag., MAG-25 (1989), pp. 4129–4131.

[23] ——, *Topological considerations in coupling magnetic scalar potentials to stream functions describing surface currents*, IEEE Trans. Mag., MAG-25 (1989), pp. 2925–2927.

[24] ——, *Topological duality in three-dimensional eddy-current problems and its role in computer-aided problem formulation*, J. Appl. Phys., 67 (1990), pp. 4717–4719.

[25] P. R. KOTIUGA AND R. GILES, *A topological invariant for the accessibility problem of micromagnetics*, J. Appl. Phys., 67 (1990), pp. 5347–5349.

[26] P. R. KOTIUGA *on* A. VOURDAS AND K. J. BINNS, *Magnetostatics with scalar potentials in multiply connected regions*, IEE Proc. A, 137 (1990), pp. 231–232.

[27] H. LAMB, *Hydrodynamics*, Dover, New York, 1932. (1879).

[28] J. C. MAXWELL, *A Treatise on Electricity and Magnetism (1891)*, Dover, New York, 1954. chap. 1, art. 18–22.

[29] A. MURPHY, *Implementation of a finite element based algorithm to make cuts for magnetic scalar potentials*, Master's thesis, Dept. of Electrical, Computer, and Systems Engineering, Boston University, Boston, MA, 1991.

[30] L. NEUWIRTH, *The theory of knots*, Sci. Am., 240 (1979), pp. 110–124.

[31] W. K. H. PANOFSKY AND M. PHILLIPS, *Classical Electricity and Magnetism*, Addison-Wesley, Reading, MA, 1962. pp. 8–10, 20–23, 125–127.

[32] H. POINCARÉ, *Analysis situs*, J. de l'École Polyt., 1 (1895), pp. 1–123.

[33] D. ROLFSEN, *Knots and Links*, Publish or Perish, Inc., Berkeley, CA, 1976. pp. 132–135.

[34] J. P. ROTH, *Existence and uniqueness of solutions to electrical network problems via homology sequences*, in Mathematical Aspects of Electrical Network Theory, SIAM-AMS Proceedings III, 1971, pp. 113–118.

[35] H. SEIFERT AND W. THRELFALL, *A Textbook of Topology*, Academic Press, New York, 1980. (Translated from the German by M.A. Goldman.) p. 337.

[36] M. SPIVAK, *Calculus on Manifolds*, W. A. Benjamin, New York, 1965. pp. 131–134.

[37] G. SPRINGER, *Introduction to Riemann Surfaces*, Addison-Wesley, Reading, MA, 1957. Section 5-8.

[38] J. STILLWELL, *Classical Topology and Combinatorial Group Theory, Second Edition*, Springer-Verlag, New York, 1993. Second Ed. pp. 172–175.

[39] J. A. STRATTON, *Electromagnetic Theory*, McGraw-Hill, New York, 1941. pp. 227–228.

[40] A. VOURDAS AND K. J. BINNS, *Magnetostatics with scalar potentials in multiply connected regions*, IEE Proc. A, 136 (1989), pp. 49–54.

MAGNETIC FIELD ANALYSIS IN INDUCTION MOTORS IN THE FIELD-ORIENTED MODE

G. Henneberger, M. Schmitz

Institute of Electrical Machines,
University of Technology Aachen,
Schinkelstrasse 4, 52056 Aachen, F.R.Germany

INTRODUCTION

The field-oriented control is a frequent used application for induction motors. This paper presents a method for simulating the magnetic field in a squirrel-cage induction motor in the field-oriented operating mode. The purpose of this simulation is to analyse the influence of eddy currents and rotor-yoke-unloading in dependence on the field oriented currents in the direct and quadrature axis.
This is done by multiple calculations of the magnetic field with a commercial CAD-finite element program. It is possible to calculate rotor fluxes and rotor currents in the direct and quadrature axis. It will be shown, that the nonlinearity of the iron and the rotor-yoke-unloading have to be considered by calculating the rotor flux or especially the rotor time constant.

FIELD ORIENTED CONTROL

The direct and quadrature theorie enables one to describe the squirrel cage induction machine in a rotor flux fixed reference frame[1]. To meet the requirements of the field-oriented mode, the torque-producing current linkage has to be perpendicular to the rotor flux. For the rotor flux linkages this leads to:

$$\Psi'_{IIA} = (1+\sigma_2)L_{1h}i'_{IIA} + L_{1h}i'_{IA} \stackrel{!}{=} L_{1h}i'_{\mu A} \qquad \text{d-axis} \tag{1}$$

$$\Psi'_{IIB} = (1+\sigma_2)L_{1h}i'_{IIB} + L_{1h}i'_{IB} \stackrel{!}{=} 0 \qquad \text{q-axis} \tag{2}$$

On condition that the stator currents are injected, the induction machine can be described as follow:

$$T_2\frac{\mathrm{d}i'_{\mu A}}{\mathrm{d}t} + i'_{\mu A} = i'_{IA} \qquad \frac{i'_{IB}}{T_2 \cdot i'_{\mu A}} + \omega = \omega_\mu = \frac{\mathrm{d}\alpha}{\mathrm{d}t} \tag{3}$$

$$M_{el} = \frac{pL_{1h}}{1+\sigma_2} i'_{\mu A} \cdot i'_{IB} = \frac{J}{p}\frac{\mathrm{d}\omega}{\mathrm{d}t} + M_{Load} \tag{4}$$

Electric and Magnetic Fields, Edited by A. Nicolet
and R. Belmans, Plenum Press, New York, 1995

$$\text{with:} \quad \frac{d\gamma}{dt} = \omega \quad \text{mech. speed} \qquad \frac{d\alpha}{dt} = \omega_\mu = \omega + \omega_R$$

In order to control the induction machine in field-oriented coordinates, the magnitude and angular position of the rotor flux has to be known. It is possible to get these magnitudes from the measured stator currents and the equation 3. This requires the knowledge of the rotor time konstant T_2. However the rotor time constant is variable:

$$T_2 = \frac{L_2}{R_2} \qquad \begin{array}{l} \leftarrow \text{depends on saturation, state of rotor-yoke-loading} \\ \leftarrow \text{depends on temperature} \end{array}$$

MACHINE MODEL

The investigated machine is a squirrel cage induction motor with a rated power of 3 kW. It has 4 poles and 36 slots in the stator and 28 in the rotor. Because of the symmetry, only one phase spacing has to be modelled. The stator and the rotor are modelled with nonlinear iron ST37, the stator winding is made of stranded copper, the rotor winding is made of aluminium. In order to investigate the rotor-yoke-loading, the shaft is modelled with linear iron and a small airgap. The used model is a first approximation, in further investigations the short-circuiting ring of the cage and the coil end leakages will be considered in an outer electrical circuit.

The model (fig. 1) consists of 1636 first order triangular elements. The constraints set the vectorpotential A to zero at the surface (unary or Dirichlet constrain) and ±A at the left and bottom side (binary constraint). For the calculation, steady-state operation and a

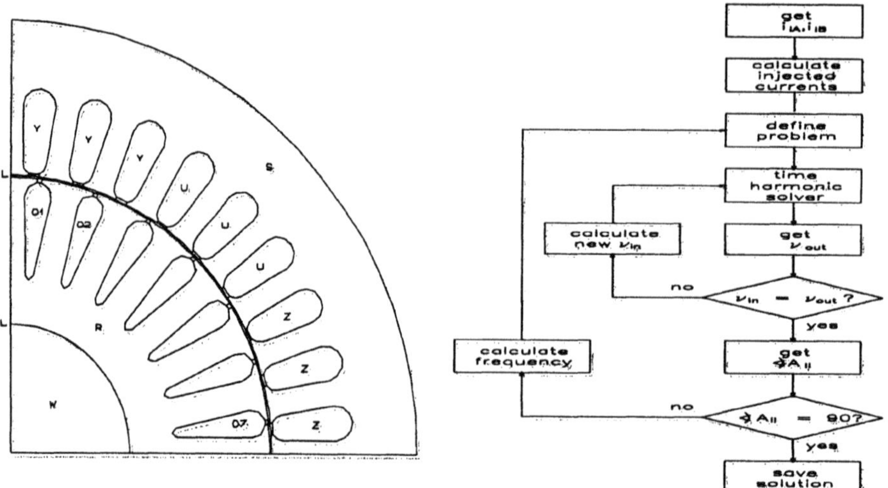

Figure 1. Finite element model **Figure 2.** Flowchart for load calculation

constant rotor velocity are assumed. If the model is transformed to a rotor fixed reference frame, the frequency of the stator current is equal to the slip frequency ω_R. In order to inject the stator currents in the model, the two-axis stator currents must be transformed into three phase currents i'_u, i'_v, i'_w:

$$\begin{bmatrix} i'_u \\ i'_v \\ i'_w \end{bmatrix} = \sqrt{\frac{2}{3}} i'_I \begin{bmatrix} \cos(\omega_R t + \varphi) \\ \cos(\omega_R t - 120° + \varphi) \\ \cos(\omega_R t - 240° + \varphi) \end{bmatrix} \tag{5}$$

$$\text{with:} \quad i'_I = \sqrt{i'^2_{IA} + i'^2_{IB}} \qquad \varphi = \arctan \frac{i'_{IB}}{i'_{IA}} \tag{6}$$

At no-load operation with $\omega_R = 0$, this leads to a static problem and can be solved with a static non-linear solver. The stator winding carries only a DC-current: $i'_{IA} = \sqrt{3}I_o$ and $i'_{IB} = 0$.

LOAD OPERATION

The load operation is solved with a 2D-time-harmonic solver for linear eddy current problems[2]. The nonlinearity of the iron is considered by a given reluctivity distribution. Because the result field and the injected current distribution lie in different directions, the reluctivity distribution does not correspond to the real flux distribution. Therefore the reluctivity of each element will be extracted after each solution and a new reluctivity will be calculated with an under relaxation procedure until the output is much as the input distribution. This algorithm is considered by the inner loop shown in the flow chart in figure 2. The phase of the rotor vector potential is calculated from this solution.

Because the vector potential and the flux linkage are perpendicular to each other, the vector potential and the A-axis have to have an angular degree of 90° to fulfil the field-oriented conditions. The magnetic field have to be turned.

In the field oriented operating mode exist 3 independent parameters: the stator currents i'_{IA}, i'_{IB} and the frequency f. The currents are given, therefore the only parameter to turn the rotor-vector-potential is the frequency. This is realized in the outer loop.

RESULTS

The field oriented mode requires, that the rotor flux linkage in the A-axis is only proportional to a magnetizing current and the rotor flux linkage in the B-axis is forced to zero (equa. 4 and 5). In figure 4 and 5 it can be seen, that the rotor flux linkage lies complete in the A-axis and thus fulfil the second condition. On the other hand the first condition can't be fulfilled, because of the coupling of the A- and B-axis in consequence of saturation the rotor flux linkage depends on the load. In addition to, the currents in the rotor direct-axis don't disappear.

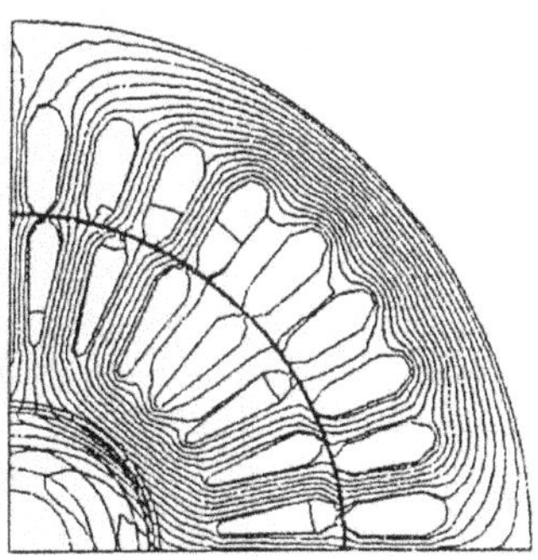

Figure 3. Flux-distribution: $i'_{IA} = 8.38\,A$; $i'_{IB} = 15\,A$; $f = 1.01\,Hz$

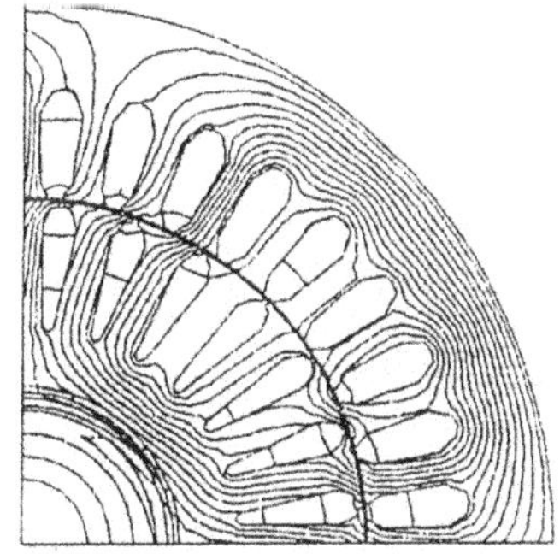

Figure 4. Flux-distribution: $i'_{IA} = 8.38\,A$; $i'_{IB} = 60\,A$; $f = 4.78\,Hz$

Whereas the influence of flux displacement at rated torque is still small (fig. 4), it is shown, that at quadruple rated torque the shaft carries only a small part of the flux (fig. 5). The effective area of the rotor-yoke is reduced, whereby the rotor inductance and the rotor flux decrease. Figure 6 shows the flux linkage Ψ'_{IIA} versus the torque at a constant

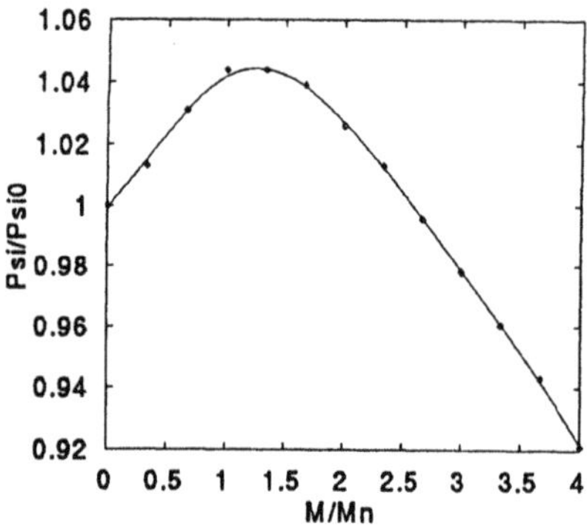

Figure 5. Rotor flux linkage

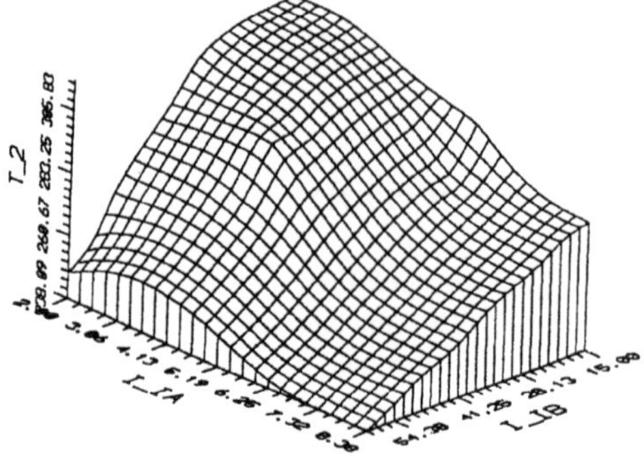

Figure 6. Rotor time constant as a function of the stator currents $T_2 = f(i'_{IA}, i'_{IB})$

magnetizing current i'_{IA}.

One can see, that after a first increase, the rotor flux decreases by 12% from rated to quadruple rated torque. This shows, that especially for high-speed motors with a thick shaft and at field-weakening operation, e.g. electric vehicles drives, a field of characteristics for $T_2 = f(i'_{IA}, i'_{IB})$ is required.

Linearization supposed, equation (6) leads to

$$T_2 = \frac{i'_{IB}}{\omega_R i'_{IA}} \quad . \tag{7}$$

Figure 6 shows, that such a field of characteristics can be created with the solution of the numerical field calculation. This method hasn't yet been proved in practise, and some more precise investigations must follow. The 4 × 4 matrix of L as a function of $i'_{IA}, i'_{IB}, i'_{IIA}, i'_{IIB}$ could lead to a new definition of L_2 and thus to a new definition of T_2 in a non-linear material.

Furthermore the torque as a function of the stator currents i'_{IA} and i'_{IB} can be calculated. A comparison of three different methods shows a good correspondence. The methods are:

1. direct calculating by the tangential force of every rotor slot $\vec{F} = I \cdot \vec{l} \times \vec{B}$,
2. calculating by the losses in the rotor bars $M = \frac{P_{Rotor}}{\omega_R}$,
3. the direct and quadrature theory leads to $M_{el} = p(\Psi'_{IIB} i'_{IIA} - \Psi'_{IIA} i'_{IIB})$.

CONCLUSIONS

The presented method enables the analyses of the induction machine in the special application of the field oriented control. Influence of saturation and field displacement in dependence on controlled variables can be displayed. Machine parameters, which are necessary for high dynamic field oriented control, can be calculated in every working point and saved in fields of characteristics.

REFERENCES

1. G. Henneberger. "Elektrische Maschinen II", Vorlesung an der RWTH Aachen, 1994.
2. P.P. Silvester, R.L. Ferrari. "Finite Elements for Electrical Engineers", Cambridge University Press, 1990.

LOSSES DUE TO ROTATIONAL FLUX IN THREE PHASE INDUCTION MOTORS

R.D. Findlay, N. Stranges and D.K. MacKay

McMaster University Hamilton, Ontario, Canada L8S 4K1

INTRODUCTION

Iron Losses under rotating flux conditions have been under study for nearly a century [1]. Although advances have been made in measurement techniques, a consensus on a standard test for predicting rotational losses in electrical steels has not been achieved. We have developed an apparatus for measuring rotational iron losses which is similar to that used by Enokizono et al., [2]. The magnetizing yoke assembly is shown in Figure 1 and the waveform generation and data acquisition system is shown in Figure 2. Additional details of this apparatus have been discussed in a previous paper [3].

It has been established, that rotational flux patterns exist in a large portion of the stator of a rotating machine. Moses and Radley [4,5] constructed a model of a large two-pole turbogenerator in order to simulate the flux and loss conditions in the stator of a rotating machine to investigate the flux density distribution using single turn search coils. The power loss distribution was measured using thermocouple techniques. They found an elliptically polarized flux density distribution at the back of the stator slots, along the inner periphery of the stator core, and an almost circular flux density polarization at the roots of the stator teeth. Core losses in these areas of the stator were found to greatly exceed those one would expect from Epstein test results.

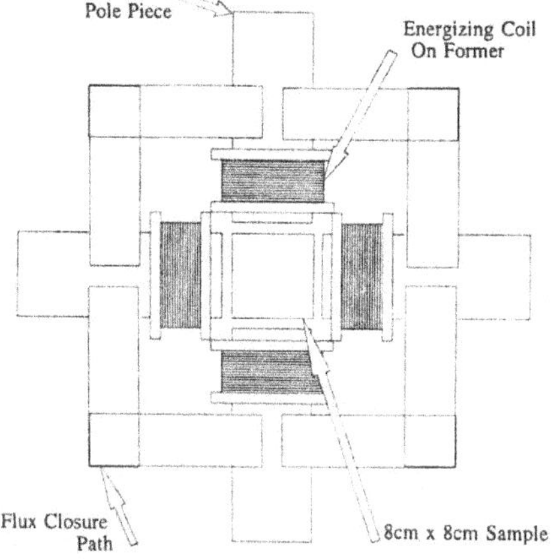

Figure 1. Magnetizing yoke asembly for measuring rotational iron losses in a sample.

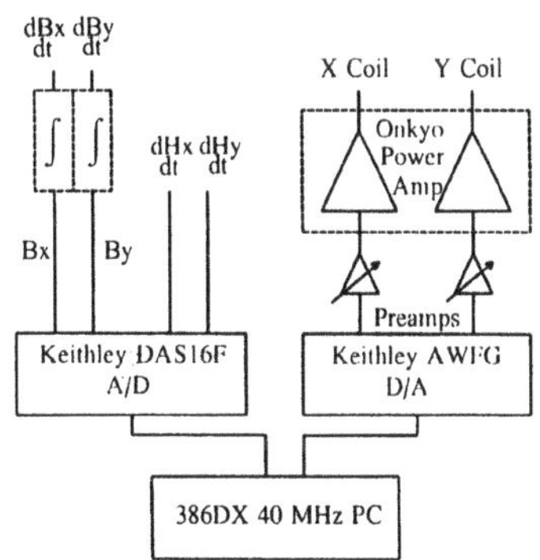

Figure 2. Waveform generation and data acquisition.

Electric and Magnetic Fields, Edited by A. Nicolet
and R. Belmans, Plenum Press, New York, 1995

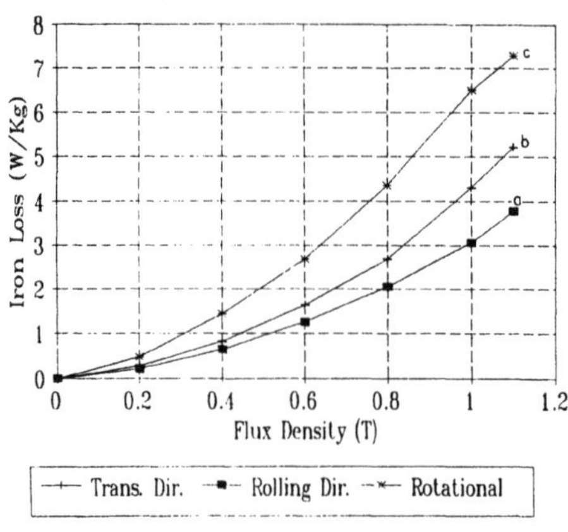

Figure 3. Rotational and alternating losses for a motor lamination steel.

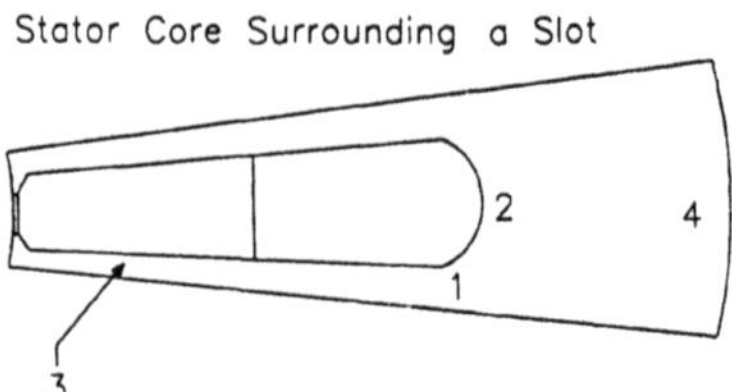

Figure 4. Regions of interest.

PROPOSED ANALYSIS USING FEM AND ROTATIONAL IRON LOSS DATA

We have discussed the possibility of using a time harmonic finite element method for determing the degree of flux polarization in an induction motor stator, in a previous paper [3]. A cross section a 3 phase, 4 pole, 5 hp squirrel cage induction motor was meshed using a first order triangular element discretization. Four regions of interest on the machine stator were investigated. These regions are show in figure 4.

Because the problem is time harmonic, the excitation currents in the stator slots must be represented as phasor quantities. We normalize the current density in all phase 'a' conductors to (1 + j0) A/mm^2. For phase 'b' conductors the current density is defined as (-0.5 + j0.867) A/mm^2. All iron paths are assumed to behave in a linear manner with a relative permeability of 1000. The magnitude of the flux density can be scaled by increasing the value of the excitation current. Since we have modelled the problem under linear conditions, the polarization pattern will not change. An excitation frequency of 60 Hz throughout the entire domain of the problem stimulates blocked rotor conditions.

The flux density was found to have only an alternating component in regions 3 and 4. In region 3 the flux was found to alternate radially along the stator tooth, while in region 4 it was found to alternate tangentially along the back of the stator core. In region 2 an elliptically polarized flux density was found, as shown in figure 5. The ellliptical flux density pattern represents a polarization of approximately 10 percent. Figure 6 shows the

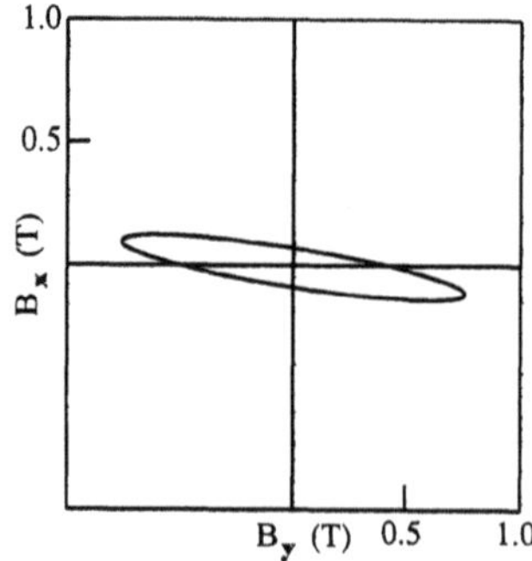

Figure 5. Flux Density Polarization in Region 2.

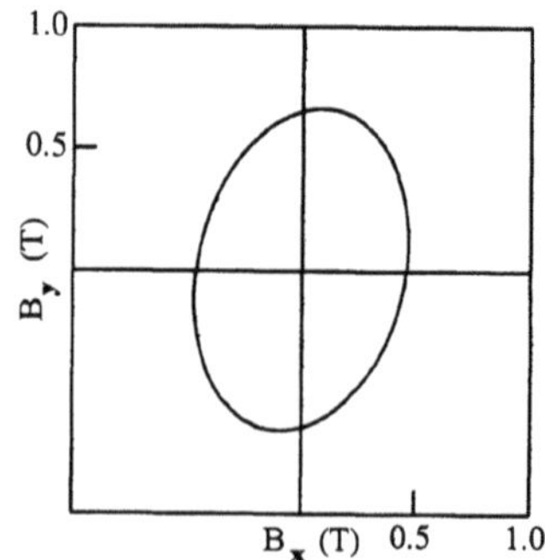

Figure 6. Flux Density Polarization in Region 1.

almost circular polarization found in region 1. A polarization of a approximately 65 percent is found in this region.

The stator of the machine is subdivided into 4 volumes. Each volume corresponds to one of the regions shown in Figure 4. The total stator core loss was calculated using two different assumptions. In the first case only the losses due to alternating flux are considered. In the second case the losses due to rotational losses are included. The values of flux density are scaled linearly so that a flux density of 1.2 T is assigned to the teeth (region 3). Resulting values of flux density for the various regions are shown in Table I. For the elliptically polarized regions, values of B_x and B_y are given. For regions where the flux is all alternating, values of B are given. Values of alternating and rotational loss for a cold rolled motor lamination steel were obtained using our test apparatus and are shown in Table II. Loss calculations were then carried out by converting the loss data into a loss/unit volume form (by multiplying by the mass density of the material) and summing the losses in each of the 4 stator volumes.

Table I Flux Density in Various Regions

Region #	Corresponding Location	Flux Density
1	Tooth Root	B_x = 0.67 T B_y = 0.44 T (Elliptical Polarization)
2	Back of Stator Slot	B_x = 0.77 T B_y = 0.077 T (Elliptical Polarization)
3	Tooth	B = 1.2 T (Alternating)
4	Core Back	B = 0.67 T (Alternating)

Table III shows the loss attributable to each region of the machine stator. When calculating the losses under the assumption of alternating flux only, the flux densities in polarized regions were taken as the values which corresponded to the major axis of the polarization ellipse in each particular region.

Table II Iron Loss Data

B_x (T)	B_y (T)	Iron Loss (W/kg)
0.67	0.44	2.18
0.77	0.077	1.69
1.2	0.0	4.09
0.77	0.0	1.61
0.67	0.0	1.26

Table III Polarized Iron Losses

Region	Losses (No Polarization) W	Losses (Polarization) W
1	1.86	3.22
2	2.83	2.97
3	9.03	9.03
4	3.31	3.31

The net iron loss for the stator considering only the alternating flux is 17.03 W. With the flux density polarization accounted for the loss is 18.53 W. This represents an increase of approximately 9 percent. Note that approximately half the losses in the stator are attributable to the losses in the teeth due to the higher flux density in that region. If the tooth losses are ignored, and only the losses attributable to the stator core are considered, the discrepancy is even greater. The net iron loss for the stator core considering alternating losses is 7.99 W. Accounting for the polarization, the net loss is 9.49 W. This represents an increase of approximately 19 percent.

Discussion

Calculating core losses using formulations similar to equation (1) may yield values which are too low. There are two fundamental problems encountered by machine designers when trying to predict additional losses due to flux polarization in a rotating machine core. The main problem results from the lack of a standardized test for measuring iron losses under rotational flux conditions. If a standard test could be agreed upon, steel producers would be able to provide rotational iron loss curves to motor manufacturers to supplement Epstein test data. It is difficult to account for increased losses due to flux polarization using Epstein test data alone. The second problem relates to the lack of analysis tools for ascertaining the degree of polarization in the various regions in the core. We have proposed a finite element method, as a first attempt at providing an analysis tool and it should be considered useful as an illustration at best. In order to make this method of analysis more accurate, certain assumption must be taken into consideration. Higher order elements might yield more accuracy since the curl of the magnetic vector potential is required to obtain the polarization of the magnetic flux density. The curl of a first order vector potential field yields a zero order flux density. The assumption of a constant permeability neglects saturation effects. The reader should also be warned that rotor motion has not been modelled since we have assumed a blocked rotor condition. We have ignored the losses in the rotor, since we are considering it strictly as a flux path for the fundamental field. Under normal operating conditions the losses in the rotor would be very small compared to the stator losses.

Conclusions

Rotational losses in MLQ steels are substantially higher than alternating losses at moderate flux density levels. At the base of the slot and in the tooth region alternating flux loss dat may be used, taking into account the appropriate rolling directions. However, at the base of the tooth and in the core area, rotational flux loss data should be used. This process is not followed in the current practices for machine design. This work should be extended to regions of higher flux density, which, we believe will have significantly higher losses. Finally, a standardized test should be agreed using the process outlined above, rather than the limited data currently used from the Epstein test.

References

[1] F.G. Baily, "The Hysteresis of Iron and Steel in a Rotating Magnetic Field", Philosophical Transactions of the Royal Society, 1986, Vol. 187, pp 715-746.

[2] M. Enokizono, T. Suzuki, J. Sievert, J. Xu, "Rotational Power Losses of Silicon Steel Sheet", IEEE Transactions on Magnetics, Vol. 26, No. 5, September 1990, pp 2562-2564.

[3] R.D. Findlay, N. Stranges, D.K. MacKay, "Losses Due to Rotational Flux in Three Phase Induction Motors", Paper # 94-115-6-EC, PES Winter Power Meeting, New York, January 30 -February 3, 1994.

[4] A.J. Moses, G.S. Radley, "Experimental Simulation of Magnetic Flux and Power Loss Distribution in the Stator Core of a Large Rotating Machine", Journal of Magnetism and Magnetic Materials, 19(1980), pp 60-62.

[5] G.S. Radley, A.J. Moses, "Apparatus for Experimental Simulation of Magnetic Flux and Power Loss Distribution in a Turbogenerator Stator Core", IEEE Transactions on Magnetics, Vol. MAG-17, No. 3, May 1981, pp 1311-1316.

A COMBINED FINITE ELEMENT - CIRCUIT MODEL OF A SQUIRREL-CAGE INDUCTION MOTOR

R. De Weerdt, U. Pahner, R. Belmans

Dept. E.E. - Electrical Energy
K.U. Leuven
Kardinaal Mercierlaan 94
B-3001 Heverlee - Leuven
Belgium

E. Tuinman

Holec Ridderkerk
The Netherlands

1. INTRODUCTION

The paper describes the results of a squirrel-cage induction motor analysis taking saturation and motor end effects into account. The main part of the induction motor analysis is formed by an iterative method earlier developed by Belmans et al.[1] allowing the inclusion of saturation in the solution. The method can be summarised as follows: The procedure starts (for current-driven mode of operation) by solving two static non-linear problems, with real and imaginary part of the stator currents as input. From these solutions an averaged reluctivity vector is generated. This vector is used with the time-harmonic solver to obtain a first estimate of the induced rotor currents. After the time-harmonic solving, two new static problems can be defined, but now with the real and imaginary part of both stator and rotor currents as input. A more accurate reluctivity vector can be extracted. This procedure is repeated until convergence is reached. For the voltage-driven mode, the procedure starts with a time-harmonic solution generating a first estimate of both stator and rotor currents. During the time-harmonic solution, end effects of the motor are taken into account. This is done by linking the finite element model with external sources and impedances [2]. The time-harmonic solver solves both the finite element model and the circuit equations from the external circuit. Temperature effects are included by a variation of the conductivity of the stator winding and the rotor conductors. Since the end-effect parameters are introduced in the model as lumped parameters, values must be assigned to them. These values are calculated from analytical formulas or using a finite-element approach.

2. DETERMINATION END-EFFECT PARAMETERS

For the estimation of the stator end winding parameters (resistance and inductance), a number of analytical formulas exist. The end winding resistance can accurately be calculated using the geometric parameters of the conductor. Formulas of the end-winding inductance can be found in the work by Nürnberg [3] and Richter [4] . The calculation of the end ring impedance can be performed using analytical or finite element techniques. here, the ring resistance is calculated analytically based on the ring geometry, the ring inductance is calculated using an axisymmetric finite element problem definition as described in [5]. The

Electric and Magnetic Fields, Edited by A. Nicolet
and R. Belmans, Plenum Press, New York, 1995

advantage of the finite element method is that the iron core, the shaft and the frame of the motor can be modelled more accurately than using an analytical approach.

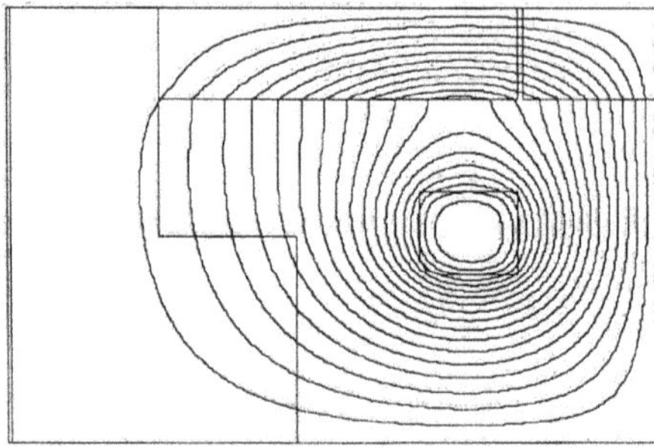

Figure 1. Flux plot end ring inductance calculation.

3. END RING INCLUSION WHEN ONLY PART OF THE MOTOR IS MODELLED

It is common practice to model only the smallest necessary part of the motor since this reduces computation time and allows a more accurate modelling of the motor. But when modelling only one pole of the motor the part of the ring that is not part of the model must not be neglected. Doing so would force the sum of the rotor bar currents to be zero and this is not correct for one pole. The solution of this problem is to enclose the finite element model with two impedances to ensure a continuity in the ring current. Assume the modelling of one pole of a 4-pole motor with 12 rotor bars (Figure 2):

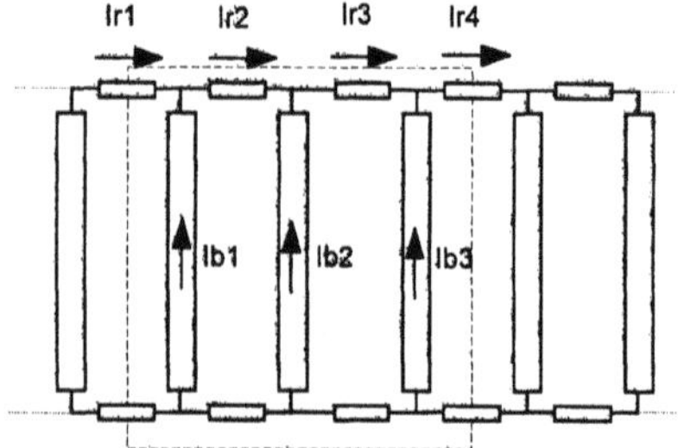

Figure 2a. Full rotor cage representation. The pole modelled is inside the dashed line.

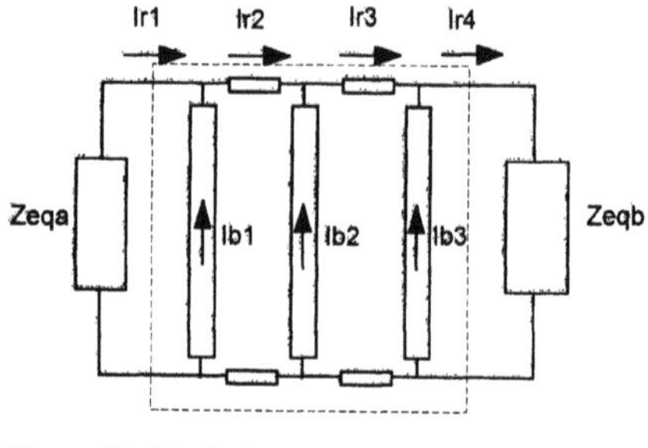

Figure 2b. Equivalent rotor cage representation.

For the currents in the bars and ring parts of a squirrel-cage induction motor, analytical expressions exist. For the equivalent rotor cage (Fig.2b) four voltage differential equations for each of the loops may be derived containing the ring- and bar currents and the equivalent impedances $\underline{Z}_{eqa}$ and $\underline{Z}_{eqb}$. Applying the analytical solution of the rotor cage to this set of equations, we find the following expressions for the equivalent impedances:

$$\underline{Z}_{eqa} = \underline{Z}_R + j * \underline{Z}_R * \text{cotg}\left(\frac{p * \pi}{N}\right) \qquad \underline{Z}_{eqb} = \underline{Z}_R - j * \underline{Z}_R * \text{cotg}\left(\frac{p * \pi}{N}\right)$$

In these formulas $\underline{Z}_R$ is the ring impedance between two adjacent bars.
Figure 3 show the effect on the bar- and ring currents when the equivalent impedances are applied or neglected (the motor used is a 4-pole motor with 40 rotor bars, one pole is modelled):

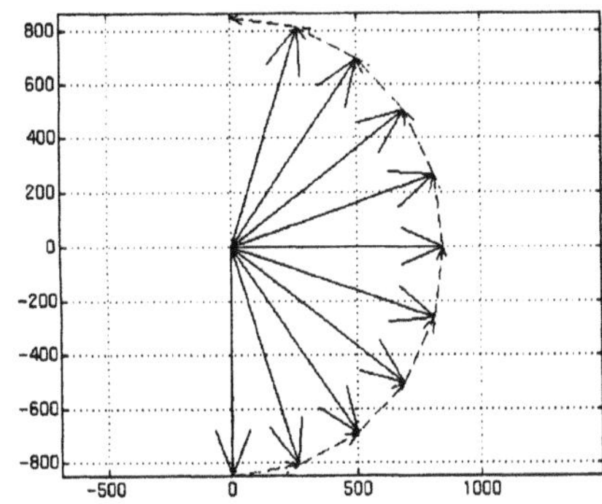

Figure 3a. Bar and ring currents when equivalent impedances are applied.

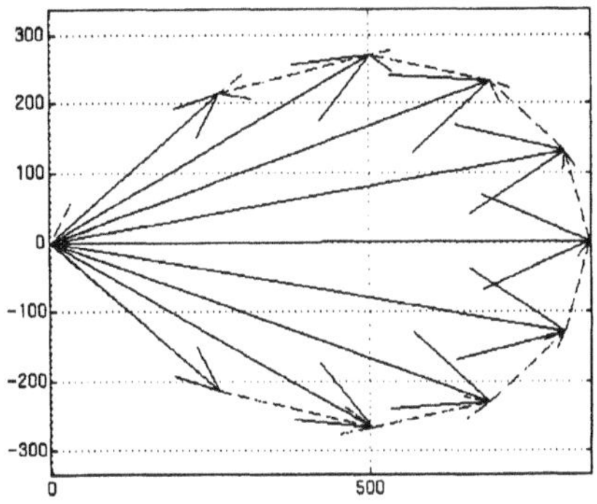

Figure 3b. Bar and ring currents when no impedances are applied

4. COMBINED FINITE ELEMENT - CIRCUIT MODEL

The above described method is applied to a 140 kW traction motor.

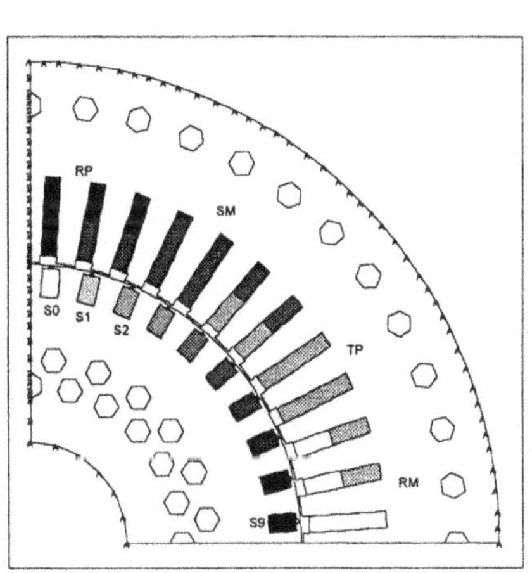

Figure 4a. Outline 2D mesh.

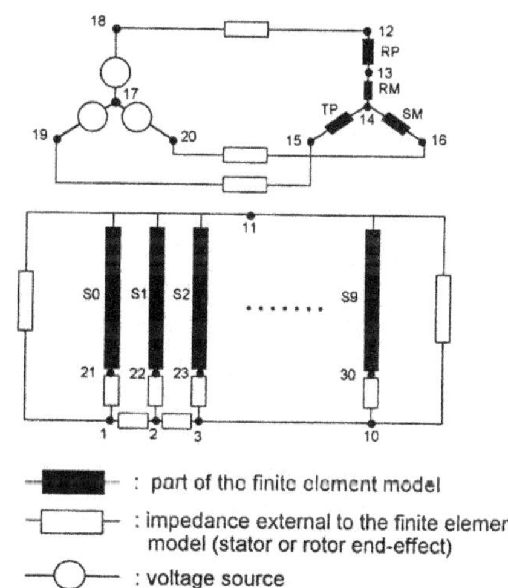

Figure 4b Layout of the combined finite element - circuit model.

5. COMPARISON BETWEEN CALCULATIONS AND MEASUREMENTS

Voltage-driven and current-driven calculations are performed for no-load, locked rotor and load conditions. The effect of temperature (by changing the conductivity) and motor end-effects can be examined since these effects are incorporated step by step in the different calculations. The five calculations have the following features:

calculation 1: conductivity at room temperature and no end-effects
calculation 2: conductivity at measurement temperature and no end-effects
calculation 3: calculation 2 and only stator end-winding impedance included
calculation 4: calculation 3 and inclusion end ring inductance
calculation 5: inclusion of all end-effects and conductivity at measurement temperature.

The following table shows the stator current and voltage for measurements and calculations at the different operating points. The current-driven calculations are compared with the same (voltage-driven) measurements.:

Table 1. Comparison between calculations and measurements.

	measured current (A)	calculation. 1 (A)	calculation. 2 (A)	calculation 3 (A)	calculation 4 (A)	calculation 5 (A)
no load	30.8	31.6	31.3	31.1		
locked	50.4	62.0	58.7	58.0	52.5	51.3
load	57.3	119.5	90.6	90.1	72.8	58.9
	Applied voltage (V)	calculation 1 (V)	calculation 2 (V)	calculation 3 (V)	calculation 4 (V)	calculation 5 (V)
no load	1626	1526	1526	1532		
locked	206.6	170.9	179.3	183.0	194.0	198.0
load	1623	700.0	914.5	920.3	1112.4	1595

Only when the end effects and the correct conductivity are included, the current driven and voltage driven calculations give comparable results. It should also be mentioned that the induced rotor currents are strongly affected by these effects in the voltage driven calculations but hardly change in the current driven calculations. This behaviour corresponds with the real motor behaviour and can be illustrated with the equivalent circuit of the induction motor. Fixing the stator current and the slip in the equivalent circuit leads to a certain rotor current. Changing the rotor resistance by 100 % hardly affects the rotor current, but substantially changes the stator voltage.

6. CONCLUSIONS

The steady state behaviour of squirrel-cage induction motors can accurately be calculated using a combined finite element - circuit model. Temperature effects and motor end effects have to be incorporated to obtain a realistic solution. The different behaviour of the induction motor when current-driven or voltage-driven is also encountered in the calculations. The current-driven motor seems much more insensitive to changes in the rotor impedance when only the rotor currents are examined. To draw the correct conclusions also the induced stator voltage has to be considered. The end-ring resistance is the most important end-effect parameter, together with the conductivity change due to temperature rise, a difference of 50 % can occur in the solutions. For the locked rotor calculation, the end ring inductance becomes more important than the end ring resistance.

ACKNOWLEDGEMENTS

The authors are indebted to the Belgian Nationaal Fonds voor Wetenschappelijk Onderzoek for its financial support for this work and the Belgian Ministry of Scientific Research for granting the project IUAP No. 51 on Magnetic Fields.

7. REFERENCES

1 Belmans R., De Weerdt R., and Tuinman E., Combined field analysis techniques and macroscopic parameter simulation for describing the behaviour of medium-sized squirrel-cage induction motors fed with an arbitrary voltage, EPE Brighton, September 1993, pp. 413-418
2 MagNet User Guide, Infolytica, 1994
3 Nürnberg W., "Die Asynchronmaschine", Springer Verlag, 1963
4 Richter R., "Berechnung elektrischer Maschinen", Birkhäuser Verlag Basel, 1967
5 Williamson S, and Mueller M.A, Calculation of the impedance of rotor cage end rings, IEE PROCEEDINGS-B, Vol. 140, No.1, January 1993, pp. 51-60

USE OF A CUBIC FINITE ELEMENT-BOUNDARY ELEMENT COUPLING METHOD IN THE COMPUTATION OF THE ELECTROMAGNETIC PARAMETERS OF A SWITCHED RELUCTANCE MOTOR

Avoki Omekanda, Christian Broche, Michel Crappe, and René Baland

Faculté Polytechnique de Mons
Bd Dolez, 31 700 Mons, Belgium

INTRODUCTION

The design optimization and behaviour prediction of a Switched Reluctance Motor (SRM) require a good knowledge of magnetic flux linkage through a motor phase and torque on the shaft[1]. In order to reach a good accuracy in the computed magnetic field, cubic curvilinear elements are used because of high precision furnished by these elements in the nodal approximation even in narrow part of the air gap. Cubic finite element formulation is applied in all internal regions of the motor. So, the magnetic non-linearity in the stator, rotor and shaft can be adequately treated. Knowledge of magnetic vector potential on the internal nodes permits the establishment of flux density lines maps over the whole motor cross-section and offers facilities in post-processing calculation of electromagnetic parameters. Cubic boundary element formulation is applied in the external air region in order to facilitate positioning of the external air outline as far as possible.

MAGNETIC FIELD EQUATION

For a SRM modelization, stationary magnetic fields are taken into account. Indeed, this stationary field assumption is valid in the context of experiments carried out for determination of static electromagnetic parameters of the machine with rotor locked. For simulation of a SRM magnetic state, the end effects and hysteresis effects are neglected.

The two-dimensional magnetostatic problem is described by the scalar non linear Poisson's equation

$$\frac{\partial}{\partial x}\left(\nu\frac{\partial A}{\partial x}\right) + \frac{\partial}{\partial y}\left(\nu\frac{\partial A}{\partial y}\right) = -J \qquad (1)$$

in which A and J are the z-components of respectively the magnetic vector potential and the current density vector, ν is the magnetic reluctivity and (x,y) are the cartesian coordinates of a local point on the considered domain. Continuity conditions along the motor's interface regions and a Dirichlet boundary condition on the external air outline[2] must be taken into account in the resolution of the partial differential equation (1).

Electric and Magnetic Fields, Edited by A. Nicolet
and R. Belmans, Plenum Press, New York, 1995

APPLICATION OF COUPLING METHOD

The combination of BEM equations (5), FEM equations (7), continuity equations along interfaces and boundary conditions on the outline external air gives a global algebraic system of equations. This system is non linear because of magnetic saturation occuring in the iron regions of the motor. The resolution of the system is carried on using Newton-Raphson method. After resolution, the vector potential at internal nodes and along interfaces, and its normal derivative at the external air outline and along interfaces are known[3]. This coupling method is applied for the computation of electromagnetic parameters of a switched reluctance motor. The main dimensions and power of the SRM prototype are : stator pole arc = 22.5°, Rotor pole arc = 30.0°, height of stator pole = 19.0 mm, height of rotor pole = 9.0 mm, length of air gap = 1.0 mm, stator core outer diameter = 229.8 mm, length of iron core = 300.0 mm, diameter of the shaft = 55.0 mm, number of turns per pole = 10 and motor power = 7.5 kW. The non-oriented silicon steel laminations are used in the stator and rotor cores.

Figure 1 (c) shows an example of cubic finite element and boundary element mesh generated from the geometric data of the SRM prototype.

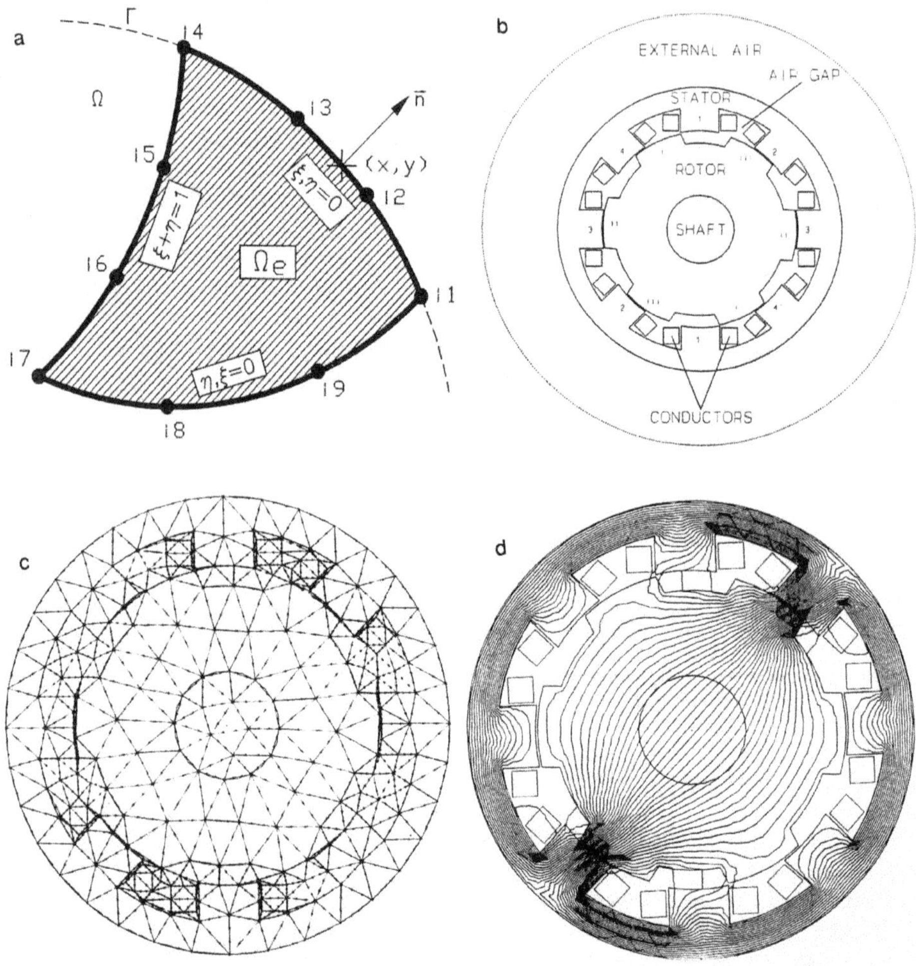

Figure 1. (a) A 9-nodes cubic finite element. (b) SRM magnetic domains. (c) FE-BE mesh of the SRM in the rotor position equal to 0°. (d) Distribution of the magnetic flux density lines.

ELECTROMAGNETIC ANALYSIS OF THE SRM PROTOTYPE

In order to obtain a complete modelization of the SRM, magnetic field was computed for several values of excitation phase current and rotor position. For each pair of these variables, a global system of equations was constituted and its resolution has given the distributions of magnetic vector potential and its normal derivative. Different kinds of magnetic maps could be established. Figure 1 (d) shows a result of the SRM magnetic simulation in the multiconduction mode of operation. In this last simulated operation, the SRM rotor is located at zero degree position while excitation currents are equal to 20 A in the phase 2-2' and 110 A in the adjacent phase 1-1'.

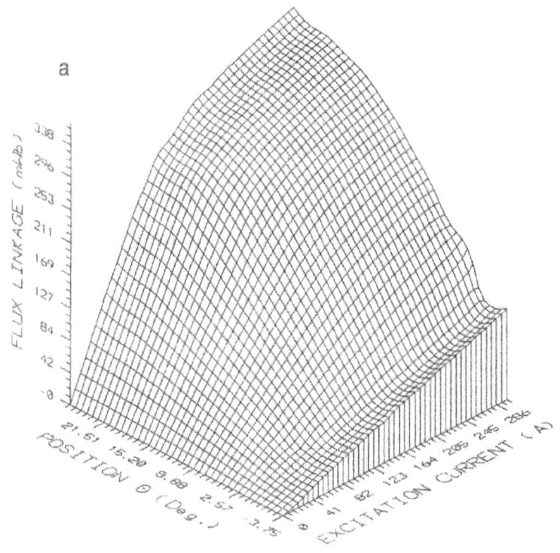

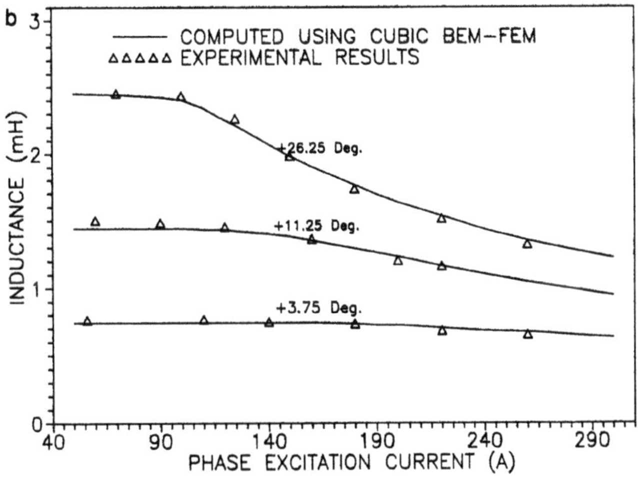

Figure 2. (a) Flux linkage versus rotor position and excitation current. (b) SRM phase inductance.

CUBIC BOUNDARY ELEMENT FORMULATION

Boundary element method (BEM) will be applied in the external air that is a non conducting (current density J equal to zero) and linear region of magnetic reluctivity ν_0. According to the differential partial equation (1) and the application of the Green's identity to this equation we have the following relationship[3]

$$C_l A_l = \sum_{e=1}^{M} \int_{\Gamma_e} \left(G \frac{\partial A}{\partial n} + A \frac{\partial G}{\partial n} \right) d\Gamma \tag{2}$$

in which Γ_e is a boundary element due to the discretization of the boundary Γ in M boundary elements and G is the Green's function. **n** is the unit outward normal vector at a current point (x,y) on the boundary Γ of the region. Coefficient C_l is equal to 1 when the observation point l is internal to the domain, but it depends on the local curvature when the point l is on the domain boundary. The cubic evolution of the magnetic vector potential on a 4-nodes one-dimensional element is

$$A = N_1^* \cdot A_{i1} + N_2^* \cdot A_{i2} + N_3^* \cdot A_{i3} + N_4^* \cdot A_{i4} \tag{3}$$

where shape functions are

$$N_1^* = -\frac{1}{16}(1-\xi)(1-9\xi^2) \; ; \; N_2^* = +\frac{9}{16}(1-\xi^2)(1-3\xi) \; ; \; N_3^* = +\frac{9}{16}(1-\xi^2)(1+3\xi) \; ; \; N_4^* = -\frac{1}{16}(1+\xi)(1-9\xi^2) \tag{4}$$

The normal derivative of vector potential and cartesian coordinates x, y of a current point on a cubic boundary element Γ_e are expressed in the same way as the vector potential (3).

Taking account of relation (2) and consideration of the cubic approximations of vector potential and its normal derivative on a 4-nodes boundary element give the following algebraic equation

$$C_l A_l + \sum_{e=1}^{M} \left[p_{1,l}^{(e)} \cdot A_{i1} + p_{2,l}^{(e)} \cdot A_{i2} + p_{3,l}^{(e)} \cdot A_{i3} + p_{4,l}^{(e)} \cdot A_{i4} \right] = \sum_{e=1}^{M} \left[q_{1,l}^{(e)} \cdot \left(\frac{\partial A}{\partial n}\right)_{i1} + q_{2,l}^{(e)} \cdot \left(\frac{\partial A}{\partial n}\right)_{i2} + q_{3,l}^{(e)} \cdot \left(\frac{\partial A}{\partial n}\right)_{i3} + q_{4,l}^{(e)} \cdot \left(\frac{\partial A}{\partial n}\right)_{i4} \right] \tag{5}$$

in which

$$p_{i,l}^{(e)} = -\frac{1}{2\pi} \int_{-1}^{+1} N_i^* \frac{\partial r}{\partial n} \frac{1}{r} j^* d\xi \quad ; \quad q_{i,l}^{(e)} = +\frac{1}{2\pi} \int_{-1}^{+1} N_i^* \ln\frac{1}{r} j^* d\xi \tag{6}$$

j^* is the jacobian of the geometric transform between the cartesian coordinates x, y and the local coordinate ξ on a cubic boundary element. r is the distance from an observation point l of the domain to a considered current point on the boundary element.

CUBIC FINITE ELEMENT FORMULATION

Taking account of the discretization of the domain and its boundary in respectively M' finite elements and M boundary elements gives

$$\sum_{e'=1}^{M'} \iint_{\Omega_{e'}} \nu \left(\frac{\partial N}{\partial x}\frac{\partial A}{\partial x} + \frac{\partial N}{\partial y}\frac{\partial A}{\partial y} \right) d\Omega_{e'} = \sum_{e=1}^{M} \int_{\Gamma_e} \nu N^* \frac{\partial A}{\partial n} d\Gamma_e + \sum_{e'=1}^{M'} \iint_{\Omega_{e'}} N J d\Omega_{e'} \tag{7}$$

$\Omega_{e'}$ is a 9-nodes triangular element as shown in figure 1(a). The nodal approximation of the magnetic vector potential on this element is

$$A = \sum_{j=1}^{9} N_j(\xi,\eta) \cdot A_{ij} \tag{8}$$

When the 9-nodes triangular element has a side in common with the boundary Γ, the approximation of the normal derivative of vector potential is

$$\frac{\partial A}{\partial n} = N_1^* \cdot \left(\frac{\partial A}{\partial n}\right)_{i1} + N_2^* \cdot \left(\frac{\partial A}{\partial n}\right)_{i2} + N_3^* \cdot \left(\frac{\partial A}{\partial n}\right)_{i3} + N_4^* \cdot \left(\frac{\partial A}{\partial n}\right)_{i4} \tag{9}$$

The exploitation of the Gauss formula[3] has permitted the calculation of total flux linkage of a motor phase. The evolution of this magnetic parameter versus rotor position and excitation current is presented in the figure 2(a). From the knowledge of flux linkage, the phase winding inductance that is an important parameter too was calculated. Figure 2(b) shows evolution of a phase inductance of the SRM prototype as a function of excitation current for different values of angular rotor position. This last figure presents a comparison between calculated and measured phase inductance. Among three usable methods for torque calculation, the Maxwell stress-tensor method has been estimated interesting for the computation of electromagnetic torque in a SRM from BEM-FEM field solution. Indeed, the tangential flux density component along an interface is directly deduced from the knowledge of the normal derivative of magnetic vector potential. The Maxwell stress-tensor formulation[4] used for calculation of the SRM torque T is

$$\vec{T} = \frac{1}{\mu_0} L_c \int_{\Gamma_R} [(\vec{r}\times\vec{B})(\vec{B}\cdot\vec{n}) - \frac{1}{2} B^2(\vec{r}\times\vec{n})]\, d\Gamma \tag{10}$$

in which Γ_R is the closed interface between the rotor and the air gap, L_c is the length of rotor iron core. Figure 3(a) gives the evolution of computed torque as a function of excitation current and angular rotor position. A comparison between computed values and experimental values of the SRM torque for excitation current equal to 70 A is presented in figure 3(b).

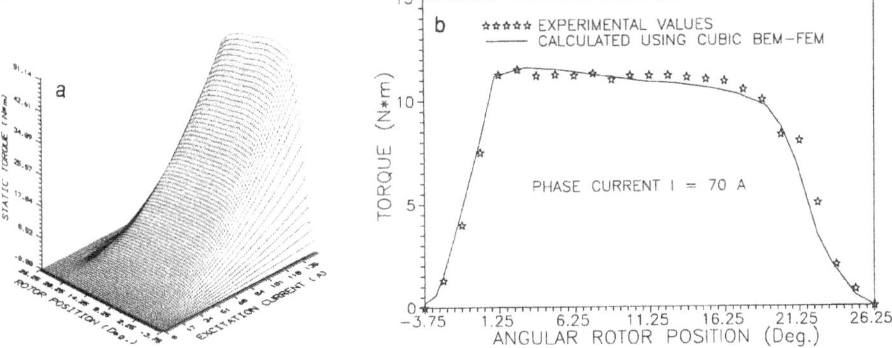

Figure 3 : (a) Torque versus rotor position and excitation current. (b) Comparison between measured and calculated torque.

CONCLUSION

A two-dimensional technique coupling cubic finite element method and cubic boundary element method has been presented. The computation of magnetic field in a switched reluctance motor has been done for different excitation current and several angular rotor positions. High precision of the cubic finite and boundary elements used in the coupling technique has permitted calculation of magnetic quantities more precisely.
Magnetic field distributions, flux linkage, inductance and torque in the given SRM prototype have been calculated. Computed magnetic quantities agree with measured values.

REFERENCES

1. W.F. Ray, P.J. Lawrenson, R.M. Davis, J.M. Stephenson, N.N. Fulton, and R.J. Blake, 'High Performance Switched Reluctance Brushless Drives', IEEE Trans. Ind. Appl., Vol. IAS-71, pp. 1769-1776, 1985.

2. A. Omekanda, C. Broche, R. Baland, 'Computation of Magnetic Field in a Switched Reluctance Motor Using a Quadratic Hybrid BIEM-FEM Method', ETEP, Vol. 2, No. 5, pp. 303-307, Sept/Oct. 1992.

3. A. Omekanda, 'Analyse du Comportement Electromagnétique d'un Moteur à Réluctance à Commutations. Utilisation d'une Méthode Hybride : Eléments Finis-Equations Intégrales de Frontière', PhD Thesis, FPMs, 1993.

4. B. Aldefeld, 'Forces in Electromagnetic Devices', COMPUMAG-78, Grenoble, Paper 8.1, Sept. 1978.

AUTOMATED 3D MESH GENERATION SUITED FOR OPTIMISATION

TB. Johansson, R. Belmans

Dept. E.E. - Electrical Energy
K.U. Leuven
Kardinaal Mercierlaan 94
B-3001 Heverlee - Leuven
Belgium

ABSTRACT

The paper deals with automation of an extrusion-based 3D mesh generator to adopt it for optimisation of electrostatic micromotors. It shows that it is possible to generate relatively complex motor models very fast and reliable. Limitations of this technique and suggestions for improvements are discussed at the end. The technique has been developed in the framework of a BRITE/EURAM project aiming for 3D optimisation of different kinds of electrostatic micromotors.

1. INTRODUCTION

Building 3D meshes is normally a time consuming process. Therefore performing optimisation in 3D is very tedious since a large number of models, slightly different from one another, must be built and analysed. For motor structures the required labour is even vaster since for every model the analysis normally must be carried out for a number of rotor positions, this in order to obtain macroscopic parameters as torque and inductance or capacitance as function of the rotor position. For most 3D FE packages a new rotor position requires a new mesh to be generated. The paper presents a technique where the fast and reliable extrusion-based 3D mesh generator can be automated so that a large range of different motor geometries can be generated, with the choice of rotor position fully arbitrarily.

2. PERIODIC GEOMETRY

All electric rotating motors inherently possess some kind of periodicity. Typically each motor has a stator and a rotor where the *geometric period*, in degrees, is defined as the pole pitch. Some parts of the motor geometry have a different period from both the rotor and the

Electric and Magnetic Fields, Edited by A. Nicolet
and R. Belmans, Plenum Press, New York, 1995

stator. The rotor shaft for instance can be described as having a *geometric period* of 360°. Each periodic geometry also has an extension within the pole pitch. This extension will in this text be referred to as the *polar extension*. For a variable capacitance motor (fig. 1) the *polar extension* of the stator teeth would be the same as the tooth width, in degrees, of the stator teeth.

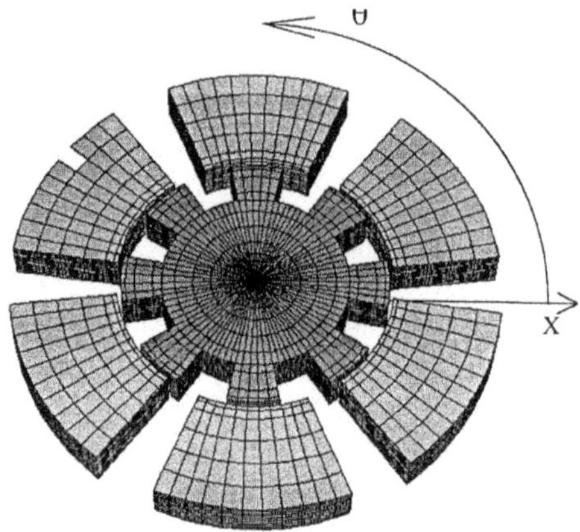

Figure 1. Mesh of a 6/8 pole electrostatic micromotor built by rotating the base-plane in positive θ direction. The reference point θ = 0 is chosen in the middle between two stator teeth.

With these two parameters, the *geometric period* τ and the *polar extension* τ', it becomes easy to describe where the sides, parallel with the rotor axis, of the stator teeth and rotor teeth are located in the motor. If the motor is scanned in positive θ direction (fig. 1), then the following equations define the location of the front-sides and the back-sides of the stator teeth.

$$\theta^1_{n,\mathrm{front}} = \left(\frac{\tau_1 - \tau'_1}{2}\right) + n \cdot \tau_1 \tag{1a}$$

$$\theta^1_{n,\mathrm{back}} = \left(\frac{\tau_1 + \tau'_1}{2}\right) + n \cdot \tau_1 \tag{1b}$$

$$\left\{ n{:}0,1,\ldots,\left(\frac{360° - \tau_1}{\tau_1}\right)\right\} \quad \text{(0 to number of poles minus one)} \tag{1c}$$

The reference point for θ is chosen in the middle between two stator teeth. By introducing the offset angle α accounting for the rotor position, the same may be done for the rotor.

$$\theta^2_{n,\mathrm{front}} = \left(\frac{\tau_2 - \tau'_2}{2}\right) + n \cdot \tau_2 + \alpha_2 \tag{2a}$$

$$\theta^2_{n,\mathrm{back}} = \left(\frac{\tau_2 + \tau'_2}{2}\right) + n \cdot \tau_2 + \alpha_2 \tag{2b}$$

$$\left\{ n{:}0,1,\ldots,\left(\frac{360^{\circ}-\tau_2}{\tau_2}\right)\right\} \quad \text{(0 to number of poles minus one)} \tag{2c}$$

Except for those parts of the motor geometry having $\tau = 360°$, the motor can be split in more then 2 periodic geometries in the general case. Subscripts (superscripts for θ) 1 and 2 indicate parameters for the stator and rotor respectively.

3. BUILDING 3D GEOMETRY-PERIODIC MODELS USING EXTRUSION

The technique, discussed in this paper, only uses rotation around the y-axis, to define the extrusion data. One vertical side of the base-plane (fig. 2) should coincide with the global y-axis. Since each copy of the base-plane only gets the θ parameter as extrusion data, one of the vertical sides of all planes also coincide, with the global y-axis, thus will the centre of the rotor axis (fig. 3).

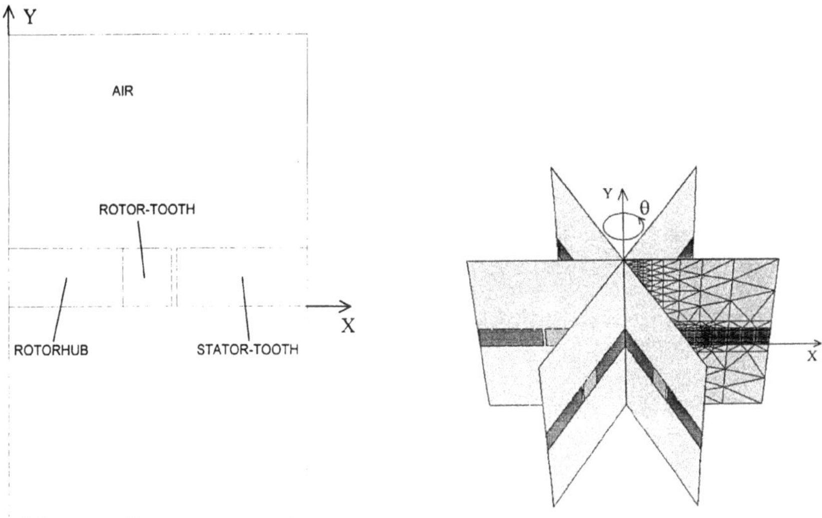

Figure 2. Outline of the base-plane used when generating model of figure 1.

Figure 3. Copies of the base-plane mesh are placed at different angles θ.

From the knowledge of the number of poles, the *polar extension* τ' of the stator and rotor, and the rotor position, and using Eqns. (1) and (2) it is easy to find where planes must be located. For each of the periodic geometries 4 different types of action (sequence of relabeling and/or constraining) are defined. Types 1 and 2 define a front and a backside respectively, on e.g. a rotor tooth. Types 3 and 4 define a plane inside or outside of e.g. a rotor tooth. These four actions define every plane for each periodic geometry. Thus the maximum number of different plane-types required to construct a model is 4^N, where N is the number of periodic geometries in the model. These planes are generated in advance and are sufficient to define all possible combinations of τ, τ' and α. This means that ones the planes are generated the remaining task is to find out at what angle which kind of plane must be located. This is performed with repetitive use of Eqns. (1) and (2) and a sorting algorithm. This sorting algorithm finds what the next plane must be from knowing the previous. A front-plane must be followed by and inside-plane, an inside-plane by a backplane and so on. The angle between two consecutive planes has to be chosen depending on the size of the elements in the base-plane and the required aspect ratio of the 3D elements, the tetrahedrons. This implies that the same kind of planes sometimes has to be repeated with the desired angle between them in order to improve the aspect ratio of the tetrahedrons. This also improves the rounded shape of the motor model. Except for that building motor models by rotating the base-plane makes it possible to generate motor models with any combination of τ, τ' and α, using only one 2D mesh, it has also a second advantage. The outline of this 2D mesh, the base-plane, tends to be very simple even for more complicated 3D motor models (figs. 4 and 5).

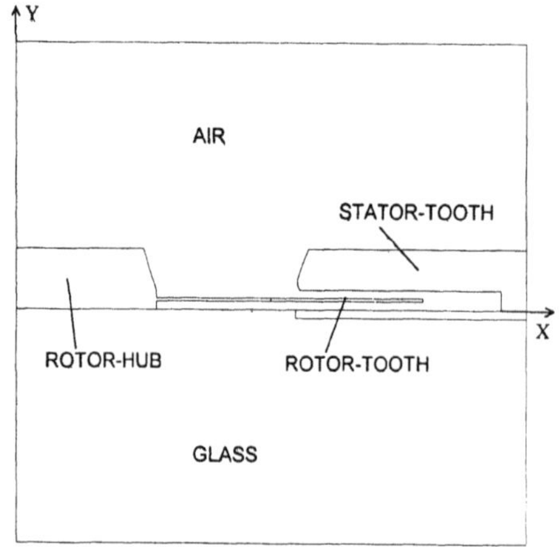

Figure 4. Outline of the base-plane used for generating model from figure 5.

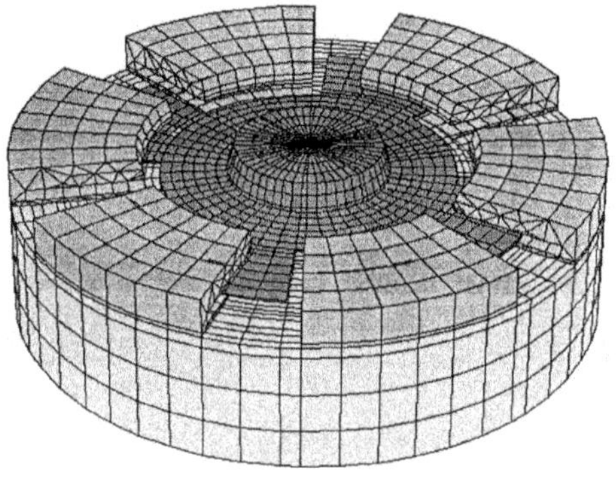

Figure 5. Mesh of a 6/4 pole axial flux electrostatic micromotor.

4. PERFORMANCE

The major advantage automating the extrusion-based mesh generator, rather then a solid modelling mesh generator, is speed. To generate a motor model with approximately 100'000 tetrahedrons takes, from that the parameters are entered to the model is built ready for the solver, on a HP715 less than 10 minutes, with the used software [1]. Once the different kind of planes are generated, it takes always less then 5 min. This should be compared with typically more then 45 min for a solid modelling mesh generator.

5. LIMITATIONS AND POSSIBLE IMPROVEMENTS

The main limitation with the used extrusion-based mesh generator is that the planes must be flat. One set of extrusion data is applied to all nodes in each plane. Therefore to mesh e.g. two crossing cylinders is very difficult. It is however possible since the used mesh generator allow the nodes to be displaced <u>within</u> each plane. With this technique it is possible to model e.g. crossing cylinders, but still, it is very difficult to get good results. The modelling possibility and capacity for such mesh generators could be improved considerably by letting the extrusion data to be different not only for each plane but also for each node or groups of nodes in each plane. A mesh generator with this possibility would together with the model-building technique presented in this paper, be very powerful.

REFERENCES

[1] MagNet Users Manual, Infolytica Corporation, Canada, 1993

FROM FEM ANALYSIS TO BACK EMF IN ROTATING MACHINES: ALGORITHMS FOR ACCURATE EVALUATION

Beatrice Mellara, Ezio Santini

Universita' di Roma "La Sapienza"
Dipartimento di Ingegneria Elettrica
Via Eudossiana 18
00184 Rome, Italy

ABSTRACT

An algorithm for the accurate evaluation of back emf at the terminals of a rotating machine is described: the starting point is the vector potential as obtained via FEM analysis. Influence of local error on the numerical solution is discussed, and strategies to minimise the influence of the local error on the integral parameters are shown. An algorithm for the polynomial approximation of back emf vs. time is proposed, and numerical examples allow to evaluate the accuracy and robustness of the methodology.

INTRODUCTION

FEM is nowadays a well established analysis technique, commonly employed in a wide range of electromagnetic problems. Results of FEM computer programs are potentials in the interpolatory nodes. On the other hand, an engineering viewpoint looks at numerical values of integral quantities (back emf, torque, losses, ...) with greater attention [1]. If these quantities have to be evaluated, results of FEM analysis have to be postprocessed. Between the machine parameters, back emf is in general the first step for the determination of the overall performances of a design, and this paper shows how it is possible to deduce accurately back emf at the machine terminals starting from FEM analysis.

The logical sequence of this technique starts from the knowledge of the vector potential: back emf is computed by means of a small number of integral data closely related to the flux through a surface. It will be evident that numerical derivation of such integral data is necessary throughout the work, and therefore a great accuracy is required in FEM computations. This means not only to refine successive FEM solutions, but also to control the evolution of such numerical parameters from one FEM analysis to another. This is not only in a single FEM analysis, but also comparatively from a solution to another, if several FEM solutions are required for various geometrical configurations of the same design (with reference to a rotating machine, for different stator-rotor relative positions).

Electric and Magnetic Fields, Edited by A. Nicolet
and R. Belmans, Plenum Press, New York, 1995

FROM FIELD ANALYSIS TO BACK EMF

A 2D section of a rotating machine is considered. The following hypotheses are adopted [2]:

- electromagnetic fields can be considered 2D;
- magnetic materials are linear, and hysteresis is not present in ferromagnetic materials (iron and PMs);
- losses in active materials are not taken into account;
- the rotor speed ω_r is constant.

High order elements were used in FEM analyses, although their use did not increase significantly the numerical results.

Flux Linkages

A FEM solution in terms of vector potential is considered arrived at. By definition, the flux linked with a surface that has in the (x, y) plane its traces in the points 1 and 2, is $\phi = A_1 - A_2$. The flux linked with a coil (ψ) is:

$$\psi=\frac{1}{S}\int A(x,y)dS \tag{1}$$

where S is the area of the transverse section of the coil and $A(x,y)$ is the vector potential (here and in the following the number of conductors in series per slot will be taken as one; the same holds for the number of pole pairs).

This is the value of the flux linked with a test coil at time t_0; values for different times can be obtained with new relative positions between stator and rotor. However, the flux linked with the test coil at time t_1 is the same flux linked at t_0 with the coil which is far an angle $(t_1 - t_0)\ \omega_r$ from the current coil, where ω is the rotating speed of the rotor. This allows to limit the number of FEM calculations to one, if the magnetic structure is isotropic. When the geometry under consideration presents a small number of coils per pole and per phase, the relevant magnetic structures present characteristics not constant in space. Therefore, the number of FEM calculations to perform is at least two: the first one when the axis of the magnets are superimposed to the axis of a slot, the second one when the axis of the magnets are superimposed to the axis of a teeth.

It is evident that evaluation of integral (1) is the basis for the analysis, and it is important to evaluate the precision reached in its computation. Let us consider the unknown, exact solution $U(x,y)$, and the approximate solution $A(x,y)$ obtained via FEM. It is possible to write: $U(x,y) = A(x,y) + h(x,y)$, where $h(x,y)$ is the unknown local error function (obviously compatible with the boundary constraints). The flux linkage can be expressed as:

$$\psi = \frac{1}{S}\int U(x,y)dS = \frac{1}{S}\int A(x,y)dS + \frac{1}{S}\int h(x,y)dS \tag{2}$$

Potential values in regions S where integral in Eq. (1) has to be calculated present in general small variations and therefore they have in general the same sign: consequently, the first integral in the second hand of Eq. (2) has the same absolute value of the integral of the absolute values of the potentials. On the other hand, local error tend to present a random behaviour: its sign can change from point to point, and in general the integral of the local error tends to cancel. It is possible therefore to write:

$$\int U(x,y)dS = \int\left|A(x,y)\right|dS + k\int\left|h(x,y)\right|dS$$

where k is a unknown factor less than one. From a numerical viewpoint, Eq. (1) can therefore be rewritten as:

$$\psi = \frac{1}{S}\int U(x,y)dS = \frac{1}{S}\int A(x,y)dS + \frac{k}{S}\int h(x,y)dS \qquad (3)$$

More generally, it is possible to say that the quantities related to the integral of the computed potential values are more exact than the potential itself. The goal to decrease as much as possible the integral of the local error is obviously associated to a decrement of the local error, that can be obtained by increasing the number of nodes in the FEM analysis (at least until when the truncation error is negligible with respect to round-off error).

The strategy which leads to an error-free solution consist in a recursive evaluate-analyze-refine procedure. Each FEM analysis has as starting point the preceding one, where local error is evaluated and adequate refinement techniques are adopted in order to minimise the local error in the next FEM computation (a common technique is to add new nodes where the local error is high). Numerical tests evidentiate that, after a limited number of iterations, the value of the integral (1) evaluated in a given region S (i.e. the linked flux) does not change even if the number of nodes is greatly increased: in practice, the maximum difference between the first and the last iteration is less than 2 %, and this means that a limited number of nodes allows to evaluate with a good precision the flux linked with a coil.

Numerical Evaluation of Back Emf

When the flux linked with a coil (or with a phase) is known, the computation of the relevant back emf is in principle a simple task, by applying the Faraday law:

$$e(t) = -\frac{d\psi(t)}{dt} = -\frac{d\psi(\theta)}{d\theta}\frac{d\theta}{dt} = -\frac{d\psi(\theta)}{d\theta}\omega_r \qquad (4)$$

where θ is the angular position, in a reference frame rigidly connected to the rotating field, of the axis of the coil, and ω_r is the angular velocity of the rotating field. From Eq. (4) it turns out evident that numerical derivation of the linked flux is the basis for the determination of the back emf.

One possibility is to calculate the back emf in a turn by means of direct numerical derivation of the flux. Accuracy reachable in this way is poor, since the linked flux is known in a small number of points of the interval; no matter what kind of numerical derivation algorithm, derivatives of degree greater than one are not taken into account. If the linked flux varies suddenly near the point under consideration, the numerical values of higher derivatives are not negligible, and this results in great numerical errors.

Linked flux can be approximated by means of analytical functions, such as: Fourier expansions, Lagrange polynomials, Tchebishev polynomials, cubic polynomial splines [3]. If a analytical approximation for the linked flux is determined, the back emf can be found by means of analytical derivation, that can be obtained without numerical errors and with no significant computational efforts. This is a general interpolation problem, where a number N of pairs of (θ, ψ) values are known in a closed interval of the θ-axis. However, adequate choice of the interval along θ-axis allows the *a-priori* knowledge of more constraints on the derivative of the linked flux: for instance, the back emf in the interpole axis (the back emf is maximum) or in the pole axis (the back emf is zero).

A Fourier approach for the interpolation seems not to be useful, since the precision in the numerical evaluation of the Fourier coefficients decreases as the order of the harmonic increases. In principle this is not a great problem, since the numerical value of the said parameter decreases when the order of the harmonics increases, and the error in the numerical value of the flux is often not appreciable. On the other hand, back emf is calculated by means of analytical derivation as in the following Eq. (5):

$$e(\theta) = -\frac{d\psi(\theta)}{d\theta} = -\sum_{k=1}^{N} k\ \psi_k\ sin(k\ \theta) \tag{5}$$

It turns out evident from Eq. (5) that the higher the order of the harmonic, the greater the weight. Numerical tests show that N in Eq. (5) should be kept small: addition of two o three harmonics to a qualitatively good $e(\theta)$ curve results often in wild jumps of the new curve that, at least in principle, should be more precise.

Lagrange polynomials are difficult to be determined if the number of interpolatory nodes is more than five or six, since this is the order of the associated Vandermonde matrix, which is ill-conditioned by its same nature. Tchebishev polynomials give better results, but their calculation requires very clever programming. Moreover, both Lagrange and Tchebishev do not insure anything about the values of the back emf and of the flux at the endpoints of the interval, where periodicity conditions are obviously required. The flux and its derivative are simply what they are, and it is not possible to constraint the time and space functions to present a given derivative.

In principle, cubic splines seem to be the most difficult technique to apply, since the N pairs of (θ, ψ) values are not enough to determine a unique set of polynomial approximations: two more numerical quantities are needed (in general, the first or the second derivative in any point of the interval). These quantities are in general not known, but adequate choice of the interval allows to know *a priori* two derivatives which have to be zero. In the problem under examination cubic splines are therefore the most suitable technique, since lack of two quantities allows two more degrees of freedom, that can be used in the control of the reached accuracy. More importantly, analytical derivation of cubic splines leads to a set of square splines, which in general do not present fast variations, and therefore functions computed in this way present a smooth behaviour.

If a cubic spline approximation is found starting from a piecewise evaluation of the back emf in the leftmost and rightmost points of the interval, the analytical behaviour of the back emf can be determined: in particular, the numerical value of the back emf in the centerpoint of the interval is known. This value must be equal (with opposite sign) to the value of the back emf at the endpoints of the interval, and it can be used to compute a new cubic spline interpolation. The process ends when the computed values of the back emf in the endpoints of the interval do not differ from the starting values.

Verification of the procedure can be obtained by means of additional FEM analyses relevant to geometry presenting relative stator-rotor positions not used in the determination of the linked flux. Linkages can be computed by means of Eq. (1) and compared with the result of the cubic spline. If the values do not agree in the bounds of the required accuracy, the accuracy in FEM analyses is not enough, and the procedure must be repeated growing the number of nodes. In numerical tests, however, this necessity was never evidentiated.

NUMERICAL RESULTS

Fig.(1) shows the 2D sketch of a 4 poles Inset Permanent Magnet machine (IPM) [4]. As evident from the figure, this machine presents a limited number of slots per pole, and this results in a irregular behaviour of the flux density in the air gap. Fig.(2) shows the results of the proposed algorithm for the determination of the back emf in the machine. In Fig. (2.a) the initial approximations for the flux linkage and for the back emf vs. the angular position of a coil are shown: the numerical value of the back emf at the endpoints of the interval was determined by a piecewise approximation. Fig. (2.b) shows the same quantities at the end of the iterative process, which ends in five iterations.

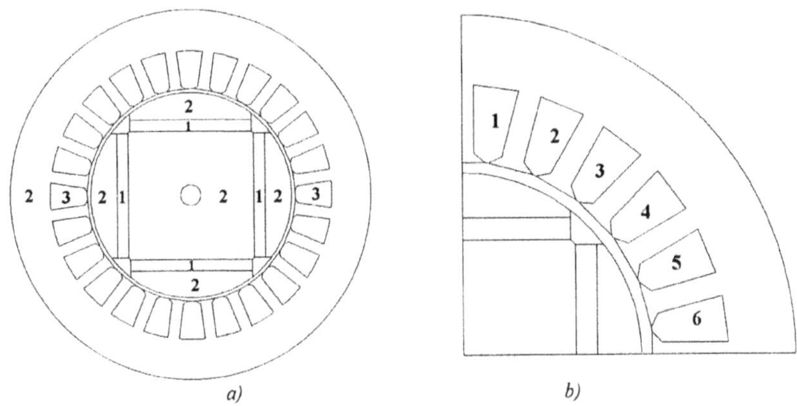

Fig. 1. - a) general layout of a IPM machine: 1 PMs, 2 iron, 3 slots; b) geometry used for the 2D analysis.

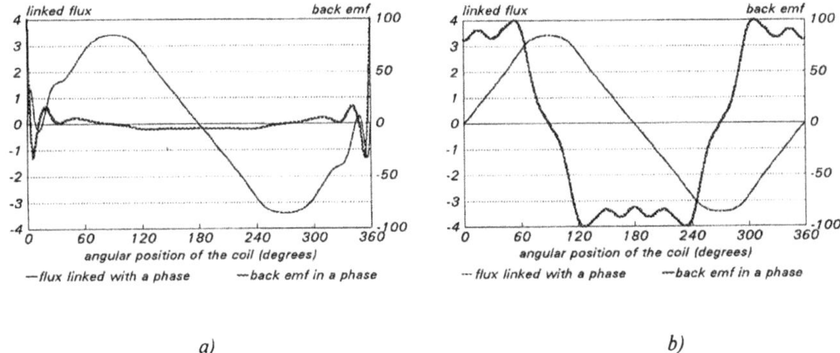

Fig. 2. - Results of the cubic polynomial spline approximation for the flux linked with a phase and for the back emf in a phase vs. the angular position of the phase. a) the numerical value of the first derivative of the linked flux at the beginning of the interval is chosen arbitrarily; b) the final spline approximation obtained with the optimisation technique described in the paper.

CONCLUSIONS

This paper presented a procedure for the numerical determination of back emf induced in a phase of an electric machine starting from FEM analysis. The procedure is rather influenced by the precision reached in FEM analysis: a strategy for the control of this precision has been described. An algorithm for the evaluation of the flux linked with a coil was discussed, and numerical derivation was reached by means of an adequate set of optimised cubic splines.

REFERENCES

[1] D.A. Lowther and P. P. Silvester. **Computer Aided Design in Magnetics**. Springer-Verlag 1985.

[2] F.Caricchi, F. Crescimbini, O. Honorati and E. Santini. **Optimum CAD-CAE Design of Axial-Flux PM Motors.** *Proceedings of the International Conference on Electrical Machines, Manchester (U.K.), 15 - 17 September 1992.*

[3] E. Santini and M. Vinci. **Fast and Efficient Techniques for Interpolation of B-H Curves** ***(invited paper).*** Electrosoft 1993, Southampton (U.K.), 4 - 6 July 1993. Edited on Software Applications in Electrical Engineering, Computational Mechanics Publications, pp. 263 - 270.

[4] A. Di Napoli, P. Pinkas, E. Santini - **CAD of IPMs** - ISTET 93: 7th International Symposium on Theoretical Electrical Enginering, Szczecin (Poland), 13 - 15 September 1993.

TORQUE CALCULATION OF A SMALL, AXIAL FLUX PERMANENT MAGNET MOTOR

M. Van Dessel, R. Belmans

Dept. E.E. - Electrical Energy
K.U. Leuven
Kardinaal Mercierlaan 94
B-3001 Heverlee - Leuven
Belgium

R. Hanitsch, E.A. Hemead

Inst. für Elektrische Maschinen
T.U. Berlin
Einsteinufer 11
D-10587 Berlin
Germany

ABSTRACT

The paper deals with the design of a disc-type motor excited by permanent magnets. For the finite element model reference is made to previous work. In this paper the calculation of the torque using the same model is described. Out of three torque calculation methods investigated, the best one is current - field interaction. The resulting torque curve when all phases are operated, is constructed. The theoretical results are compared to values obtained from measurements.

1. INTRODUCTION

In computer peripherals, office equipment and storage devices, the trend is towards small packages, low power consumption, high efficiency and low cost. Consequently, these types of drives need to have a small rotor inertia and a small volume. In a number of applications the disc-type motor is the best choice due to its short axial length. In portable devices where energy is supplied by a battery, there is crucial demand for high efficiency. To attain these goals, permanent magnets are used for the excitation and an electronic commutation can be combined with speed control as shown by Hanitsch et al. [1]

2. MOTOR DESIGN

The design has been described in previous work [2]. Figure 1 shows a cross section of the prototype. The main dimensions are: outer diameter about 45 mm, axial length about 15 mm. The rotor is a disc of sintered permanent magnet material, magnetised in the axial direction in a multipole arrangement. Magneto-resistive sensors are used for the detection of the rotor position.

Electric and Magnetic Fields, Edited by A. Nicolet
and R. Belmans, Plenum Press, New York, 1995

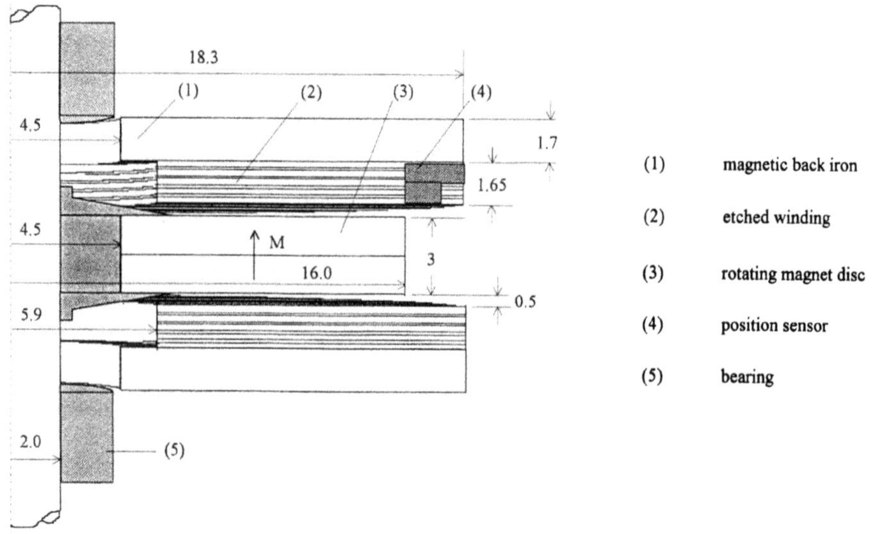

Figure 1. Main dimensions of the prototype

The stator consists of two disks of ferromagnetic back iron material with two airgap windings. On each stator side, 8 windings are installed and connected forming two phases (figure 2). The prototype has symmetrical planar windings, manufactured using etching techniques. The required torque makes it necessary to employ a four layer winding to establish a sufficiently high current layer.

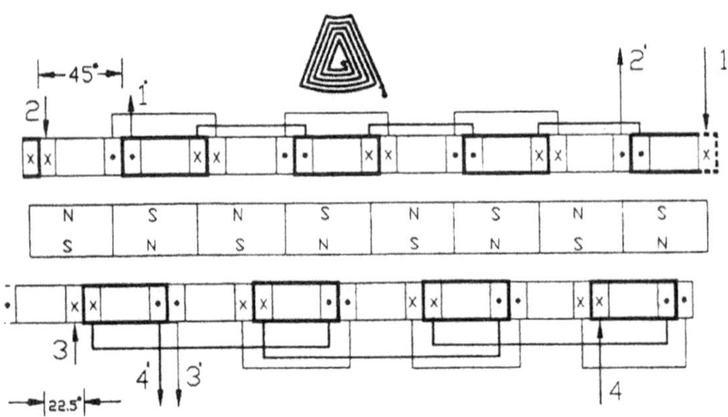

Figure 2. Winding arrangement

3. TORQUE CALCULATION

3.1 Finite Element Analysis

Due to its special geometry, this type of motor can not be analysed using a two dimensional approach, as generally found in modern design procedures. Previous work [3] describes the design of a 3D finite element model for the motor. The characteristics obtained from the model are the flux and induced voltage in one phase, the inductance of one phase, and the variation of the flux density at the magnet surface on load. This paper treats the torque calculation using the same model.

3.2 Calculation Methods

a) Interaction between Current and Flux Density

The torque on the stator is obtained by summation of the torque contributions of the forces on the individual conductors:

$$\vec{T} = \int_0^1 \vec{r} \times (I d\vec{l} \times \vec{B}) \tag{1}$$

Only the z-component of the stator torque is calculated. The motor torque is the opposite of the torque acting on the stator.

b) Virtual Work

The torque is given by the derivative of the co-energy with respect to the rotor position.

$$T = \left. \frac{dW_{co}}{d\theta} \right|_{i=cte} \tag{2}$$

c) Maxwell Stress

Integrating the Maxwell stress tensor on a surface enclosing the rotor yields the torque.

3.3 Conditions for the Torque Calculation

- The calculation is done in the steady state: the developed torque is balanced by the load torque. The field is calculated under time-invariant conditions. Therefore, eddy currents are neglected, and a magnetostatic analysis is used.
- Under normal operating conditions two coils carry current: one in the top and one in the bottom stator. Due to linearity, this can be treated as the superposition of two cases with one active coil. Linearity is found by checking the operating points in the B-H characteristic of both the magnets and the back-iron. Therefore, only the torque of one coil is calculated by the finite element analysis. The torque curves belonging to the other three coils can be obtained using a time shift.
- The current in the winding is constant and equals 0.5 A.

3.4 Results

The best results are obtained using interaction between current and flux density. Figure 3 shows the four torque curves resulting from operating the coils over one period. The basic curve is the torque T_1 obtained from the current - field interaction method.

$$T_1 = -2.466 \sin(4\theta) \quad [\text{mNm}] \tag{3}$$

Figure 3 also gives the resulting torque when switching is done for a counter-clockwise rotation. This means that at every position, two coils giving a positive torque are active. The properties of the overall torque are:

- periodicity 22.5° (1/4 of motor periodicity)
- average torque 3.14 mNm
- peak to peak ripple 1.02 mNm.

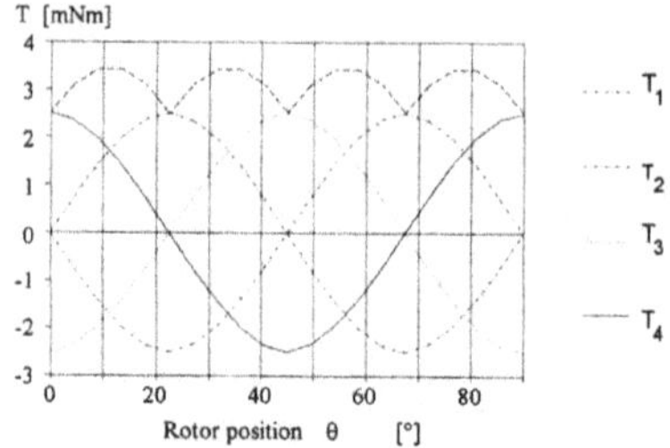

Figure 3. Torque curves for the disc-type motor (Windings supplied with 0.5 A)

4. COMPARISON WITH MEASURED TORQUES

A considerable problem for the prototype is the mounting of the rotor, that has to be centred between upper and lower stator to have equal airgap lengths. To get this centring a mechanical pre-stress is applied to the bearings. Since this causes considerable friction, the directly measured torque is not comparable with the calculated electromagnetic torque. The electromagnetic torque has to be estimated from measurements. For a permanent magnet motor the amplitude of torque and induced voltage in one phase are:

$$T_A = C_T\, i \tag{4}$$

$$U_A = C_V\, \omega \tag{5}$$

C_V is measured from the induced voltage curve at 1000 rpm.

$$C_V = \frac{U_A}{\omega} = \frac{0.62\,\mathrm{V}}{\left(\frac{2\pi.1000}{60\mathrm{s}}\right)} = 5.92\ \ 10^{-3}\,\mathrm{Vs} \tag{6}$$

With $C_T = C_V$ and $i = 0.5$ A, the estimated value of the torque amplitude is $T_{A,est} = C_T\, i = 2.96$ mNm. This is comparable to the calculated amplitude of 2.466 mNm.

REFERENCES

1. C.S.Park and R.Hanitsch, "Novel 10 W brushless DC motor of the pancake type", IEE Conference Publication 324, 1990, pp. 435 - 439.
2. D.S.Choi and R.Hanitsch, "Disc - Type Motor with etched windings and magneto-resistive position sensing", Proceedings of PCIM '93, Nurnberg, Germany, June 1993, pp. 456 - 461.
3. M.Van Dessel, R.Belmans, D.Verdyck, W.Geysen, R.Hanitsch, and E.A.Hemead, "Three dimensional finite element analysis in axial flux permanent magnet motors", Proceedings of the SPEEDAM 92, Positano, Italy, May 19 - 21, 1992, pp. 43 - 46.

TORQUE OPTIMIZATION OF A BURIED P.M.S.M BY GEOMETRIC MODIFICATION USING F.E.M.

C. Marchand, Z. Ren and A. Razek

Laboratoire de Génie Electrique de Paris
U.R.A. 127 C.N.R.S., Univ. Paris VI-Paris XI, Ecole Supérieure d'Electricité
Plateau du Moulon, 91192 Gif-Sur-Yvette, Cedex, France

INTRODUCTION

Permanent Magnet Synchronous Machines are often used in variable speed drives or servo drives. In this type of applications, some features are particularly appreciated: dynamic behaviour, position precision, low torque ripples...

To achieve these operating objectives, it is necessary to consider all the elements of the drive: "converter - machine - control system" and the machine structure can not be dissociated from the current fed wave form. For example, previous studies for a given machine have shown that torque optimisation (maximal torque without torque ripple) can be obtained by the introduction of current harmonics.

However the implementation of this optimisation torque approach requires a high performance of digital control system.

An alternative solution to improve the system performances is presented in this paper. It consists of the modification of the structure and the determination of a suitable geometry of the machine for a classical current fed (sinusoidal or rectangular wave form).

The numerical procedure is based on the Finite Element Method where local magnetic material saturation is considered. The electromagnetic force distribution and torque are calculated from the virtual work principle.

In this paper a parametric study on the optimization of critical geometric dimensions is performed. The process is automatic (geometry modification, mesh and calculation) from a modification of coordinates. The influence of a buried P.M.S.M pole geometry on the electromagnetic torque (mean value and pulsations) for a given current fed wave form is presented.

Electric and Magnetic Fields, Edited by A. Nicolet
and R. Belmans, Plenum Press, New York, 1995

PRESENTATION

In synchronous servo motors the classical phase current control takes only into account the e.m.f wave form of the machine. Consequently, two current wave forms are commonly applied: sinusoidal current wave form when the e.m.f. is sinusoidal or rectangular wave form when the e.m.f. is trapezoid. However, for some applications this classification is not adapted. For example, when the torque must be as smooth as possible, the structure of the machine must be more considered. It is the case of machines with flux enhancement arrangement like the buried magnet machine. These ones have an important rotor magnetic saliency, which can provide an interesting reluctance torque but also creates torque pulsations when the phase current wave forms are sinusoidal or rectangular. For these machines, torque optimisation and minimisation of torque ripple lead to impose non-sinusoidal current fed[1]. The current wave form can be predetermined for each operating point by numerical method from local field calculation based on finite element technique(F.E.M)[2,3]. This instantaneous current can be also calculated on line by the use of a torque estimator. In this case a precise model of the machine is used. The instantaneous torque is estimated and compared to a reference torque in order to create the current reference.

The two methods are not easy to implement and require high performance control microprocessor like D.S.P (Digital Signal Processor). For the first, the predetermined current is implemented under the form of harmonic decomposition where the coefficients depend on the operating point. The experimental results show an interesting reduction of the torque ripple but also that the method is very sensitive facing the coefficient determination. The second method (on line) can be employed only in a speed range. At high speed the calculations of current derivatives increase noise, at low speed the imperfections of the inverter and the division by the estimated speed disturb the torque estimation.

These statements show that to increase the performance of a variable speed drive ("converter-machine-control system"), each element of the system must not consider independently. To simplify the control system, a classical sinusoidal current is applied. The developed torque will be perfected if the machine is well adapted to this current. Consequently in the aim to satisfy torque optimisation criterion, we search to adapt or modify the geometry of the machine.

A parametric study to the determination of a suitable machine geometry by the modification of some parameters is performed. The studied synchronous machine had buried permanent magnets. The initial structure is shown in figure 1.

ELECTROMAGNETIC TORQUE DETERMINATION

The knowledge of the flux density **B** and the application of the virtual work principle permit the determination of the local force distribution with the consideration of the magnetic material saturation. **B** is determined by the numerical solution of Maxwell equations using Finite Element Method [4]. The variation of magnetic energy in a domain due to a virtual displacement of a geometric point when the flux is hold constant gives the electromagnetic force on the considered point.

With F.E.M. the force calculation is applied on every node of the mesh. In the interior of the magnetic material the force values are negligible compared with the force values on the air gap surface. The torque is thus due to the last one. Figure 2 shows the force distribution due to the rated sinusoidal current on a pole of the rotor for the initial position ($\theta = 0$).

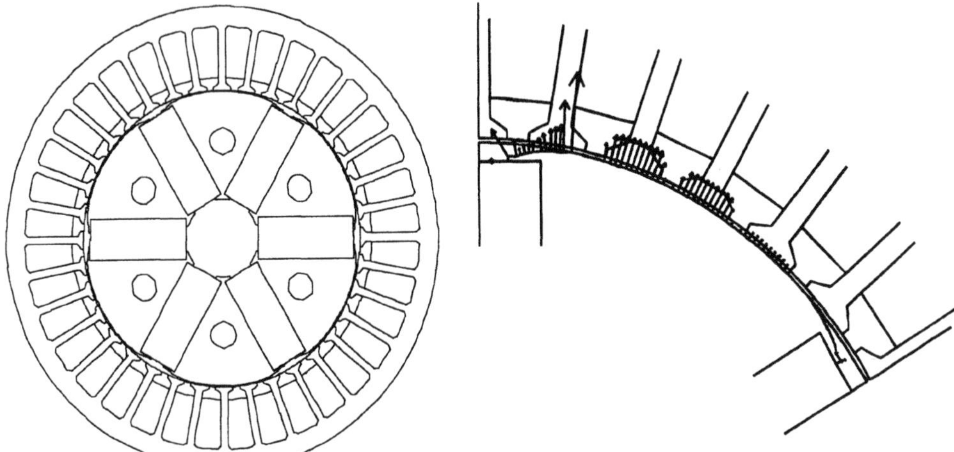

Figure 1. Structure of the studied P.M.S.M.

Figure 2. Electromagnetic force distribution on the rotor

OPTIMISATION PROCEDURE

Optimisation of electromagnetic devices using F.E.M. is more and more employed[5,6,7]. The flowchart of the procedure is presented in Figure 3.

The objective function is the optimisation of the torque: minimisation of the torque ripple and increasing of the mean value. The structure modification is carried out to reach this objective. The presented procedure is the first step of the work and it is based on an parametric approach. Three variables a, b and c are chosen as shown in figure 4.

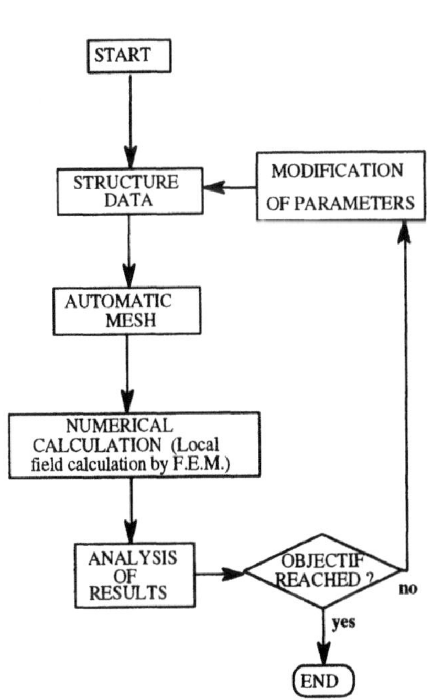

Figure 3. Flowchart of an optimisation process

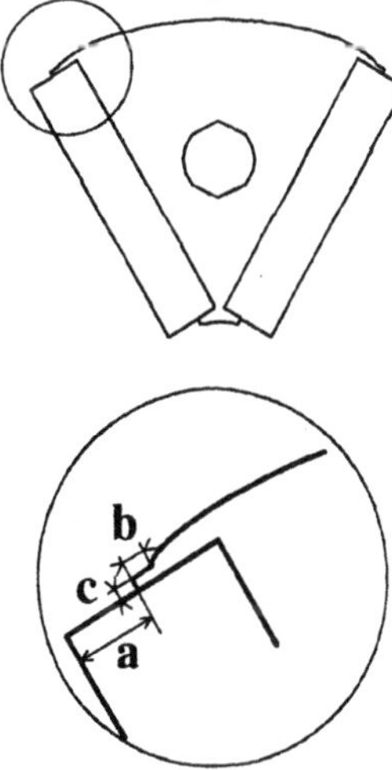

Figure 4. The detail of the structure variables

In order to minimise the calculation time, only the modified region is remeshed for each iterati A constant mesh density is maintained.

The results of the optimisation are shown in figure 5 by the comparison of the calculated tor before and after the optimisation. The mean value of the torque is increased and the variation magnit is decreased but stays important. Actually, an automatic procedure with a numerical algorithm wh take into account the objective function but also the geometrical constraints is implemented.

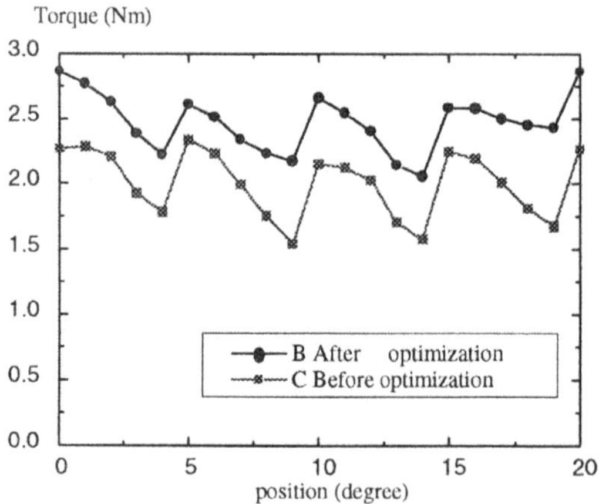

Figure 5. Torque variations as function of the rotor position

CONCLUSION

In this paper, the influence of the rotor structure on the electromagnetic torque is presented. T use of a parametric procedure associated with F.E.M. permits an improvement of the torque variati as function of the rotor position.

REFERENCES

1. Marchand C. and Razek A., 'Optimal torque operation of digitally controlled permanent magnet synchronous m drives', IEE Proc-B, Vol. 140, No.3, pp. 232-240, 1993
2. Colamartino F., Marchand C., Ren Z. and Razek, A., 'Using of electromanetic modelling in the digital control permanent magnet synchronous motor', Second International Workshop on Electric and Magnetic Fields , 18-20 M 94, Leuven.
3. Colamartino F., Marchand C., and Razek, A., 'Estimation and minimisation of electromagnetic torque ripple in a bu permanent magnet synchronous motor', ICEM 94, 5-8 septembre94, Paris.
4. Ren Z., Besbes M. and Bouktache S., 'Calculation of local magnetic forces in magnetized materials', First Internatio Workshop on Electric and Magnetic Fields Liege, 27-29 September 1992
5. Subramaniam S. and Hoole S.R., 'Optimization of a magnetic pole face using linear constraints to avoid jag contours', Record of the 9th COMPUMAG conference on the computation of electromagnetic fields, pp. 548-5 MIAMI, October1993
6. Magele C. A., Preis K., Biro O., Richter K.R., 'Different strategies in the optimisation of electromagnetic devic IMAC'S 91 13th World congress on computation and applied mathematics Vol.4, pp. 1594-1595, july 22-26 19 DUBLIN.
7. Kadded K., Saldanha R. R. and Coulomb J. L., 'Minimization of the induction space harmonics of a P.M. synchron machine using finite element method and penalty method', Journal de physiqueIII France3, pp. 413-422, March 19

FLUX-WEAKENING OPERATION OF PERMANENT MAGNET SYNCHRONOUS MOTORS FOR ELECTRIC VEHICLE APPLICATION

G. Henneberger, J.R. Hadji-Minaglou, R.C. Ciorba

Institute of Electrical Machines,
University of Technology Aachen,
Schinkelstrasse 4, D-52056 Aachen, Germany

Abstract

The permanent magnet synchronous motor (PMSM) is as well applicable for the electrical vehicle as the asynchronous motor but needs a special design. The requirements for a large power to weight ratio combined with an extended torque/speed characteristic and a high efficiency is a considerable challenge for the motor designer. This paper is concerned with the design and the comparison of two PMSM and the test results have been built for the adequate variant.

The design of the motors

The modeled motors have the same stator with a three-phase winding placed in 48 nuts and two layers with a $\frac{11}{12}$ fractional-pitch winding. The number of pole pairs is p=2. Each pole contains 13 NdFeB rectangular magnets, radially magnetized. The two extremes magnets are higher to minimize the demagnetizing effect. The cut-outs in the rotor poles, different in the two motors are designed to decrease the inertia of the rotor and to increase the saliency ratio $\xi = \frac{L_d}{L_q}$. The rotor shaft is nonmagnetic. Figure 1 presents the fluxplot in one pole section for each variant.

The machine model

The steady state equation per unit in d-q axis theory are:

$$u_d = r \cdot i_d - n \cdot x_{q0} \cdot i_q \tag{1}$$

Electric and Magnetic Fields, Edited by A. Nicolet
and R. Belmans, Plenum Press, New York, 1995

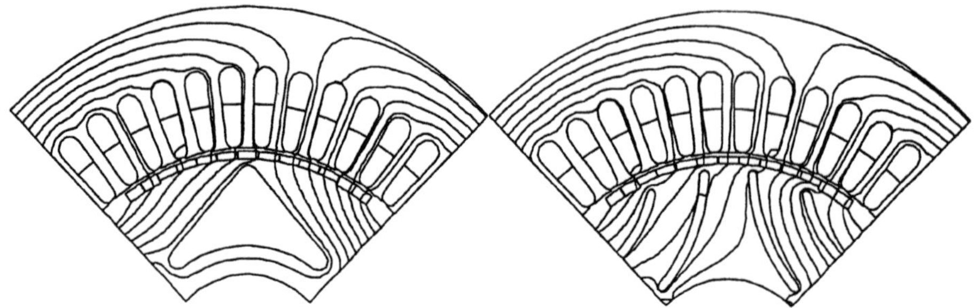

Figure 1. PMSM1 design PMSM2B design

$$u_q = r \cdot i_q + n \cdot x_{d0} \cdot i_d + n \cdot U_{p0} \tag{2}$$

$$m = (U_{p0} - (x_{q0} - x_{d0}) \cdot i_d) \cdot i_q \tag{3}$$

$$n = \frac{u_q - r \cdot iq}{U_{p0} + x_{d0} \cdot i_d} \tag{4}$$

The rated speed, neglecting cooper losses ($r = 0, \Psi = 0$), is:

$$n_N = \frac{n_0}{\sqrt{1 + (x_{q0} \cdot i)^2}} \tag{5}$$

and the maximal speed is n_{max}:

$$n_{max} = \frac{n_0}{1 + x_{d0} \cdot i} \tag{6}$$

where n_0 is no load speed. An optimal flux weakening area can only be reached with a negative direct current if the saliency ratio ξ is high. For this reason a large airgap in the q-axis combined with flux barriers have been designed for the two prototypes PMSM1 and PMSM2.

Analysis and Simulation

These special motors have been calculated by the program package MagNet. The parameters L_d and L_q were calculated with the energy method based on FEM-analysis and the results are presented in figure 2a. The PMSM2 has a better saliency ratio ξ than PMSM1 at $i < 2.6 I_N$ (fig2b).

To derive with high accuracy the induced voltage versus time from the field vector potential, ten distinct positions between the rotor and the stator are used for each model. The pole width on an arc contour in the air gap has been divided in 120 segments. With 12 slots per pole this results in 10 segments per slot pitch. Each rotor position differs 0.75^o from the other and that permits, together with symetrical current supply, to obtain 120 values for the induced voltage at a rotation with an electrical angle of 2π. As we expected, the induced voltages for both motors are nearly the same.

Next step was the simulation of the motors over the whole speed range with different loads. In figure 3b the torque-speed characteristics of both motors are shown.

PMSM1 has a smaller weight and is easier to manufacture. With these preliminary considerations the PMSM1 has been built and tested.

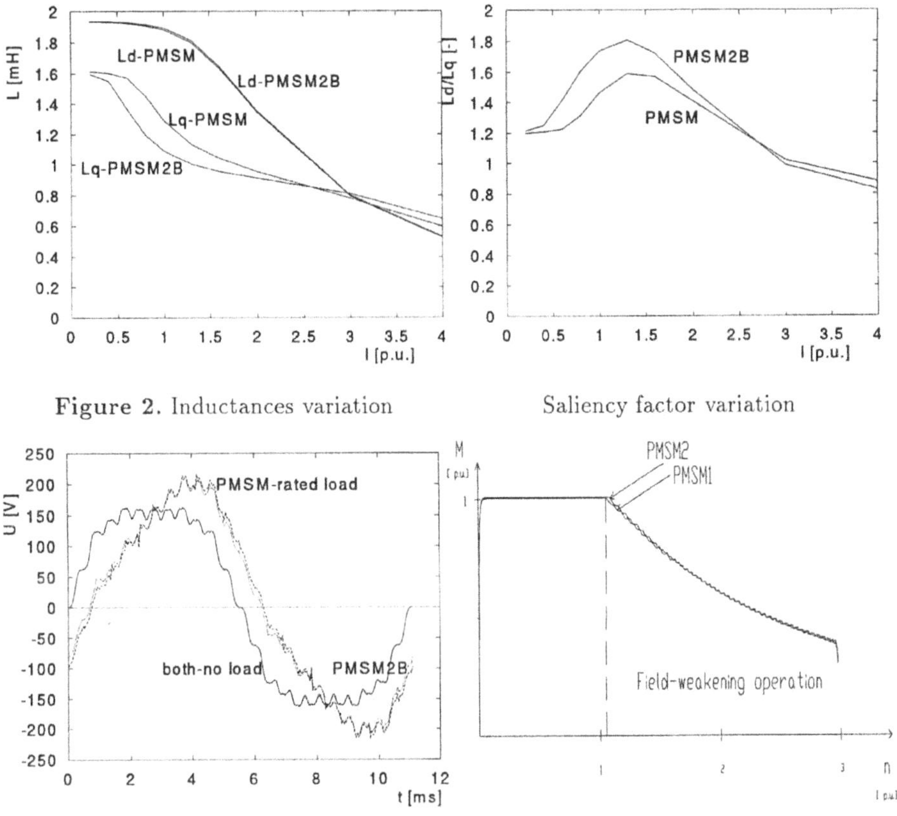

Figure 2. Inductances variation Saliency factor variation

Figure 3. a) Calculated induced voltages. b)Torque speed characteristics.

The torques are estimated with the Maxwell stress tensor method based on FEM-analysis. In figure 4a the torque variation with rotor mechanical angle are shown. The torques are calculated at rated current (I_N) for different current angles (Ψ):

- P1 corresponds to $\Psi = 0^o$,
- P2 corresponds to $\Psi = -10^o$,
- P5 corresponds to $\Psi = -45^o$
- P7 corresponds to $\Psi = -75^o$.

A Fourier analysis of the induced voltage in the PMSM1 is shown in figure 4b. The aim was to minimize the harmonics with adequate winding of the armature.

Laboratory test results

A series of tests has been performed on PMSM1 to confirm the design. This includes open circuit and load tests as generator, heat run test, load tests with constant load torque over whole speed range as motor.

The results for the induced voltages versus time, measured in generator mode, correspond with accuracy to the calculated values in FEM-Analyse (fig.5a).

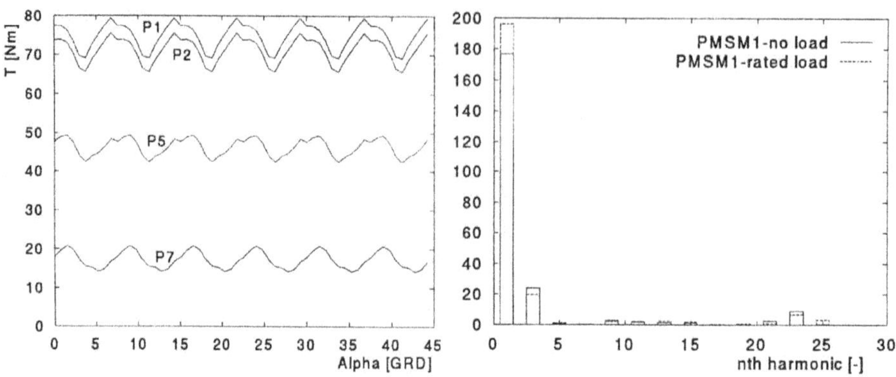

Figure 4. a)Torque variation with angle b)Voltages harmonics of PMSM1

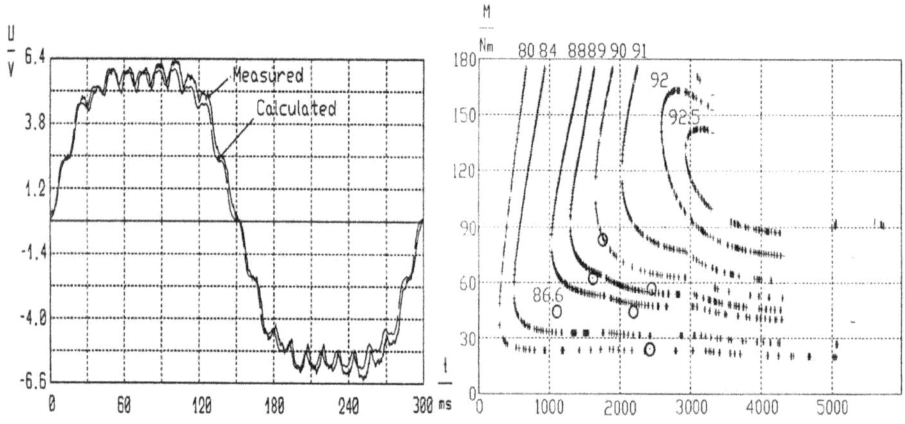

Figure 5. a)Induced voltages versus time. b)Shell curves for efficiency

An important subject is the efficiency. It is calculated for many values of torque and speed and is represented in figure 5b. In the same graphic some experimental values of the efficiency are represented with circles.

References

[1] W.L.Soong and T.J.E.E. Miller *Theoretical limitations to the field-weakening performance of the five classes of brushless synchronous AC motor drive* Electrical Machines and Drives Conference sep.1993 pp 127-132.

[2] Langheim J. *Einzelradantrieb fuer Elektrostrassenfahrzeuge* Dissertation RWTH Aachen 1993.

[3] Henneberger G., Hadji-Minaglou JR. *Entwurf und Vergleich von verschiedenen Motortypen fuer den Einsatz im Elektrofahrzeug* 4. Aachen Colloquium, Aachen, Oct. $5-7^{th}$ 1993.

USING OF ELECTROMAGNETIC MODELLING IN THE DIGITAL CONTROL OF A PERMANENT MAGNET SYNCHRONOUS MOTOR.

C. Marchand, F. Colamartino, Z. Ren and A. Razek

Laboratoire de Génie Electrique de Paris
U.R.A. 127 C.N.R.S., Univ. Paris VI-Paris XI, Ecole Supérieure d'Electricité
Plateau du Moulon 91192 Gif-Sur Yvette, Cedex, France

INTRODUCTION

In classical controls of synchronous motor drives, flux distributions are supposed sinusoidal with the angular rotor position. Usual control is performed in the d-q rotating frame bound to the rotor where all parameters and electromagnetic variables are considered constant. But these assumptions depend on motor structure. Inset permanent magnet synchronous machines with flux enhanced arrangement have a non-sinusoidal field distribution and parameters varying in d-q frame. Motor produces torque ripple with an usual sinusoidal fed current which leads to vibrations and mechanical instabilities.

In order to reduce these pulsations, electromagnetic modelling of the machine by Finite Element Method (F.E.M.) is used. In this paper two techniques are proposed. The first is based on optimised predetermined current wave forms and the second concerns on-line torque estimations.

In the first case, from numerical calculation, the machine is modelled. With a direct torque calculation based on the virtual work principle, we predetermine the current wave form for a given torque reference. We can also calculate model parameters of the machine (inductances and e.m.f.) and then torque can be calculated analytically and the instantaneous current can be optimised for a given torque. To verify the validity of these techniques, we have built an experimental system where the digital control is based on a Digital Signal Processor (DSP32C) which permits to implement complex control laws. The torque measurement is performed from strain gages force sensors on the floating mounted stator.

Due to the aim of reducing the number of calculations and predeterminations, the second solution consists in on-line torque estimation from measured electrical values and based on the machine model. This estimation allows to control the torque in a closed loop.

MODELLING AND CURRENT PREDETERMINATIONS

For a permanent magnet synchronous machine electromagnetic relations in the stator stationary frame (a,b,c) can be written:

Electric and Magnetic Fields, Edited by A. Nicolet
and R. Belmans, Plenum Press, New York, 1995

$$[V] = R[I] + [L_{abc}]\frac{d[I]}{dt} + \frac{d[L_{abc}]}{dt}[I] + \frac{d[\Phi_f]}{dt} \qquad (1)$$

$$\Gamma = n_p\left(\frac{1}{2}[I]^t\frac{d[L_{abc}]}{d\theta}[I] + [I]^t\frac{d[\Phi_f]}{d\theta}\right) \qquad (2)$$

where n_p is the number of pole pair, θ is the electric angle, $[L_{abc}]$ is the phase inductance matrix, $[V]$, $[I]$ and $[\Phi_f]$ are respectively the voltage, the current and the permanent magnet flux vectors. Γ is the electromagnetic torque.

If rotor flux and armature magnetomotive force distributions are assumed sinusoidal each term of the phase inductance matrix is sinusoidal with the angular position. In these conditions, the electromagnetic torque (2) is constant when currents are sinusoidal and all values in (1) are sinusoidal.

The figure 1 shows the experimental measured torque with impressed sinusoidal currents on a buried P.M.S.M. The torque is not constant and contains harmonics multiple of 6θ. In fact the assumptions described above are not verified for this type of machine. The field distributions are non sinusoidal and the pulsations are due to the rotor saliency and the inductances variations[3].

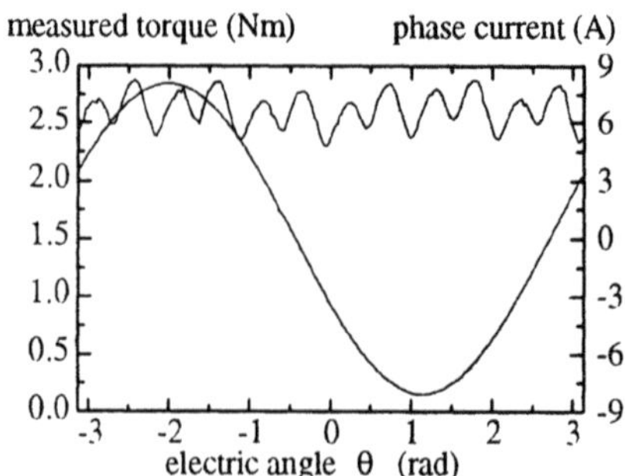

Fig. 1. measured torque with sinusoidal phase current fed.

To cancel the torque ripple, we can impose a current wave form which contains harmonics[1,2,3]. The determination of this current wave form can be obtained from numerical modelling of the machine based on F.E.M. Two techniques have been developed: direct an indirect methods.

Indirect method

This method has been already presented[1,3]. F.E.M in 2D is employed to determine the e.m.f. of the machine (fig. 2a) and the phase inductances variations with the rotor position[4] (fig. 2b). An optimisation procedure permits to determine the optimal current wave form (fig 2c) which gives the optimal torque (a perfect constant torque without pulsation with a minimal instantaneous current). In this method, several calculation steps are needed to determine the current. Consequently, the time computation is important and intermediary values lead to numerical errors. The second method avoids this inconvenience.

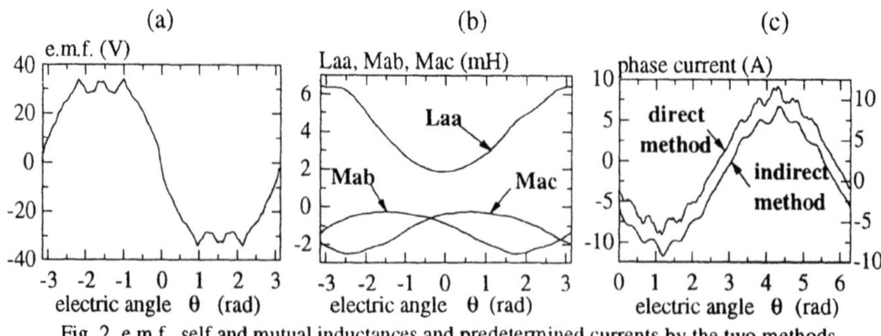

Fig. 2. e.m.f., self and mutual inductances and predetermined currents by the two methods.

Direct method

The modelling of an electromagnetic device by F.E.M. allows the determination of the magnetic state with the consideration of the real geometry and the magnetic saturation of the material. Knowing the vector potential distribution, the application of virtual work principle permits to calculate the magnetic force on each node of the finite element mesh [3]. For an electrical machine, the torque is obtained from the force density distribution on the pole of the rotor.

Virtual work principle. From a virtual displacement, the variation of the magnetic energy with constant magnetic flux gives the force. Applying this principle to a finite element mesh, the nodal force can be obtained by the derivative of the energy in elements with respect to a virtual nodal displacement[5].

The current wave form is calculated by an iterative procedure based on F.E.M. For each rotor position θ the needed current is determined to obtain a given reference torque. The flowchart of the method is given in fig 3. The resulting current for an operating point is shown in figure 2c and compared with the previous current (determined with indirect method). The two current wave forms are very similar.

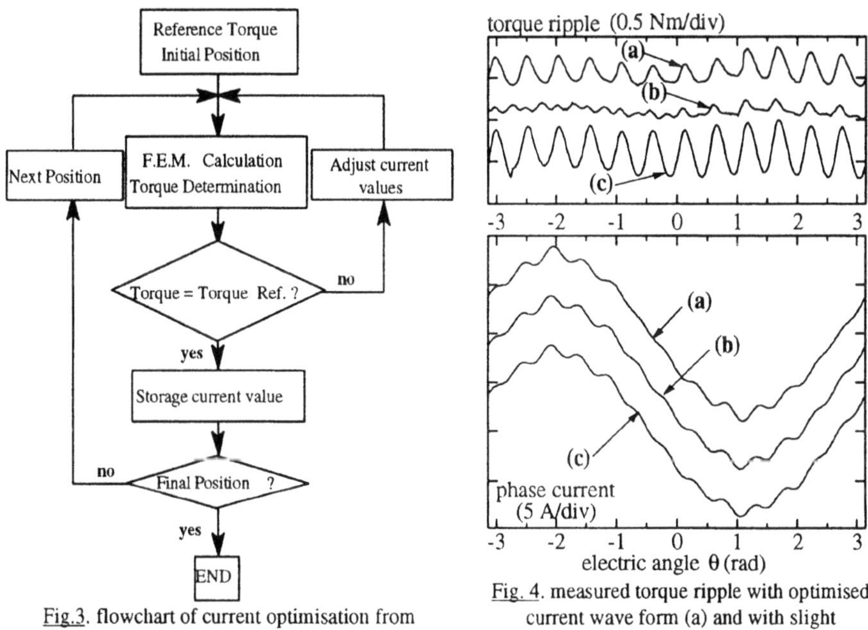

Fig.3. flowchart of current optimisation from direct torque calculation by F.E.M.

Fig. 4. measured torque ripple with optimised current wave form (a) and with slight modification (b) and (c)

Experimental results. The predetermined phase current wave form reference is implemented in DSP board as sum of harmonics:

$$I_a(\theta) = \sum\left[I_{cp}\cos(p\theta) + I_{sp}\sin(p\theta)\right] \quad \text{with} \quad p = 1,5,7,11,13 \tag{3}$$

This expression and current regulation is performed on the torque measurement system at low speed. The measured torque shows that the 6th torque's harmonic is removed but the 12th harmonic remains important (see figure 4, curve a).

The experimental system permits also to observe the influence of current parameters (equ. 2). A slight modification of current harmonics leads to a great variation of the torque ripple like it is shown in figure 4 (curves b and c) where the current seems scarcely modified. Consequently, the method which consists to predetermine the current requires a great precision on parameter determination and is too sensitive. Another solution to minimise the torque pulsations can be envisaged: the direct torque regulation.

ON-LINE TORQUE ESTIMATION

The torque can also be estimated from voltage values, currents, position, speed and a model of the machine[6]. In fact from equations (1) and (2) the torque can be rewrote:

$$\Gamma = \frac{n_p}{2}[I]^t\left\{\frac{1}{\omega}\left([V]-R[I]-[L_{abc}]\frac{d[I]}{dt}\right)+\frac{d[\Phi_f]}{d\theta}\right\} \quad (4)$$

We can show by simulation that if only the first harmonic of inductances and permanent magnet flux is used, the estimation is satisfy. This estimator is implemented in DSP and the experimental result is shown on figure 5 in comparison with measured torque. But this estimation becomes bad at low speed due to the derivation by electric pulsation ω=dθ/dt and the imperfections on voltage values. Currents derivation leads also to noise on the estimate. This estimation can be used to regulated the instantaneous torque.

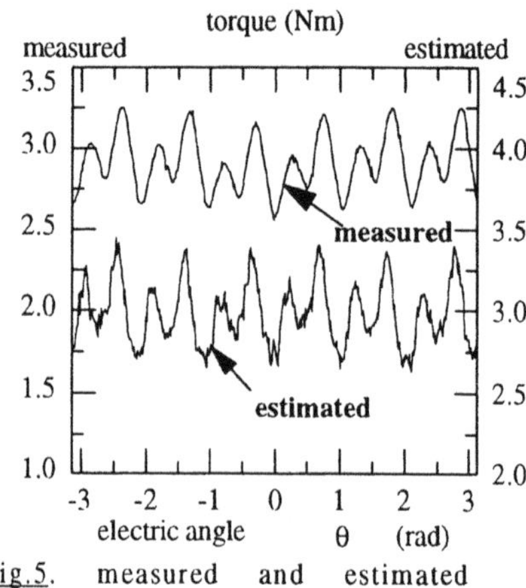

Fig.5. measured and estimated instantaneous torque.

CONCLUSION

In this paper we show by using of the modelling based on finite element method and experimentally that the classical assumptions of sinusoidal field distribution is not valid in buried PMSM. This kind of motor have electromagnetic torque ripple with a sinusoidal fed current. To reduce this ripple we determine optimised current wave form from numerical calculations. However this technique requires a lot of calculations and a great precision to give good results. For these reasons we propose a direct torque regulation using an on-line torque estimator. This estimation is valid only in a range of speed.

REFERENCES

1. Marchand, C. and Razek, A.: 'Electromagnetic modelling to optimise low speed and position control in servo motors', International Workshop on electric and Magnetic Fields, Liège 1992
2. Favre E. and Jufer M., 'Current shapes leading to a constant torque', IMACS-TC'90, Nancy, 1, pp. 135-140, 1990.
3. Marchand, C. and Razek, A.: 'Optimal torque operation of digitally controlled permanent magnet synchronous motor drives', IEE Proc.-B, 1993, 140, (3), pp.232-240
4. Piriou, F. and Razek, A., 'Calculation of saturated inductances for numerical simulation of synchronous machines', IEEE Trans. Magn., MAG-19, pp. 2628-2631, 1983.
5. Ren Z., Besbes M. and Bouktache S., 'Calculation of local magnetic forces in magnetized materials', First International Workshop on Electric and Magnetic Fields, Liege 1992,
6. Colamartino F., Marchand C. and Razek A. 'Estimation and minimisation of electromagnetic torque ripple in a buried permanent magnet synchronous motor." International Conference on Electrical Machines, I.C.E.M. 94, Paris, 5-8 September 1994

AN ENERGY EFFICIENT BRUSHLESS DRIVE SYSTEM FOR A DOMESTIC WASHING MACHINE

Harmer K, Howe D, Mellor P H, Riley C D, Mitchell J K

Dept. Electronic & Electrical Engineering
University of Sheffield
PO Box 600, Sheffield, S1 4DU, UK

ABSTRACT

The paper describes the design, analysis, simulation and testing of a brushless motor for a front loading domestic washing machine. The motor employs bonded NdFeB magnet materials and drives the washing machine drum directly without the need for conventional gearing systems. The key stages in the design are highlighted and test results are presented.

INTRODUCTION

For technical and economic reasons, the drum drive systems employed in domestic washing machines have traditionally been based on a combination of a high speed geared motor and a pulley and belt, energy performance not being a primary design consideration. However, improved detergents and modern fabrics allow low temperature washing, making the energy consumed by the drum drive more significant and providing impetus for more efficient designs.

The paper explores the feasibility of incorporating rare-earth permanent magnet brushless drive technology into front-loading domestic washing appliances. The intention is to improve performance, efficiency and reliability, whilst reducing constructional complexity and without significantly increasing the cost. One of the most radical design possibilities has been chosen in a shift from conventional geared motor systems to a direct drive configuration. The drive motor is mounted on the rear of the washing tub and its rotor is coupled directly to the rotating washing drum. This eliminates the need for motor gearing and supplementary mounting hardware.

DRIVE SYSTEM PERFORMANCE REQUIREMENTS

There are essentially four distinct operational 'modes' which are used to provide a varied range of washing programmes on modern washing machines, viz, wash, rinse, distribute and spin. It has been shown experimentally that two distinct operating points satisfy the torque/speed requirements of each of the above modes. The torque/speed specification for each operating point is shown in Table 1 along with a target efficiency which provides an improvement in motor efficiency of approximately 30% over conventional motors.

Table 1 Motor target specifications (referred to washing drum)

	Low Speed (Tumble)	High Speed (Spin)
Torque (Nm)	10	3
Speed (RPM)	50	1500
Power (W)	52	471
Target Efficiency (%)	60	90

DESIGN PROCEDURE FOR DIRECT DRIVE BRUSHLESS DC MOTOR

The design procedure selected for the direct drive brushless dc washing machine motor is outlined in Figure 1. The following sections elaborate on the techniques employed at each stage.

Electric and Magnetic Fields, Edited by A. Nicolet
and R. Belmans, Plenum Press, New York, 1995

Initial Estimation of Motor Dimensions

Initial estimates of the machine dimensions were made using the general torque expression

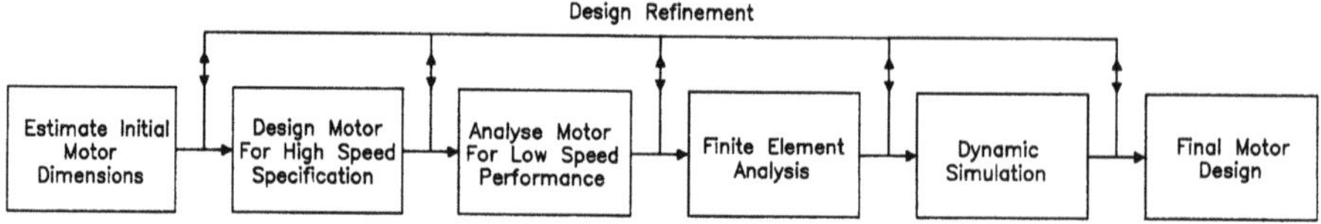

Figure 1 Flow diagram of motor design procedure.

$$T = \frac{\pi}{2} D^2 LBQ$$

where T is rated torque, D is rotor diameter, L is rotor axial length, B is magnetic loading and Q is electric loading. Values of Q and L were obtained from the rated specification of the machine, its method of cooling and the available washing machine cabinet dimensions. Using the standard expression

$$B = \frac{B_r}{1 + \mu_r \frac{l_g}{l_m}}$$

an estimate of B was made by selecting a value of l_g/l_m to give economic use of magnet material.

Design Synthesis and Analysis

Use was made of CAD software developed at the University of Sheffield comprising a suite of programs linking lumped parameter electromagnetic design synthesis with subsequent cogging torque calculations, finite element modelling, detailed iron loss analysis and dynamic simulation of selected designs [1]. Due to the need for optimisation of the design at both low and high speed operating points, a two stage design procedure was adopted.

Stage 1. Parameter scanning and graphical optimisation techniques were used in conjunction with the high speed motor specification to examine a wide range of feasible designs.

Stage 2. Designs from stage 1 were analysed to determine their performance under low speed operating conditions.

This procedure short-listed many feasible designs. The design which satisfied the two torque/speed requirements, approached the target efficiency specifications, made most efficient use of magnet materials and was compact and low in mass, was selected and is shown in Figure 2.

Figure 2 Rotor and stator assemblies of prototype motor.

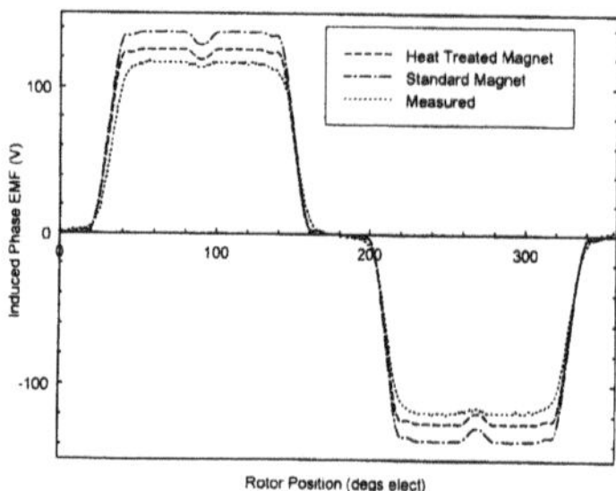

Figure 3 Motor back EMF waveforms.

This 3-phase machine has an external rotor, 20 poles and 30 slots, and is equipped with concentrated stator windings which have short end turns offering low copper loss. The overall diameter is 200mm, the axial length 40mm and the total mass is 3.4kg (excluding bearings and mounting hardware) of which 0.2kg is bonded NdFeB magnet material.

Finite Element Analysis

Finite element analysis was employed in order to check for excessive saturation of the magnetic circuit, derive a back-emf waveform for use during subsequent dynamic simulation, calculate winding inductances and electromagnetic torque, and check the degree of permanent magnet demagnetisation.

The analysis showed that with rated winding currents saturation is well controlled and the demagnetisation margin is acceptable considering the intentionally small quantity of NdFeB permanent magnet material used.

Winding inductances were calculated over a range of rotor positions and with varying winding currents. Due to the low level of saturation, only small variations in inductance were observed with position and excitation, and constant values of self (2.8mH) and mutual (1.35mH) inductance were used during subsequent time stepped dynamic simulation.

The motor back-emf, tooth flux and yoke flux waveforms (Figures 3, 4 and 5) were predicted using the remanence value of 'standard' magnet material together with a lower value to assess the degradation in magnetic properties due to the heat treatment of the magnets required during construction of the machine.

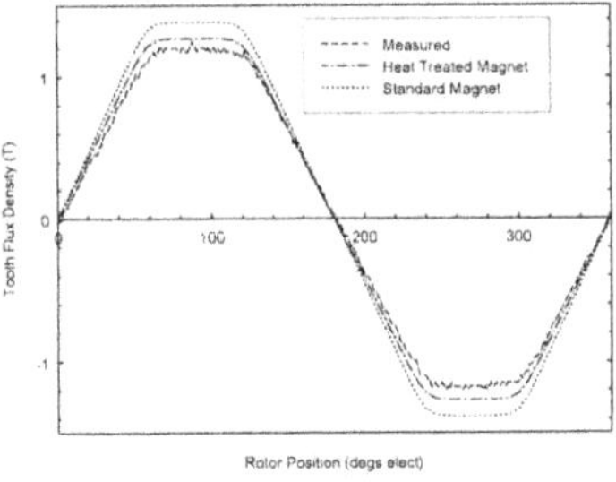

Figure 4 Motor tooth flux density waveforms.

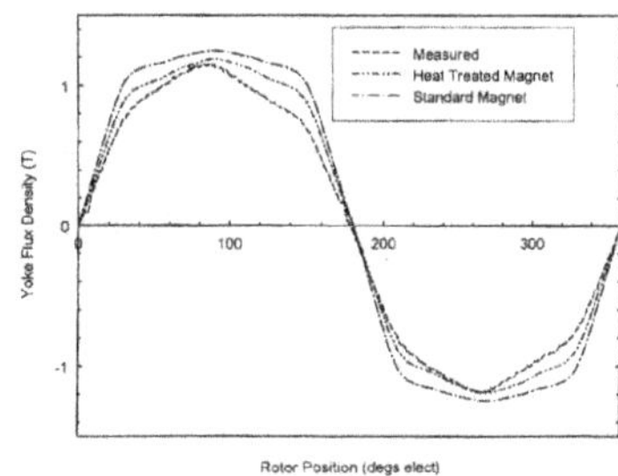

Figure 5 Motor Yoke flux density waveforms.

Electromagnetic torque was calculated using both Maxwell Stress and Lorentz techniques. The results are shown in Figure 6. There is good agreement with the 10Nm maximum torque specification of the motor. As the Lorentz method does not account for reluctance torques, the slight difference between the two methods would be expected.

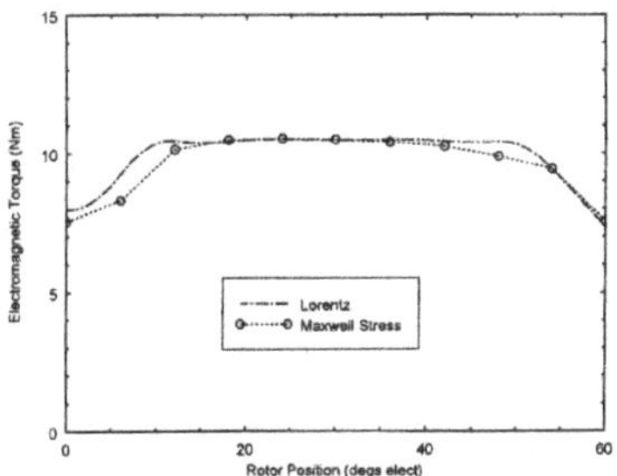

Figure 6 Electromagnetic torque.

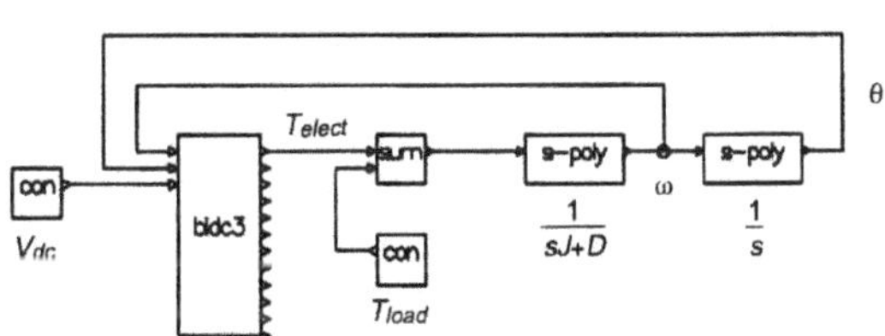

Figure 7 Time stepped simulation network.

Time Stepped Dynamic Simulation

Dynamic simulations were performed under steady-state conditions, at the two specified operating points. Simulations were performed assuming a constant DC link voltage, although this will be extended to account for the mains input rectifier and filter. The network used for the simulations is shown in Figure 7.

Estimation of Motor Losses

In order to accurately estimate overall motor efficiency it is imperative that the motor loss mechanisms are modelled precisely. Under high speed operating conditions iron losses are dominant. Estimating iron losses in brushless dc motors using traditional data (which is based upon sinusoidally varying fluxes) may result in unacceptable errors since the flux waveforms are essentially trapezoidal as shown in Figures 3, 4 and 5. For this reason, a more appropriate iron loss estimation technique [2] has been employed, utilising loss data measured from single sheet material samples. A summary of motor loss components under high speed operating conditions is shown in Table 2. For comparison, three iron loss estimation techniques are tabulated. The most simple method assumes sinusoidally varying fluxes whilst the most sophisticated technique is based upon finite element modelling.

Under low speed operating conditions copper losses are dominant and may be easily computed once the RMS motor phase currents have been established from the dynamic simulation.

Table 2 Summary of predicted high speed loss components and estimated efficiencies.

	'Standard' Magnets			Heat Treated Magnets		
	Iron Loss (W)	Copper Loss (W)	Efficiency (%)	Iron Loss (W)	Copper Loss (W)	Efficiency (%)
Sinusoidal	60	3.9	88	50	4.9	90
Piecewise Linear	68		87	57		88
Finite Element	79		85	65		87

Table 3 Summary of predicted low speed loss components and estimated efficiencies.

'Standard' Magnets			Heat Treated Magnets		
Iron Loss (W)	Copper Loss (W)	Efficiency (%)	Iron Loss (W)	Copper Loss (W)	Efficiency (%)
0.45	44	54	0.37	52	50

PROTOTYPE MOTOR

Construction

The stator lamination stack was fabricated using CNC wire erosion techniques from 0.5mm gauge silicon iron material (Losil 450) which was subsequently annealed to relieve rolling and machining stresses thus improving magnetic performance.

The rotor was constructed from 20 pre-magnetised NdFeB magnet arc segments produced from flat blocks by a thermal deforming process. Unfortunately, the thermal conditions imposed upon the rotor magnets during shaping has a degrading effect upon the magnetic properties of the material resulting in a reduction in remanence from 0.68T to 0.62T.

An initial study has demonstrated that it would be feasible to multi-pole magnetise a complete rotor assembly (ie the back iron and injection moulded magnet) using capacitor discharge magnetisation techniques [3]. This eliminates the thermal deformation process and alleviates the problem of handling individual pre-magnetised magnet arcs, thus reducing production time and cost.

Testing

The experimentally measured back-emf, tooth flux and yoke flux waveforms shown in Figures 3, 4, and 5 are in good agreement with those predicted for heat treated magnets.

At 1500RPM, an open circuit iron loss figure of 70W was measured which compares well with the predicted data in Table 3.

Initially, an overall motor efficiency of 82% was measured at the high speed operating point of 3Nm at 1500RPM and at the low speed operating point an efficiency of 35% was recorded. These figures are lower than predicted and are, in part, due to the effects of non-ideal commutation and the additional resistance of the winding connection leads.

ACKNOWLEDGEMENTS

The authors would like to acknowledge EPSRC for the provision of a CASE studentship for Mr K Harmer and the industrial collaborators Cookson Group PLC.

REFERENCES

[1] BOLTE, E, ACKERMANN, B, HALFMANN, J, HOWE, D, ZHU, Z Q, JENKINS, M K, EVISON, P R, MITCHELL, J K; 'Computer-aided design and analysis of permanent magnet brushless drive systems', Proc of 6th Int Forum on CAD, UK, Sept 1991, p54-73.

[2] ATTALAH, K, ZHU, Z Q, HOWE, D; 'Flux waveforms and iron losses in permanent magnet brushless dc machines', Proc 12th Rare-Earth Magnet Conference, Canberra, 1992, p109-120.

[3] JEWELL, G W, HOWE, D, BIRCH T S; 'Simulation of capacitor-discharge magnetisation', paper BQ-01 presented at IEEE Intermag Conference, Brighton, April 1990, and published in IEEE Trans Magnetics, MAG26(5), 1990, p1638-1640.

TORQUE CALCULATION APPLIED TO OPTIMIZATION METHODS OF PERMANENT MAGNET SYNCHRONOUS MOTORS BY FINITE ELEMENT ANALYSIS

G. Henneberger, S. Domack

Institute of Electrical Machines
University of Technology Aachen
Schinkelstrasse 4, 52056 Aachen, Germany

INTRODUCTION

Permanent magnet excited synchronous motors are applied for robotics and machine tools. Excellent dynamic behavior, high position precision and low torque ripples are requested. To reduce the motor volume, high energy rare-earth magnets are used. The three- phase stator windings are fed sinusodially and the induced air gap flux density has a rectangular shape. Minimal current consumption is achieved with a special current control method, [1] which uses the rotor reluctancy to economize the expensive magnet material. Therefore asymmetrical rotor geometries $X_q > X_d$ are used in combination with a negative current component.Two different asymmetrical rotor designs will be presented as an example in this study and be compared with a conventional symmetrical motor type. Special optimization technics have to be applied to work out, which rotor is the most effective. For that an accurate torque calculation is necessary. Due to the saturation a finite element package is used.

ROTOR DESIGNS AND CURRENT CONTROL METHOD

The Motor M77 in Figure 1 (a) has a conventional symmetrical rotor with radially magnetized magnets on the rotor surface. Rare earth SmCo is used. This model will be the reference to the other models. The stator layout and the nominal stator currents will be kept the same throughout the study.

In general the armature current has only a quadrature component. The reactance in the quadrature and direct axis have the same value, because the whole rotor surface is covered with magnets. It is possible to achieve additional torque with a construction, which has $X_q > X_d$.

Electric and Magnetic Fields, Edited by A. Nicolet
and R. Belmans, Plenum Press, New York, 1995

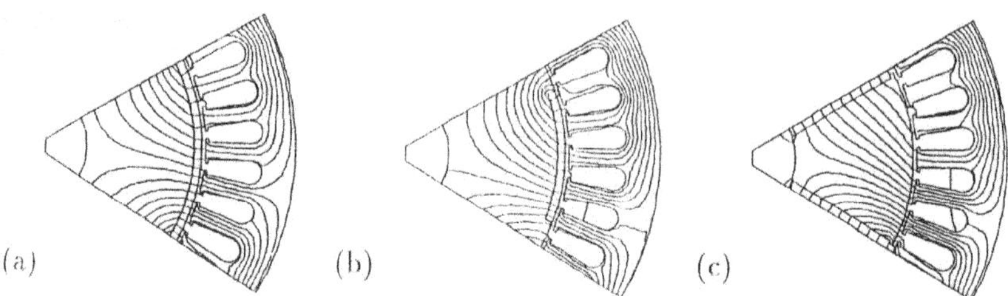

Figure 1. Flux plots of rotor designs: (a) M77, (b) SC, (c) VERT

A simple method to achieve an asymmetrical rotor geometry is by changing the outside magnet pieces of a pole pitch of the M77 into iron (Figure 1 (b)). The permeability of the permanent magnets is approximately equal to the value of free space. A second asymmetrical rotor layout is demonstrated in Figure 1 (c). The interior radial magnets extend to the shaft. The reluctance in the direct axis is higher, as the flux lines have to pass the magnets, thus causing X_q to exceed X_d.

$$M = \frac{3p}{\omega} \left(U_{p0} - I_d \left(X_q - X_d \right) \right) I_q \ , \tag{1}$$

$$I_d = I \cdot \sin \psi \ , \ I_q = I \cdot \cos \psi \ . \tag{2}$$

The rotor reluctance in combination with a negative current component $\psi < 0$ provides additional torque for a better dynamic behaviour [2].

TORQUE CALCULATION

There are different ways of torque calculation by finite elements. Investigations on the maxwell stress tensor and the virtual displacement method [3] have shown, that the latter delivers more accurate results with the same discretisation geometry. Therefore this method will be compared with a newly developed flux method.

The flux calculated out of the vector potential delivers the induced voltage. The effect of the chorded winding and skewed stator slots can then be taken into account. Therefore the induced voltage of a phase has to be calculated as a function of time. In the 2D- field calculation the vector potential of a slot A_{slot} has to be applied, which will be the average of the slot area. The teeth of the stator are parallel and the influence of the position of the coil in the slot can be neglegted, even at high current levels. The motor has six slots per pole pitch and a effective motor length of l_{FE} (laminated rotor sheets). Using the symmetry conditions of one pole pitch the flux of a single phase for a point of time t_1 due to the slot m with a chorded winding $s/\tau_p = 5/6$ can be calculated as:

$$\Psi_{phase}(t_1) = w_{phase} l_{FE} \left(\frac{1}{2} A_{slot_{m-1}} + A_{slot_m} + \frac{1}{2} A_{slot_{m+1}} \right) \tag{3}$$

If the five other slots are taken into account, six time points for one rotor position can be recieved with symmetry conditions of the pole pitch. Additional rotor positions have to be considered for a more precise function of flux and voltage. The voltage U_i of N

time points can be determined with a discret fourier transformation. The speed ω is assumed to be constant:

$$U_i(t) = \frac{d\psi(t)}{dt} , \tag{4}$$

$$\Rightarrow U_i(n) = \Delta\omega \sum_{k=0}^{N} b_k \cos(2\pi \frac{nk}{2N} - \phi_{uk}) . \tag{5}$$

The skewed stator can be integrated with factors:

$$U_i(n)_{skewed} = \Delta\omega \sum_{k=0}^{N} b_k \chi_k \cos(2\pi \frac{nk}{2N} - \phi_{uk}) , \chi_k = \frac{\sin(k\frac{\pi}{6q})}{k\frac{\pi}{6q}} . \tag{6}$$

If the speed ω is constant the torque M can be calculated with the three phases u, v, w:

$$M(n) = \frac{1}{\omega} \sum_{k}^{u,v,w} U_{ik}(n) \cdot I_k(n) , n = 1, \cdots N \tag{7}$$

The application of the virtual displacement method with a skewed stator requires more calculation steps. The motor has to be intersected several times with the length of the motor. The torque of every intersection has to be calculated with a different load angle ψ.

The comparision of the flux method and the virtual displacement method delivers very good results, shown in Figure 2. There is a small deviation of the values for the model VERT. The reason has to be found in the principle of the virtual displacement method. The offset appears, if the distorted layer of finite elements surrounding the movable part has elements, whose boundary is in direction with the polarisation of the close magnet element. A small descretisation error of the vector potential effects the result of the torque.

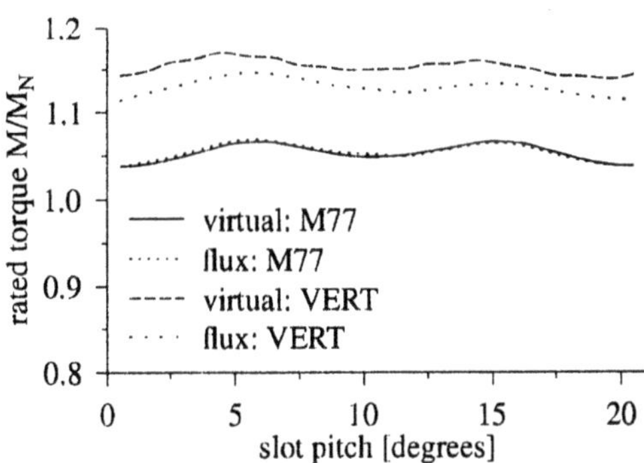

Figure 2. Comparison of virtual displacement and flux method with skewed stator slots

The presented flux method has been proved by measurements. The designs VERT and SC will now be applied to special optimization methods. Therfore the torque calculation will be tested with a Fuzzy- method and a Newton- Lagrange- method.

APPLICATION

The magnet volume of the motors SC and VERT has to be minimized. The torque and current consumption have to be the same as for the motor M77. The motor length of all variants remains the same. For the maximal torque the load angle ψ has also to be estimated.

The Newton- Lagrange- method uses for the estimation of the Hessian matrix the BFGS- formular [4]. The unequaltity restrictions are considerd by the method of the active sets [4].

The Fuzzy- method applies the experience, that the torque is mainly dependent on the magnet volume and uses the deviation of torque as a prediction for the volume.

Both methods have been applied and the results are given in Table 1. Two rotor positions have been taken into account for the torque calculation. For the motor SC the magnet covering for the pole pitch α_i and the hight of the magnet h_M has to be optimized. The magnet dimension for the motor VERT are length l_M and hight h_M.

Table 1. Results of optimization with the Fuzzy- (FZ) and Newton- Lagrange- (NL) method

motor	M/M_N	l_M [mm]	α_i	h_M [mm]	ψ_{opt} [degrees]	saving %
SC (FZ)	1	-	0.78	2.19	-15.5	14.2
SC (NL)	1	-	0.78	2.22	-16.3	14.2
VERT (FZ)	1	26.6	-	2.98	-17.9	11.2
VERT (NL)	1	26.6	-	3.02	-17.8	10.1

The dimensions for the magnet of both optimization methods correspond very well, which confirms the torque calculation with the induced voltage (or flux) method.

CONCLUSION

Investigations of optimization technics with Fuzzy and Newton- Lagrange have shown, that this torque calculation method with the flux or induced voltage can easily be applied and delivers accurate results, which are proved by measurements. The method may also be used in 3D field calculation and it is independent on geometry designs. A corse mesh refinement can be used, which reduces the calculation time and skewed stator slots can easily be taken into account.

REFERENCES

1. G. Henneberger,Dynamic Behaviour and Current Control Method of Brushless DC-Motors with Different Rotor Designs, *EPE* , 1531:1536 (1989)
2. G. Henneberger, S. Domack, Design and Control Methods of Permanent Magnet Excited Selfcommutated Synchronous Machines, *SM 100*, 1124:1129 (1991)
3. J.L. Coulomb, G. Meunier, Finite Element Implementation of Virtual Work Principle for Magnetic or Electric Force and Torque Computation, *IEEE Trans. on Magnetics* 1894:1896 (1984)
4. R. Fletcher,"Practical Methods of Optimization",John Wiley & Sons, Chichester, (1991)

DIFFERENT PERMANENT-MAGNET STRUCTURES AND THEIR INFLUENCE ON THE TORQUE OF SMALL DC MOTORS

M. Rizzo,[1] A. Savini,[2] and J. Turowski[3]

[1] Department of Electrical Engineering, University of Palermo, Italy
[2] Department of Electrical Engineering, University of Pavia, Italy
[3] Technical University of Lodz, Poland

INTRODUCTION

The design of permanent-magnet DC micromotors implies solving a number of problems. A few of them belong to all DC micromotors, i.e. size and number of rotor slots, width of air gap, external stator and internal rotor diameter, ratio of length to rotor diameter etc.[1]. One problem, certainly not the lowest in importance is, however, specific[2], i.e. the choice of the most convenient permanent magnet for prescribed specifications of the micromotor. To date the progress in the production of permanent magnets has made it possible to work out different structures of motors and, for each structure, the most suitable permanent magnet for specified applications. Rare-earth permanent magnets and, in particular, those based on Samarium-Cobalt are so far rather expensive; this suggests their use in small volumes and especially in those cases where high values of torque are required.

The aim of the work has been that of comparing, assigned all other variables, the torques developed by three different models of micromotor, considering, for each model, two different types of magnet.

MOTORS

A comparison is made of the performances of 3 motors having the same geometrical dimensions and the same volume of the magnet (see Table 1).

Table 1. Main characteristics of the motors examined.

Number of pole pairs	2
Number of parallel path pairs	$a = 2$
Outer diameter of rotor	$D_r = 20$ mm
Outer diameter of stator	$D_s = 42$ mm
Air-gap	$\delta = 0.6$ mm
Number of armature slots	$Z = 13$
Number of armature turns	$N = 1820$

Two types of permanent magnets are considered, i.e. Ferrite FB-3 and Samarium-Cobalt; their B-H characteristics are depicted in Fig. 1.

Electric and Magnetic Fields, Edited by A. Nicolet and R. Belmans, Plenum Press, New York, 1995

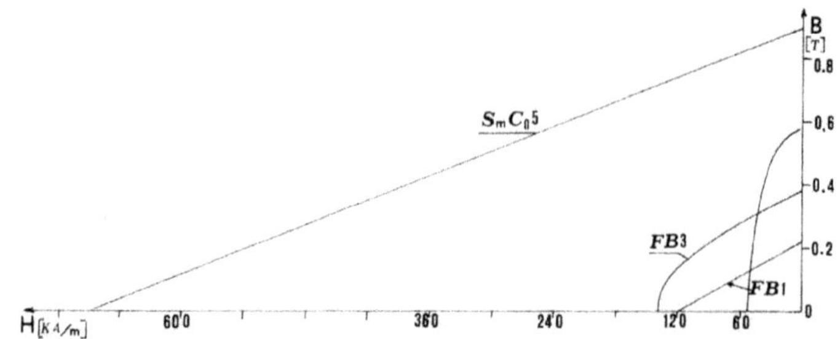

Figure 1. Magnetic characteristic of permanent magnets.

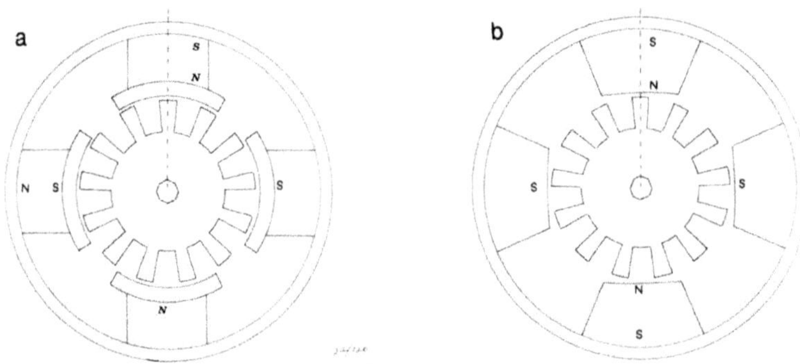

Figure 2. Motors with radial magnetization and: a) polar shoe; b) variable air gap.

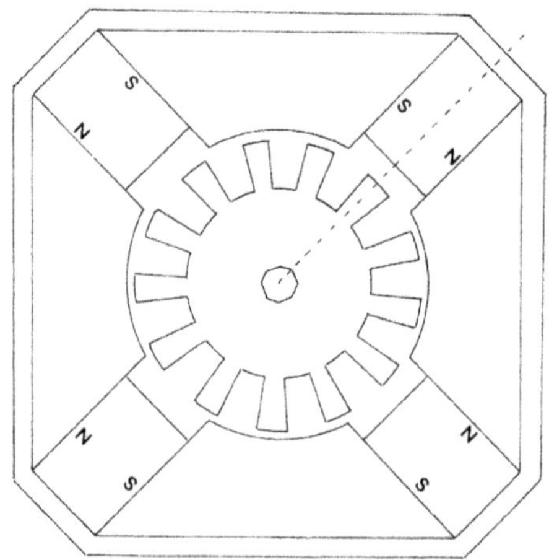

Figure 3. Motor with magnet in angular position and parallel magnetization.

For the two motors in Fig. 2 a radial , while for the motor in Fig. 3 a tangential magnetization is assumed for the magnet.

The same rotor with an odd number of slots is taken for all motors, i.e. more exactly, a copper rotor (large slots and thin teeth) for the Ferrite magnet and an iron rotor (thin slots and large teeth) for the Samarium-Cobalt magnet[3] (Fig. 4 a, b).

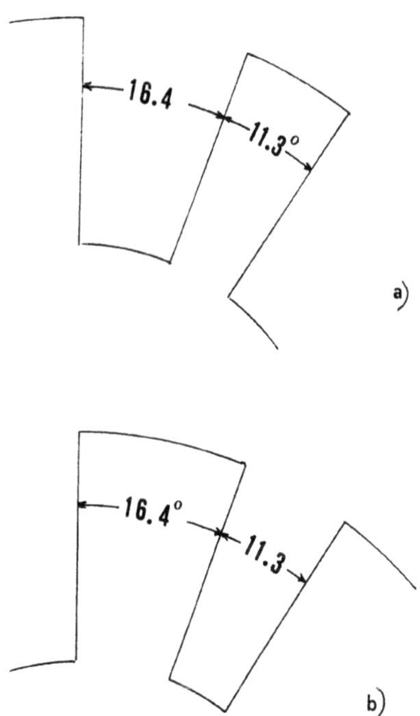

Figure 4. Tooth pitch for: a) copper rotor; b) iron rotor.

A few simplifying assumptions are made: a two-dimensional model of the motor is considered; the slots are supposed to be open and have rectangular cross-section; a winding with a single layer is considered even if, as a rule it should have two layers; the magnet is supposed to have hard magnetization, edge effects and eddy currents are neglected as well as hysteresis effects; uniform current density is supposed to be applied to each slot.

The analysis, carried out by the finite element method[4], makes it possible to determine the values of normal component B_n and tangential component B_t of flux density at air gap. Knowing the distribution of magnetic field on the surface of the rotor, by the application of the stress tensor method, it is possible to compute force F

$$\mathbf{F} = \iint_s (1/\mu_o)\ ((\mathbf{B.n})\ \mathbf{B} - 0.5\,|\mathbf{B}|^2\ \mathbf{n})ds =$$

$$= (\iint_s 0.5(1/\mu_o)(B_n^{\ 2} - B_t^{\ 2})\ ds)\mathbf{n} + (\iint_s (1/\mu_o)B_n B_t\ ds)\mathbf{t} =$$

$$= F_n\ \mathbf{n} + F_t\ \mathbf{t} \quad (1)$$

where

$$\mathbf{B} = B_n\ \mathbf{n} + B_t\ \mathbf{t} \quad (2)$$

and then torque M

$$M = F\ .\ D_r/2 \quad (3)$$

The value of torque (N . m 10^{-3}) is calculated for six different angular positions of rotor, with reference to the symmetry axis of the magnet within a tooth pitch , that is 0°, 5.65°, 9.75°, 13.85°, 17.95°, 22.05°. Both no-load and full-load operations are considered. The values for each motor are reported in Tables 2,3,4.

Table 2. Torque for the motor in Fig. 2 a).

	FB - 3		Sam - Co	
position	no load	full load	no load	full load
0.	68.3	133.0	171.0	300.9
5.65	67.6	23.6	67.7	266.5
9.75	40.31	14.55	42.6	270.2
13.85	102.8	160.2	112.6	136.7
17.95	11.45	107.53	183.0	323.3
22.05	103.5	186.0	223.9	308.5
average value	65.66	104.15	133.47	267.68
peak value	103.5	186.0	223.9	308.5

Table 3. Torque for the motor in Fig. 2 b).

	FB - 3		Sam - Co	
position	no load	full load	no load	full load
0.	82.32	178	116.5	321.9
5.65	65.2	25.57	138.8	267.6
9.75	31.7	85.0	171.6	284.8
13.85	98.0	140.0	60.0	166.3
17.95	84.4	130.0	218.9	370.5
22.05	100.0	205.0	229.8	358.6
average value	76.94	127.26	155.9	294.95
peak value	100.0	205.0	229.8	358.6

Table 4. Torque for the motor in Fig. 3.

position	FB - 3 no load	FB - 3 full load	Sam - Co no load	Sam - Co full load
0.	39.0	72.0	46.8	84.0
5.65	4.96	8.3	97.5	168.0
9.75	2.89	6.2	72.2	114.0
13.85	9.0	24.0	70.0	125.0
17.95	6.0	19.0	63.2	90.0
22.05	7.1	12.0	23.0	42.0
average value	11.49	23.58	62.12	103.83
peak value	39.0	72.0	97.5	168.0

REMARKS AND CONCLUSION

The values of torque reported in the above Tables lead us to note that the motor with variable air-gap (Fig. 2) and radial magnetization both with FB-3 magnet and Samarium-Cobalt magnet provides better performances compared with the other ones. In fact, the variable air-gap offers lower losses than those implied by the pole shoe of the motor in Fig. 3 with radial magnetization as well; losses are lower also with respect to those of the motor in Fig. 4 with parallel magnetization. This can be explained by the fact that, with radial magnetization and variable air-gap, flux lines are forced to develop within the magnet throughout the air gap; therefore, thanks to the high value of reluctance met by flux lines,the effect on the distribution of main flux are less pronounced than those occurring in other configurations or for other types of magnetization. At given load, moreover, the motor in Fig. 2 requires a volume of Samarium-Cobalt magnet 57% lower than that of FB-3 magnet. As concerns the motor in Fig. 4, we can remark that the angular position of the magnet turns out to be more suitable for the Samarium-Cobalt magnet than for the FB-3 magnet in the case of parallel magnetization; also for this motor, at given load, the volume of Samarium-Cobalt magnet can be reduced to 43% of the volume of the corresponding FB-3 magnet.

REFERENCES

1. J. Turowski, K. Komeza, A. Pelikant, S. Wiak, M. Rizzo and A. Savini, Electromagnetic field problems in small electronically controlled motors, in "Proc Int. Conf. ACEMP", Kusadasi, Turkey (1992).
2. M. Rizzo, A. Savini, J. Turowski and S. Wiak, The influence of permanent magnets in brushless DC motors, in " Proc. CEFC '92 Conf"., Los Angeles (1992).
3. T. Kenjio and S. Nagamori, "Permanent Magnet and Brushless DC Motors", Clarendon Press, Oxford (1985).
4. M. Rizzo, A. Savini, and J. Turowski, The shaping of flux density at the air gap of small DC motors with different permanent-magnet poles (companion paper)

COMPARATIVE ANALYSIS OF THREE CLASSES OF EXPERIMENT DESIGN APPLIED TO OPTIMIZATION OF PM DC MACHINES

Kostadin Brandisky, Uwe Pahner, Ronnie Belmans

Lab. EMA, Electrical Engineering Department
K.U.Leuven, Kard. Mercierlaan 94, 3001 Leuven, Belgium

INTRODUCTION

The combination of numerical field calculation with the response surface methodology permits computer simulations to be extended to the design stage, rather than to be used in the analysis of a system. The necessity of using response surface methodology and simplified analytic models stems from the fact that the numerical field computation is very time-consuming, especially for 3D or transient analysis of complicated devices. This makes the direct use of such computer models in optimization algorithms unattractive, because most of them require many evaluations. When using the design of the experiment, a comparatively small number of computer model evaluations are required in obtaining an approximate polynomial model. Moreover, this approach is inherently parallelizable using massively parallel computers.

Second-order polynomial models are commonly used in the response surface methodology, e.g. the central composite designs[1], the Box-Behnken designs[2] and the small composite designs[3]. Here, these experiment designs are compared by their coverage of the investigated region using variance dispersion graphs, by accuracy of the fit, prediction and number of required runs in the optimization of a permanent magnet dc machine with seven design variables.

DESCRIPTION OF THE INVESTIGATED SECOND ORDER DESIGNS

Three-level complete and factorial designs are commonly used for exploring second-order response surfaces. The complete factorial design requires $N=3^k$ runs, where k is the number of variables (also called factors). When the number of factors increases, investigating 3^k factorials become too expensive, e.g. for k=7, $N=3^7=2187$. Furthermore, these designs do not give equal precision for the fitted responses at points that are at equal distance of the center of the factor space (rotatability property).

Electric and Magnetic Fields, Edited by A. Nicolet
and R. Belmans, Plenum Press, New York, 1995

Two designs, using more efficiently the experimental runs than the 3^k factorial experiments, are the central composite design and the Box-Behnken design. Both are fractions of the 3^k factorials and can be made rotatable. Also small composite designs are economic variants of the central composite designs.

The central composite design (CCD) is constructed in the following way[1]:

a) a cube part, consisting of 2^k vertices of a cube;
b) 2k axial points along the co-ordinate axes, (the star), at axial distance α from the center;
c) n_0 center points.

The total number of experiments in a central composite design, based on a complete 2^k factorial is $N=2^k+2k+n_0$. Usually n_0 is $2 \leq n_0 \leq 5$. The central composite design is rotatable if $\alpha=(F)^{1/4}$, where F is the number of factorial points ($F=2^k$ for complete factorial). If k is large, a fraction of the 2^k factorial is taken. For the design with 7 variables a half fraction is taken, $F=2^{k-1}=2^6=64$, and the total number of runs is $N=2^6+2.7+5=83$.

When $\alpha=1$ is chosen, i.e., the star points are at the faces of the cube formed from 2^k factorial, the design is called face-centered cube design (FCC). It has the advantage of requiring only 3 levels, whereas other CCD's with $\alpha \neq 1$ require 5 levels.

The Box-Behnken's designs (BBD) are formed by combining 2^k factorials with incomplete block designs[2]. The result is a design that is more economic than CCD and also rotatable or nearly so. BBD requires 3 levels for each factor and is preferable to the other 3-level design - FCC, because it requires fewer runs and is rotatable.

There are several kinds of small second-order composite designs[3] (SCD). The Draper's small composite design for seven factors is used due to its simplicity. It is constructed as follows:

a) for the cube portion, 7 columns of the 28-runs Plackett-Burman's design are used;
b) 2k=14 axial points are added, with $\alpha=2.8284$ to ensure rotatability;
c) add center points as necessary to ensure a good prediction variance near the center of the region ($n_0=5$ is chosen here).

Thus, the total number of runs of the used small composite design is N=28+14+5=47.

COMPARISON USING VARIANCE DISPERSION GRAPHS

The variance dispersion graph[4] is increasingly used in assessing the design prediction capability over region of interest. It is a two-dimensional plot displaying the maximum, minimum and spherical average prediction variances on various spheres through the region of interest. The resulting plot shows where the design predicts well or not. It also displays where the design's support is inconsistent, i.e. where the difference between maximum and minimum prediction variance is relatively large.

On Figure 1, the absolute values of the prediction variances of CCD, BBD and FCC are compared. Maximum, minimum and average prediction variances are practically the same for CCD and BBD because both designs are rotatable. For $R<0.5$ the BBD exhibits less prediction variance than CCD. CCD performs better near the boundary of the region. The overall performance of BBD is better because it requires less runs (62 versus 83). Because FCC is non-rotatable, it exhibits some difference between its maximum and minimum variance for $R>0.3$. This can jeopardise the accuracy.

Figure 2 displays maximum, minimum and average variance for SCD. There is a considerable dispersion between the maximum and the minimum variances for $R>0.5$. For $R<0.5$ the absolute values of the prediction variances are comparable with these of BBD and CCD.

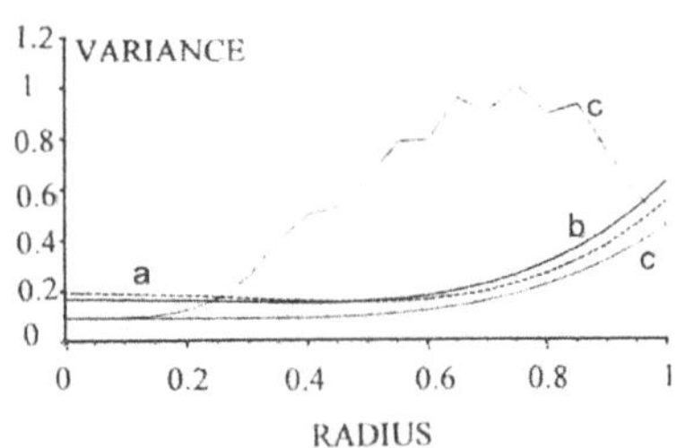

Figure 1. Variance dispersion graphs for a seven-variable CCD, BBD and FCC:
(a) CCD with N=83, α=2.8284 and n_0=5;
(b) BBD with N=62, n_0=6;
(c) FCC with N=79, α=1 and n_0=1.

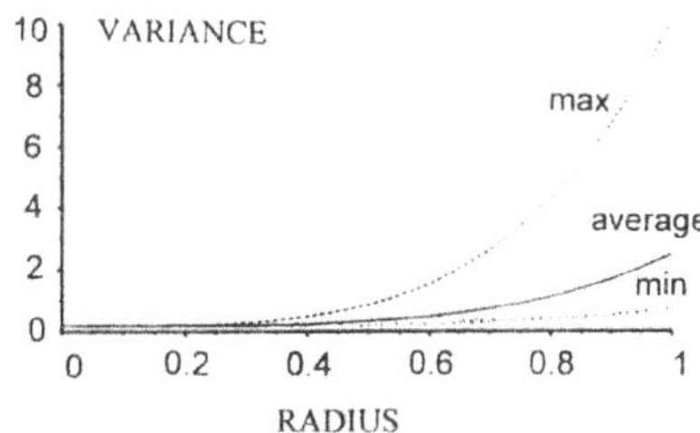

Figure 2. Variance dispersion graph for a Draper's seven-variable SCD : N=47, n_0=5, α=2.8284.

A PM DC MACHINE AS AN EXAMPLE

The designs under investigation are used for constructing approximating polynomials and performing multiresponse optimization of a permanent magnet dc motor. The aim of the optimization is minimum material cost and field strength in the permanent magnets, maximum torque and output power. The optimal settings of seven design variables are sought for - yoke thickness, air gap length, permanent magnet height, outer rotor radius, length of the motor, number of conductors in a slot and wire diameter[5]. To perform the necessary computer experiments a parametrized input shell around a CAD-finite element package[6] is programmed. Then, the least squares method is applied to fit full quadratic polynomials to the machine performances. Desirability functions are formed from the responses and evolution strategy optimization[7] is applied to find the optimum point.

The designs are compared by accuracy of the fit (RSD - relative standard deviation, i.e. an integral estimate of the distance of the observations from the accepted shape of the response surface), difference between the predicted and the verified optimum performances (Diff.) and number of computer experiments (Table 1).

Table 1. Accuracy comparison

Design	Cost		Torque		H_{max}		Power		Runs
	RSD [%]	Diff. [%]	RSD [%]	Diff. [%]	RSD [%]	Diff. [%]	RSD [%]	Diff. [%]	[-]
BBD	0.025	0.135	0.71	2.7	0.99	1.86	0.26	1.5	57
CCD	0.089	0.099	1.78	0.8	0.47	1.30	0.75	0.54	79
FCC	0.058	0.019	0.346	1.85	1.92	2.64	0.20	0.09	79
SCD	0.1	0.52	2.15	4.1	0.56	4.27	0.85	2.44	43

Table 2. Performances for the optimum variants

No.	Design	Cost [BF]	Torque [N.m]	H_{max} [kA/m]	Power [W]
1	BBD	51.89	0.44	172.2	120.6
2	CCD	50.34	0.451	171.8	118.3
3	FCC	53.03	0.443	159.3	110.1
4	SCD	50.31	0.44	170.4	121.2

In Table 2, the optimum values of the performances for the different optimum points are shown.

As seen from Table 1, the SCD has the largest RSD values. CCD and SCD exhibit larger RSD than BBD and FCC, except for H_{max}.

The differences between the predicted and verified responses (Diff.) are local errors near to the optimum points found using the different designs. Typically they are 2-5 times larger than RSD, with some exceptions. The smaller real errors in CCD are due mostly to the fact, that the optimum point is closer to the center of the region, while for the other designs the optimum point is near the boundaries where the prediction is worse. Again, the error is larger for the SCD. CCD and FCC are more accurate.

As a whole, the optimum point performance errors are comparatively small, typically 1-2 %, with exception of the SCD where they reach 4.1-4.27 % in the torque and H_{max}-values. The optimum points differ considerably in their coordinates: many optimum points having nearly equal performances exist in the real response surface. The different experiment designs approximate in a different way the real surface, so with every design different optimums are fitted and found by the optimization program.

CONCLUSIONS

Comparison shows that the central composite design gives the best accuracy of the fit and the prediction, but requires most computer experiments. The small composite design is most economic, but it has the largest prediction errors. The Box-Behnken design is in between in terms of prediction error and required runs. These results are in agreement with the variance dispersion graphs.

The overall conclusion is that the BBD is a good compromise between accuracy and computation, and can be recommended for computer experiments. Also, BBDs exist for a comparatively large number of factors k=3,4,5,6,7,9,11,12 and 16, while SCDs exist for k=5,6,7 and 9. SCD is preferable in experiments requiring large computing time.

ACKNOWLEDGEMENTS

The authors are grateful to the Belgian Nationaal Fonds voor Wetenschappelijk Onderzoek for its financial support of this work and the Belgian Ministry of Scientific Research for granting the IUAP No. 51 on Magnetic Fields.

REFERENCES

1. G.E.P. Box, N.R. Draper, "Empirical Model-Building and Response Surfaces", John Wiley & Sons, (1987).
2. G.E.P. Box, D.W. Behnken, Some new three level designs for study of quantitative variables, *Technometrics*, vol. 2, No. 4, pp. 455-475, (1960).
3. N.R. Draper, Small composite designs, *Technometrics*, vol. 27, no. 2, pp. 173-180, (1985).
4. G G. Vining, A computer program for generating variance dispersion graphs, *Journal of Quality Technology*, vol. 25, no. 1, pp. 45-58, (1993).
5. K. Brandisky, R. Belmans, U. Pahner, Optimal design of a segmental pm dc motor using statistical experiment design method in combination with numerical field analysis", submitted to *ICEM'94*, Paris, France, (1994).
6. E.M Freeman, "MagNet 5 Users Guide", Infolytica Corp., (1993).
7. H.-P. Schwefel, "Numerical Optimization of Computer Models", John Wiley & Sons, (1981).

3D FINITE ELEMENT ANALYSIS FOR EDDY CURRENT COMPUTATION IN RF PLASMA DEVICES WITH METAL COOLING SYSTEM

F. Z. Louai, D. Benzerga and M. Féliachi

LRTI / GE44, CRTT, Boulevard de l'Université
44602, Saint Nazaire cedex, France

ABSTRACT

In this paper, we present an analysis of the electromagnetic phenomena that occur in Radiofrequency plasma devices. To compute the eddy current occurring as well in the plasma as in the metallic cooling system, we have used the three-dimensional AV formulation. As the working frequency is high (3MHz), it is difficult to discretize the areas because the skin effect is very strong. To avoid this problem, we have recourse to the use of the Impedance Boundary Condition (IBC). The effects of the cooling system on the plasma are analysed.

INTRODUCTION

Over the last years, many researches have developed mathematical models for the RF plasma device, to analyse the physical phenomena that occur in it[1,2,3]. The most models are developed in the 2D case. The main difficulty is that the plasma is one phenomenon characterized by the dependency between the electromagnetic, thermal and fluid equations. The dependency of the electrical conductivity on temperature is taken into account by coupling electromagnetic, fluid flow and thermal[2,3] equations. The studied plasma device has 3D structure when the cooling system is present (see Figure 1). On the other hand, the eddy current are characterized by a strong skin effect. Previously, a 2D mathematical model has been developed in the case of cold crucible furnaces by Mühlbauer[4], when the following assumptions:

$$c \ll R_0 \quad \text{and} \quad b \ll c \tag{1}$$

are checked (see Figure 1). Furthermore, this model have been extended and applied to plasma devices with cooling system[5].

As the assumptions (1) are not checked in the case of the studied device, we propose to apply a 3D model using an AV formulation. To compute the eddy current on the surface of the metallic segments an Impedance Boundary Condition is used[6].

Electric and Magnetic Fields, Edited by A. Nicolet
and R. Belmans, Plenum Press, New York, 1995

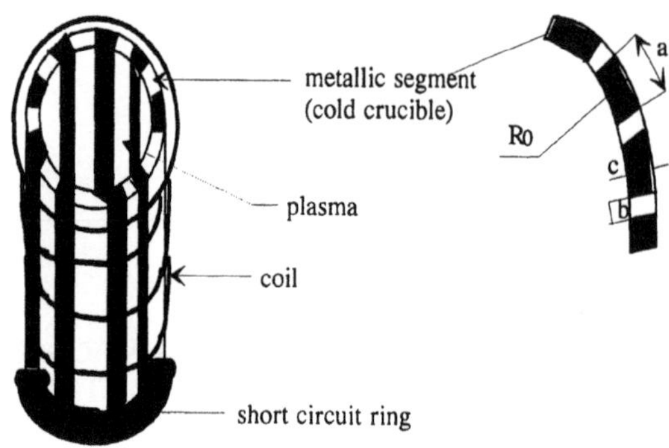

Figure 1. Induction plasma device scheme

FINITE ELEMENT FORMULATION

In order to generalize the study of such a device, we propose to apply a three dimensional modeling based on the AV formulation, to compute the eddy current as well in the metallic segments as in the plasma. In the conducting region, after integrating by parts and using Galerkin projective method, we solve the following equations:

$$\left\{\begin{aligned}&\int_{\Omega}\left[\mathbf{curl\,N}_i\,.\,\nu\,\mathbf{curlA}+\nu_p\,\mathrm{div}\,\mathbf{N}_i\,\,\mathrm{div}\,\mathbf{A}+\mathbf{N}_i\sigma\left(j\omega\mathbf{A}+\mathbf{gradV}\right)\right]d\Omega\\&-\int_{\Gamma_c}\mathbf{N}_i\left(\mathbf{H}\times\mathbf{n}\right)d\gamma_c-\int_{\Gamma_c}\mathbf{N}_i\,\mathbf{n}\,\nu_p\,\mathrm{div}\,\mathbf{A}\,d\gamma_c=\int_{\Omega}\mathbf{N}_i\,\mathbf{J}_s\,d\Omega\end{aligned}\right. \tag{2}$$

$$\int_{\Omega}\mathbf{grad}\alpha_i\,\sigma\left(j\omega\mathbf{A}+\mathbf{gradV}\right)d\Omega=0 \tag{3}$$

$\mathbf{A}$ is the magnetic vector potential and V is the electric scalar potential.
ν : magnetic reluctivity; ν_p: penalty coefficient; σ : electric conductivity; $\mathbf{J}_s$: source current density ; $\mathbf{N}_i$: vector weighting functions and α_i: scalar weighting functions.
Γ_c : the surface of the cooling system segments and $\mathbf{n}$ is the normal to the surface Γ_c.

In the free space, only the magnetic vector potential is used. The uniqueness of the solution is obtained by imposing a Coulomb gauge introduced in (2) as penalty term. So, the second term of (2) on Γ_c vanishes[8].

The high value of the frequency (3MHz), permits to assume the eddy currents be on the surface of the cooling system segments. The current density in the metal varies according to an exponential law[6]: $\mathbf{J}(z)=\mathbf{J}_0\,e^{-\left(\frac{1+j}{\delta}\right)z}$; where $\mathbf{J}_0$ is the current density at the surface, note that the direction z is the normal direction and δ is the skin depth defined by: $\delta=\sqrt{\dfrac{2\,\nu}{\omega\,\sigma}}$, ω: is the angular frequency.

Then, the superficial current density is given by $K=\int_0^{\infty}\mathbf{J}\,dz$, this makes the tangential component of the magnetic field $\mathbf{H}$ discontinuous along Γ_c. Therefore, the magnetic field $\mathbf{H}$ is linked to the electric field $\mathbf{E}$ by the following relation:

$$\mathbf{H} \times \mathbf{n} = \frac{1}{Z_s}(\mathbf{n} \times \mathbf{E}) \times \mathbf{n} \tag{4}$$

With $\mathbf{E} = -j\omega\mathbf{A} - \mathbf{grad}V$ and $Z_s = \frac{1+j}{\sigma\delta}$. Then, the first surface integral on Γ_c is replaced by the expression (4) in (2). So, the equation (2) will be expressed in terms of the unknowns **A** and V.

NUMERICAL RESULTS

The implementation of the 3D model is carried out using the Flux-Expert package[7] and compared with 2D model, previously developed[4] and applied to plasma torch[5]. Without load, the obtained results with this model are in good agreement with measurements operated by Mühlbauer[4] in the case of the cold crucible. In this case of the 2D model, the effect of the cooling system is taken into account by a current representing the cooling system. When the assumptions (1) are verified, The corresponding current density is given in terms of vector potential :

$$\mathbf{J} = \frac{k}{\mu_0}\frac{\partial^2 \mathbf{A}}{\partial z^2} \tag{5}$$

The parameter k depends on the geometrical characteristics of the cooling segments:

$$k = \frac{2\Pi c R_0}{b\,m} \tag{6}$$

In (7), m is the number of segments and the coefficients c, R_0 and b are defined in Figure 1, $\mu_0 = 4\Pi 10^{-7}$ H/m. Figure 2,3,4, show the variation of the magnetic induction following symmetrical axis direction for various values of the parameter k. When the conditions (1) are checked, the relative error is small (see Figure 2) and becomes great (see Figure 4) in the opposite case. These results show how a 3D model is necessary to use.

In the presence of the load (plasma), the effect of the fluid and thermal phenomena is taken into account within the electric conductivity distribution determined from a 2D modeling[3]. Figure 5 shows the effect of the cooling system on the electric power induced in the plasma.

CONCLUSION

The validity of the proposed 3D model is obtained when comparing the results to the Mühlbauer experimental ones. Then, the method is applied to study the effect of the cooling system on the electric power density induced in the plasma.

ACKNOWLEDGEMENT

The scientific support of F. Bouillault from the L. G. E. P lab. is gratfully acknowledged.

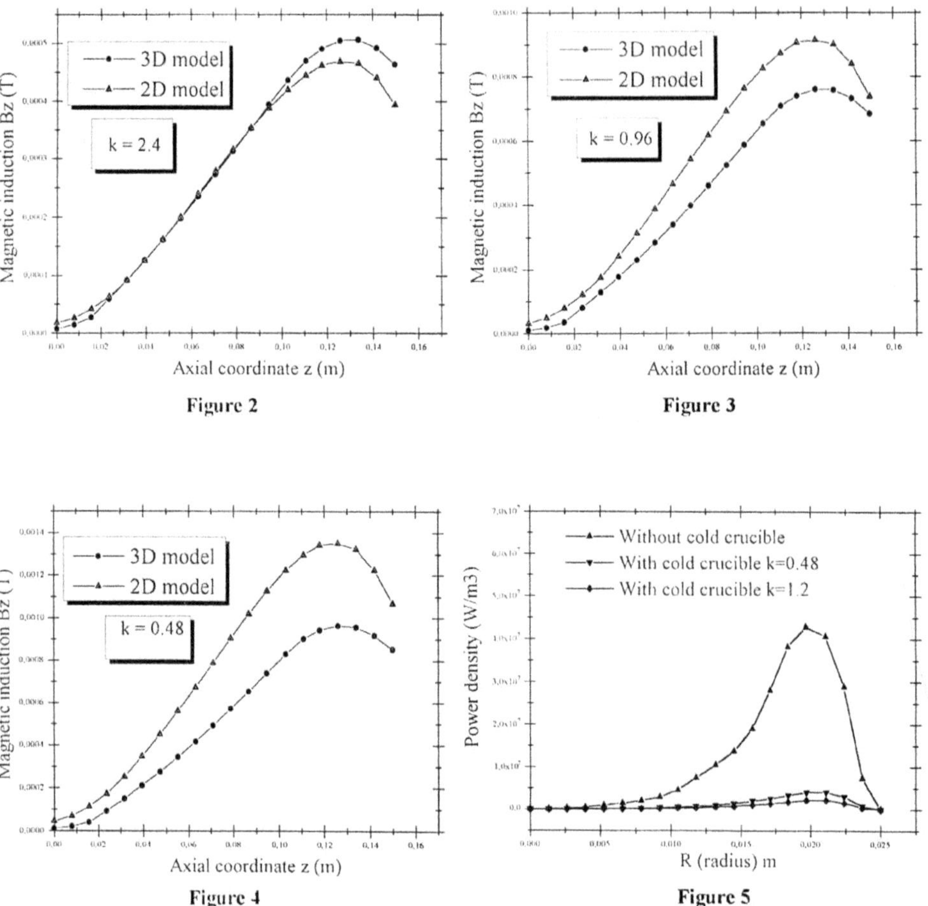

Figure 2 **Figure 3**

Figure 4 **Figure 5**

REFERENCES

1. M. I. Boulos,"Temperature and flow field in the fire ball of the inductively coupled plasma," *IEEE. Trans. Plasma Sci*, PS-4, p.28, (1976).
2. J. Mostaghimi and M. I. Boulos," Two dimensional electromagnetic field effects in induction plasma modelling", *Plasma Chemistry and Plasma Process*, Vol 9, n°1, pp 24-29, (1989).
3. M. R. Mékidèche and M. Féliachi", Finite element induction plasma modelling", 19th . *Inter. Conf. On Plasma Science*, Tampa, Florid, USA, June (1992).
4. A. Mühlbauer, A. Muznicks and A. Jakowitscki", Modelling of the electromagnetic field in induction furnaces with cold crucible", *Elctrowärm International*, n°49, (1971).
5. M. R. Mékidèche", Contribution à la modélisation numérique de torches à plasma d'induction", *Thèse de Doctorat*, Université de Nantes, Octobre (1993).
6. S. Ratnajeevan H. Hoole", The natural finite element formulation of the impedance boundary condition in shielding structures", *J. Appl. Phys.* 63 (8), 15 April (1985).
7. Ph. Massé," Analyse méthodologique de la modélisation numérique des équations des milieux continus à l'aide de la méthode des éléments finis: Flux-Expert un système d'aide à la construction des logiciels", *Thèse de Doctorat ès-Sciences*, INPG, Grenoble, (1983).
8. P. Robert, M. Ito and T. Takahashi", Numerical solution of three dimensional transient eddy current problems by the A-ϕ method" *IEEE. Trans On Mag*, Vol 28, n°2, March (1992).

COMPARISON OF 2D AND 3D EDDY-CURRENT CALCULATIONS WITH RESPECT TO INDUCTION FURNACES

G. Henneberger, S. Dappen, W. Hadrys

Institute of Electrical Machines
University of Technology Aachen
Schinkelstrasse 4, 52056 Aachen, Germany

ABSTRACT

This paper deals with the comparison of 2D and 3D eddy-current calculations especially in the case of induction furnaces. These machines are widely used in industries to heat up and melt metals for further metallurgical processes. The magnitude of these furnaces is in the range from some kilograms up to many tons of melt within the pot. Nowadays the induction furnaces are supplied with resonance inverters with a typical frequency range from 70 to 1000 Hz. Typical induction furnaces consist of five main parts, some of them are very important for the electromagnetic effects as the distribution of power losses in the windings, the efficiency or the melting power. These are the furnace's coils, the yokes and the melt. Less important for electrical effects are the melting pot and the mechanical suspension of the furnace.

All calculations, which have been done in the past used twodimensional models not only for the calculation of electrical parameters but also for coupled problems as fluid flow phenomenons and their free surface behaviour [1]. For dealing with these problems the usage of axisymmetrical twodimensional eddy-current calculation has approved its applicability. Now the question is, if there are typical threedimensional effects in the construction of induction furnaces, which will lead to limitations in geometrical magnitude or the maximum electrical power.

CALCULATION OF POWER LOSSES IN THE WINDINGS

For dealing with the above mentioned questions an 2t induction furnace is taken into consideration with 10 windings working with a frequency of 50 Hz and power-supply voltage from mains. The melt is steel at 1500°C temperature and the yoke consists of laminated iron.

Fig. 1 shows on the left side these electromagnetically relevant regions in a 18°-segment which could be used because of the furnace's twenty-fold symmetry. Fig. 1 also shows on the right hand side some other construction parts as screws, a beam and a metal platform lying above the coil and the yokes and being used as an access for workers to the melting pot. In

Electric and Magnetic Fields, Edited by A. Nicolet
and R. Belmans, Plenum Press, New York, 1995

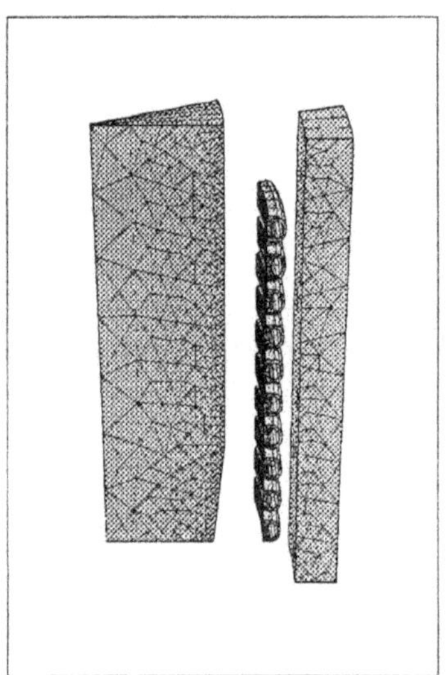

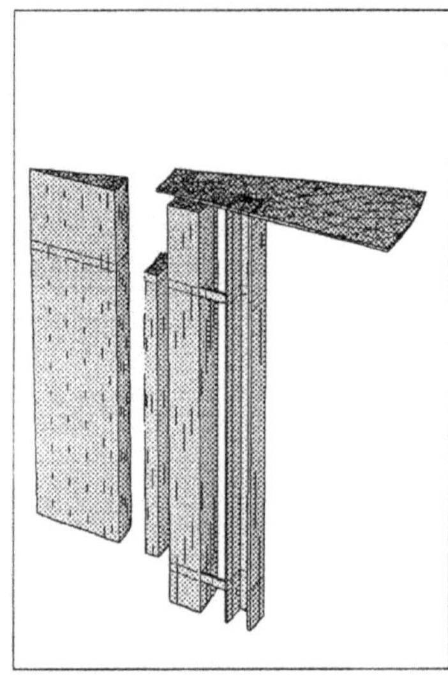

Figure 1. 3D geometry with major construction parts on the left and mechanical suspension parts on the right

high power furnaces the power losses within these construction parts limitate the maximum furnace power. But in this work there will be regarded and compared at first the calculated power losses in the coil's windings when using a classical twodimensional model in contrast to threedimensional eddy-current calculations.

For getting comparable results the yokes of the furnace are modelled as a cylinder also in the threedimensional model. This is a principle neglect in all twodimensional approaches. In a third model being also threedimensional this neglect of a cylindrical yoke will be revised. So in this model there are 10 yokes being put in equal distances around the coil.

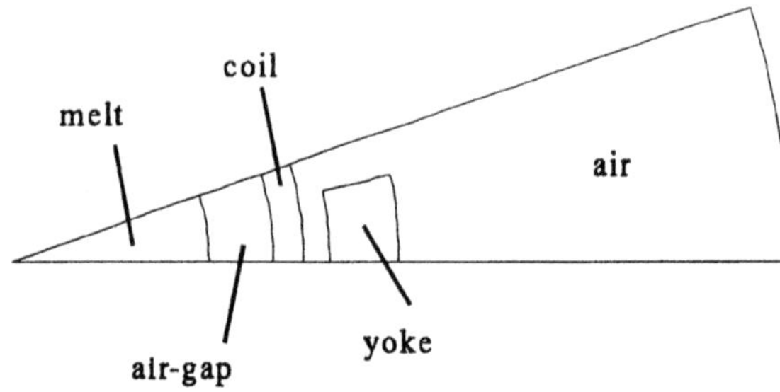

Figure 2. view from top on the model with separated yokes

Fig. 2 shows this geometrical distribution with a view from the top. The twodimensional axisymmetric calculation uses the vector potential approach whereas the threedimensional approach utilizes the scalar-vector method to get the results shown in Fig. 3.

The models compared consist of approximately 4100 nodes solved with second order approximation functions (2D) and 245880 tetrahedral elements with 46051 nodes and first order trial functions (3D). The main consequence of Fig. 3 is the possibility to neglect the non-cylindrical form of the yokes without disturbing the results of the current-density distribution and power losses in the melt and the coil windings although they are positioned nearby the yokes. The twodimensional approach is sufficient for these tasks and its main advantages are the low computing time and main memory requirements which make calculations possible even on smaller computer systems.

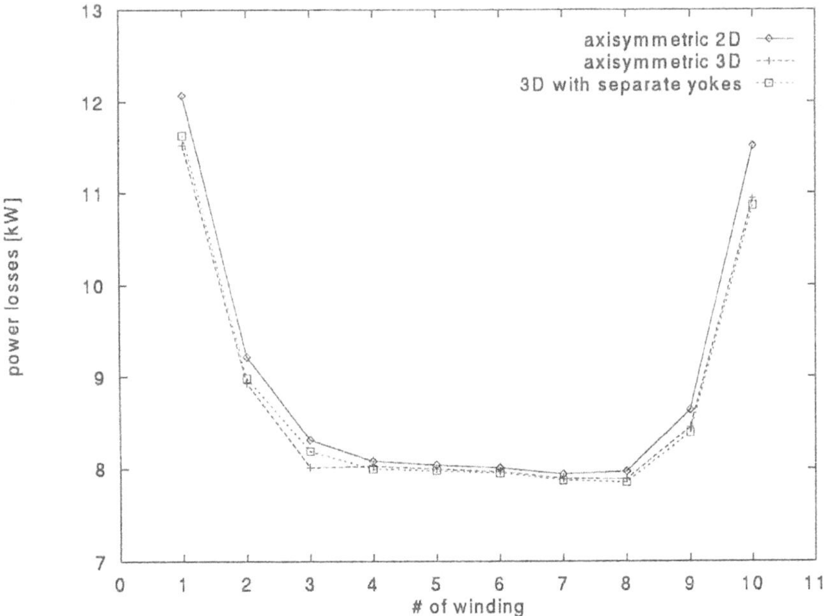

Figure 3. Comparison of different calculation approaches

The distribution of power losses (Fig. 3) is typical for induction furnaces. The windings in the middle of the coil are only coupled to an axial magnetic field but the end windings are coupled to axial and radial field components. This results in a stronger proximity-effect in these regions and as a further consequence in higher power losses.

POWER LOSSES IN MECHANICAL SUSPENSION PARTS

A typical threedimensional electromagnetic problem concerning the construction of induction furnaces are the losses in the mechanically relevant parts. These are the screws, beams and metal the platforms. The calculation of losses can be regarded as a first step to the calculation of temperature distributions or other coupled problems within these parts [2]. As it can be seen from Fig. 1 there is no reduction possible to a twodimensional calculation with respect to any symmetry. For example the screws and the beams destroy both, an axisymmetrical and a 2D-cartesian problem reduction. The results for the induction furnace mentioned above are

$$P_{loss,beam} = 9\,\mathrm{W} \quad , \quad P_{loss,platform} = 19\,\mathrm{W} \quad , \quad P_{loss,screws} < 1\,\mathrm{W}.$$

In this case of a furnace working with power-supply voltage from mains and a frequency of 50 Hz, these losses can be neglected but in the cases of higher frequency supply this method gives the opportunity to optimize the mechanical construction parts in their geometry and position within the furnace in order to avoid superheating of these parts being able to endanger the service staff.

YOKE- AND LEAKAGE-FIELD

It is more and more important to know the leakage-field of high power induction furnaces exactly. This is because of reduced limit values of leakage-fields in service staff's working areas. As a result of this the prediction of the field outside the coil, especially in the region of the yokes is imperative. The twodimensional model cannot be used because of its cylindrical yoke modelling, but a threedimensional analysis using the model shown in Fig. 2 answers this question. Some results obtained by this model are shown in Fig. 4 where 3 cuts in the radial-circular plane at different axial values have been done.

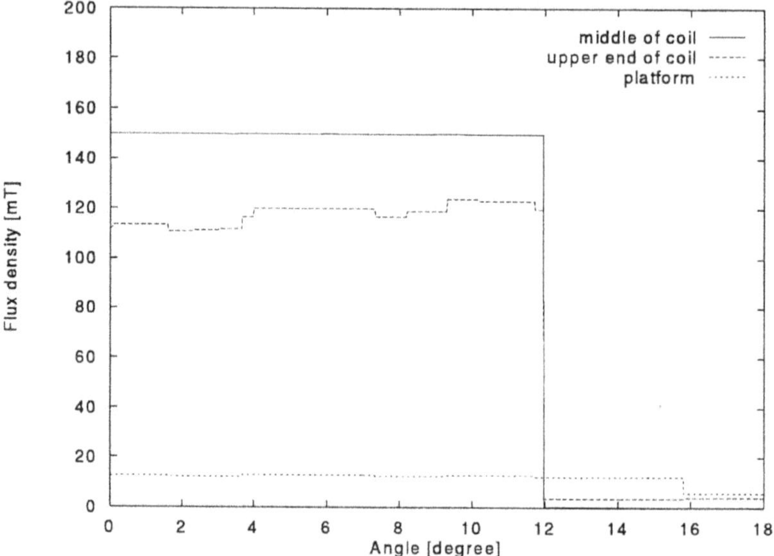

Figure 4. yoke- and leakage-field in different axial cuts

It is shown the flux-density distribution at a line lying in the middle of the yoke in the 18°-segment. From 0° to 12° there is placed the yoke and from 12° to 18° an air-gap remains in the furnace. In the middle of the yoke all flux runs through it and there remains less than $1mT$ of flux-density in the air-gap. The results change when taking into consideration a cutting plane being placed above the yokes. Here the flux-density distribution is much more smoothed compared to the middle of the yoke but nevertheless it can be seen that the flux tries to go into the yoke's iron. A main result of this leakage-field analysis is the fact that the yokes are able to take up nearly all of the flux in their present construction and that only a very low magnetic leakage-field will be nearby the servive staff although they are not build cylindrically closed around the coil.

References

[1] G. Henneberger, Ph.-K. Sattler, D. Shen, W. Hadrys: *Coupling of Magnetic and Fluid Flow Problems and its Application in Induction Melting Apparatus*, IEEE Trans. Magn., Vol. 29, pp. 1589-1594, March 1993.

[2] G. Henneberger, W. Hadrys, W. Mai: *Threedimensional Calculations of Mechanical Deformations caused by Magnetic Load*, Proceedings of the second International Workshop on Electric and Magnetic Fields, Leuven 1994.

INDUCTORS MODELLING AND OPTIMIZATION IN COOKING INDUCTION HEATING SYSTEMS

D. Leschi[1], N. Burais[1], J.Y. Gaspard[2]

[1]CEGELY - URA CNRS 829
Département d'Electrotechnique - Ecole Centrale de Lyon
BP 163 - 69131 Ecully, Cedex, France
[2]CEPEM-TEM
18 rue du 11 Octobre - BP105
45142 St Jean de la Ruelle, Cedex, France

INTRODUCTION

Induction heating is widely used in metallurgy. Its application in cooking systems is recent. Induction cooking presents more advantages than classical heating systems (resistance or gas): direct heating up of pans without thermal inertia, electronic adjustable power control. Although the induction principle has been known many years ago, the design of such systems necessitates the study of the three parts of an induction system: resonant converter, inductor, and pan.

Because of the use of different pan sizes with various ferromagnetic properties, the following contrains have to be taken into account : homogeneous thermal distribution in pan, constant equivalent inductor impedance, high power efficiency, minimum electromagnetic interference.

In this system (figure 1), an inductor (coil) is powered by a resonant converter. Magnetic field time variation induce eddy currents in a pan placed above this inductor. The heating of the pan is related to eddy current distribution which is dependent on the following parameters :

- geometrical structure of the whole inductor-pan,
- electromagnetic properties of the pan (electrical conductivity σ, permeability μ),
- current frequency and rms value.

In the resonant converter (figure 1), fixed capacitance value is adapted to an equivalent impedance value of the inductor-pan system in order to obtain rms current value and frequency (f = 25 kHz). But this equivalent impedance takes into account eddy current distribution in pan (as a secondary part of an equivalent transformer) and owner impedance of the inductor (primary part of this "transformer"). So, we can notice that the three parts of the induction system (converter, inductor, pan) are very closed together. Although many

Electric and Magnetic Fields, Edited by A. Nicolet
and R. Belmans, Plenum Press, New York, 1995

publications about electronic power and induction applications exist, only a few number of them are devoted to the study of converter-inductor system where the authors[1,2] use lumped parameter transformer model of inductor-pan.

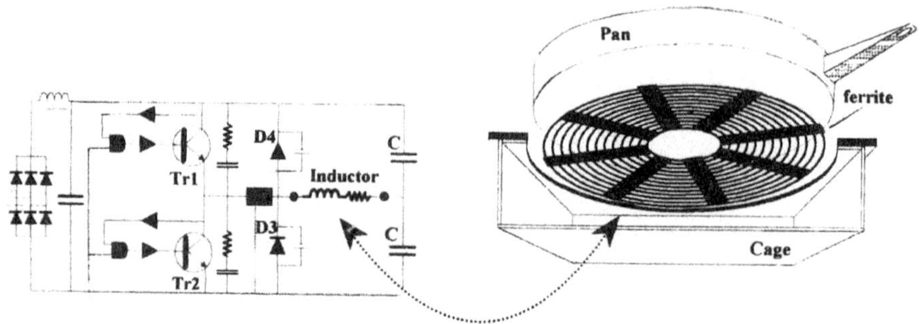

Figure 1. Cooking induction heating system (resonant converter, inductor and pan)

ELECTROMAGNETIC FORMULATION AND FIRST RESULTS

In order to develop optimal induction cooking systems, it is necessary to study eddy current and heating distributions, and to calculate equivalent impedance of the whole inductor-pan. Physical phenomena in this system can be simulated by solving Maxwell's equations. The system geometry is 3D. Because of the axisymetric structure of the inductor and the pan, an axisymetric solution (figure 2) is possible using an equivalent reluctance and permeability for the 3D magnetic core.

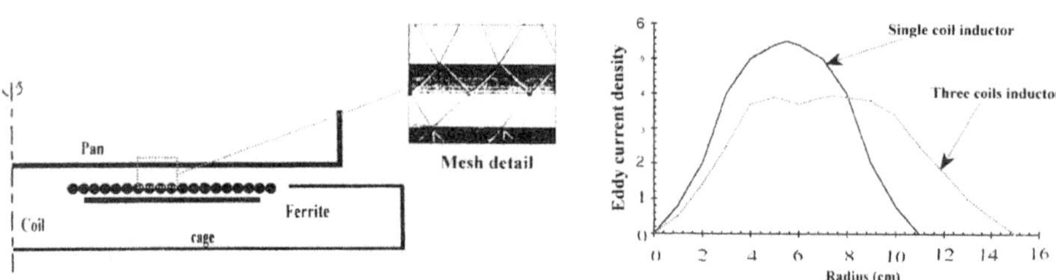

Figure 2. Classical axisymetric study plane **Figure 3.** Eddy current distribution along radius of large pan

Using the magnetic potential vector **A** defined by **B** = **Curl A**, electromagnetic phenomena are modeled by the classical magnetodynamic equation :

$$\mathbf{Curl}\,(1/\mu.\mathbf{Curl\ A}) + \sigma.\delta\mathbf{A}/\delta t = \mathbf{J_{ex}} \quad (1)$$

with J_{ex} is the current density in the inductor.

This equation is solved with Finite Element Method and is implemented in package FISSURE developed in CEGELY.

A first study has been made with various parameters as, inductor and pan sizes, supporting structure, and magnetic core. A very good correlation between numerical and experimental results is obtained. But this study shows a non homogeneous thermal distribution called 'corona' effect in large diameter pans (figure 3), and a high level of electromagnetic interferences for small diameter pans (figure 4).

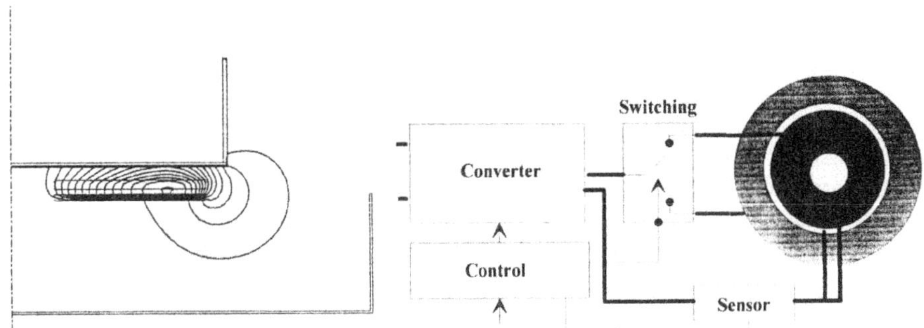

Figure 4. Flux lines distribution with small pan and single coil

Figure 5. Successive power inductor

DEVELOPMENT AND MODELISATION OF AUTO-ADAPTIVE INDUCTORS

The authors thus have developed auto-adaptive inductors with concentric coils, which allow to obtain more uniform thermal distributions in different size pans, with constant equivalent impedance in order that these inductors can be supplied by the same resonant converter. Various coils systems[3] have been tested :

- two parallel coils,
- two coils supplied in different power cycles (figure 5),
- three coils with various connection states (serial and/or parallel, one or two coils in short-circuit,...) (figure 6).

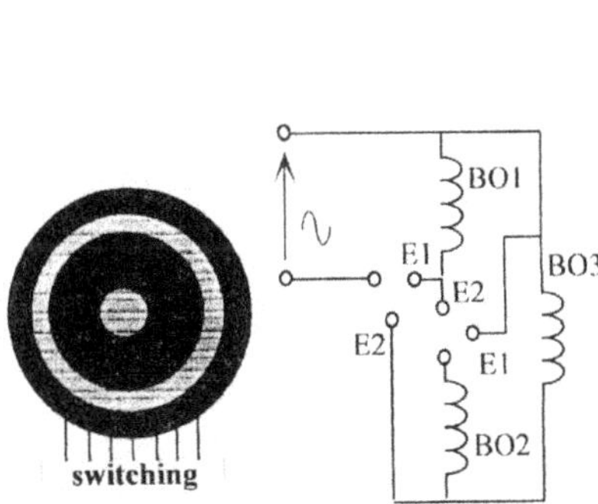

Figure 6. Three coils inductor

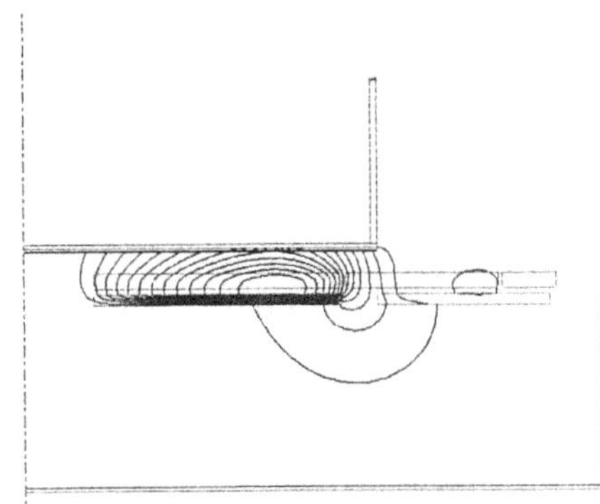

Figure 7. Flux lines with small pan for 3 coils inductor

In order to determine currents in different coils, equivalent impedance and eddy current distribution in pan, a special package has been developed in FISSURE to calculate particularly self and mutual reactances Xij of inductors coils.
For each coil i, equation (1) is automatically solved with :

$J_{ex} = J_i$ for one coil i

$J_{ex} = 0$ for others coils j

and all equivalent reactances and resistances are calculated from flux distribution and energy consideration.

Simulation of the whole system (power converter - inductor - pan) is made by associating power electronic circuit simulation package (SICOS) and field calculation package (FISSURE) developed in CEGELY. These packages allow to determine current rms value and frequency, and magnetic field distribution in the system. Performances of these differents inductors can be thus easily compared :
- Two parallel coils system is interesting for medium and large pan diameter, but not adapted for small diameter.
- Two successif powered coils system is adapted for small and medium diameter. It implies that the two concentric coils have the same equivalent impedance. This system is more simple because no mutual inductance must be taken into account in the current calculation. Current switching between the coils must be realised at zero current.
- Three coils system has more complicated states but is adapted to all diameters. For small pan diameter (figure 7), state 1 = BO1 supplied, BO2 +BO3 short-circuited; for large diameter, state 2 = (BO1+BO2)//BO3.

For example, figure 3 shows eddy current distribution in three coils system, compared with distribution obtained with only one classical coil.

CONCLUSION

Good correlation between numerical and experimental results are obtained. These new inductors satisfy international interference standards and give more homogeneous thermal distribution with various diameter pans. They are actually tested in professional cooking induction systems. These study is an example to show how classical electromagnetic modelisation allows to design and optimize an electromagnetic system taking into account all physical and technical constrains.

ACKNOWLEDGMENTS

This study is supported by CEPEM (Orléans) for domestic appliances and BONNET (Villefranche/Saône) for professional appliances.

REFERENCES

1. G. Morizot, "Etude d'une table de cuisson par induction et convertisseur MF, 2 kW", Thèse CNAM (1981).
2. S. Carthy, "State plane analysis of induction hob generators", Ph D Dublin (1994).
3. J.Y. Gaspard, "Modélisation et réalisation de nouveaux systèmes de cuisson par induction", Thèse de Doctorat, Ecole Centrale de Lyon (1993).

FINITE ELEMENT ANALYSIS OF INDUCTIVE REHEATING FACILITIES FOR THE STEEL STRIPS PRODUCTION

P. Costa, M. Santinelli

ENEA, Associazione EURATOM-ENEA sulla Fusione
C.R.E. Frascati
C.P. 65 - 00044 Frascati, Rome (Italy)

I. THE PROBLEM

The production of steel strips and plates is usually performed with processes during several hours in plants more than 1400 meters long. In order to reduce investment and production costs, a continuous process during only 15 minutes has recently been set in Italy in an innovative plant (8 times shorter than usual) where in addition power requirements and ecologic impact are lighter. Such a compact arrangement has been possible by means of inductive reheating facilities, that maintain the strip *hot*, i.e. with both a liquid core and in fully solidified state. The new line-up (thin slab cast rolling machine, coiler-furnace, four-high stands finishing mill) is easily adaptable to different steels and enables a cost favourable hot strip production even with reduced quantities of product. thanks to the inductive reheating facilities. This paper will present the FEA of these facilities. Figure 1 describes flux concentrator core and two windings coil of the facility fore-and-aft section.

II. FINITE ELEMENT ANALYSIS (FEA) OF THE INDUCTIVE HEATING FACILITIES

Heating facilities design is a complex job involving the interaction of different skills. The authors have been charged of theoretical aid, while a prototype reheating module was arranged. Their investigations started with a study of the designate prototype electromagnetic field, in order to optimize its geometry. Then the heat diffusion through the strip cross section, from the skin depth to the strip core and to the radiating surface, has been described with a 1-D-FEM. The thermic network analogy method has after been employed, to calculate how the temperature distribution changes in the time. In this phase, computer aid has been provided by ATP (Alternative Transient Program) to analyse the hot strip transient reheating.

After these computations:

Electric and Magnetic Fields, Edited by A. Nicolet
and R. Belmans, Plenum Press, New York, 1995

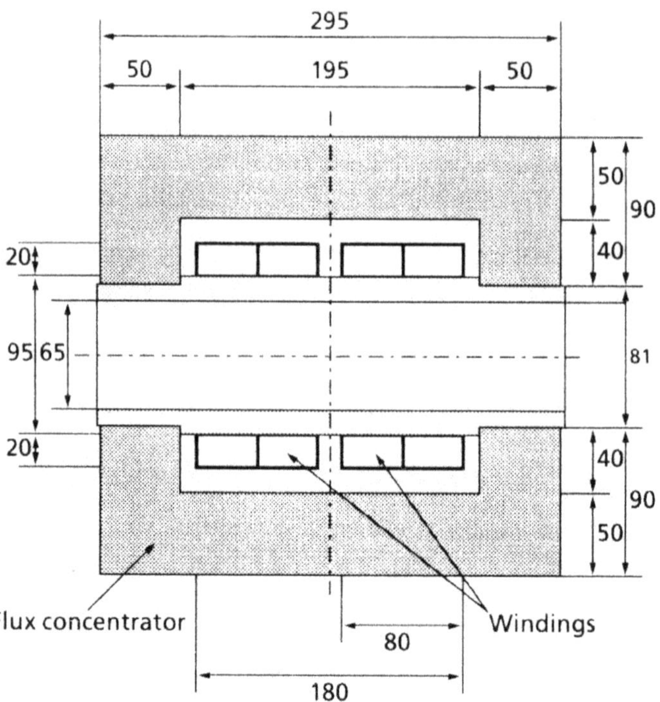

Figure 1. Inductive reheating facility fore-and-aft section.

- the prototype reheating module has been redefined (by increasing power supply frequency, to cope with the strip thickness, and by changing flux concentrators shape, to widen the current distribution);
- FEA has been adopted to give the most realistic description of the process.

The hot strip temperatures map is established by the balance between skin depth heating and external surface cooling. The heat distribution depends on the eddy current density one, which in turn is given by the magnetic flux density distribution. The authors have utilised a FE code (ANSYS) able to carry out a coupled analysis, in the time domain, of the three fields: electric, magnetic and thermal. First, a preliminary 3-D-FEM of a single module has been developed. The 1820 elements model represents one eighth of the prototype operating at 6.7 KHz. The outputs map the magnetic field and the current density induced in the strip. Luckily the edge effects haven't come out to be very significant. The authors have afterward employed a 2-D-FEM to describe a quarter of strip, core and coil prototype cross sections: the results were not very different from the 3D ones. This allowed them to carry on the analysis with the 2-D-FEM and avoid expensive computations.

III. THE MOVING FEM

During the process, the strip heats and cools as it enters and leaves all the inductive reheating facilities. The authors have realized a 2-D-FEM that simulates the strip moving at a speed of 6 m/min through four consecutive facilities. The fore-and-aft strip section is divided in 26 successive segments. Every ANSYS iteration two segments come along and go in the reheating facility, while the two ones that were in it, come out and cool themselves. After few iterations, the first two segments come in the second facility, while the last ones are still in front of the first. In this way, the strip goes on moving through all the facilities and the segments cross different values of magnetic flux and heat the strip by their eddy currents, in different way. A transient dynamic analysis has been performed to take into account the effects on the strip temperature, caused by its moving through the different consecutive heating conditions. But the cooling conditions also change in and out the facilities. Cooling, in fact, is chiefly caused by the strip radiation, i.e. outside the facilities, it depends on the ambient temperature, and on the facility one, inside it.

IV. THE EXPERIMENTAL VALIDATION

The temperatures distribution at the surface of a test strip passing through the prototype module has been monitored with two pyrometers, placed at the input and at the output. Thermocouples have been put in the test strip at different depths, to measure the penetration of the inductive heating. The measures have been acquired by a computer and related to the simulation results. The experimental data have outcome to be similar to the analysis results. The maximum difference has been less than 10%. The comparison between experiment and simulation is described in Figure 2.

V. THE TOOL

The little difference between experiment and simulation enabled us to assess the qualitative and quantitative validity of our FEA, that has come out to be a tool, useful to evaluate in few hours and at low cost the effects of possible improvements on the heating facilities, such as:

- power supply optimization: e.g. different values of the coil current;

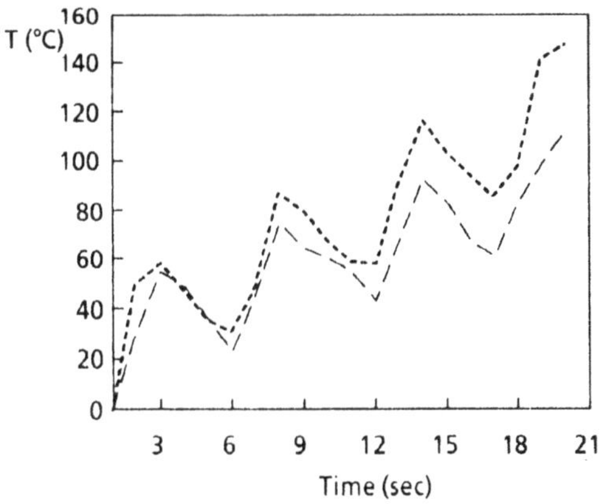

Figure 2. Comparison between experiment (dotted line) and simulation (dashed line).

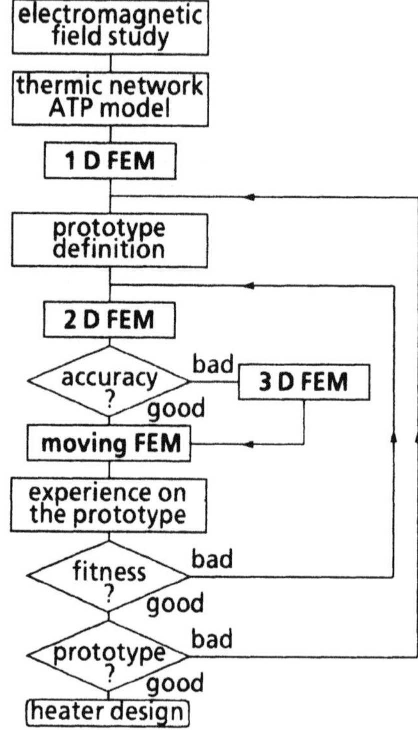

Figure 3. FEA contributions to the heating devices design.

- heating depth in the strip: e.g. different values of power supply frequency;
- magnetic field concentration in the strip: e.g. different shapes and dimensions of the magnetic core;
- efficiency, by diminishing the strip heat losses: e.g. installation of thermal shields.

In Figure 3, the investigations, carried out during the prototype reheating module was realized, have been put in a sequence, proceeding from the coarse to the fine approach, in order to avoid the unnecessary use of expensive methods. This flow-chart could be developed in a process control, suitable to improve the production line adaptability to the different qualities or quantities of product.

3-D NON LINEAR MODELLING OF MICROWAVE HEATING PROCESS USING FINITE ELEMENT METHOD

Abdelkrim Sekkak, Lionel Pichon and Adel Razek

Laboratoire de Génie Electrique de Paris
U.R.A. D0127 CNRS
Universités Paris 6 et Paris 11- Ecole Supérieure d'Electricité
Plateau du moulon, 91192 Gif sur Yvette- France

INTRODUCTION

The study of the microwave heating structures has become an interest of many industries. Actually the use of these structures is rapidely growing. Thanks to the brief start-up time and internal heating due to penetration of the wave they become an attractive energy source for heating. But the design of these structures, for example microwave ovens, remains largely an empirical process: it takes a long conception time and is commercially too expensive for the designers. Many authors have been working with different numerical methods[1,4] to analyse the microwave heating process.

Our work was to develop a coupled model using the finite element method technique, in order to study the microwave heating process and to provide an efficient tool to assist the designer for optimisation and building the structures[5,6]. The 3-D coupled magneto-thermal model is based on solving a deterministic electromagnetic driven problem using edge elements and the heat conducting problem using nodal elements. It allows us the determination of the electromagnetic field distribution and the temperature of the heated load. The temperature dependance of the electromagnetic and the thermal parameters is taken into account (ρ,C_p,K,ε',ε''). In this paper we have used this model to study the influence of the dielectric (ε' ε'') and the thermal parameters (ρ,Cp,K) on the temperature variation.

Food products have considerable losses in general and so microwave heating processes are widely used in food industries. We present in this paper the analyse of several foodstuffs.

MATHEMATICAL EQUATIONS

We use Maxwell's time harmonic equations with the following boundary conditions :

$$\text{curl}\,(\text{curl}\,e) - \omega^2\mu_0\varepsilon\, e = 0 \quad \text{in} \quad \Omega \qquad (1)$$

$$n \wedge e = 0 \quad \text{on} \quad \Gamma \qquad (2)$$

$$n \wedge h = n \wedge h_s \quad \text{on} \quad S \qquad (3)$$

Where $\varepsilon = \varepsilon_0\varepsilon_r = \varepsilon' - j\,\varepsilon''$ is the complex permittivity, μ_0 and ε_0 are the vacuum permeability and permittivity, h_s is the magnetic field source on S.

The temperature T of the heated object V (fig.1) is determined by the following heat equation with the boundary condition :

Electric and Magnetic Fields, Edited by A. Nicolet
and R. Belmans, Plenum Press, New York, 1995

$$\rho(T)C_p(T)\frac{\partial T}{\partial t} = \mathrm{div}(K(T)\mathrm{grad}T) + Q(T) \quad \text{on } V \quad (4)$$

$$h(T)(T - T_a) = -K(T)\frac{\partial T}{\partial n} \quad \text{on } \partial V \quad (5)$$

ρ, Cp, K, h are respectively the density, the specific heat capacity,the thermal conductivity and the convective transfert coefficient (h is the air convective coefficient).T_a is the room temperature, Q is the heating source density and δV is the boundary of the heated load.

Equations (1) and (4) accounting for boundary conditions are solved by iterations (see flowshart fig. 2). In the heating process the temperature changes are much slower than the variation of the electromagnetic field thus we consider that the time harmonic steady state is obtained for each temperature distribution. The coupling is taken into account through the dissipated power density per volume unit :

$$Q(T) = \omega\varepsilon_0\varepsilon''_r\, |e|^2/ 2$$

NUMERICAL RESULTS

The coupled model is used to analyse a rectangular cavity (short-circuited standard waveguide : WR 340) containing a lossy dielectric material constituted by foodstuff (Fig. 3). The convective boundary condition is applied on the face in front of the electromagnetic wave. The faces against the waveguide walls are supposed to be thermally isolated.The same mesh is used for the electromagnetic problem and the thermal problem (Fig.4).

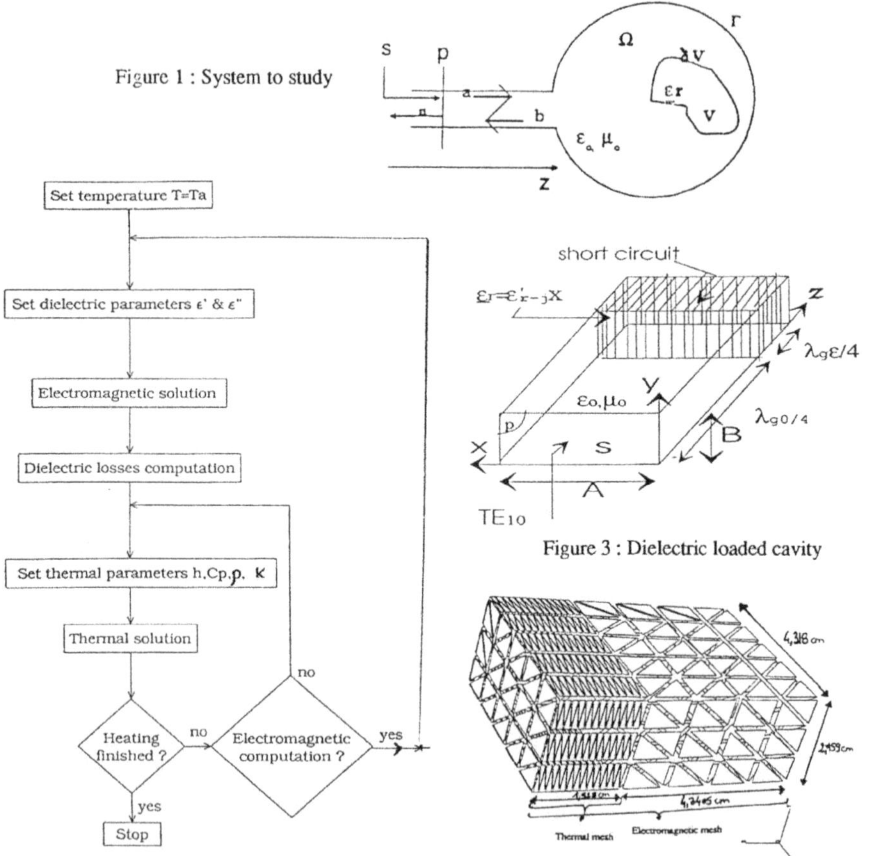

Figure 2: Flowshart of the non-linear problem

Figure 3 : Dielectric loaded cavity

Figure 4: Mesh of the quarter of the loaded cavity

The cavity is excited by a TE_{10} mode at 2.45 Ghz. The power generated by the exciting source is about one kW at the beginning of the heating. The initial temperature is 25°C. The convective heat parameter is taken constant for all the study $h = 50\ W/m^2°C$ (air).

In order to study the influence of the thermal or the dielectric parameters on the evolution of the heating inside the products, we have performed our analysis for several foodstuffs.

Influence of the thermal parameters

We present here the heating of "french fried potatoes" for which the dielectric parameters are given by Ohlsson and Al.[8]. In Fig.5 we show the evolution of the temperature distribution inside a quarter of the product. First the thermal parameters are taken as those of water (Figure 5a). The term $[\rho C_p / K]$ which is proportional to the time constant of the first order heating equation varies between 6 and 7.65 ($*10^6\ s^{-1}$) when the temperature T grows from 0 to 140 °C . In a second step the value of $[\rho C_p / K]$ has been doubled (Figure 5b). In the two cases the maximum of temperature is in the middle of the sample. It may be noted that the maximum computed electric power corresponds to this hot region. We can conclud that the temperature is growing more rapidely when the constant time is enhanced. The hot region stays at the same place.

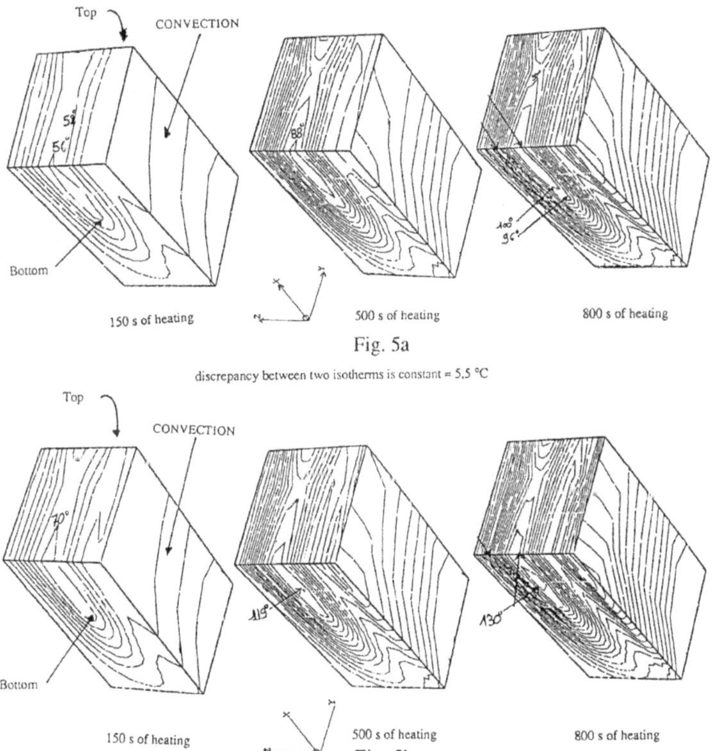

Fig. 5 : Influence of the thermal parameters on the heating of "French fried potatoes"

Influence of the dielectric parameters :

In order to analyse the effects of the dielectric parameters two products with the same thermal parameters[3] are studied: "Pizza baked dough"[$3 < \varepsilon'_r < 6.5$ - $0.35 < \varepsilon''_r < 1$] and "the stuffing pizza" [$7.5 < \varepsilon'_r < 12.5$ - $1.5 < \varepsilon''_r < 6$]. On Fig. 6, we can compare the evolution of temperature at different times of heating. We noted that both the postion of the hot region or the temperature distribution obtained are different. The study of other different examples[6] shows that the difference of temperature depends essentially on the imaginary part of the permittivity (losses). The position of the hot region is governed by the real part of the permittivity.

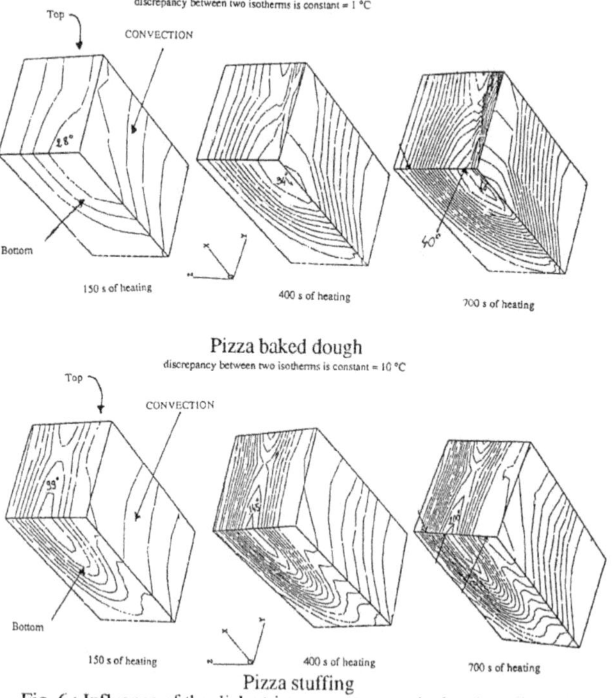

Fig. 6 : Influence of the dielectric parameters on the heating of the Pizzas

CONCLUSION

We present a 3-D finite element coupled model which provides an efficient tool for predicting runaway heating effects during microwave heating.

It allows us to say that the variation of the thermal parameters only affects the growing time heating of the products. The real part of the permittivity is the cause of the maximum of losses position. Finally the imaginary part of the permittivity affects essentially the intensity of the dissipated power.

REFERENCES

1. Z. Shouzheng, J.B. Davies, "Non linear modeling of microwave heating problem", IEE International Conference on computation in electromagnetics, pp. 86-89, (Nov. 1991).
2. C.T.M. Choi, A. Konrad, "Finite element modelling of the RF heating process", *IEEE Trans. on Magnetics*, Vol. 27, no.5, pp. 4227-4230, (Sept 1991).
3. K.G. Ayappa, H.T. Davis, E..A. Davis and J. Gordon, "Analysis of microwave heating materials with temperature dependant properties", *Aiche journal*, Vol.37, no.3, pp.313-322, (March 1991).
4. R.A. Desai, A.J. Lowery, C. Christopoulos, P. Naylor, J.M.V. Blanchard and K. Gregson, "Computer modelling of microwave cooking using the transmission line model", *IEE Proc. A*, Vol.139, no.1, pp. 222-223, (Jan 1992).
5. A. Sekkak, L. Pichon, A. Razek " 3-D FEM magneto-thermal analysis on microwave ovens" to be published in *IEEE Trans. on Magn.*, (Sept. 1994).
6. A. Sekkak " Modélisation 3-D de structure en hyperfréquences par la méthode des éléments finis d'arêtes. Application au chauffage par micro-ondes", These de l'université Paris XI., (Sept.. 93)
7. F. M. Clark, " Insulating materials for design and engineering practice", John Wiley and Sons, (1963).
8. TH. Ohlsson and N.E. Bengtsson, "Dielectric food data for microwave sterilization processing", *J. of Microwave Power*, Vol. 1, no.1, pp. 93-107, (1975).
9. Weast, "Handbook of chemistry and physics", The chemical ruber and Co, 47th edition, (1966-1967).

STUDY OF MAGNETOELASTIC PROBLEMS BY STRONG COUPLING MODEL

Mondher Besbes, Bogdan Ionesco, Zhuoxiang Ren and Adel Razek

Laboratoire de Génie Electrique de Paris, U.R.A.127 CNRS, Universités Paris6/11, ESE, Plateau du Moulon, 91192 Gif sur Yvette cedex, France

INTRODUCTION

Design of electromagnetic systems such as electrical machines, transformers, magnetic levitators requires analysis of different coupled physical phenomena and particularly mechanical and magnetic phenomena. The importance of their coupling differs with applications.

In the case of electrical machines, for example, deformations induced by magnetic forces are so small that they can not modify the magnetic field distribution. So, solving these two problems separately may be sufficient. The model is thus non-coupled.

However, in some other applications (high field applications) the deformations (or displacements) are quite important, then the magnetic field distribution will be changed as well as the magnetic forces. These phenomena are strongly coupled. In consequence, it is necessary to solve them simultaneously. This leads to a strong coupling model[1].

The purpose of this paper is to present an application of this model using the finite element method in the case of elastic deformation[2]. This model allows to find magnetic field distribution and the mechanical deformations simultaneously. Compared to a non-coupled model and a weakly coupled model, the interest of the present model is underscored.

FORMULATIONS AND MODELLING

Using finite element method, the discretised equations relative to magnetostatic field with a formulation[3] in term of magnetic vector potential (**a**) and to static elastic problem with a formulation[4] in term of displacement (**u**) are respectively (1) and (2) :

$$S\,A = J \tag{1}$$

$$K\,U = F \tag{2}$$

where A is the unknown magnetic vector potential, J the source vector, S the magnetic stiffness matrix , U the unknown displacement vector, F the force vector and K the mechanical stiffness matrix.

This system of equations is coupled through magnetic force F. The force is calculated using the virtual work principle which states that the magnetic force is equal to the derivation of the magnetic energy W with respect to configuration parameter u while the magnetic flux is held constant.

Using the reference element, the contribution of one element to the force acting on a giving node is expressed as follows[1] :

$$F(a,u) = -\int_{\hat{\Omega}} (\frac{v}{2} \partial_u (\nabla a^t . \nabla a) \det(J) + W(a,u) \partial_u (\det J)) d\hat{\Omega} \quad (3),$$

where W(a,u) is the magnetic energy density in a given element, det(J) represents the determinant of the Jacobian matrix J and v is the reluctivity.

In the case of small deformations, solving (1) and (2) separately gives the good results because these deformations don't influence the magnetic field distribution. This is known as a non-coupled model. But when deformations are non negligible the change of mechanical structure modifies the magnetic field as well as the magnetic force distribution. So, a strong coupling model is necessary to solve the magneto-mechanical problem [2,5]. The solution giving with this model corresponds to a point where the equilibrium state between the magnetic energy and the elastic energy is established.

The coupled system is non-linear because of the saturation phenomena and the dependence of the magnetic force on the magnetic vector potential and the displacement. In consequence, an iterative procedure should be used. We note that only the elastic deformations are considered in this study.

Using Newton-Raphson's method, the initial equation system is transformed to the following algebraic equation for unknown ΔA and ΔU:

$$\begin{bmatrix} S_i + \partial_A S_i & \partial_u S_i A_i \\ -\partial_A F_i & K_i - \partial_u F_i \end{bmatrix} \begin{bmatrix} \Delta A_{i+1} \\ \Delta U_{i+1} \end{bmatrix} = \begin{bmatrix} J_i - S_i A_i \\ F_i - K_i U_i \end{bmatrix} \quad (4),$$

$$\text{with: } \partial_u F_i = -\int_{\hat{\Omega}} (\tfrac{1}{2} \partial_u v \, \partial_v (\nabla a^t . \nabla a) \det(J) + W(a,u) \partial^2_{u,v} (\det J)$$
$$+ \tfrac{v}{2} \partial_u (\nabla a^t . \nabla a) \partial_v (\det J) + \tfrac{v}{2} \partial_v (\nabla a^t . \nabla a) \partial_u (\det J)) \, \partial\hat{\Omega} ,$$
$$\text{and } \partial_A F_i = -A_i^t \partial_u S_i ,$$

where ∂ means either the derivative versus A at u constant, or the derivative versus u at A constant and v is a direction of displacement (not necessarily identical to u).

The obtained global system is symmetric and definite positive. One can solve this system with a Cholesky's method using an adaptive profile storage. Solving this system allows to find A and U simultaneously.

EXAMPLE AND DISCUSSIONS

The present model is applied to a 2D magnetic field and a plane deformation problem using first order Lagrange elements. It consists of a thin ferromagnetic shell of infinite length on an uniform magnetic field $\mathbf{b}_0$ (fig.1).

According to symmetries presented in this example, only one forth of the shell is modelled with the adequate boundary conditions.

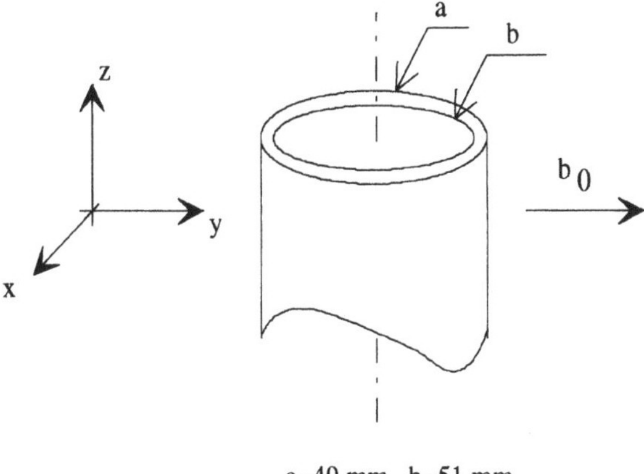

a=49 mm , b=51 mm

Figure 1. Geometric characteristic of the studied shell.

For a small values of $\mathbf{b}_0$, the same results are obtained with the non-coupled model, the strong coupling model and the weakly coupled model. In this case, the deformations are so small that they can not affect the magnetic field distribution.

But, when the external magnetic field increases the global system diverges with the weak coupling model and clear differences between the results of the non-coupled model and the strong coupling model appears (fig.2). This can be explained using a simplified model of the equation (2) where the mechanical stiffness matrix is reduced to a scalar k. The solution of this equation is presented in the figure 3 which shows clearly the inexactess of the non-coupled model solution. The convergence with the weak coupling model can be reached only in the case of rigid displacement where the mechanical stiffness matrix is not changed.

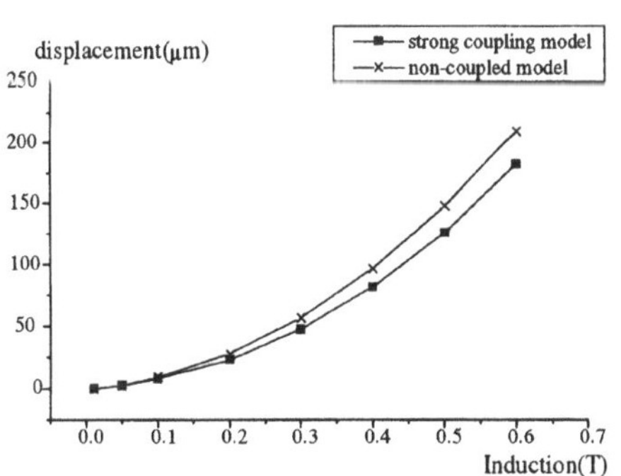

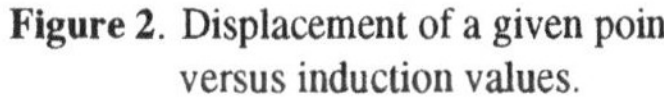

Figure 2. Displacement of a given point versus induction values.

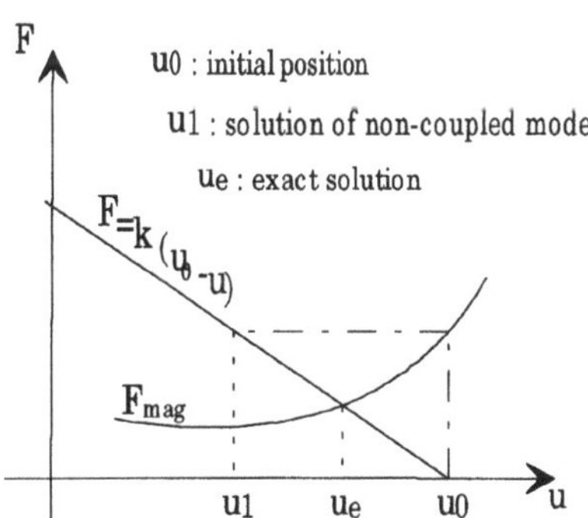

Figure 3. Solution of the coupled problem.

Figures 4 and 5 present respectively the distribution of magnetic forces and the resulting deformations when $\mathbf{b}_0$ is equal to 0.3 Tesla.

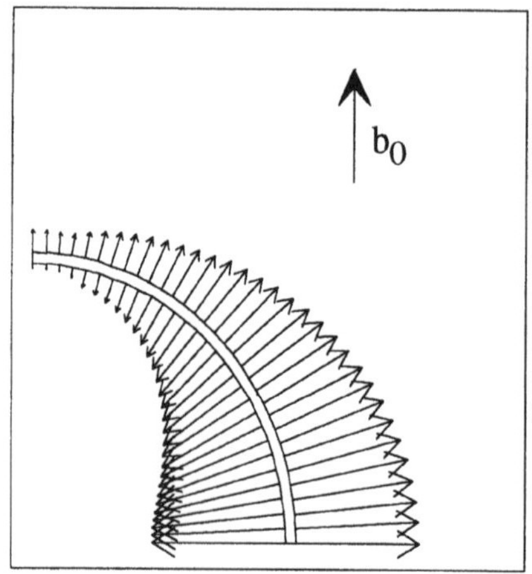

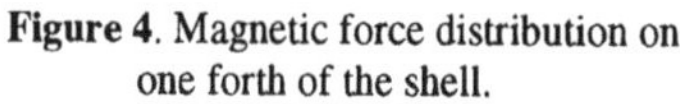

Figure 4. Magnetic force distribution on one forth of the shell.

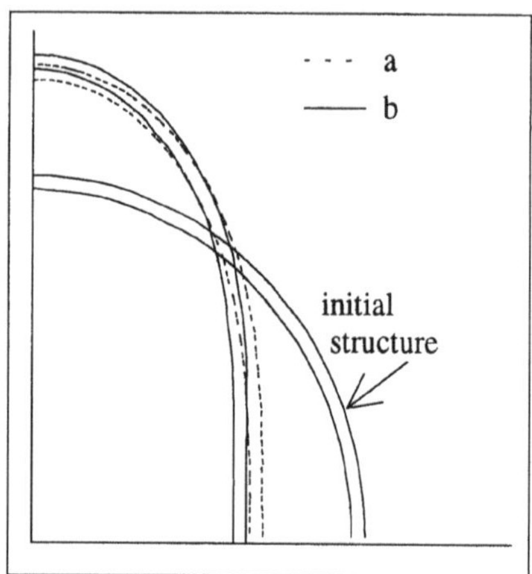

Figure 5. Computed structure deformations (a):strong coupling model, (b): non-coupled model (100 × real deformations)

The global system (4) depends on two different physical characteristics . This induces in some cases an ill-conditioned matrix. In consequence, the convergence is slow and sometimes we have an oscillation of the error. It may be therefore interesting to apply the Uzawa's method[6].

CONCLUSION

A strong coupling model is applied to study magneto-elastic problems. It consists of solving magnetic field equations and elastic equations simultaneously in order to take into account the influence of the structural deformations on the magnetic field distribution. This technique is interesting for high field applications such as magnetic forming.

In addition, it can be easily applied to treat problems of transient magneto-structural coupling.

REFERENCES

1. Z. Ren and A. Razek," A strong coupled model for analysing dynamic behaviours of non-linear electrimechanical systems", Compumag 93, Miami, Nov. 1993, to appear in IEEE Trans. on Mag., Sept.1994
2. Z. Ren, B. Ionescou, M. Besbes and A. Razek, "Calculation of mechanical deformation of magnetic materials in electromagnetic devices", CEFC'94, Aix-les-Bains, France, July 5-7,1994.
3. Z. Ren, M. Besbes and S. Boukhtache," Calculation of local magnetic forces in magnetized materials", Proceeding of International Workshop on Electric and Magnetic fields from Numerical Models to Industrial Applications, pp 64.1-64.6, 1992.
4. P. Karasudhi," Foundations of solid Mechanics.", K. Luwer Academic Publishers, 1991.
5. F. C. Moon. "Magneto-solid mechanics," Theoretical and applied mechanics, Cornell university, Itheca,Newyork (1984).
6. P. G. Ciarlet, "Introduction à l'analyse numérique matricielle et à l'optimisation", Masson 1985

COUPLED MAGNETO-THERMAL FINITE ELEMENT COMPUTATION OF LOSSES AND AMPACITY IN UNDERGROUND SF_6 INSULATED CABLES

Dimitris Labridis and Vassilis Hatziathanassiou

Division of Electrical Energy
Department of Electrical and Computer Engineering
Aristotelian University of Thessaloniki
Thessaloniki, GR 54006

INTRODUCTION

The solution of the coupled magneto-thermal field in three-phase gas insulated cables, using a finite element formulation, is presented in this paper. The proposed method takes into account the real geometry and the real electromagnetic and thermal properties of the involved materials. The given quantity is the rms current flowing through each conductor and the result is the computation of the magnetic vector potential (MVP) and temperature field distributions. From the MVP distribution, the current density distribution and the cable losses are obtained. From the temperature distribution, using a given maximum sheath temperature as a limitation, the ampacity of the cable is also determined.

THE FIELD EQUATIONS

The two-dimensional steady-state electromagnetic diffusion problem for the z-direction components of the MVP, A, and of the total current density vector, J, is described by the system of equations[1]

$$\begin{aligned} \frac{1}{\mu_0 \mu_r}\left[\frac{\partial^2 A}{\partial x^2}+\frac{\partial^2 A}{\partial y^2}\right]-j\omega\sigma(T)A + J_s = 0 \\ -j\omega\sigma(T)A + J_s = J \end{aligned} \tag{1}$$

subject to a Dirichlet boundary condition on the limit C of the region of interest S. In the above equation, μ_0 is the magnetic permeability of vacuum, μ_r is the relative magnetic permeability, ω is the angular velocity, σ is the electrical conductivity (which is a function of temperature T), J_{sz} is the z-direction component of the uniformly distributed source current density and I_{rms} is the rms of the current flowing through each conductor.

The two-dimensional steady-state heat conduction problem is described by the differential equation

$$-k\left[\frac{\partial^2 T}{\partial x^2}+\frac{\partial^2 T}{\partial y^2}\right]=\dot{q} \tag{2}$$

and the corresponding Dirichlet and Neuman boundary conditions on C. In the above equation, T is the temperature, $\dot{q}$ is the rate of heat generated per unit volume per unit time and k is the thermal conductivity.

CALCULATION OF THE EFFECTIVE THERMAL CONDUCTIVITY OF SF_6

The temperature drop across a concentric fluid gap, in the case where one conductor is located inside a sheath, may be expressed as

$$Q=\frac{2\pi k_{eff}\left(T_c-T_s\right)}{\ln\left(R_i/R_e\right)} \tag{3}$$

where Q is the heat flow per unit length, k_{eff} is the effective thermal conductivity of SF_6, T_c is the mean conductor temperature, T_s is the mean sheath temperature, R_i is the inner sheath radius and R_e is the outer conductor radius. The effective thermal conductivity k_{eff}, including the effects of free convection and radiation, may be calculated from (2) if we calculate the overall heat flow Q per unit length of the cable

$$Q=Q_r+Q_c \tag{4}$$

where Q_r and Q_c indicate heat flow per unit length due to radiation and convection respectively. If the problem is related to a three-phase one-sheath cable, we may consider an imaginary cylindrical heat dissipation surface of radius R_e, which is an envelope of the three conductors. The heat flow due to radiation is given by Stephan-Boltzman law, while the heat flow due to convection is given by a formula recommended by Doepken[2].

COMPUTATION OF LOSSES AND AMPACITY

In order to take into account the temperature dependence of the electrical conductivities, the iterative solution of the coupled set of partial differential equations (1) and (2) is necessary. An additional complication of the problem is that in the heat transfer through SF_6 both convection and radiation are important. Because these two heat transfer modes depend on the conductor's and sheath's temperatures, it also leads to an iterative solution for the thermal problem, in order to compute k_{eff}. Therefore, the problem is solved using a very fast iterative procedure. The outline of this procedure is as follows: by solving the steady-state heat conduction problem, the temperature distribution in the cable and in the surrounding soil is calculated. The input for the heat conduction problem is the ac loss due to the imposed and induced currents in the conductors and sheaths. By solving the steady-state electromagnetic problem, the field distribution in the cable cross-section is obtained. The input for the electromagnetic field calculation is the rms measurable current flowing through each conductor.

The iterative solution leads to the computation of the nodal values of the MVP A and of the temperature T, as well as of the source current density J_{si} in conductor i (where $i=1,2,3$). The average loss density $\dot{q}^e$ for the finite element e may be calculated from[1]

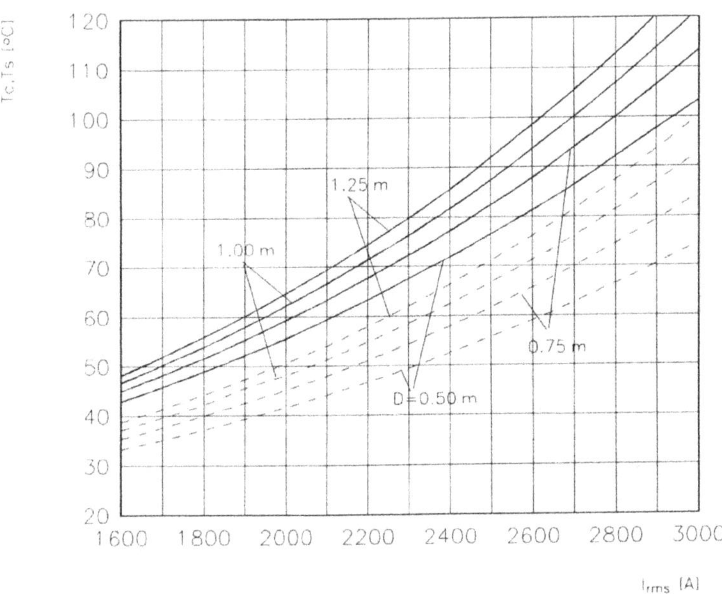

Figure 1. Sheath (dashed lines) and conductor (solid lines) mean temperatures T_s and T_c respectively vs. rms load current, for different depths of lay D.

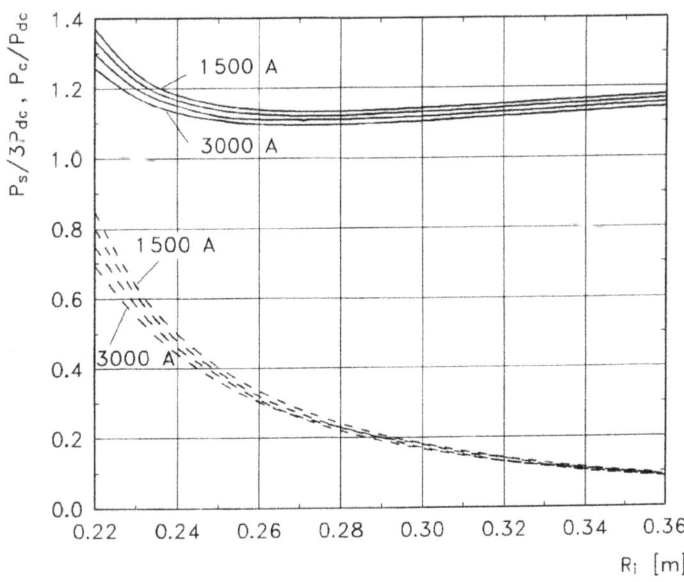

Figure 2. Conductor (solid lines) and sheath (dashed lines) ac/dc loss ratios vs. sheath internal radius R_i, for different load rms currents I_{rms}.

$$\dot{q}^e = \frac{J^e J^{e*}}{\sigma^e} \tag{5}$$

where J^e and σ^e are the total current density and the electrical conductivity of element e respectively. The mean value of the losses per unit length of element e will be obtained by the integration of $\dot{q}^e$ in the cross-section of this element S^e. The total loss per unit length of conductor i is finally obtained from the summation of the element loss contribution of this conductor, and the total loss per unit length of the sheath from the summation of the element loss contribution of the sheath.

The ampacity of a gas insulated cable is limited only by the maximum sheath temperature. The mean temperatures of each finite element are used, in order to calculate the mean temperatures T_{ci} and T_{si} of conductor i and sheath i respectively. After the termination of the iterative procedure, these mean temperatures may be plotted against the cable current. If we define the maximum conductor and sheath temperature that is allowed, the cable ampacity is estimated. It should be noted that due to the thermal proximity effect in the isolated phase arrangement, the temperatures of the central conductor and sheath cable will be higher than those of the other two cables. So, these temperatures determine the cable ampacity.

RESULTS

The validity of the method was confirmed by comparing with existing solutions as well as with measured data[3], and the agreement was found to be excellent. In Fig.1 the sheath and conductor mean temperatures T_s and T_c respectively are shown vs. rms load current, for different depths of lay D of a three-phase gas insulated cable, with geometrical and physical properties defined in (18) of[3]. Finally, in Fig.2, the conductor and sheath ac/dc loss ratios are shown vs. sheath radius R_i, for the same cable and for different load currents.

CONCLUSIONS

A finite element formulation has been used for the calculation of the ampacity and losses of three-phase gas insulated cables. The sheath and conductors mean temperatures and loss ratios may be computed, using a variety of physical and geometrical data. The most important parameters that influence the cable ampacity are the soil thermal conductivity, the burial cable depth and the ambient air temperature. Other parameters such as emissivities of cable materials or ambient soil temperatures have confined influence. The ac cable losses depend mainly on the geometrical and not on the thermal parameters. The method is capable to take into account any configuration of underground cables (using any geometrical and physical properties), surrounded by materials with different thermal conductivities as well as the effect of forced and seasonal cooling.

REFERENCES

1. V. Hatziathanassiou and D. Labridis, Coupled magneto-thermal field computation in three-phase gas insulated cables. Part 1: Finite element formulation, *Arch. Elektrotech.* 76:285 (1993).
2. H. Doepken, Calculated heat transfer characteristics of air and SF_6, *IEEE Trans. Power Appar. Syst.* 89:1979 (1970).
3. V. Hatziathanassiou and D. Labridis, Coupled magneto-thermal field computation in three-phase gas insulated cables. Part 2: Calculation of ampacity and losses, *Arch. Elektrotech.* 76:397 (1993).

COMPARISON OF POTENTIAL DUAL FORMULATIONS DEVELOPED WITH DIFFERENT ELEMENTS

N. Gasmi,[1], S. Bouissou,[2] and F. Piriou[1]

[1] Laboratoire de Génie Electrique de Paris E.S.E., U.R.A. 127 C.N.R.S.
Universités de Paris VI et XI, Plateau du Moulon, 91192 Gif-sur-Yvette, France
[2] UNELEC, Route de Guise 02322 Saint-Quentin, France

INTRODUCTION

To extend the field application of numerical simulation, it is necessary in many applications to consider the electric circuit. In 3D for the coupling between magnetic and electric circuit equations, the formulations in terms of potential are very attractive[2,4,8]. The electric formulation (**A**) can be used in many applications but it is necessary to determine three components. It may be noted that the necessity to impose a gauge condition permits to reduce the vector potential to two components if we use[3,8] (**U**.**w**=0) .

With the magnetic formulation (**T**-Ω) we can reduce the unknown number but there is a problem to consider the inductors. However the formulation (T_0-Ω) permits to take into account the inductor when the direction of current density is known [1,5,7].

In this communication to solve magnetic equation with the finite element method, we propose to study and compare both formulations (**T**$_0$-Ω and **A**). For space discretisation we consider different elements (tetrahedron, prism and hexaedron)[3]. As example of application we study a cubic conductor spanned by a current density. An other example concerns the simulation of a current transformer where the numerical solutions are compared with experimental one.

MAGNETIC AND ELECTRIC EQUATIONS

To-Ω formulation

For thin conductors the direction of density current in the inductors is known. So it becomes possible to choose a direction for the electric vector potential $\mathbf{T}_0$ so that:

$$\mathbf{curl}\,\mathbf{T}_0 = \mathbf{J}_0 \tag{1}$$

where J_0 is the current density in the inductor. So we can write the following Maxwell's equation:

$$\mathrm{div}\,\mu(\mathbf{T}_0 - \mathbf{grad}\Omega) = 0 \tag{2}$$

where μ is the magnetic permeability and Ω the magnetic scalar potential. The weighting residual technique and the Galerkin method are applied to equation (2). Then by using vectorial relations we get:

Electric and Magnetic Fields, Edited by A. Nicolet
and R. Belmans, Plenum Press, New York, 1995

$$\int_D \mu \mathbf{grad}\Omega' \mathbf{grad}\Omega dv - \int_D \mu \mathbf{grad}\Omega' . \mathbf{T}_0 dv + \int_S \Omega' \mathbf{B} . \mathbf{ds} = 0 \tag{3}$$

where Ω' represents the nodal shape function and **B** the magnetic induction. The boundaries conditions are ensured by vanishing the surface integral in equation (3).

If we consider an iron core coil, constituted of thin conductors, connected to an electric circuit, the relation between the voltage U and the current I is given by:

$$U = RI + \frac{d\Phi}{dt} \tag{4}$$

where R is the ohmic resistance and Φ the flux linkage in the winding. The coupling between equation (3) and (4) is performed by introducing a wire density **K** which permits to express the electric vector potential as function of current I.and the flux density Φ as the function of current I and magnetic scalar potential[7] Ω. Then the flux linkage can be written:

$$\Phi = \int_D \mu \mathbf{K} . (\mathbf{K}I - \mathbf{grad}\Omega) dv \tag{5}$$

A formulation

In magnetostatic Maxwell's equations with a magnetic vector potential formulation can be written under the form:

$$\mathbf{curl}(\nu \mathbf{curl} \mathbf{A}) = \mathbf{J}_0 \tag{6}$$

where ν is the reluctivity and **A** the magnetic vector potential. To solve this equation weighting residual technique and the Galerkin method are used. Then we get the following equation :

$$\int_D \nu \mathbf{curl} \mathbf{A}' . \mathbf{curl} \mathbf{A} dv - \int_D \mathbf{A}' . \mathbf{J}_0 dv + \int_S \nu \mathbf{A}' . \mathbf{curl} \mathbf{A} ds = 0 \tag{7}$$

where **A'** represents the weight function. As in equation (3) the boundaries conditions can be impose with the surface integral. To ensure the uniqueness of the solution we use the gauge condition **A.w** = 0. The coupling with electric equation (4) is obtained by expressing the flux density in function of the potential vector and the current density from the current[8].

FINITE ELEMENT DISCRETISATION

To solve magnetic and electric circuit equations the studied domain is discretized by edge elements for the vector **A** and nodal element for Ω formulation. In our study we have considered tetrahedron, prism and hexahedron.

For those three types of elements, the classical nodal shape functions are used.

Concerning the interpolation function associated to edge , a first order edge elements which unknowns are the circulation of the field ,are used[6].

APPLICATIONS

The first application concerns a cubic conductor of one meter of side spanned by 10^7 A/m^2 current density[3]. It is a linear material and the current density is known. The analytical solution of stored energy is known and given 2.208 MJ. Figure 1 shown the flux density distribution obtained from A-formulation when the discretisation is made by tetrahedral elements.We can see figure 2 the energy values obtained with both formulations, and two kinds of elements (tetrahedron and hexahedron). The results are compared with analytical solution for different meshes. In table 1 we can see and compare for different meshes the results obtained with both formulation and both elements (tetrahedron and hexahedron).

The second example concerns a current transformer built by the "Société UNELEC" which is represented in figure 3. It consists of a primary conductor fed by a known sinusoidal current and of two secondary windings (4000 turns) connected to an electric circuit constituted of a resistance.In figure 5 we can compare the secondary current wave form obtained with two types of elements using the **A**-formulation. The validity of our models can be observed in figure 6 where, for a primary current of 8440A, we compare the experimental result with those obtained by T_0-Ω and **A**-formulations.

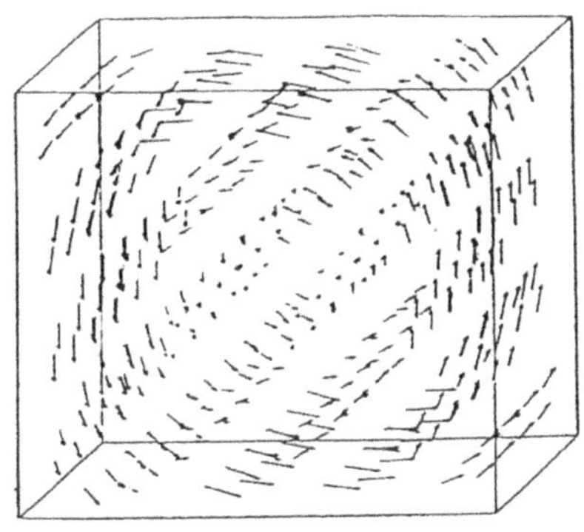

Figure 1. Flux density distribution for electric formulation in the case of tetrahedral edge elements.

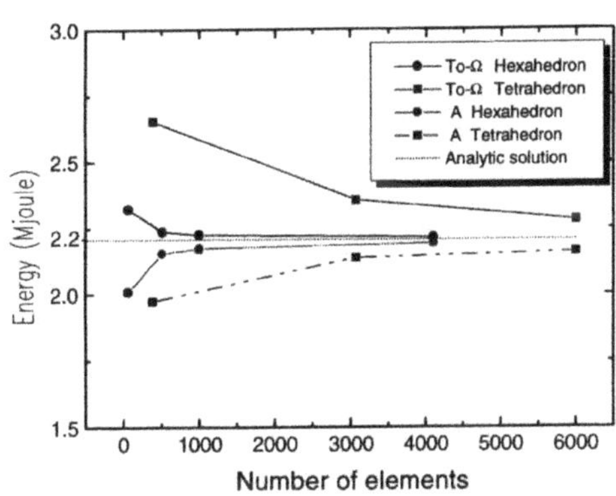

Figure 2. Comparison of the energy computed by the hexahedral and tetrahedral elements for electric and magnetic formulations.

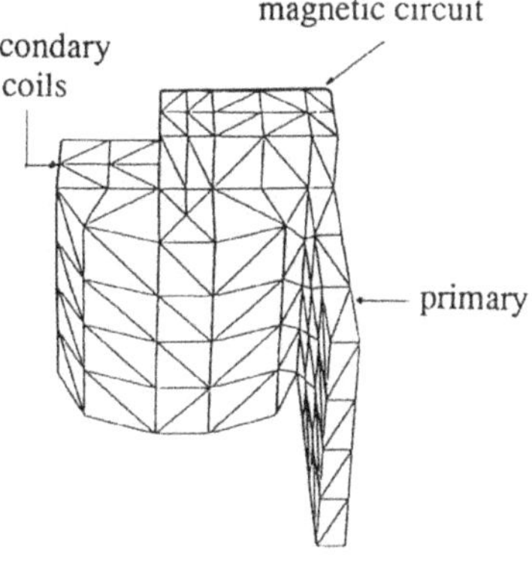

Figure 3. Representation of the studied current transformer

Table 1

degrees of freedom	125	729	1331	4913
Energy (Mj) Hexa	2.322	2.237	2.226	2.215
Energy (Mj) Tetra	2.655	2.355	2.280	2.253
Error (%) Hexa	5.16	1.31	0.81	0.31
Error (%) Tetra	20.2	6.65	3.26	2.03

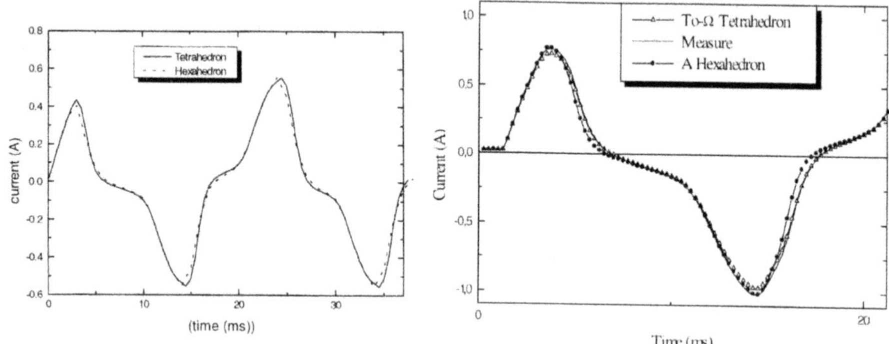

Figure 4 . Secondary current distribution computed with tetrahedral and hexahedral elements

Figure 5. Comparison between experimental and computed secondary current distribution.

CONCLUSION

In this paper two types of formulation, A and To-Ω, taking into account the electric circuit equations have been compared. They have been used with two kinds of elements. Using tetrahedral elements and for a given number of unknowns, the best accuracy is obtained with the A formulation but the To-Ω formulation is the most perform in term of time calculation. On the other hand, using hexahedral elements and the To-Ω formulation, we obtained a better accuracy in less computational time. Finaly, regardless the formulation used, hexahedral elements seem to be the most suitable.

REFERENCES

[1] T. Nakata, N. Takahashi Fujiwara and Y. Okada, "Improvment of T-Ω method for 3D eddy current analysis", I.E.E.E. Trans., vol. Mag. 24, pp 94-97, 1988.

[2] P. J. Leonard and D. Rodger, "Voltage forced coils for 3D finite-element electromagnetic models", I.E.E.E. Trans, vol. Mag. 24, pp 2579-2581, 1988.

[3] R. Albanese, G. Rubinacci, "Magnetostatic field computations in terms of two components vector potentials", International Journal for Numerical Method in Engineering, vol. 29, pp 515-532, 1990.

[4] T. Nakata and K. Fujiwara, " Recent progress in numerical analysis for electromagnettic devices", I.E.E.E. Trans., vol. Mag. 27, pp 4221-4227, 1991.

[5] A. G. Kladas and J. A. Tegopoulos, "A new potential formulation for 3-D magnetostatics necessitating no source field calculation," IEEE Transactions Magnetics, vol. 28, pp 1103-1106, March 1992.

[6] P. Dular, A. Genon, J.-Y. Hody, W. Legros, J. Mauthin, A. Nicolet, "Calculation of 3D eddy current using a coupling between edge elements, nodal finite element and boundary elements", Int. Workshop on Electric and Magnetic Fields, Liège 28-30 September 1992.

[7] S. Bouissou and F. Piriou, "Study of 3D dual formulation in potential for coupling magnetic and electric circuit equations", Int. Workshop on Electric and Magnetic Fields, Liège 28-30 september 1992.

[8] F. Piriou and A. Razek, "Finite element analysis in electromagnetic systems accounting for electric circuit", IEEE Transactions Magnetics, vol. 29, pp 1669-1675, March 1993.

THREE DIMENSIONAL CALCULATIONS OF MECHANICAL DEFORMATIONS CAUSED BY MAGNETIC LOAD

G. Henneberger, W. Hadrys, W. Mai

Institute of Electrical Machines,
University of Technology Aachen,
Schinkelstrasse 4, 52056 Aachen, F.R.Germany

ABSTRACT

The subject of this contribution is the electromagnetic mechanical coupling of dynamic structure deformations which are caused by magnetic field forces. These deformations can cause both the wear on the used materials and high emission of noise. Therefore the prediction of these deformations is very important. The calculation process deals with the evaluation of electromagnetic field- and force- relations. These forces are both electromagnetic Lorentz forces in eddy current regions and surface force-densities on ferromagnetic materials. The finite element method is used for the process of calculation [2]. The calculation of the force distribution, the mechanical modelling of a three dimensional induction furnace and the evaluation of the deformation are presented. Finally the calculations are compared with measurements.

LORENTZ FORCE DENSITY

The Lorentz force density $\vec{f}_L$ is given by (1)

$$\vec{f}_L = \frac{1}{2}(\vec{J} \times \vec{B}) \qquad (1)$$

$$\frac{1}{\mu}\vec{\Delta}\vec{A} - j\omega\sigma\vec{A} = -\vec{J}_{Source} \qquad (2)$$

for the alternating part of a time harmonic field distribution. In this formula $\vec{J}$ means the current density and $\vec{B}$ the flux-density. For a complete representation of the effects the introduction of a complex force vector field in eddy current regions is necessary. Because of the product in (1) the angular frequency of the force is twice the angular electrical frequency. To calculate the flux-density, the induction furnace is assumed as axisymmetrical. Therefore only a twodimensional section is considered. The linear complex differential equation with a time harmonic source current $\vec{J}_{Source}$ for the vector potential $\vec{A}$ is given in (2). The field distribution and the distribution of the Lorentz forces following (1) in the coil and melt is shown in figure 1.

Electric and Magnetic Fields, Edited by A. Nicolet
and R. Belmans, Plenum Press, New York, 1995

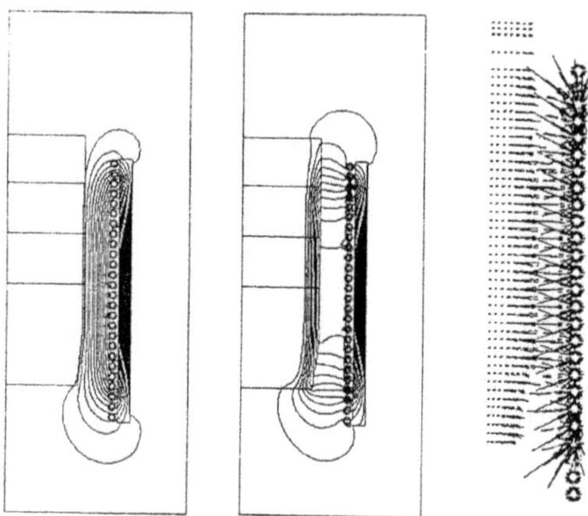

Figure 1. Lorentz forces in the coil

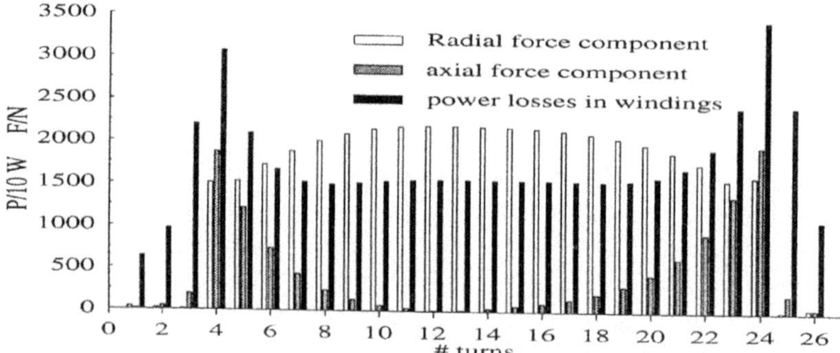

Figure 2. Distribution of forces and losses

SURFACE FORCE DENSITY

The alternating part of the surface force density $\vec{\sigma}$

$$\vec{\sigma} = \frac{1}{2}(B_n(H_{1n} - H_{2n}) - (w_1{}' - w_2{}')) \cdot \vec{n}_{12} \tag{3}$$

is imperative for the force calculation on magnetic materials especially in this case for the force densities on the yokes [1]. B_n means the normal component of the flux-density, H_{1n} and H_{2n} are the normal components of the magnetic field intensity on each side of the contact surface of the material 1 and 2. $w_1{}'$ and $w_2{}'$ are the co-energies and $\vec{n}_{12}$ the normal surface vector.

The equations for the magnetic field are:

$$\nabla \times \vec{H} = \vec{J}_{Source} \; ; \quad \nabla \cdot \vec{B} = 0 \; ; \quad \vec{H} = \frac{1}{\mu(\vec{H})} \cdot \vec{B} \tag{4}$$

A three dimensional model is necessary. Because of the twenty-fold symmetry the finite-element net is reduced to a 18°–segment. The yoke is halved lengthwise. It consists of the yoke, coil and air.

The force on each node is assigned by integrating the surface force density over the neighbouring surface triangles. The surface force density consists of normal components only, with the exception on the edges where the normal vectors of neighbouring surfaces are different. The results of the integrated surface force densities are in a range of approximately 2% compared to the Lorentz–Forces.

THE MECHANICAL MODEL

Fundamental for the displacement calculation is the following equation as a derivative of Hamiltons principle:

$$(\underline{K} - \omega_{mech}^2 \cdot \underline{M}) \cdot \underline{\hat{D}} = \underline{\hat{R}} \tag{5}$$

Here $\underline{K}$ means the stiffness matrix, $\underline{M}$ the mass matrix, $\underline{\hat{D}}$ the vector of displacements for each node, $\underline{\hat{R}}$ the force vector and ω_{mech} the angular frequency beeing twice the angular electrical frequency for the considered harmonic. Because of the complex representation of the force distribution following (1) a complex time vector of the deformation is the consequence. For this reason it is possible to calculate waves and shock effects in the mechanical structures.

This demands a complete three dimensional mechanical model which contains in contrast to the electrical models all mechanical relevant conditions.

Therefore the model is not limited to the region up to the yoke. The outer construction, like beams and cover sheets also the wall of the pot and the material between the turns, must be taken into account.

Because of the twenty–fold symmetry this model is also reduced to a 18°–segment. The surface of the meshed model is given in figure 3 [3]. The forces of both electrical models are converted to the mechanical model and normalized.

The principle of mechanical modelling allows only fixed connections between materials. But the furnace consists of some loose connections like the transition from the melt to the wall of the pot. Therefore the force in the melt is set to zero in the period of time, in which the forces remove the melt from the wall. To use the time harmonic equation (5) the resultant force function is distributed to the harmonics by using the Fourier analyis. Because of the linear process of calculation the results are superposed.

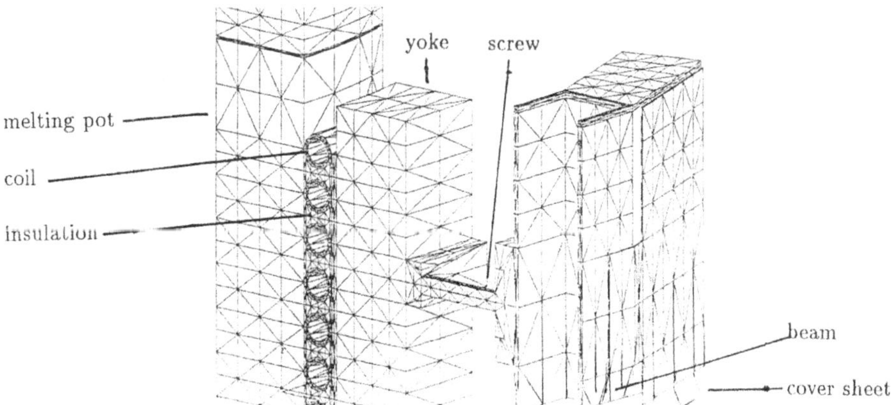

Figure 3. The three dimensional mechanical model

Another difficulty concerns the mechanical coupling of the different regions in the mechanical model. When modelling all regions as connected, no friction is possible between them. This results in a very stiff structure. One possibility for correction is to introduce an anisotropy not in the equations of elasticity but in the driving force distributions.

In this work we got the best results when taking into account the coil and their insulation as stiff regions and all other regions coupled with friction. The following factors were found for the r - direction: 9.48 or 15.45 (dependent upon the insulation's condition) and the axial - direction: 7.23. All these values are found theoretically with respect to the temperature dependency of the elastic moduli.

DEFORMATION

In fig. 4 the real and imaginary component of the deformed and undeformed furnace are shown on the load conditions $f_{el} = 250\text{Hz}$, $J = 5684\text{A}$ and $P = 1600\text{kW}$.

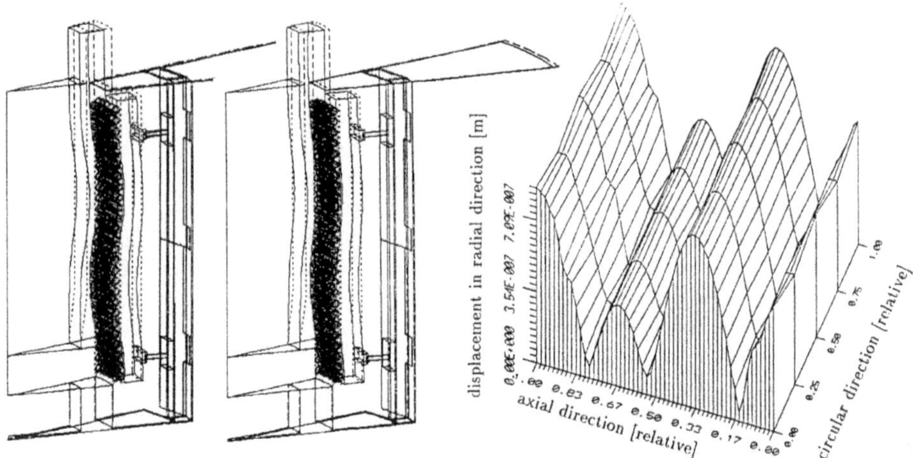

Figure 4. Deformed and undef. structure of the ind. furnace

Fig. 5 displays the comparison of the calculated results and measured values on the outside of the yoke. It has been chosen a line in z - direction beginning with the first winding at the lower part of the furnace and ending with the last one. The measurements have been done using accelerometers connected to a signal analyzer.

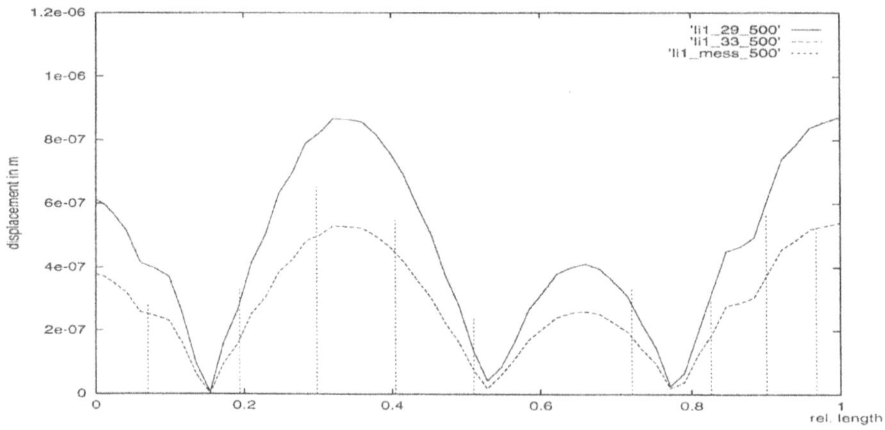

Figure 5. Comparison of calculation and measurement

References

[1] J.R. Melcher, Continuum Electromechanics, MIT Press, Cambridge MA 1981.

[2] O.C. Zienkiewicz, The finite element method, McGRAW-HILL, London 1989.

[3] W. Mai, Berechnung des dreidimensionalen strukturdynamischen Verhaltens eines Tiegelinduktionsofens unter elektromagnetischer Belastung mittels Finiter Elemente, Diploma-thesis, IEM RWTH Aachen 1994.

ON THE USE AND INTERPRETATION OF ELECTRICAL VALUES WHEN COUPLING ELECTRIC CIRCUIT AND ELECTROMAGNETIC FIELD EQUATIONS

P. Lombard

Cedrat Recherche,
10 chemin du pré carré, Zirst 4301
38943 Meylan, Cedex, France
lombard@cedrat-grenoble.fr

INTRODUCTION

The aim of this paper is to present how to interpret properly the results when modelling electromagnetic devices. We will focus especially on electrical values such as current, voltage, electrical power, copper losses in stranded or solid conductors. The coupling between electromagnetic field and electric circuit equations in 2D is becoming well known. Many people have studied the topic. Two approaches have been developed: the indirect coupled approach[1] and the direct approach. In the latter case, there are mainly two methods: the integro-differential method[2] and the direct method[1,3,4]. The latter method seems to be more efficient(in terms of speed and memory requirements)[5]. After developing this coupling, the way of applying it is not so obvious: the choice of the geometry to study, according to the symmetry, is sometimes difficult; the interpretation of the electrical power in solid and stranded conductor is not clear.

To analyse these problems, we propose first to redefine properly the electrical values and their effects on the symmetries. Then, we will study and interpret electrical values using two examples, computed with the program FLUX2D: a transformer and an induction machine.

DEFINITION

The equations describing the coupling between electric and magnetic phenomena have already been described. In this part, we intend to show the use of solid and stranded conductors and to define the electrical power and the copper losses.

Solid Conductors

These conductors can develop eddy currents. The current density is linked to the way the flux penetrates the solid conductor. An example of a solid conductor is given in Figure 1.

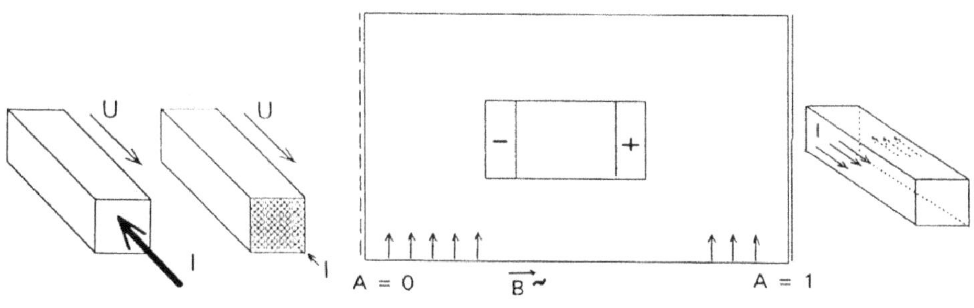

Figure 1. Solid conductor **Figure 2.** Stranded conductor **Figure 3.** Specific geometry

Electric and Magnetic Fields, Edited by A. Nicolet and R. Belmans, Plenum Press, New York, 1995

They are characterised by the value of their resistivity. Typical uses of solid conductors include modelling bars in a squirrel cage for induction machines or to imposing a null current in a magnetic shield.

Stranded Conductors

These conductors cannot develop eddy-current. The main hypothesis is that the current density is constant over the entire cross section. One stranded conductor consists of one side(in 2D but not axisymetric) of a coil with N turns. The current I is the current in one turn. The configuration is given in figure 2. Stranded conductors are characterised by the number of turns, the fill factor and the resistivity. Typically, stranded conductors are used to model coils. In fact, for a coil we need two stranded conductors plus two resistances and two inductances for the end part of the coil which is not represented in the finite element domain. The ending resistances can be determined easily, either with a direct D.C. resistance calculation(by approximation with sample geometry and the formula $R = \rho \, . \, \ell \, / \, S$) or by measurement(if we know the total value for the coil, we can subtract from it the D.C. resistance of the two coil sides); In many cases the ending inductances are insignificant. Otherwise, their values can be found by using a 3D program (such as FLUX3D).

Symmetries

Symmetries with coils can easily be used to reduce the number of unknowns and the time of computation. For solid conductors, there are some limitations where we cannot use symmetry. An example deals with a magnetic shield problem. There is one conductor and a plate of iron in the air. The total current in the cross section of the iron plate is null. The geometry presents a symmetry in the middle of the iron. But the difficulty is that if we study only half of it, we don't know what to impose for electric boundary conditions on the iron plate. Indeed, there is no reason to impose a null current on only half of the cross section. Furthermore, if we don't impose any boundary condition, by default, we impose a null voltage(due to Maxwell's equation). This is not the right condition because if the total current is null, there is actually a small non-zero voltage.

Electrical Power and Copper Losses

The aim of this section is just to remember the definition of the electrical power P_e and the copper losses P_c on a conductor characterised by its current I and its voltage U :

$$P_e(t) = U(t) \, . \, I(t) \tag{1}$$

$$P_c(t) = \int_S \rho \, . \, J(t)^2 \, dS \tag{2}$$

where ρ is the resistivity of the conductor and S the cross sectional area of the conductor. We can show one example so as to illustrate the difference. We consider a piece of copper, one meter long(so we can study it in 2D), which is in an alternating field. This can be studied as a magneto dynamic problem with the proper boundary conditions. The geometry is given in Figure 3. We impose a null current on the copper piece. The alternating field creates a variation of flux on the copper piece. The variation of flux generates a voltage. As the total current is null, there is a part of the current goes in and part comes out. Note that the electrical power is null(because of the null current) but copper losses are important(due to induced current).

TRANSFORMER

Description

Now we will study a transformer. The primary coil is a winding fed by an A.C. voltage supply. The secondary coil is short-circuited. Due to symmetry, the geometry is limited to one quarter of the device. The electric circuit and the flux lines are given respectively in Figure 4 and 5:

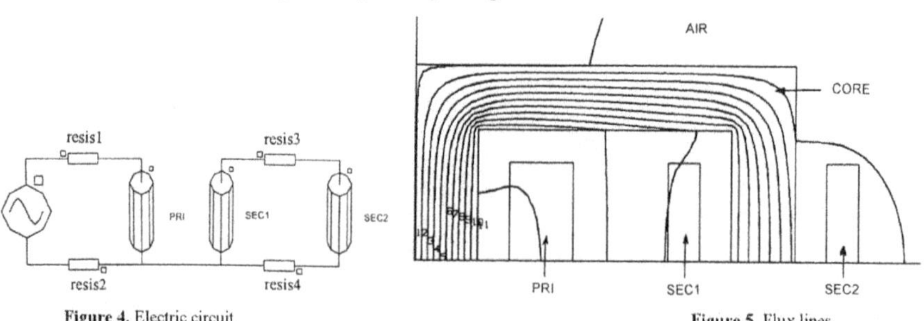

Figure 4. Electric circuit

Figure 5. Flux lines

A power balance has been computed for the active power. Four different cases are investigated:

a the secondary is short-circuited

b the secondary circuit is opened (R3 = 10 ^{6} ohms for example)

c the secondary part of the second coil is replaced by one fictitious turn
the corresponding region behaves like vacuum

d the secondary part of the second coil is replaced by an equivalent resistance and inductance. The corresponding region behaves like vacuum.

Results

Table 1. Analysis of input and output power(W)

	conductor	Case a	Case b	Case c	Case d
primary circuit	source	-25.47	-1.67	-41.39	-25.51
	resis1	.0299	.00822	.0708	.03
	resis2	.0299	.00822	.0708	.03
	pri	25.42	1.65	41.25	25.45
copper losses	pri	5.98	1.643	14.16	5.99
	sec1	9.62	6.E-7	26.56	9.72
	sec2	9.62	6.E-7	0	9.65
	resis3	.0962	6.E-9	.27	.096
	resis4	.0962	.006	.27	.096
	Σ P	25.41	1.65	41.26	25.55
secondary circuit	sec1	-9.85	-.0061	-.53	-9.84
	sec2	9.66	.0001	0	9.65
	resis3	.096	0	.265	.096
	resis4	.096	.006	.265	.096

Interpretation

From the table, we can see that the sum of all electrical power in the whole electric circuit is null, which is normal. The electrical power represents the power that goes through the conductor. For example, the electrical power of the primary coil (25.42 W) in Case a is equal to the sum of the copper losses in the whole device(5.98+9.62+9.62+0.096+0.096 = 25.41 W).

Another important remark is that the first part of the secondary coil behaves as an electric source for the secondary circuit(negative value of the electrical power). What is interesting is that the two parts of the same coil do not have the same utility. As shown in Cases c and d, the second part of the secondary coil is not necessary.

INDUCTION MACHINE

Description

We propose to study an induction machine at locked rotor. The stator is fed by a three phase A.C. source. The rotor bars are connected at their end rings, each segment of which possesses both resistance and inductance. Due to symmetry, only one quarter of the machine has been studied. The electric circuit and the flux lines of the stator are given respectively in Figure 6 and 7.

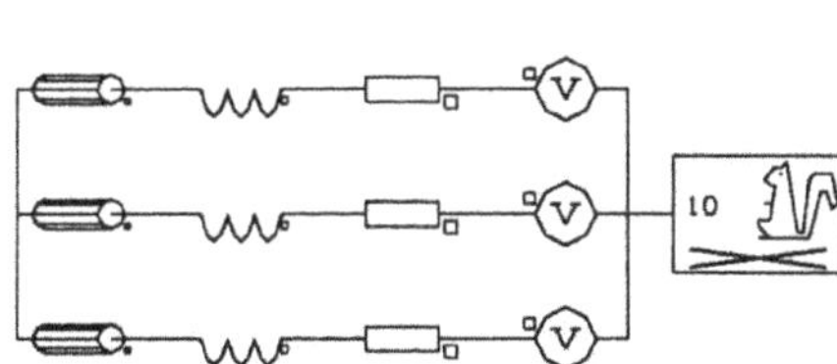

Figure 6. Electric circuit

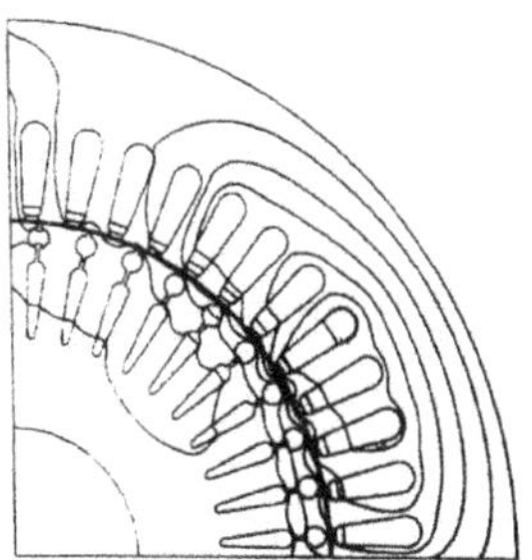

Figure 7. Flux lines

Results

We make a power balance of the whole machine. Results are given in Table 2. Our goal is to understand how to interpret the power balance, not to compare with measurements.

Table 2. Analysis of input and output power

	conductor or region	active power (W)	reactive power (VAr)
electric circuit of the stator	sources	-3199.8	-7855.1
	resistance of head-winding	196.48	0
	inductance of head-winding	0	375.72
	coil	3003.3	7479.4
copper losses	coil	186	1221.7
	bars	2757.8	3051
	resistance of end ring	59.9	0
	inductance of end ring	0	68.4
	other regions	0	3139.2
	Σ P or Σ Q	3003.7	7480.3

Interpretation

The sum of all electrical power for the whole electric circuit is still null, which is as expected. We can also notice that the active electrical power transmitted through the three coils (3003.33W) is dissipated in copper losses in the whole machine(coil 186.01 + bars 2757.85 + external resistance of end ring 52.95 = 3003.78). The same remark is valid for the reactive electrical power.

We can note that for the squirrel cage and for the given position, some bars are considered as electrical sources of current whereas others are just dissipating power.

CONCLUSION

Thus, we have recalled the definition and the use of solid and stranded conductors. The case of symmetries and their limitations have been studied with their limit. Then, we analysed the difference between the electrical power and the copper losses. Through various examples, we see that copper losses correspond to the dissipated power. The electrical power represents the power which is transmitted through a conductor. In fact, for a given conductor, the copper losses are always a part of the electrical power(in absolute value).
Another interesting point, noticed in the transformer example is that two parts of the same coil do not have the same utility. In the transformer, the inside part of the secondary coil behaves as a source of current(which is what we need). The outside part of the secondary coil, however, dissipates the power gained by the inside part. The analysis of the electrical values allows the designer to determine which part of the device will be really useful.

REFERENCES

1. F. Piriou. "Numerical Simulation of Electromagnetic Systems Considering Electric Circuit Equations", International Workshop on Electric and Magnetic fields, Vol. N°1, 1992, pp. 5.1-5.6

2. A. Konrad. "Integro-Differential Finite Element Formulation of Two-Dimensional Steady State Skin Effect Problems", IEEE Transactions on Magnetics, Vol. MAG-18, N°1, 1982, pp. 282-292

3. P. Lombard, G. Meunier. "Coupling Between Magnetic Field and Circuit Equations in 2D", International Workshop on Electric and Magnetic fields, Vol. N°1, 1992, pp. 7.1-7.6

4. S. Salon, M. DeBortoli, R. Palma. "Coupling of Transient Fields, Circuits, and Motion Using Finite Element Analysis", Journal of Electromagnetic waves and Applications, Vol. 4, N°11, 1990, pp. 1077-1106

5. H. Lindfors, J. Luomi. "A General Method for the Numerical Solution of Coupled Magnetic Field and Circuit Equations", Proc. of the Int. Conf. on Electrical Machines, pp 141-146, 1988

FINITE ELEMENT ANALYSIS OF TEMPERATURE DISTRIBUTION IN AN INDUCTION MOTOR

A. Bousbaine, M. McCormick, W.F. Low

Department of Electronic and Electrical Engineering
De Montfort University, The Gateway
Leicester, LE1 9BH, U.K.

INTRODUCTION

Modern motors are now being operated much nearer to their overload limits because of the stringent high torque to inertia requirements of mechanical systems. Consequently, the risk of adverse thermal conditions increases and the motor must be designed to ensure that the temperature of the motor components remain within their designated limits. The success of any design hinges on an advance knowledge of the likely temperature rise in the machine. It is difficult to develop mathematical approximations for thermal processes in electrical machines because of their complex construction and uncertainties in the characterisation of the composite material used. Early attempts at the prediction of temperature rise were largely based on empirical formulae and trial and error methods. These were, in general, not accurate[1] because the effect of insulation and air flow on the thermal performance of the machine were not fully understood. Lumped parameter thermal methods have since been developed to predict the temperature rise for a given machine geometry, iron loss density distribution, and a set of thermal characteristics. Although these models give a reasonable accurate temperature distribution on a 'macroscopic' level localised temperature can be inaccurate unless the thermal network is further refined to represent the model more closely.

The introduction of numerical techniques, such as finite elements and boundary elements, to the field analysis has made it possible to predict accurate localised temperature distributions. This has allowed detailed studies on, and the characterisation of, the various iron loss components. However, despite the presence of these powerful techniques their success is still contingent upon the availability of reliable information on loss density distribution, thermal data and characteristics of the composite materials used in the construction of the machine.

It has been shown theoretically that the loss density distribution can be determined using the temperature time method [2]. Bousbaine [3] extended this method to investigate the loss distribution over the volume of a standard 3-phase, 4kW, 4-pole totally enclosed fan cooled induction motor under various load conditions and showed the empirical methods of assigning local loss densities, used by others researchers, were not correct despite good overall agreement. The aim of this contribution is to corroborate the measured loss density distribution by comparing the measured temperature distribution with that obtained by finite element analysis.

FINITE ELEMENT FORMULATION

For 2-dimensional steady state heat flow, the governing partial differential equation is given by,

$$\frac{\partial}{\partial x}\left(k_x \frac{\partial \theta}{\partial x}\right) + \frac{\partial}{\partial y}\left(k_y \frac{\partial \theta}{\partial y}\right) + q = 0 \qquad (1)$$

Electric and Magnetic Fields, Edited by A. Nicolet
and R. Belmans, Plenum Press, New York, 1995

where k_x and k_y are the thermal conductivities in the x and y directions, and q is the loss density.

To ensure a unique solution sufficient boundary conditions must be imposed. The more common of these, shown in figure (1), are:

Dirichlet boundary conditions defined by

$$\theta\,(x,y) = \theta_0 \tag{2}$$

Neumann boundary conditions defined by,

$$k_n \frac{\partial \theta}{\partial n} + q'_{S_2} = 0 \tag{3}$$

where n is the outward normal direction to a surface of the region, k_n is the thermal conductivity in the normal direction, q'_{S2} the heat flux from the boundary.

The solution of equation (1) can be obtained by formulating the scalar potential temperature problem in variational terms by an energy functional, and minimising with respect to a set of trial functions. Rao[4] has shown that a functional which, when minimised, satisfies equation (1) is given by,

$$\xi = \frac{1}{2} \iint_R \left[k_x \left(\frac{\partial \theta}{\partial x}\right)^2 + k_y \left(\frac{\partial \theta}{\partial y}\right)^2 - 2\,q\,\theta \right] dx\,dy + \int_{S_2} q'\,\theta\,d\,S_2 \tag{4}$$

The double integral is taken over the region of interest R, the last term is introduced to account for Neumann conditions. The solution for the temperature distribution is obtained by minimising the functional given by equation (4), with respect to the nodal temperatures. This leads to a set of simultaneous equations which, when combined with similar equations for the remaining elements in R, results in a set of n simultaneous equations which can be expressed in matrix form as

$$[G]\,[\theta] = [q] \tag{5}$$

where [G] is the coefficient matrix, [θ] is the unknown temperature vector, [q] is the loss density vector, and n is the number of nodes in the domain R.

The analysis of thermal field distribution thus reduces to one of solving an $n \times n$ matrix equation. In order to obtain a unique solution the prescribed boundary conditions are imposed in equation (5). [G] is usually very large and sparse since non-zero coefficients will exist only when nodes are coupled by finite element mesh. The sparsity of the matrix [G] can be exploited by using one of the sparse algorithms described by Low [5].

Figure 2 shows a finite element for one pole pitch of the induction. To simplify the finite element analysis the following assumptions are necessary,

1. Thermal conductivity is constant over a linear element.
2. The loss density, if present, is also constant within an element.
3. Temperature distribution is assumed to be periodic over a pole pitch.
4. The temperature and loss density distribution are independent of rotor position.
5. A 2-dimensional finite element model is used so heat flow is assumed to be entirely radial and is unaffected by the end effects.

ASSIGNMENT OF LOSS DENSITIES

Accurate temperature prediction can only be achieved if the correct loss density is assigned. Since the distribution of iron losses in the machine is not fully understood, various authors have used two dimensional magnetic field distribution [1,6] and others have proposed empirical ratios for assigning the losses [7]. Using the temperature time technique a more accurate distribution was determined and is shown in Table 1.

For the 2-dimensional model copper losses in the stator are assigned values according to the ratio of the length of the embedded windings to the total length of the windings. The rotor copper losses, however, are assigned according to the ratio given by Alger [8], i.e.

$$\frac{R_e}{R_b} = \frac{C_2 D_r}{2\pi C_e L_b p^2} \tag{6}$$

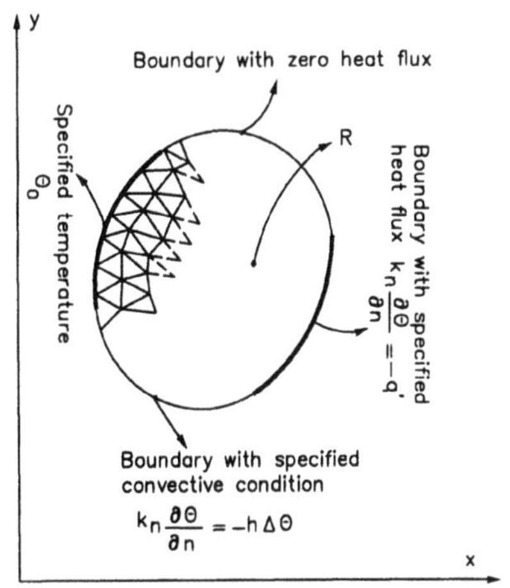

Figure 1. A generalised 2D region showing the discretisation boundary conditions.

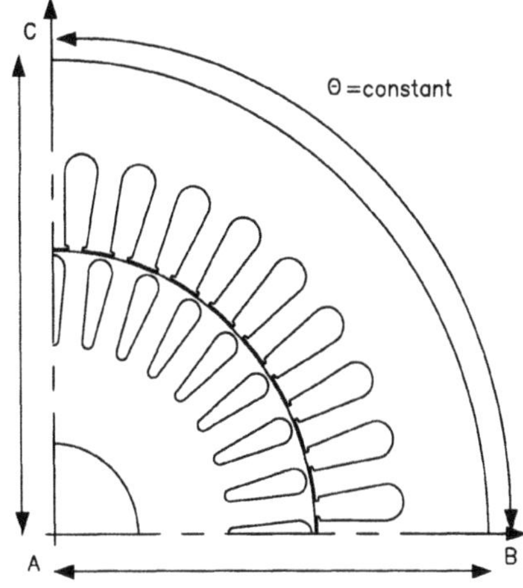

Figure 2. Section of an induction motor used in the finite element analysis.

where R_e is the end rings resistance, R_b is the bars resistance, D_r is the diameter of the end ring, C_2 is the total cross section of all bars, L_b length of each bar, C_e is the cross section of one end ring, and p is the number of pair poles.

Table 1. Various loss density distribution used by authors.

	Sarkar (W/m^3) [6]	Mellor [7]	Temperature time (W/m^3) [3]	
Stat. back iron	0.0389×10^6	0	0.0527×10^6	69.7%
stator teeth	0.0392×10^6	50%	0.1060×10^6	30.3%
Rotor	-	50%	-	0%

RESULTS AND DISCUSSIONS

Figure (2) shows one pole pitch of the machine. This has been selected on the basis of geometrical and thermal symmetry. Boundary conditions applicable to this thermal problem are also shown in figure (2), i.e. AB periodic to AC and BC specified temperature (Dirichlet). Figure (3) shows the discretised finite element domain created using a 2-D thermal analysis program. Figure (4) shows a typical equi-thermal contours distribution for the assigned losses. Table 2 shows a comparison between experimental results and those obtained using finite element analysis for various loss density distributions.

A comparison of the measured and the finite element results shows good agreement in the stator section for both loss distributions. In the rotor, the correlation with the measured results are not as good and this is probably due to the difficulty in obtaining reliable readings from the rotor. However, despite the discrepancies the results show that considerable emphasis must be put on specifying the loss density correctly since this has a marked effect on the localised temperatures.

CONCLUSIONS

The two dimensional, steady state, finite element technique for the thermal analysis of induction motors has been described. Using finite element to predict the temperature in electrical machines is an accurate method provided that the thermal properties and loss densities are available to the required accuracy. This contribution has shown that the temperature time technique is an accurate method for calculating the iron loss distribution in machines. The method is fast, inexpensive and lends itself to immediate visual pictures of the temperature patterns in the induction motor. Moreover, the two dimensional finite element method has shown good agreement with experimental and lumped thermal model (LTM) results when the loss density distribution based on the temperature time method is used.

Table 2. Comparison of the temperatures for different methods used.

Components	Temperatures (^oC)			
	Measured	Loss density distribution		
		FE [3]	FE [7]	L TM [3]
Stator iron	72.2	72.20	72.20	69.53
Stator teeth	77.0	79.10	80.76	75.05
Emb. winding	102.1	100.58	101.47	94.33
End winding	110.0	-	-	109.15
Air gap	-	118.40	144.87	124.69
Rotor winding	-	156.94	193.43	151.11
Rotor iron	128.8	156.94	193.43	150.40
Endcap air	71.7	-	-	83.89

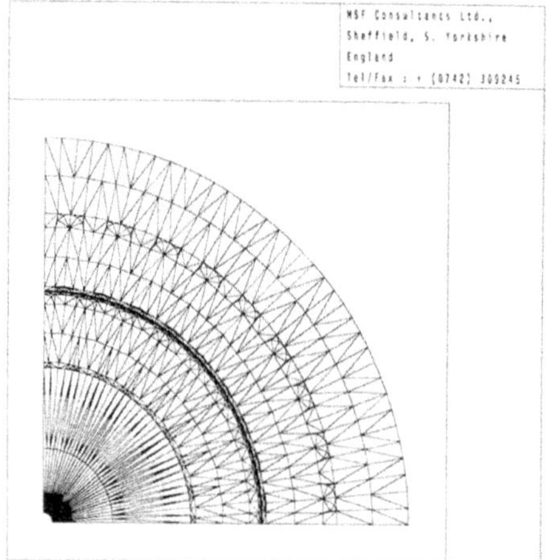

Figure 3. Two dimensional finite element mesh.

Figure 4. Equi-thermal contours distribution.

REFERENCES

1. T.J. Roberts, ''The Solution of Heat Flow Equations in Large Electrical Machines'', Proc. IME, Vol.184, Pt.3E, 70:83(1968-70).
2. A. J. Gilbert, '' A Method of Measuring Loss Distribution in Electrical Machines'' Proc. IEE, Vol.108A, 239:244(1961).
3. A. Bousbaine, W.F. Low, M. McCormick and N. Benamrouche, ''Thermal Modelling of Induction Motors Based on Accurate Loss Density Measurements'', ICEM, 53:957(1992).
4. S.S. Rao,''The Finite Element Method in Engineering'', Pergamon Press Plc., (1989).
5. W.F. Low, '' The Computation of Magnetostatic Fields in Permanent Magnet Devices'', Ph.D. Thesis, University of Sheffield, (1985).
6.D. Sarker, P.K. Mukherjee and S.K. Sen, ''Use of 3-Dimensional Finite Elements for Computation of Temperature Distribution in the Stator of an Induction Motor'', Porc.IEE, Vol. 138, 75:86(1991).
7. P.H. Mellor, D. Robert and D.R. Turner,'' Lumped Parameter Thermal Model for Electrical Machines of TEFC Design'', Proc. IEE, Vol. 183, No.5, 205:218(1991).
8. P.L. Alger. '' Induction Machines'', Gordan and Breach Science Publishers, (1970).

A FINITE-ELEMENT SIMULATION OF AN OUT-OF-PHASE SYNCHRONIZATION OF A SYNCHRONOUS MACHINE

Silvio I. Nabeta, Albert Foggia, Marcel Ivanes, Jean-Louis Coulomb and Gilbert Reyne

Laboratoire d'Electrotechnique de Grenoble (INPG) URA CNRS 355
BP 46, 38402 Saint-Martin-d'Hères Cedex, France

INTRODUCTION

A very common and important stress for a synchronous machine arises during an out-of-phase synchronization.

An out-of-phase synchronization is a faulty 3 phase synchronizing that occurs when the generator is switched to the system and the conditions of voltage, frequency and phase are not respected.

In this paper we assumed that voltage and frequency are similar for the generator and the system and only the phase angle differs.

Detailed simulations are proposed for two specific conditions:

1. Out-of-phase synchronization with a 180° of angular lag. In this case the armature currents reach the maxima values.
2. Out-of-phase synchronization with 120° of angular lag. In this second case the torque reachs the maximum.

THEORETICAL OVERVIEW

Consider a synchronous generator unloaded. The stator voltages per phase can be expressed as[1]:

$$\begin{aligned} V_{a0} &= c.\sin(\omega t + \theta_0) \\ V_{b0} &= e.\sin(\omega t + \theta_0 - \frac{2\pi}{3}) \\ V_{c0} &= e.\sin(\omega t + \theta_0 - \frac{4\pi}{3}) \end{aligned} \qquad (1)$$

As stated above only the phase equality is not respected. So the system voltage may be written as:

$$\begin{aligned} V_{a1} &= e.\sin(\omega t + \theta_0 - \lambda) \\ V_{b1} &= e.\sin(\omega t + \theta_0 - \lambda - \frac{2\pi}{3}) \\ V_{c1} &= e.\sin(\omega t + \theta_0 - \lambda - \frac{4\pi}{3}) \end{aligned} \qquad (2)$$

where λ is the phase difference at the switching instant.

Applying the Park transformation to these two equation sets we have:`

machine: $\quad V_{d0} = 0, \quad V_{q0} = e \qquad$ system: $\quad V_{d1} = -e.\sin\lambda, \quad V_{q1} = e.\cos\lambda$

Electric and Magnetic Fields, Edited by A. Nicolet
and R. Belmans, Plenum Press, New York, 1995

The voltage variation in the machine terminal is:

$$\Delta V_{d0} = V_{d1} - V_{d0} = -e.\sin\lambda = -2e.\sin\frac{\lambda}{2}.\cos\frac{\lambda}{2}$$
$$\Delta V_{q0} = V_{q1} - V_{q0} = e.\cos\lambda - e = -2e.\sin^2\frac{\lambda}{2} \tag{3}$$

using the operational notation we have:

$$\frac{\Delta V_{d0}}{p} = -p\Delta\phi_d - \omega.\Delta\phi_q - r_a.\Delta I_d$$
$$\frac{\Delta V_{q0}}{p} = -p\Delta\phi_q + \omega.\Delta\phi_d - r_a.\Delta I_q \tag{4}$$

with
$$\Delta\phi_d = L_d(p).\Delta I_d$$
$$\Delta\phi_q = L_q(p).\Delta I_q \tag{5}$$

Eliminating ΔI_d and ΔI_q in equations (4) we reach the system below:

$$\begin{bmatrix} \frac{\Delta V_{d0}}{p} \\ \frac{\Delta V_{q0}}{p} \end{bmatrix} = \begin{bmatrix} -\left(p+\frac{r_a}{L_d(p)}\right) & -\omega \\ \omega & -p\left(p+\frac{r_a}{L_q(p)}\right) \end{bmatrix} \cdot \begin{bmatrix} \Delta\phi_d \\ \Delta\phi_q \end{bmatrix} \tag{6}$$

Solving (6) we find :

$$\Delta\phi_d(p) = \frac{1}{p(p^2 + 2\alpha p + \omega^2)} \cdot \left[-\Delta V_{d0} \cdot \left(p+\frac{r_a}{L_d(p)}\right) + \Delta V_{q0}.\omega \right]$$
$$\Delta\phi_q(p) = \frac{1}{p(p^2 + 2\alpha p + \omega^2)} \cdot \left[-\Delta V_{d0}.\omega - \Delta V_{q0} \cdot \left(p+\frac{r_a}{L_q(p)}\right) \right] \tag{7}$$

where: $\frac{1}{\alpha} = T_a$ is the armature time-constant.

Introducing (7) in equation (5) it gives the current variations:

$$\Delta I_d(p) = \frac{\Delta V_{d0}}{L_d(p)(p^2 + 2\alpha p + \omega^2)} + \frac{\omega.\Delta V_{q0}}{pL_d(p)(p^2 + 2\alpha p + \omega^2)}$$
$$\Delta I_q(p) = \frac{\Delta V_{d0}}{pL_q(p)(p^2 + 2\alpha p + \omega^2)} - \frac{\Delta V_{q0}}{L_d(p)(p^2 + 2\alpha p + \omega^2)} \tag{8}$$

Considering that the out-of-phase phenomenon is harmful in the first instants we can neglect all resistances[1]. Moreover assuming that $X''_d = X''_q$ we obtain in the time-domain terms:

$$I_d(t) = \Delta I_d(t) = \frac{2e}{X''_d}.\sin\frac{\lambda}{2}.\left[\sin\left(\omega t+\frac{\lambda}{2}\right) - \sin\frac{\lambda}{2}\right]$$
$$I_q(t) = \Delta I_d(t) = -\frac{2e}{X''_d}.\sin\frac{\lambda}{2}.\left[\cos\left(\omega t+\frac{\lambda}{2}\right) - \cos\frac{\lambda}{2}\right] \tag{9}$$

And the phase A current, in per unit values, can be written as:

$$I_a(t) = \frac{2e}{X''_d}.\sin\frac{\lambda}{2}.\left[\sin\left(\omega t-\theta_0-\frac{\lambda}{2}\right) - \sin\left(\theta_0-\frac{\lambda}{2}\right)\right] \tag{10}$$

The torque equation is obtained by using:

$$C_e = \frac{e^2}{X''_d}.\omega\left[\phi_d.I_q - \phi_q.I_d\right] \tag{11}$$

In our case we obtain in per-unit values:

$$C_e = \frac{e^2}{X''_d}\left[\sin\lambda - 2\sin\frac{\lambda}{2}.\cos\left(\omega t + \frac{\lambda}{2}\right)\right] \tag{12}$$

It can be seen that the maximum current occurs when:

$$\lambda = \pi, \qquad \theta_0 = 0, \qquad \omega\tau = \pi$$

and we obtain:

$$I_{amax} = \frac{4e}{X''_d} \tag{13}$$

For the torque the maximum is reached with:

$$\lambda = \pm\frac{2\pi}{3}$$

and:

$$C_{emax} = \frac{3\sqrt{3}\ e^2}{2\ X''_d} \tag{14}$$

NUMERICAL SIMULATION

The numerical simulation was carried out by using a time-stepped Finite-Element method coupled with electrical circuits[2,3] and the moving-air-band technique[3] that allowed the representation of the rotor motion.

Figure 1 shows the machine geometry and the corresponding armature electrical circuit used.

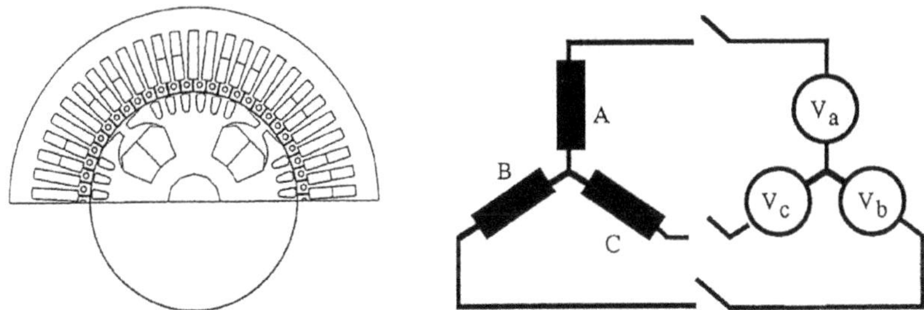

Figure 1. Machine geometry and electrical circuit for the armature

To allow the out-of-phase synchronizing simulation it has been necessary to reach the generator steady-state at unloaded condition. This has been done by a time-stepped simulation with the field winding fed with a DC voltage source and the rotor speed at 1500 rpm.

Afterwards, switching with adequate angle, the out-of-phase synchronization to the system represented by the generators V_a, V_b and V_c was executed.

Synchronizing angles were chosen at 120° and 180° corresponding to the maxima of the torque and the armature current respectively.

RESULTS

Figure 2 shows the phase A current and the torque for the out-of-phase synchronizing at 180° and 120°.

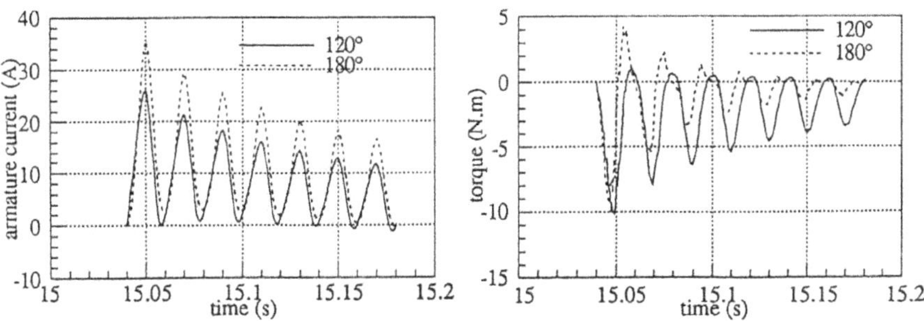

Figure 2. Phase A current and torque

Table 1 presents the comparison between computed and analytical values of maximum current and maximum torque during the sub-transient period.

Table 1. Maxima values comparisons

Parameter / Method	I_{amax} (A) 180°	C_{emax} (N.m) 120°
Analytical	40.29	-8.07
Simulation	35.50	-10.20

It can be seen from computed curves that simulations take into account current and torque decrements due to armature and rotor resistances.

We observe from table 1 that for λ=180° the current analytical value is greater than simulated one. This difference can be explained by the fact that neglecting all resistances in the analytical expressions the current decrements are not taken into account which leads to an overstimated result.

Nevertheless analytical torque is smaller than simulated one. We can argue that neglecting the resistances we also neglect the unidirectional torques due to the ohmic losses in the armature, field and damper circuits[4,5].

Moreover, assuming $X''_d=X''_q$ we neglect the second order torque due to the pole saliency which can contribute to increase the torque result.

CONCLUSION

A numerical analysis of an out-of-phase synchronization has been performed by a Finite-Element simulation.

Computed results accord with theory and have shown that maxima currents and torque occur at angular lags of 180° and 120° respectively.

Furthermore, numerical simulation allows to predict with less simplifications the out-of-phase synchronizing transient.

Indeed, it takes into account complex phenomena as the unidirectional torques while analytical expressions can hardly deal with.

ACKNOWLEDGMENTS

The authors acknowledge the CNPq - Brazilian Research Council for the financial support during this work.

REFERENCES

1. P. Barret, "Régimes Transitoires des Machines Tournantes Electriques", Editions Eyrolles, Paris, (1982).
2. P. Lombard, G. Meunier, A general purpose method for electric and magnetic combined problems for 2D, axisymmetric and transient systems *CEFC 92*, 3-5 aug , Claremont, USA. *IEEE Transactions on Magnetics*, vol. 29, N° 2, pp. 1737-1740, march 1993.
3. E. Vassent, G. Meunier, A. Foggia, G. Reyne, Simulation of induction machine operation using a step by step finite-element method coupled with circuit and mechanical equations, *Fifth joint MMM-Intermag Conf.* Pittsburgh, USA, 18-21 june 1991. *IEEE Transactions on Magnetics*, vol. 27, N° 6, nov. 1991
4. C. Concordia, "Synchronous Machine - Theory and Performance" John Wiley&Sons, London 1951.
5. H.S. Kirshbaum, Transient electrical torques of turbine generators during short circuits and synchronizing, *AIEE Transactions* , vol. 64, pp. 65-70, feb. 1945.

THE PREDICTION OF LOSS DENSITY DISTRIBUTION IN ELECTRICAL MACHINES USING A DIRECT 'INVERSE FIELD' FINITE ELEMENT TECHNIQUE

W.F. Low, A. Bousbaine, M. McCormick

Department of Electronic and Electrical Engineering
DeMontfort University
The Gateway
Leicester LE1 9BH, U.K.

INTRODUCTION

The increasing awareness in energy conservation has led to renewed enthusiasm for the design of efficient and reliable machines. The availability of improved analytical and numerical techniques has made it possible to design electrical machines to exacting standards. This is particularly true in the design of magnetic circuits where established numerical techniques, such as finite elements, can be applied to study the magnetic field distribution in very great detail. Thermal design, however, has tended to lag behind its electromagnetic counterpart because of the considerable difficulties associated with the correct assignment of loss density distribution within the volume of the machine. These uncertainties arise as a result of the lack of total understanding about the nature and origin of iron losses. Whilst an incorrect loss density distribution may not affect the overall thermal distribution, localised temperatures could be grossly inaccurate. There have been attempts to quantify and characterise the losses using empirical methods[1], and experimental methods such as the calorimetric method[2], and the temperature-time method[3], but these have proved difficult and cumbersome and cannot be readily incorporated into general design methodologies. Furthermore the lack of information on the thermal characteristics of materials, especially when these are combined to form a composite mass, for example the copper windings which consist of copper, lacquer, air, and insulation, has compounded the problem of accurate thermal modelling.

This contribution describes an accurate and novel method to alleviate some of the problems associated with studies on, and the prediction of, the steady-state loss density distribution in electric machines. The method is based on a 2-dimensional direct 'Inverse Field' finite element technique whereby a set of prespecified subsidiary conditions, ie the temperature distribution, is coupled to the standard finite element matrix equation. By solving the resultant 'weakly coupled' system it is possible to determine, in one iterative cycle, the loss density distribution which gives rise to the prescribe conditions. However, unlike other finite element based optimisation techniques which scan all the field sources to obtain the best combination, this new method utilises a Newton-Raphson algorithm so that convergence to the correct solution is very rapid. This technique was applied to the study of the loss density distribution in a 3-phase, 4kW induction motor where the loss density and temperature distribution were measured experimentally.

FORMULATION OF THE PROBLEM

The steady-state 2-dimensional heat conduction problem is governed by Poisson's equation, ie

$$k_x \frac{\partial^2 T}{\partial x^2} + k_y \frac{\partial^2 T}{\partial y^2} = -Q \qquad (1)$$

Electric and Magnetic Fields, Edited by A. Nicolet
and R. Belmans, Plenum Press, New York, 1995

where T is the temperature, k_x and k_y are the thermal conductivities, and Q is the loss density. A finite element solution to Eq. 1 requires the discretisation of the domain into a set of elements within which the temperature is assumed to vary in some predefined manner, for example

$$T = \sum N_i T_i \tag{2}$$

where N_i is the shape function, and i is associated with the nodes defining the element.

The solution to the partial differential equation can be obtained by minimising an energy functional respect to the space function defined in Eq. 2. The required functional is

$$\xi = \frac{1}{2} \int_R \int k_x \left\{ \frac{\partial T}{\partial x} \right\}^2 + k_y \left\{ \frac{\partial T}{\partial y} \right\}^2 - 2\, Q\, T \quad dR \tag{3}$$

and it can be shown[4] that the Euler equation to the functional is given by Eq. 1.

For a first order triangular finite element the variational procedure is well documented. For an assemblage of elements in a domain with n nodes, the result of the variational procedure is a matrix equation of the form

$$[S][T]=[Q] \tag{4}$$

where $[S]$ is an nxn sparse coefficient matrix, $[T]$ is the temperature vector, and $[Q]$ is the loss density vector.

Eq. 4 can be inverted to give $[T]$ but convergence can be improved by utilising the Newton-Raphson algorithm[5]. Assuming that the variables are T and Q , the resulting matrix equation to be solved is thus

$$\left[\frac{\partial S}{\partial T_n} \mid \frac{\partial S}{\partial Q_n} \right] \begin{bmatrix} \Delta T \\ \Delta Q \end{bmatrix} = [S]\,[T] - [Q] = [R] \tag{5}$$

where the required unknowns are now the increments $[\Delta T]$ and $[\Delta Q]$. The coefficient matrix is no longer square but it can be made invertible if sufficient subsidiary conditions are coupled to it. These conditions directly dictated the source distribution.

SUBSIDIARY CONDITIONS

The subsidiary conditions relate to the required temperature distribution. For example, if a temperature, $T_{spec(i)}$ is required at position (i) , figure (1), then following the solution for the temperature distribution there will be an error function at (i) given by

$$\xi_{(i)} = T_{spec(i)} - T_{calc(i)} \tag{6}$$

where $T_{calc(i)}$ is the calculated temperature at node (i), and $T_{spec(i)}$ is the required temperature for node i.

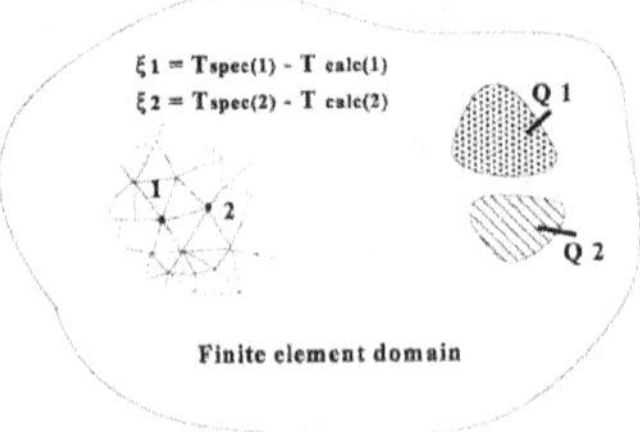

Figure 1. Application of subsidiary conditions

Unless the source distribution has converged to its correct value the error function will be non-zero and in the context of the Newton-Raphson method this is given by

$$\xi_{(i)} = \frac{\partial T_{calc(i)}}{\partial T} \Delta T$$

Multiple subsidiary conditions can be defined in a similar manner and coupled to Eq. 5 so that the resultant matrix to be solved is now given by

$$\begin{bmatrix} \frac{\partial S}{\partial T_n} & \frac{\partial S}{\partial Q_n} \\ \frac{\partial T_{calc}}{\partial T} & 0 \end{bmatrix} \begin{bmatrix} \Delta T \\ \Delta Q \end{bmatrix} = \begin{bmatrix} R \\ \xi \end{bmatrix} \tag{7}$$

Inversion of Eq. 7 yields the corrections [ΔT] and [ΔQ] so the improved solutions are now given by

$$\begin{aligned} [\,T\,]_{k+1} &= [\,T\,]_k + [\Delta T\,] \\ [\,Q\,]_{k+1} &= [\,Q\,]_k + [\Delta Q\,] \end{aligned} \tag{8}$$

where k is the iteration count. The iterations continue until the corrections are less than some preset tolerance. From Eq. 7 it is clear that any number of unknown sources can be determined provided the same number of subsidiary conditions are defined.

For any specified temperature distribution the loss density distribution is, in general, not unique. Therefore in specifying the locations for applying the subsidiary conditions care must be taken to select positions which give an unambiguous source distribution. A general rule of thumb for avoiding this problem is to select sites which are within the region where the loss density is required and where the expected temperature gradient is high.

FINITE ELEMENT MODEL

The new method was used to predict the loss density distribution in a 3-phase, 4 kW induction motor for which loss density and thermal information are available[3]. Figure 2 shows the cross-section for one slot-pitch of the motor, chosen on the basis of thermal symmetry, and the corresponding boundary conditions. Figure 3 shows the sites for the losses, and the locations at which the subsidiary equations are applied. Thermal conductivities were obtained from manufacturer's data and equivalent thermal conductivities for composite masses were calculated using established techniques[3].

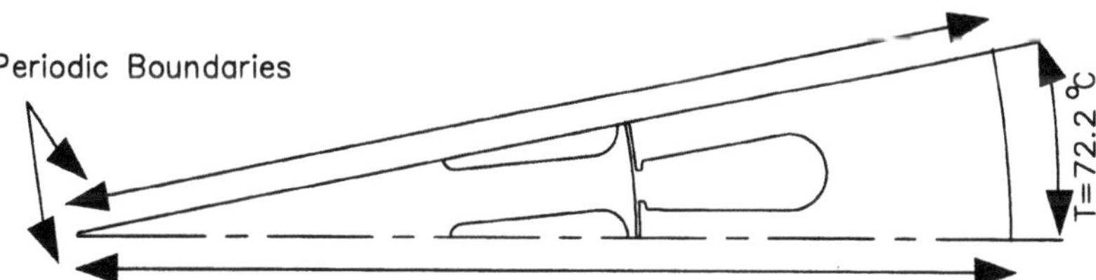

Figure 2. Cross-section of 1 slot-pitch

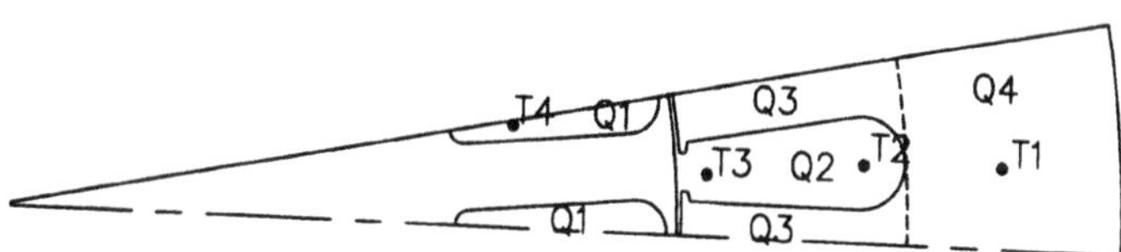

Figure 3. Sources and subsidiary conditions

VERIFICATION OF THE METHOD

Figure 4 shows the resulting thermal field distribution using the new method.

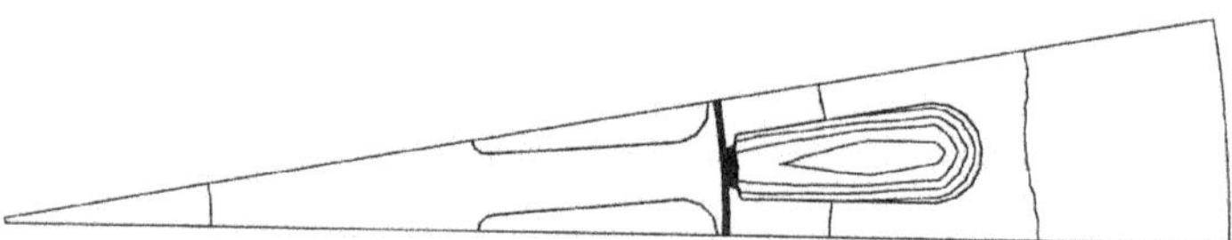

Figure 4. Thermal field distribution

Table 1 compares the predicted loss density distribution with those measured. The temperature had converged to the correct solution as indicated by the negligible error. Good agreement was also obtained for the loss density distribution. The percentage error was relatively high in some regions and this was attributed to the non-uniqueness of the source distribution. These errors can be reduced by refining the number of regions in which losses are imposed. This allows more subsidiary conditions to be defined and helps to ensure a unique solution for the loss distribution.

Table 1. Comparison of the loss density distribution

Loss Density, $x\,10^6\ W m^{-3}$			$\sum (T_{calc} - T_{spec})^2 < 1.0\,x\,10^{-6}$
	T-T Measurements	Finite Elements	| % Difference |
Q_1	1.0735	1.0636	0.97
Q_2	0.6564	0.6532	0.48
Q_3	0.1060	0.1326	25.0
Q_4	0.0527	0.0602	14.0

CONCLUSIONS

A finite element method for the direct solution of the inverse thermal field problem has been implemented. The method involves coupling a set of subsidiary conditions to the standard finite element equations and solving the resultant matrix iteratively. The results show good agreement with measured data but care has to be taken to ensure that the subsidiary conditions are applied at suitable locations to guarantee a unique solution. Further research on the characterisation and distribution of iron losses is now be possible since these can now be readily quantified using this method.

REFERENCES

1. Subba Rao, V. "Losses in ferromagnetic laminations due to saturation", Proc IEE, Vol 111, No. 12, Dec 1964, pp2111-2117
2. Turner, D.R., Binns, K.J., and Shamsadeen, B.N. "Accurate measurement of induction motor losses usinga balanced calorimeter", Proc IEE, Vol 138, No. 5, Sept 1991, pp 233-242.
3. Bousbaine, A. "An Investigation into the Thermal Modelling of Induction Motors", PhD thesis, University of Sheffield (1993).
4. Rao, S.S. "The Finite Element Method in Engineering", 2nd. Edition, Pergamon Press, (1989).
5. Low, W.F. "The Computation of Magnetostatic Fields in Permanent Magnet Devices", PhD thesis, University of Sheffield (1985).

A COMPUTATION OF THE TRAJECTORIES OF PARAMAGNETIC PARTICLES

E. Nava, G. Vinsard, A. Mailfert.

Green, URA 1438, 2 Av. de la forêt de Haye
54516 Vandoeuvre-les-Nancy, France

INTRODUCTION

The magnetic filtration of small paramagnetic particles consists in making a high static magnetic field, which is created by a supraconductor inductor and magnetizes many iron wires that are embedded in a fluid flow. The magnetic field, in the vicinity of the iron wires, comports high gradients that trap the paramagnetic particles carried by the fluid. The modelling of the magnetic filtration is very complex[1] because the fluid flow must be computed accurately at both the scale of the iron wires and particles to take into account the many interactions. Moreover the interaction between the fluid and particles is difficult to state. The magnetic analysis is also complex even if we only treat the 2D problem, which is issued from the assumption that the iron wires are infinite cylinders. In fact the magnetic field can be computed by using its scalar potential as state variable, and the magnetic force is proportional to a double derivative of this potential whose values are difficult to obtain numerically[2]. The aim of this paper is to show the results that we have obtained by using a re-interpolation of the double derivative of the potential on a triangular mesh that is generated from the finite element mesh from which we compute the potential.

ANALYSIS IN THE UNCOUNTABLE DIMENSION

Let us consider a general magnetostatic problem: if $\mathbf{h}_s$ is the source magnetic field, which would exist in the space without any magnetic wires in the space, and $\mathbf{h}_t$ the total magnetic field, which exist in presence of the wires, then we can defined the reaction magnetic field $\mathbf{h}_r$ as the difference between these two fields. $\mathbf{h}_r$ is the finite-energy non-rotational field that maximizes the coenergy functional[3]

$$W(\boldsymbol{h};\, \boldsymbol{h}_s) = \int_{R^3} \mu_0 \boldsymbol{h}_s.\boldsymbol{h} - \frac{1}{2}\bar{\bar{\mu}}(\boldsymbol{x},|\boldsymbol{h}_s+\boldsymbol{h}|)(\boldsymbol{h}_s+\boldsymbol{h})^2 \;\; d\boldsymbol{x} \tag{1}$$

where the non linear magnetic behavior of the wires D_w is described by the space function

Electric and Magnetic Fields, Edited by A. Nicolet
and R. Belmans, Plenum Press, New York, 1995

$$\bar{\bar{\mu}} : R^3 \times R \rightarrow R \; ; \quad x \times |h| \rightarrow \left| \begin{array}{l} \bar{\bar{\mu}}(|h|) \; in \; D_w \\ \mu_0 \; in \; D_a = R^3/D_w \end{array} \right. \tag{2}$$

in which the scalar functions

$$\bar{\bar{\mu}}(|h|) = 2/|h|^2 \int_0^{|h|} u \, \bar{\mu}(u) \, du \; ; \; \bar{\mu}(u) = 1/u \int_0^u \bar{\mu}(v) \, dv \tag{3}$$

represent : the energy ($\bar{\bar{\mu}}$), the total ($\bar{\mu}$) and the differential (μ) permeabilities.

Now we assume that the reaction field is very low behind the source field because the wires are strongly saturated ; then the functional (1) can be approximated by its two-order Taylor development near the source field

$$W_2(h;\, h_s) = \int_{R^3} \tfrac{1}{2}(\bar{\bar{\mu}}(x,|h_s|)h_s^2 + \bar{\mu}(x,|h_s|)(h_s \times h)^2/|h_s|^2 + \mu(x,|h_s|)(h_s.h)^2/|h_s|^2) \\ + (\bar{\mu}(x,|h_s|)-\mu_0)h_s.h \; dx \tag{4}$$

which is linear with respect to $\mathbf{h}$ and provides a single-step algorithm to compute the reaction field $\mathbf{h}_r$. Another interest to use this kind of superposition method is that the (small) values of $\mathbf{h}_r$ are not hidden by those of $\mathbf{h}_s$. This formulation can also be see as the first step of the Newton method where the initial field is the source field $\mathbf{h}_s$.

If the source field is uniform, the force $\mathbf{f}$ on a single paramagnetic particle, which if embedded in D_a and whose volume, magnetic susceptibility and mass are v,χ and m, can be written as

$$f = \mu_0 \chi v \; \nabla h_t^2/2 = \mu_0 \chi v \; \nabla(h_s.h_r) + o(h_r^2) \tag{5}$$

We neglect the term in square of $\mathbf{h}_r$, which is small, and then the force is proportional to the first derivative of the magnetic field $\mathbf{h}_r$. Another important point is that the force derives from a potential energy. In the case where there are no supplementary conservative forces, we obtain an invariant of the motion of the particle, the Hamiltonian that is easy to compute and furnishes a good mean to control the trajectory of the particle.

THE DISCRETIZATION

The minimization problem (4) can be formulated in term of magnetic scalar potential -the major reason is that we have in mind to treat 3 dimensional problems later and the state variable of the magnetic scalar potential formulation is easy to implement. Then the non rotational reaction magnetic field can be written as

$$h = \nabla\phi \tag{6}$$

The functional (4) can be discretized on a Lagrange P1 triangular finite element basis and the scalar potential ϕ_r is computed as a piecewize-linear function on the elements. Consequently the magnetic field $\mathbf{h}_r$ is computed as a piecewize-constant function on the elements and its derivative -and the force- in the direction of $\mathbf{h}_s$ is a Dirac distribution on the edges of the elements. These forces are impulsions on the edge of the elements and it is very easy to compute the trajectories of the particles, which are straight segments inside each element. However, we have tried this approach without any success ; the trajectories were wrong. To overcome this drawback, some (non exhaustive) possibilities exist:

0. The magnetic particle size is larger than the elements size. Probably the edge distribution can be used straightforwardly, however in many applications such a situation will lead to a too large mesh. We will not consider this situation.

1.The scalar potential is computed with the boundary element method rather finite element method. Then the first derivative of $\mathbf{h}_r$ is continuous and probably we can obtain a robust algorithm. However, if there are many wires with arbitrary shapes then the boundary element method leads to a non band-width matrix, which can be difficult to solve.

2.The finite element method is used to compute the field ; a region, the Green region, is defined on the boundary of which the potential is obtained by interpolation. The Green formula is used in this region to obtain a smooth potential that can be derived twice[2]. This is an elegant method, which shows that the high order derivatives of a numerical field can be found almost exactly. However the Green region must have a shape such that the Green function is known in closer form. In our case this region must be in the exterior of the wires and if we have more than a single wire, we can not find such a region. Then, if we use this technic we must move the Green region because the particle move in the space and this motion will be arbitrary.

3.The Hermit finite elements, for which the potential and its derivatives are simultaneously computed, are perhaps the solution. However, their use is not easy and, moreover, if an high continuity of the potential is required in the exterior of the wires, this is just the opposite on the boundary of the wires.

These technics, except perhaps 3, have a drawback. However the technic 2 shows[2] that the accuracy on the double derivative of the potential exists in the Lagrange P1 interpolation of the potential. To attempt to find this accuracy, a possibility is to re-interpolate the potential energy (scalar product $\mathbf{h}_r.\mathbf{h}_s$), on an another triangular mesh, which can be easily constructed by using the Delaunay procedure on the set of point whose elements are the middles of the triangles of the first mesh. On the new mesh, we have called it the element mesh, the potential energy becomes piecewize-linear and the force piecewize-constant by elements. Moreover the re-interpolation smoothes locally the potential energy in a physical sense because the elements mesh only depends on the positions of the

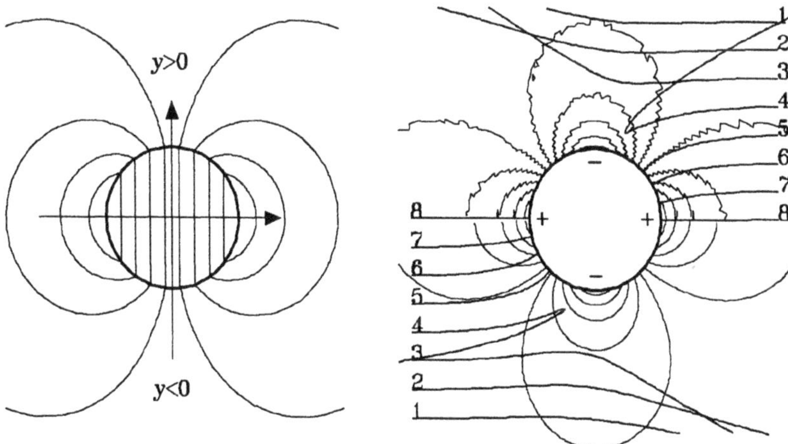

Figure 1. Left: the magnetic potential lines. right: the potential energy, down : analytical, up: numerical, on the elements mesh ; the trajectories of particles: down: analytical, up: numerical.

elements. This technic is almost like those presented by Bastos[4], except that we work here with unstructured meshes.

RESULTS

To test the accuracy of the above technic, we have computed the trajectories of particles around a single wire problem for which an analytical solution of the magnetic potential is available. The field $\mathbf{h}_s$ is uniform, oriented on the x direction, the field $\mathbf{h}_r$ is computed: (1) by using the form

$$\phi_s = h_0 x \quad ; \quad \phi_r = h_r \left| \begin{array}{l} x \ in \ D_w \\ xR^2/(x^2+y^2) \ in \ D_a \end{array} \right. \tag{7}$$

where R is the radius of the wire D_w and h_r is computed by introducing the form (7) in the functional (4), (2) by using the Lagrange P1 finite element method on a mesh sufficiently large for approximate the infinity. The magnetic scalar potential (Figure 1) is the same -the maximum difference is less than 10^{-6} h_0R between the two cases. The energy potential is represented: (1) the analytical computation in the inferior half plane (y<0), (2) the numerical computation in the superior half plane (y>0). The maximum difference is 10^{-2} h_0^2, but this difference is not significant because the gap is localized in all the plane. The symbol + and - represent the sign of the potential energy, i.e. the attractive or repulsive polarity.

Eight particles are positioned as indicated in the figure ; they have an horizontal speed and a mass sufficiently low for that the wire has a strong effect on their trajectories. We have made three computation for each of the particle: (1) the analytical expression is straightforwardly introduced in a Runge-Kutta procedure, the time step being very small, (2) the analytical expression is used to compute the values on the nodes of the elements mesh and the computation of trajectories is analytical piecewize by element, (3) the values on the elements mesh are the numerical ones, the computation of the trajectories are the same as (2). There is no difference between (1) and (2), which are represented on the left down in the figure. The numerical trajectories (3) are represented on the right up. There is difference with (1) and (2) when the particle are rejected by the wire (particle 4 essentially), due essentially to the lack of accuracy of the numerical procedure (3). However, if we are interested by global quantities, as the capture section of the wire, the results are almost the same for the three procedures.

CONCLUSION

The numerical method that we have used is not very accurate but the trajectories are not absolutely wrong and can be used to the computation of global quantities. In our opinion the re-interpolation technic can be improved, for example by smoothing with another re-interpolation.

REFERENCE

1. D. Fletcher, "Fine particle high gradient magnetic entrapment", IEEE trans on Magn., Vol 27, No 4, (July 1991).
2. P. P. Silvester, D. Omeragić, "Differentiation of finite element solutions of the Poisson equation", IEEE trans on Magn., Vol 29, No 2,(March 1993).
3. E. Durand, "Magetostatique", Masson, (1968) (in french).
4. J.P.A. Bastos, N. Ida, R.R. Fereira, "Virtual elements: a smoothing technique to provide continuity in FEM discretized results", Proc. Compumag 93, Miami, (1993).

REDUCTION OF EDDY CURRENT LOSSES IN PIPE-TYPE CABLE SYSTEMS

R.D. Findlay and J.H. Dableh

Power Research Laboratory
McMaster University
Hamilton, Ontario, Canada, L8S 4K1

INTRODUCTION

The most dominant designs for underground systems of 132 kV and above are self-contained oil-filled cables [1] and pipe-type cables [2]. However, incremental losses occur in both the pipe itself, due to the induced eddy current, and in the cables in reaction to the effect of the pipe. The pipe is exposed mainly to a circumferential time-varying magnetic field created by the three phase current in the cables inside the pipe. These undesirable eddy currents generate additional heat, limiting the current carrying capacity of the cables. Moreover, due to the close proximity between the cables themselves on one hand and between them and the pipe on the other, the ac resistance of each cable is increased and higher ohmic losses result.

In a previous paper [3] we described an electromagnetic field shaper to reduce eddy current losses on the walls of pipe-type cables. This paper presents the results of the first set of tests performed on a pipe-type cable arrangement with a single phase conductor in conjunction with an aluminum shield. Two shields were tested, one designed to fit on the inner surface of the pipe, while the second shield was constructed in the same way and out of the same material as the first, but with a smaller diameter leaving a gap between the outer surface of the shield and the inner surface of the pipe.

DESCRIPTION OF THE SHIELDING ARRANGEMENT

We propose to reduce eddy current and hysteresis losses by interposing an electromagnetic shield between the cables and the pipe. Two aluminum shields were built using aluminum strips of width 1.25 cm., noting that the penetration depth in aluminum at 60 hertz is approximately 1.1 cm. The thickness of the strip chosen for this first generation of shield prototypes was selected to be 0.6 mm. The aluminum band was wound spirally on a 0.9 meter length inert former for support and shape. The spiral was in the same direction as the stranded conductor used inside the pipe. Two diameters of shield were tried, 19.7 cm and 10.8 cm, respectively.

As the aluminum strip was wound on the form, an insulating strip was used to separate each individual spiral, as shown in Fig. 1. This provided a means to ensure that

Electric and Magnetic Fields, Edited by A. Nicolet
and R. Belmans, Plenum Press, New York, 1995

the eddy current path in the axial direction was indeed limited to the width of one strip of the shield. The first shield, with larger diameter, was wrapped in paper to provide an insulating layer between the inner surface of the pipe wall and outer surface of the shield, since this shield is designed to fit tightly inside the pipe.

DESCRIPTION OF THE EXPERIMENTAL SET UP

The electrical supply was drawn from a 230-V, single-phase, 60-Hertz supply via a variable autotransformer and an isolation transformer at 1 to 46 turns ratio. The secondary of the transformer was rated to supply up to 1200 A at 5 V. The simulated pipe-type cable system consisted of a 2.4 meter long conductor placed along the centre of a 2.4 meter long section of a carbon-steel pipe. The cable used is an aluminum stranded conductor, diameter of 4.8 cm., suspended by structural braces which were isolated electrically from the electrical cable by an insulating washer. The carbon-steel pipe had an inside diameter of 20.2 cm and a thickness of about 1.25 cm.

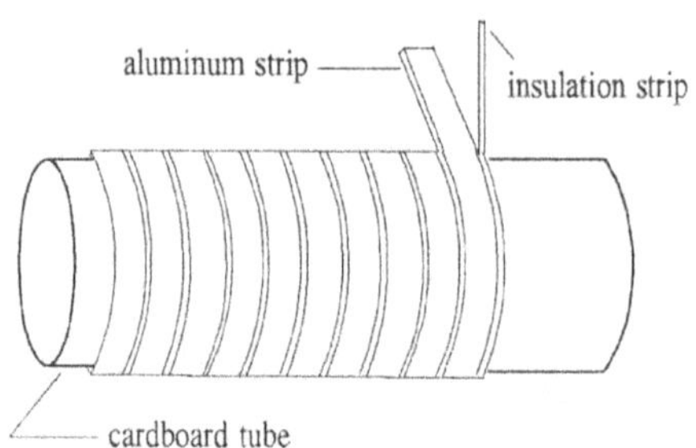

Figure 1. Construction of Aluminum shields.

During each experiment at every current setting, the heat generated in the pipe was monitored, at specific time intervals, by measuring the temperature using thermocouples [4] placed around the pipe, as illustrated in Fig. 2. Readings from each of the numbered thermocouples were recorded until the temperature reached a steady state level. The conductor was tensioned to 1100 kg.

Experiments

A series of experiments were performed at various current levels, in three different simulated system configurations, to verify the validity of the hypothesis and quantify the amount of possible loss reduction for each shield. The simulated configurations were as follows:

a) pipe without any shield
b) pipe with the large diameter shield
c) pipe with the small diameter shield

In each of the above configurations, the conductor was positioned along the axis of the pipe to ensure that all the heating effects were uniformly distributed around the wall of the pipe.

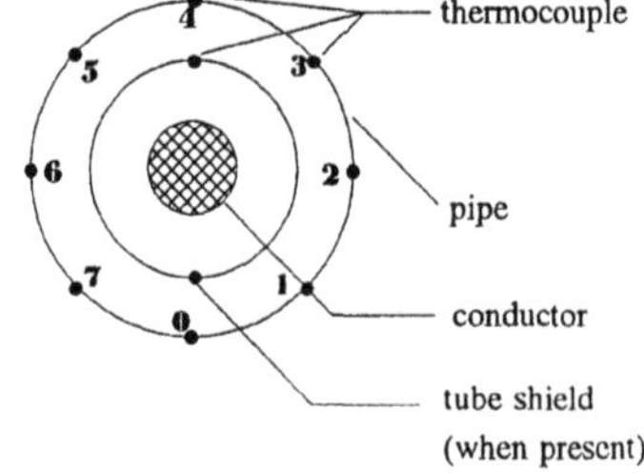

Figure 2. Locations of thermocouples

DISCUSSION OF THE EXPERIMENTAL RESULTS

For each of the configurations listed, readings from the eight pipe thermocouples and two thermocouples placed on the top and bottom of the aluminum shield were recorded every eight minutes, over a period of at least two hours. These measurements were repeated for four current levels at 200, 400, 600 and 700 amperes.

Examination of the data collected for the pipe without any field shaper, indicated that there was a temperature gradient between the top and bottom of the pipe, with the maximum temperature on the top of the pipe. Fig. 3 presents the temperature measurements at that location as a function of time, for the four current levels. Examination of the data collected for the two other configurations with an aluminum shield indicates similar temperature rise profiles at lower values. At each current level, the curve follows a typical heating pattern, except that there is some evidence of the effects of the nonlinear pipe material.

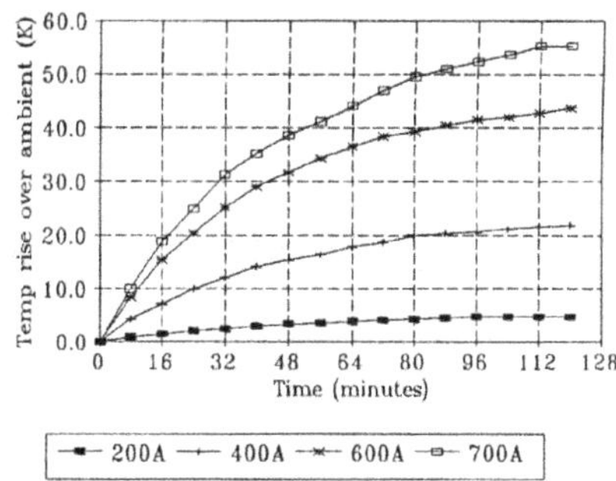

Figure 3. Pipe Temperature versus time for various currents, unshielded, top of pipe.

Pipe Temperature With and Without the Shield

A comparison of the temperature profile around the wall of the pipe, for the three configurations listed above, at the various current levels was performed. Fig. 4 presents the temperature at the top of the pipe for each of the three configurations for a current level of 600 amperes.

This figure clearly shows that the temperature of the pipe is lowest when the aluminum shield with a small diameter is used. In the steady states, the use of this shield reduced the temperature at the top of the pipe by 6.6 degrees. This is equivalent to 15 percent of the total temperature rise for the case of unshielded pipe.

As for the case of the larger shield, where the outer diameter of the shield was separated from the inner surface of the pipe by only a thin layer of paper, there was a reduction of the temperature rise of about 5.5 percent. These results are well in line with theoretical expectations. When the aluminum shield is placed closer to the cable, the induced current in it is higher, and consequently the net field as seen by the pipe outside the shield is lower. This will result in lower magnetization of the pipe and lower induced eddy current in it.

The influence of the cable current on the amount of heat generated in the system, as well as the effects of each shield on the temperature at the top of the pipe, are summarized in Fig. 5. The temperature readings plotted for each configuration at each current level were taken 32 minutes after the system was energized. This time was sufficient for the steady state temperature to

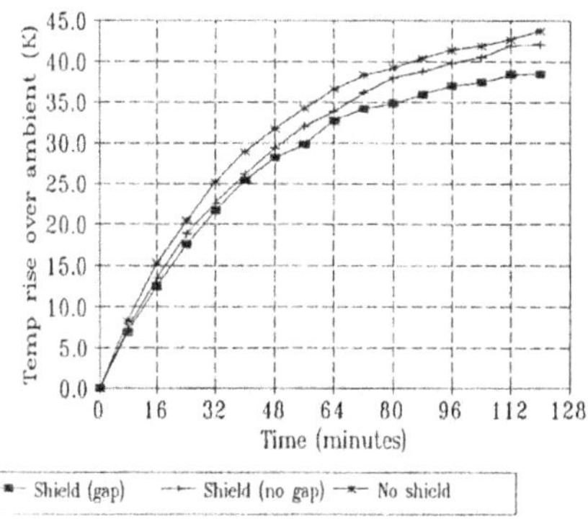

Figure 4. Temperatures at the top of the pipe with and without the aluminum shields, at 600 A.

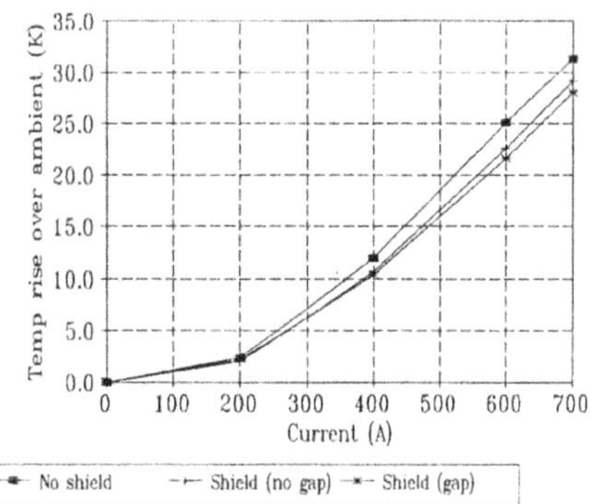

Figure 5. Pipe temperatures at 32 minutes.

expected, the highest temperatures occurred when there was no shield, with the lowest occurring with the smaller shield.

Power Requirements

The heat power flux on the inner surface of the pipe and on the surface of the shield can be estimated from the formula:

$$q = \rho_m c_w (2\pi r \tau) \frac{d\theta}{dt}\Big|_{t=0}$$

where q = heat flux in W/m, r = inner pipe radius in m, θ = temperature difference from ambient in kelvins, τ = thickness of material in m, c_w = specific heat in J/kg·K, and ρ_m = mass density in kg/m^3.

The results are given in Table I. All results are given for 600 A at the top of the pipe. The heat power flux on the shields was also determined. It was in the order of 5.3 W/m length axially along the pipe, or about 1% of the heat flux of the pipe. Consequently, the aluminum shield did not contribute substantially to the power loss loading in the cable system.

Table I

Case	Heat flux, W/m
No shield	587
Shield, small diameter	500
Shield, large diameter, no gap	488

CONCLUSIONS

Due to the difficulty of analysis, no attempt has yet been made to determine quantitatively the losses in the pipe and the shield. However, using the temperatures of the shield and the pipe, we can establish that the shield will result in lower losses. The following conclusions can be drawn from the data presented:

1. Heating due to eddy currents was decreased using a spiral aluminum shield.
2. In the practical case, it is easier to install a shield at the pipe inner surface. However, this would require more material and perform only marginally more efficiently than a shield with a smaller diameter.
3. The eddy current losses incurred by the shield are very substantially less than those saved by using the shield.
4. The shield is more effective at higher currents.
5. Savings in losses of the order of 17% were found at 600 A. Higher currents resulted in higher savings.

REFERENCES

[1] C.A. Arkell, B. Gregory, G.J. Smee, "Self-Contained Oil-Filled Cables for High Power Circuits", IEEE Trans., V PAS-97, March/April 1978, pp. 349-357.

[2] J.H. Dableh and R.D. Findlay, "An Annotated Summary of Analysis and Design Techniques for Pipe-Type Cable Systems", IEEE Trans. on Power Apparatus and Systems, Vol. PAS-103, No. 10, October 1984, pp. 2786-2793.

[3] J.H. Dableh, R.D. Findlay, R.T.H. Alden and B. Szabados, "A Novel Electromagnetic Field Shaper For Pipe-Type Cable Systems", IEEE Transactions on Power Delivery, Vol. 5, No. 1, January 1990, pp 1-8

[4] EPRI Report, "Determination of AC Conductor and Pipe Loss in Pipe-Type Cable Systems", EL-2256, Feb. 1982.

THE FINITE ELEMENT ANALYSIS OF THE MAGNETIC VIBRATIONS IN THE INDUCTION MOTOR

Pawel Witczak

Institute of Electric Machines and Transformers
Technical University of Lodz
18 Stefanowskiego St.
90-924 Lodz, Poland

INTRODUCTION

The magnetic vibrations of induction motors have been investigated for over 70 years. A great amount of theoretical work and manufacturing practice resulted in establishing rules of the slotting choice[1] to get a noiseless machine. These rules may be briefly presented as follows:

- the time and space distributions of electric currents and magnetic field in the air-gap should be as close to the mono-sinusoidal ones as possible;
- the frequency spectra of low-order stator natural vibrations and magnetic field excitations should be shifted in frequency domain one from another as far as possible.

The modern PWM controllers, which by their definition use deformed voltages and currents, have brought once more the question of magnetic vibration to the present-day investigations.

The paper deals with a theoretical mechanical model of the induction motor. Exploiting the finite element technique, there were attemps to add some new explanations of the vibration behaviour of the motor under magnetic radial forces loading. The example analysis of this problem, limited to 2D approach, was presented by Henneberger et al.[2]

The fundamental of each theoretical analysis is the decision which geometrical and material properties of the origin may be neglected or simplified in the model. The actual questions are:

What is the influence of stiffness and mass of stator windings?

What is the difference between elastically supported and rigidly mounted motor?

What is the relation between results of 2D and 3D analysis?

All calculations concern 7.5 kW, four-pole squirrel-cage motor with following main geometrical data: outer/inner diameter of the stator lamination - 104/67 mm, frame thickness - 6 mm, core length - 140 mm. This motor seems to be a typical one for the whole family of low-voltage induction motors. Therefore, the conclusions of the analysis may be widely extended, up to the motors of rated power of about 100 kW.

Electric and Magnetic Fields, Edited by A. Nicolet
and R. Belmans, Plenum Press, New York, 1995

TWO-DIMENSIONAL MODELS

The cross-section of the stator consisting of lamination sheet, windings and frame with feet has been modelled by means of 2D elements, which is shown in fig.1.

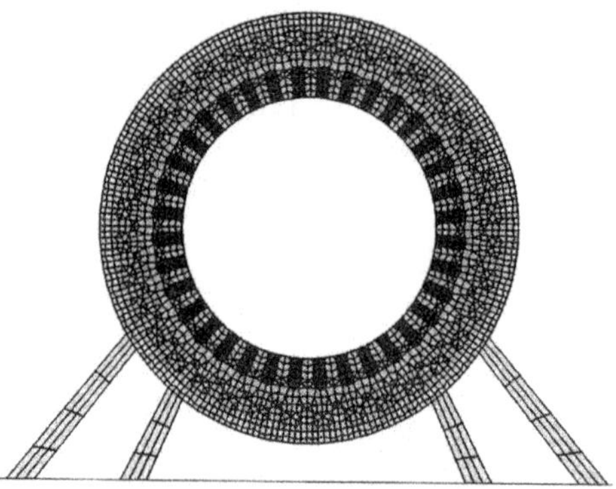

Figure 1. The two-dimensional model of the stator of the induction motor.

Inserting the different values of material properties into particular parts of this mesh, a group of diversified models has been obtained. The composite area of windings has the equivalent Young modulus E=0.095 MPa. The results of calculations of natural frequencies and vibration modes are presented in table 1.

Table 1. The list of results of natural frequency analysis for various types of two-dimensional models of the induction motor.

Order of vibration	Natural frequency, Hz			
	Model a[1]	Model b[2]	Model c[3]	Model d[4]
2	1475	1780	1688	1860,2080,3320
3	3892	4635	4379	5092, 5440
0	7233	7173	6830	> 9000
4	6866	8133	7668	8230, 8570

[1] lamination + frame, elastically supported;
[2] lamination + windings (slot part) + frame, elastically supported;
[3] lamination + windings (slot & end part) + frame, elastically supported;
[4] lamination + windings (slot & end part) + frame, rigidly supported;

While analysing models a and b one can see that stiffness of winding has the prevailing influence comparing to its mass - the values of natural frequencies are increasing in spite of the addition of significant amount of the material. When only mass density is increased (models b and c) we obtain the uniform decrease of frequency values, as it has been expected. The introduction of rigid mounting conditions brings about a substantial growth of each frequency and also the appearance of different modes of the same order.

THREE-DIMENSIONAL MODEL

The simplest way to get the 3D model is to expand the mesh shown in fig.1 in third direction. This approach has one but serious disadvantage - it causes a very significant increase of the number of degrees of freedom in the model. It has been estimated that the 3D model of the analysed motor, which would preserve the slotting geometry, must have over 30,000 degrees of freedom. The solution of such a mesh is possible but the amount of computing time seems to be too extensive at present. To avoid these difficulties an equivalent thick-ring model of the stator has been established. The equivalence means that two kinds of conditions were simultaneously fulfilled (with the accuracy about 2%):
- the static deflections resulted from radial loading of order $r < 6$ (sinusoidal in space) are the same;
- the natural frequencies of modes of order $r < 6$ are the same.

Assuming that the outer diameter of this ring is the same as in core laminations, it has been found that the data of equivalent thick-ring model are: $d_{eq}=28.5$ mm, $E_{eq}=1.49*10^5$ MPa. It should be noticed that yoke height and stiffness are: $d_y=21$ mm, $E_y=1.95*10^5$ MPa.

Taking into account the above mentioned simplification, the 3D finite elements model has been constructed and it is presented in fig.2.

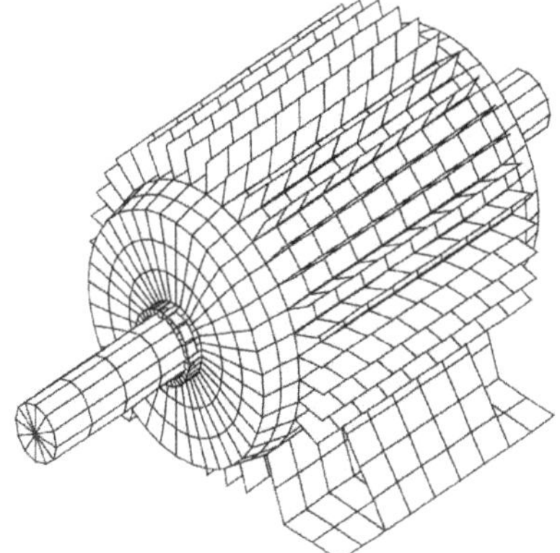

Figure 2. The three-dimensional model of the induction motor.

The radial magnetic forces acting on the stator teeth surface were calculated in usual manner from magnetic stress tensor T_{nn}

$$T_{nn}(\alpha,t) = \frac{1}{2\,\mu_o} B^2(\alpha,t) \quad , \tag{1}$$

where the flux density distribution of magnitude $B_m=1$ T has the form

$$B(\alpha,t) = B_m \sin(\omega t - p\alpha) \tag{2}$$

These forces with pole pair number p altered from 1 to 3, were applied both to two- (model d) and three-dimensional models under static conditions. The obtained values of residual deflections are compared in table 2. The surface distributions of von Mises reduced stresses in 3D model are presented in fig.3. To improve the readability of these distributions, the cooling ribs have not been shown in the picture and the deflection drawing scale has been enlarged.

Table 2. The comparison of static deflections in two- and three-dimensional models of the induction motor with different order of the magnetic force excitation.

Type of model	Static deflections, μm		
	p=1	p=2	p=3
2D	1.57	0.56	0.42
3D	2.20	0.62	0.44

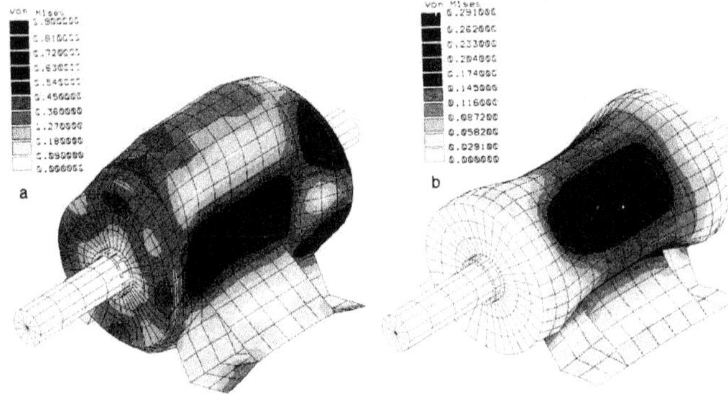

Figure 3. The reduced stresses (values in MPa) and deflection surface distributions resulted from magnetic forces. a. for 2-pole flux-density wave; b. for 4-pole flux-density wave.

The good agreement of results in both models for p>1 is illustrated by undeflected and nearly free of stresses end-shields (fig.3b).

CONCLUSIONS

The finite element analysis of mechanical behaviour of induction motor from the point of view of magnetic excitations, which was briefly presented in this paper, leads to following general remarks.

1. The three-dimensional approach is only necessary when second-order force waves are expected. The modern motors are designed to avoid these effects, therefore the 2D model is sufficient in most cases. However, some vibration modes, which may be important when the rotor eccentricity appears, can be correctly calculated in 3D model only. The limited volume of this paper does not allow to give here a more detailed description.

2. The stiffness of winding material and also the rigid mounting conditions have a significant influence on deflection level and natural frequency spectrum. Therefore, they cannot be omitted or considerably simplified.

REFERENCES

1. H.Jordan "Geräuscharme Elektromotoren", Verlag W.Giradet, Essen 1950
2. G.Henneberger, Ph.K.Sattler, W.Hadrys, D.Shen, Procedure for the numerical computation of mechanical vibrations in electrical machines, IEEE Trans.,Vol.MAG-28,No.2,March 1992,pp.1351-1354

ANALYSIS OF INDUCTION MOTORS BY COUPLING OF TRANSIENT ELECTROMAGNETIC FIELD EQUATIONS, CIRCUIT EQUATIONS AND MOTION EQUATIONS USING FINITE ELEMENTS METHOD

V.N.Savov, E.S.Bogdanov, Zh.D.Georgiev

Department of Theoretical Electrical Engineering,
Technical University of Sofia, 1756 Sofia, Bulgaria

INTRODUCTION

A mathematical model for transient and steady-state electromechanical analysis of cage indication motors is proposed. The model proposed has been applied for investigation of a given cage indication motor with a 3-phase stator winding, which is connected in a "star" and the star points are not connected. Only the stator line voltage is a known input quantity.

DESCRIPTION OF THE MATHEMATICAL MODEL

It is assumed that: a) the electromagnetic field is quasistationary in the motor and is two-dimensional in the active zone of the motor; b) there are no eddy currents in the stator phase windings; c) the vector current density over the cross section of each coil (section) of the stator winding is constant.

Let the axis of the investigated motor coincide with the z axis of the fixed coordinate system ($Oxyz$, **i, j, k**). The active zone of the motor has a length l. In this zone the magnetic vector potential $\mathbf{A}=A\mathbf{k}$ and the electric scalar potential Φ of the electromagnetic field satisfy the nonlinear differential equation

$$rot\nu\, rot\mathbf{A} + \sigma\frac{\partial \mathbf{A}}{\partial t} = \mathbf{J}^{\epsilon} - \sigma\nabla\Phi + \sigma\mathbf{v}\times rot\mathbf{A} \tag{1}$$

where $\mathbf{J}^{\epsilon} = J^{\epsilon}\mathbf{k}$ is the electric current density in the stator windings, ν and σ are the reluctivity and conductivity of the medium and $\mathbf{v}$ is the linear velocity of a given rotor point.

It is convenient to use for the rotor points a moving coordinate system which is fixed to the moving rotor whereupon the vector product $\mathbf{v}\times rot\mathbf{A}$ disappears in equation (1).

Because of the periodicity (or antiperiodicity) of the electromagnetic field in the space, eq.(1) is solved in only one sector (S) of the motor's cross section. Let the motor's cross section contain N_S such sectors and the number of the rotor bars which intersect the sector (S) be b.

For further investigation it is convenient to introduce the functions

Electric and Magnetic Fields, Edited by A. Nicolet
and R. Belmans, Plenum Press, New York, 1995

$$\beta_p^s(x,y) = d_p^s \frac{N_{pc}}{S_{pc}} \quad (p=1,2,3), \qquad \beta_q^r(x,y) = d_q^r \quad (q=1,2,\dots,b) \quad , \tag{2}$$

where N_{pc} is the number of turns in a coil c (section c) of the stator phase winding p, and S_{pc} is the area of each one cross section (coil side) of the same coil c. The coefficient d_p^s takes the values d_p^s =1, resp, d_p^s= -1, when the point (x,y) belongs to the positively, resp. the negatively oriented cross section[1] of the coil c of the stator phase winding p, and d_p^s=0 otherwise. The index c is identified successively with the numbers of these coils of the stator phase winding p, which have a cross section with the solution sector (S). The coefficient d_q^r takes values d_q^r=1, when the point (x,y) belongs to the cross section of the rotor bar q and d_q^r=0 otherwise.

For the space discretization of eq.(1) finite elements mesh on the region (S) is generated, which has m elements (S_e) $(e=1,2,\dots,m)$ and n nodes. By applying the method of Galerkin to eq.(1) the following system of field equations is obtained

$$[\mathbf{S}]\{\mathbf{A}\}+[\mathbf{T}]\frac{\partial}{\partial t}\{\mathbf{A}\}=[\mathbf{D^s}]^T[\mathbf{K}]^T\{\mathbf{i^{so}}\}+[\mathbf{D^r}]^T\{\mathbf{u^r}\} \quad , \tag{3}$$

where $\{\mathbf{A}\}=[A_1 \; A_2 \; \dots \; A_n]^T$ is a column matrix of the nodal values of A, $\{\mathbf{i^{so}}\}=[i_1^s i_2^s]^T$ is a column matrix of the independent currents of the stator phase windings, $\{\mathbf{u^r}\}=[u_1^r u_2^r \dots u_b^r]^T$ is a column matrix of the voltages of the rotor bars, and $[\mathbf{K}]_{(2\times3)}$ is a matrix having elements $K_{11}=K_{22}=1$, $K_{21}=K_{12}=0$, $K_{13}=K_{23}=-1$. The local (elemental) coefficients of the matrices $[\mathbf{S}]_{(n\times n)}$, $[\mathbf{T}]_{(n\times n)}$, $[\mathbf{D^s}]_{(3\times n)}$ and $[\mathbf{D^r}]_{(b\times n)}$ are obtained in the following way

$$S_{ij}^e = \iint_{(S_e)} \nu\nabla\alpha_i^e.\nabla\alpha_j^e ds \;,\quad T_{ij}^e = \iint_{(S_e)} \sigma\alpha_i^e\alpha_j^e ds \;,\quad D_{pi}^{se} = \iint_{(S_e)} \beta_p^{se}\alpha_i^e ds \;,\quad D_{qi}^{re} = \frac{\sigma}{l}\iint_{(S_e)} \beta_q^{re}\alpha_i^e ds \;,$$

where $\alpha_i^e, \alpha_j^e, \dots$ are the shape functions of the element (S_e).

Let R^s be the resistance of each stator phase winding and L_e^s be the end-region inductance of each stator phase winding. In this case it can be proved that the following system of field-circuit equations for the stator is valid

$$[\mathbf{C}]\{\mathbf{V^s}\}=R^s[\mathbf{K}][\mathbf{K}]^T\{\mathbf{i^{so}}\}+L_e^s[\mathbf{K}][\mathbf{K}]^T\frac{d}{dt}\{\mathbf{i^{so}}\}+lN_s[\mathbf{K}][\mathbf{D^s}]\frac{\partial}{\partial t}\{\mathbf{A}\} \;. \tag{4}$$

where $\{\mathbf{V^s}\}=[V_1^s V_2^s V_3^s]^T$ is a column matrix of the stator line voltages. The matrix $[\mathbf{C}]_{(2\times3)}$ has elements $C_{11}=C_{12}=C_{22}=0$ and $C_{13}=C_{21}=C_{23}=-1$.

Let R^r be the resistance of each rotor bar, R_e^r and L_e^r be the resistance and the inductance of each end ring segment (inter-bar segment) of the rotor cage. In this case it can be proved that the following two systems of field-circuit, resp. circuit equations for the rotor are valid

$$[\mathbf{D^r}]\frac{\partial}{\partial t}\{\mathbf{A}\}-\frac{1}{l}[\mathbf{M}]^T\{\mathbf{i^c}\}-\frac{1}{lR^r}[\mathbf{1}]\{\mathbf{u^r}\}=0 \;, \tag{5}$$

$$[\mathbf{M}]\{\mathbf{u^r}\}=2R_e^r[\mathbf{1}]\{\mathbf{i^c}\}+2L_e^r[\mathbf{1}]\frac{d}{dt}\{\mathbf{i^c}\} \;. \tag{6}$$

where $\{\mathbf{i^c}\}=[i_1^c i_2^c \dots i_b^c]^T$ is a column matrix of the currents of the end ring segments. The matrix $[\mathbf{M}]_{(b\times b)}$ has elements $M_{ii}=1$ $(i=1\text{-}b)$, $M_{i,i-1}=-1$ $(i=2\text{-}b)$, $M_{1b}=-1$, $M_{ij}=0$ in the remaining cases and $[\mathbf{1}]_{(b\times b)}$ is the unit matrix.

Equations (3), (4), (5) and (6) are discretized in the time domain by means of backward difference scheme and the resulting equations are assembled into a global coupled system of equations. By solving this system the unknown quantities $\{A\}_{t+\Delta t}$, $\{i^{so}\}_{t+\Delta t}$, $\{u^r\}_{t+\Delta t}$, $\{i^e\}_{t+\Delta t}$ (at the instant $t+\Delta t$) are determined provided the quantities $\{A\}_t$, $\{i^{so}\}_t$, $\{u^r\}_t$, $\{i^e\}_t$ (at the instant t) are known.

The coupled system is nonlinear and the Newton-Raphson method is used for its solution whereupon the set of equations giving the correction vector at the iteration step $(N+1)$ is

$$\begin{bmatrix} [P]^{N+1}_{t+\Delta t} & -[D^s]^T[K]^T & -[D^r]^T & [0] \\ -[K][D^s] & -[G^s] & [0] & [0] \\ -[D^r] & [0] & \frac{\Delta t}{lR^r}[1] & \frac{\Delta t}{l}[M]^T \\ [0] & [0] & \frac{\Delta t}{l}[M] & -[E^r] \end{bmatrix} \begin{bmatrix} \{\Delta A\}^{N+1}_{t+\Delta t} \\ \{\Delta i^{so}\}^{N+1}_{t+\Delta t} \\ \{\Delta u^r\}^{N+1}_{t+\Delta t} \\ \{\Delta i^e\}^{N+1}_{t+\Delta t} \end{bmatrix} = - \begin{bmatrix} \{f^f\}^{N+1}_{t+\Delta t} \\ \{f^s\}^{N+1}_{t+\Delta t} \\ \{f^r\}^{N+1}_{t+\Delta t} \\ \{f^c\}^{N+1}_{t+\Delta t} \end{bmatrix} , \tag{7}$$

where the upper index $(N+1)$ (resp. N) denotes the iteration step number and

$$\{f^f\}^{N+1}_{t+\Delta t} = ([S]^N_{t+\Delta t} + \frac{1}{\Delta t}[T])\{A\}^N_{t+\Delta t} - [D^s]^T[K]^T\{i^{so}\}^N_{t+\Delta t} - [D^r]^T\{u^r\}^N_{t+\Delta t} - \frac{1}{\Delta t}[T]\{A\}_t ,$$

$$\{f^s\}^{N+1}_{t+\Delta t} = -[K][D^s]\{A\}^N_{t+\Delta t} - [G^s]\{i^{so}\}^N_{t+\Delta t} + [K][D^s]\{A\}_t + [H^s]\{i^{so}\}_t + [C^s]\{V^s\}_{t+\Delta t} ,$$

$$\{f^r\}^{N+1}_{t+\Delta t} = -[D^r]\{A\}^N_{t+\Delta t} + \frac{\Delta t}{lR^r}[1]\{u^r\}^N_{t+\Delta t} + \frac{\Delta t}{l}[M]^T\{i^e\}^N_{t+\Delta t} + [D^r]\{A\}_t ,$$

$$\{f^c\}^{N+1}_{t+\Delta t} = \frac{\Delta t}{l}[M]\{u^r\}^N_{t+\Delta t} - [E^r]\{i^e\}^N_{t+\Delta t} + \frac{2L^r_e}{l}[1]\{i^e\}_t ,$$

$$[G^s] = \frac{R^s\Delta t + L^s_e}{lN_s}[K][K]^T,\quad [H^s] = \frac{L^s_e}{lN_s}[K][K]^T,\quad [C^s] = \frac{\Delta t}{lN_s}[C],\quad [E^r] = \frac{2(R^r_e\Delta t + L^r_e)}{l}[1]$$

The matrix $[P]^{N+1}_{t+\Delta t}$ results from the local (elemental) matrix $[P^e]^{N+1}_{t+\Delta t}$ with entries

$$P^{e,N+1}_{ij,t+\Delta t} = S^{e,N}_{ij,t+\Delta t} + \frac{1}{\Delta t}T^e_{ij} + 2\sum_p\sum_q[\iint_{(S_e)} \frac{\partial v^{e,N}_{t+\Delta t}}{\partial B^2}(\nabla\alpha^e_i.\nabla\alpha^e_p)(\nabla\alpha^e_j.\nabla\alpha^e_q)ds]A^N_{p,t+\Delta t}A^N_{q,t+\Delta t} ,$$

where $\sum_p$ and $\sum_q$ indicate summation over all of the nodes of the element (S_e).

The mechanical motion equations are discretized by means of forward difference scheme and with the aid of the resulting equations the rotor position at each time step is calculated.

RESULTS OBTAINED

A given cage induction motor has been analyzed . The basic data of the motor are: Power-45 kW, Stator line voltage-380 V, Frequency-50 Hz, Number of pole pairs-2, Rotor diameter-213.6 mm, Air gap length-0.7 mm, Stator winding-3-phase, connected in a "star" with isolated star point, Number of parallel paths-4. The external torque and the coefficient of viscous friction are neglected. The solution region (S) has been subdivided into 2933 triangular elements with 1651 nodes. A variable time step Δt=(0.25-0.5).10^{-4} s has been used. Fig.1 shows the dependence of the stator currents i^s_1, i^s_2, electromagnetic torque T_{em} and angular velocity of the rotor Ω on time t. Fig.2 shows the magnetic vector potential isolines at two different instants of time t_1=0.015 s and t_2=0.2 s with corresponding rotation angles θ_1=0.156 rad and θ_2=24.654 rad.

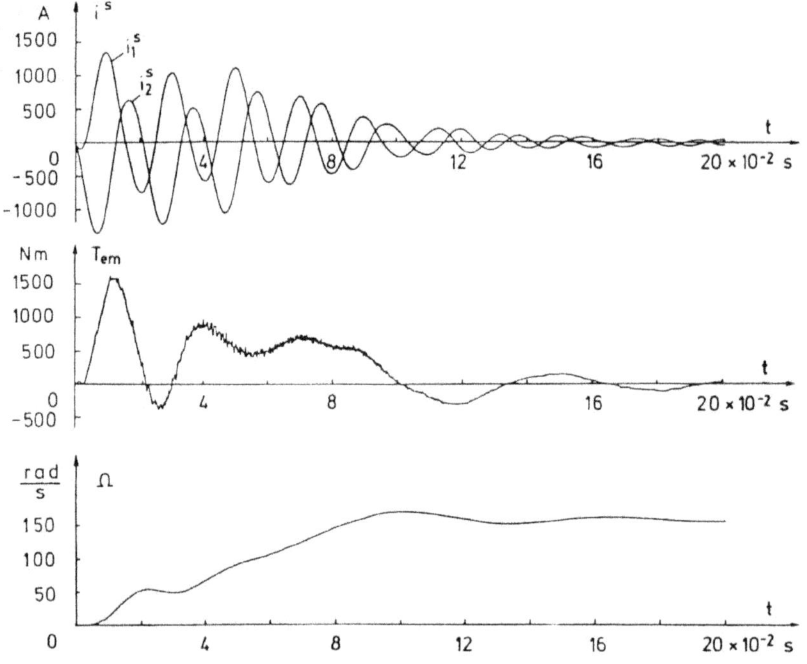

Figure 1. Graphs of the functions $i_1^s(t)$, $i_2^s(t)$, $T_{em}(t)$, $\Omega(t)$.

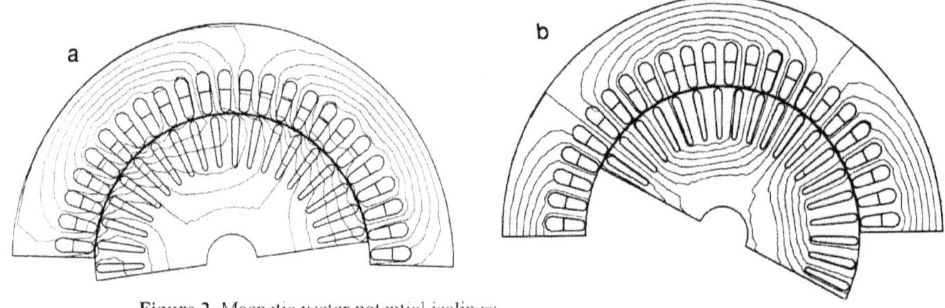

Figure 2. Magnetic vector potential isolines:
a) at the instant t_1=0.015 s (θ_1=0.156 rad); b) at the instant t_2=0.2 s (θ_2=24.654 rad)

REFERENCES

1. Strangas E.G., Coupling the circuit equations to the non-linear time dependent field solution in inverter driven induction motors, *IEEE Trans. Magn.* MAG-21:2408 (1985)
2. Meunier G., D.Shen, J.-L. Coulomb, Modelisation of 2D and axisymmetric magnetodynamic domain by the finite elements method, *IEEE Trans. Magn.* MAG-24:166 (1988)

ON THE USE OF THE GENERAL PURPOSE CODE ANSYS TO SOLVE ELECTRIC AND MAGNETIC COUPLED PROBLEMS UNDER NON-LINEAR CONDITIONS

A. Geri, M. La Rosa and G. M. Veca, IEEE Senior Member

Dipartimento di Ingegneria Elettrica
Università degli Studi di Roma "La Sapienza"
Via Eudossiana n. 18, 00184 Roma, Italy

INTRODUCTION

This work describes a field-circuit method to solve non-linear massive conductor systems with a coupled method. The approach developed by authors has been studied in such a manner to permit industrial designers to front most common problems, involving saturable massive conductor lines, without using specialised numeric codes. In order to verify the capability of the power grid to supply non-linear devices (e.g., a spot-welding machine), having characteristics rapidly and greatly variable in time, we have developed a numerical method to evaluate the effects due both to transverse electric and magnetic fields in the network multiphase lines. We have approached this problem defining an equivalent circuit for saturable lines. All parameters have been fixed by means of a 2D finite element analysis (FEA). The coupled problem is solved by means of an iterative procedure, where general purpose FEA code and network solvers can be used. In this very simple form the model is able to simulate short lines, where propagation effects can be neglected without problems. For more complex situations, the line can be divided in short tracts and everyone of these can be represented using this model.

THE MODEL

Evaluation of the circuit parameters

The approach developed by the authors is based on the idea to represent each line as a series branch including a time varying ideal voltage generator linked to the Ohm's resistance of the conductor. The voltage generator, E, represents all induced effects (i.e., skin and proximity ones) related to the transverse total magnetic field, while the ohmic resistance, R, takes in to account Joule losses. Then effects of electric transverse line field are approximately taken into account by means of lumped transverse elements, conductance G and capacitance C, at the two line ends, obtaining an equivalent circuit similar to the traditional Π one. Using a 2D approach, the classic

Electric and Magnetic Fields, Edited by A. Nicolet
and R. Belmans, Plenum Press, New York, 1995

formulation[1,2,3,4,5,6], concerning the solution of electric and magnetic coupled problems, is founded on the following fundamental equation

$$-(\nabla \cdot \upsilon \nabla)\vec{A} + \sigma \nabla \cdot V + \sigma \frac{\partial \vec{A}}{\partial t} = 0 \tag{1}$$

that for the j-th conductor of the system, having cross section S_j, length l_j and conductivity σ_j, becomes

$$\Delta Vj = \frac{I_j \cdot l_j}{\sigma_j S_j} + \frac{l_j}{S_j} \cdot \int_{S_j} \frac{\partial \vec{A}}{\partial t} \cdot d\vec{s} \tag{2}$$

and then

$$\Delta V_j = R_j I_j + E_j \tag{3}$$

The cross parameters can be determined using the static field analysis techniques. In particular, we only need to determine one group of parameters (e.g., the capacitances) because the supposed TEM mode of propagation allows to deduce the other values (e.g., the conductances) by $\sigma[C]=\varepsilon[G]$.

Field analysis

Electric field analysis. The axial symmetry of the line section permits to study only half system, with suitable boundary conditions. The enclosure has been supposed equipotential (this assumption is acceptable for the most of the industrial plants). Thus the capacitive parameters can be easily evaluated by means of the ANSYS code.

Magnetic field analysis. The line magnetic field computation is needed in order to fix E_j in (3) at every time instant. The line electromagnetic field can be determined by solving an open boundary 2D symmetrical problem (see Fig. 1a). In order to solve (1) the 2D domain, corresponding to busbar cross section and to its surrounding region, is divided in two subdomains (see Fig. 1a): a central one, Ω_f, that is shaped by means of ordinary elements (FE's), and a peripheral one, Ω_b, that is framed by infinite boundary elements[7] (IBE's).

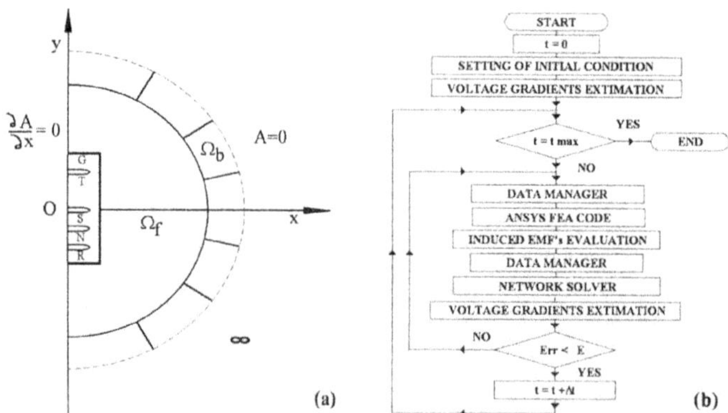

Figure 1. (a) Global arrangement of the busbar system and partitioning of the mesh area used for FEA. (b) Flow chart of the proposed method.

The inner region magnetic field is fixed by means of Galerkin's method, while the field in the peripheral zone is estimated using an integral approach. This system can be studied by means of the ANSYS code, using voltage gradients defined by external circuit and taking into account skin and proximity effects in massive conductors as well as the non-linear magnetic characteristics of the busbar steel enclosure. Numerically, the non-linear transient field problem is solved by means of the Newmark Method that is applied using the Newton-Raphson Iterative Procedure[7].

Coupled analysis

We have described a line circuit transient model where all the parameters can be computed before becoming plant analysis, except for the E_j bipoles. These values can be obtained by means of an iterative procedure (see Fig. 1b). The transient non-linear behaviour of busbar system starts estimating, by means of a network analysis, the voltage gradients in each conductor for null induced effects. These values are the initial conditions for FEA. Imposing the estimated set of voltage gradient values, we solve (1) by means of the ANSYS code. Then using a suitable post-processor we fix the E_j values in (3). Thus, using the circuit analysis, it's possible to evaluate the voltage gradients for successive attempts. The FEA is restarted for each voltage gradient updating, until a fixed accuracy is reached. Next a subsequent time instant can be processed. The circuit analysis can be performed by means of a generic network solver, such as SPICE or EMTP. In order to simulate the electrical network, for many simple plant configurations, it is enough to implement, by means of ANSYS parametric design language (APDL), traditional algorithms[8].

APPLICATIONS

The authors have applied the proposed method to analyse two typical situations. Table 1 summarised the geometrical values of each configurations considered, the electrical property of all materials and the non-linear magnetization characteristic of the enclosure. The first test refers to one phase spot-welding machine[9] controlled by an electronic SCR device during early time instants after welding firing. An ideal voltage sources, Vp, simulates the device which is studied performing a linear armonic analysis. Initial conditions are evaluated considering a rest. The system analysis is started at a zero passage of welding machine supply voltage. The computed load current is shown in Fig. 2a. This curve is in a good agreement with industrial reserved experimental data referring to similar configurations. Then, we have analysed, a solid compact insulating busbar system, during one phase-earth short circuit. The computation has been performed considering the short when the voltage is trespassing the zero value ($\phi=0$), and when the voltage is near to its maximum ($\phi=\pi/2$). The remaining bar are assumed to be unloaded, according to the usual hypotheses. The transient has been studied in the time domain to take into account the saturation of steel enclosure (in this case a strong induced field spread out of the line).

Table 1. Geometrical and electrical data for all simulations executed.

Test I	Sections:	Phase=300 mm^2			Neutral=220 mm^2			Enclosure=650 mm^2	
Test II	Sections:	Phase=600 mm^2			Enclosure=272 mm^2				
	Insulation:	Resistivity=10^{11} Ωm			Relative dielectric constant=5				
Resistivity		Steel=0.172 $\mu\Omega$m			Copper=0.018 $\mu\Omega$m				
H [As/m]	300	600	780	900	1200	1560	2280	2580	3000
B [T]	0.118	0.313	0.420	0.482	0.604	0.719	0.850	0.882	0.921

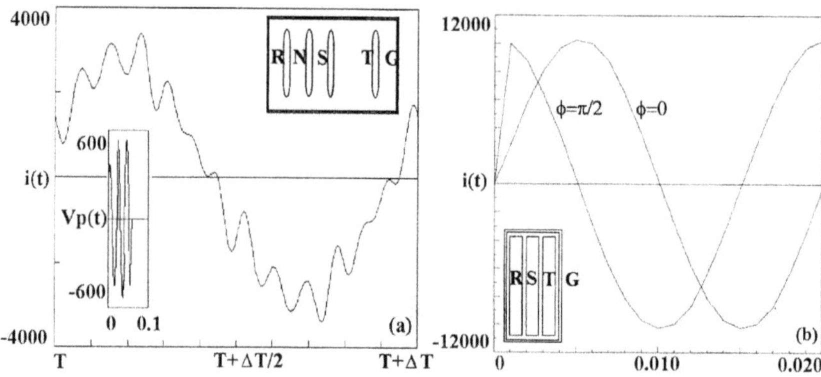

Figure 2. (a) Computed load current of spot-welding machine during early time instants after welding firing in an effective industrial plant[9] [ΔT=0.020 s]. (b) Computed short circuit currents in a busbar compact system geometry for ϕ=0 and $\phi=\pi/2$. [i(t) is expressed in Ampere, t is expressed in seconds].

FEA analysis is performed by means of ANSYS, while the equivalent network is solved using SPICE. The computations results are plotted in Fig. 2b. Although the computed values of derived currents are more relevant in the second test than that in the first one, they are always negligible in comparison with the line current. For this reason, analysing this configuration, the transverse parameter may be removed. The ANSYS code can also be advantageously applied to evaluated the electromagnetic interference around the busbar system.

CONCLUSIONS

This work has showed that ANSYS industrial FEA code and general purpose network solver can be integrated to solve field-circuit problems in the time domain and in the frequency domain. The proposed approach has permitted the development of a simple and affordable tool, even for industrial designer. In addition, the use of a general purpose FEA code permits an easy simulation of various configurations and a interesting evaluation of the influence of secondary effects. This last characteristic is very useful for great amplitude transients, where the computation of electrodynamics stresses or inducted effects can be needed.

1. A. Konrad, Integrodifferential finite element formulation of two-dimensional steady state skin effect problems, *IEEE Trans. Mag.*, Vol. MAG-18, NO. 1, 284 (1982).
2. J.R. Brauer, Finite element calculation of eddy currents and skin effects, *IEEE Trans. Mag.*, Vol. MAG-18, NO. 2, 504 (1982).
3. T.H. Fawzi, P.E. Burke and T.C.H. Lau, BIE analysis of eddy current losses in rectangular busbars in nonuniform fields, *IEEE Trans. Mag.*, Vol. MAG-18, NO. 6, 1061 (1982).
4. P. Belforte, M. Chiampi and M. Tartaglia, Fem analysis and modelling of busbar systems under ac conditions, *IEEE Trans. Mag.*, Vol. MAG-21, NO. 6, 2284 (1985).
5. I.A. Tsukerman, A. Konrad, G. Meunier and J.C. Sabonnadière, Coupled field-circuit problems: trends and accomplishments, *IEEE Trans. Mag.*, Vol. 29, NO. 2, 1701 (1993).
6. P. Lombard and G. Meunier, A general purpose method for electric and magnetic combined problems for 2D, axisymmetric and transient systems, *IEEE Trans. Mag.*, Vol. 29, NO. 2, 1737 (1993).
7. ANSYS, User's manual, Revision 4.4a, Swanson Analysis System, Inc.
8. Leon. O. Chua and Pen_men Lin. "Computer-aided analysis of electronic circuits," PRENTICE-HALL, INC., Englewood Cliffs, New Jersey (1975).
9. A. Geri, M. La Rosa and G.M. Veca, Modelling and analysis of electric and magnetic coupled problems under non-linear conditions, in press on *J. App. Phys.*, April, (1994).

FEM THERMAL MODELING OF AN INDUCTION MOTOR

Matjaž Plejić, Viktor Goričan, Božidar Hribernik

University of Maribor, Faculty of Technical Sciences
Smetanova ul. 17, 62000 Maribor, Slovenia

INTRODUCTION

The circuit approach is still prevailing in thermal calculations of electrical machines. However, the finite element method (FEM) is gaining ground due to its advantages. The modeling by FEM is fast and flexible, the parameters once obtained are generally valid, the results cover the entire problem area and they can be clearly presented. The majority of commercial CAD-FEM programs includes thermal solvers which makes possible coupled electromagnetic-thermal analysis on the same model.

The exact description of thermal processes in electrical machines involves three dimensional modeling and fluid dynamics. The modeling is time-consuming and computer times necessary to solve such problems usually still exceed acceptable limits. There is an obvious need for simplified models that will allow predicting the thermal behavior of electrical machines already in the design phase and will yet produce results good enough for practical purposes.

The aim of our work was to study thermal processes in an fan-cooled induction motor by means of its radial and axial cross section and to obtain in this way the reliable information on the temperature field in the motor without 3D modeling.

THERMAL MODELS AND PARAMETER IDENTIFICATION

The intensity of the heat flow from the outer surfaces of the motor to the ambient air is determined by the local Nusselt's numbers from which the convection coefficients are derived. The procedure adopted in practical calculations is to introduce the average coefficient, usually given as a function of the air speed in the ribs. In this way the speed reduction along the ribs due to the radial dispersion is neglected, which is acceptable only for low speeds, as found by Di Gerlando and Vistoli (1993). In our study the thermal calculation was carried out for different speeds of the motor and Table 1. shows the identified average convection coefficients for the radial and axial model. Since we measured the local air speeds in the ribs with a hot wire anemometer, in the next step the convection coefficient in the axial model was defined as a function of the axial position. This increased accuracy, but speed profiles are hard to predict, as they depend on the rib geometry, axial length and initial speed. The air speed varies also among different ribs (up to 40 % !), which can all cause considerable errors in the calculations.

Table 1. The identified convection coefficients [W/m^2 K]for the radial and axial model together with the measured and calculated temperatures in the windings

speed [n/min]	631	1389	1681	2270	2845
α_{radial}	72	110	115	150	185
α_{axial}	60	95	100	135	165
$T_{measured}$	116.2	108.5	92.1	80.3	74.6
T_{radial}	119	111	93	83	77
T_{axial}	117	111	91	80	76

The identified average convection coefficients were used to predict the temperature behavior of several motors of similar geometry and size and they produced satisfying results. However, they depend on the geometry and for different machines they have to be adapted.

The use of the radial cross section shown in Figure 4 is based on the presumption of the infinite length of the motor and it is therefore suitable for motors with long stacks. The actual convection coefficient needs to be corrected to account for the limited length. The shorter the stack, the higher the correction coefficient. The ratio between the length of the stack (L) and stator outer diameter (D) proved to be a rather gcod criterion for this purpose. The graph in Figure 2 was constructed on the base of measurements and calculations of several motors of different lengths and it enables the approximate determination of the convection coefficient for the radial model.

To obtain the more accurate value of the convection coefficient the calculation in the axial cross section has to be carried out. The combining of two different models was a necessity, because only by introducing the axial model the axial heat flows in the motor and different side effects can be taken into account. The axial model gives more information, as it is basically a 3D model, due to the defined axial symmetry of the problem. Nevertheless, some details such as the rib geometry and heat flow between the slots are included only by the radial model.

The reference point for both models is the calculation in which the ribs in the radial model are replaced by the cylindrical surface. By comparing the results with those obtained with the modelled ribs, the "rib factor" (ε), which is the ratio of both areas is obtained:

$$\alpha' = A / A' \alpha = \varepsilon \alpha$$

By multiplying the actual convection coefficient with the rib factor, the convection coefficient for the axial model is obtained.

The correction coefficient for the radial model is the ratio between the convection coefficient of the radial model without ribs and identified convection coefficient of the axial model, when both models yield the same results.

The internal convective exchanges were represented by the conduction. The *effective thermal conductivity* λ_{eff} of the air gap, that depends on the geometry and Reynolds number (Hatziathanassiou 1994), was introduced. For the calculated conditions the value 0.1 W/mK was used.

The intensity of the convective heat transfer from the end windings and rotor to the internal air depends mainly on the peripheral speed. In our case the internal air was modeled as a conductive material with the conductivity 0.4 W/mK.

To avoid the time-consuming discretization of the slots and to reduce the total number of elements, the equivalent thermal conductivity of the slot λ_{sl} was employed in both models (Plejic

1992). A series of calculations with different disposition of the materials in the slots enabled the construction of the graph in the Figure 3, from which the λ_{sl} can be obtained. The solution sensitivity to this parameter is approximately 1K / 0.01W/mK for the radial model and 0.5K / 0.01W/mK for the axial model.

The losses were divided according to the location of their origin and their contributions to the heating of the motor were studied separately, so the discrepancies between both models were easier to observe. If the parameters are correct, the temperature distributions of both models coincide.

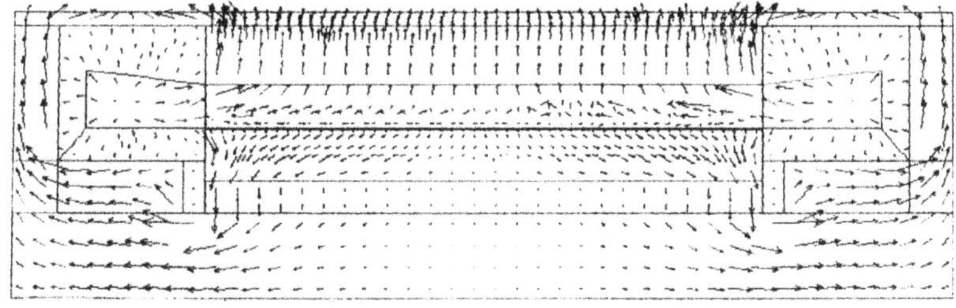

Figure 1. The axial cross section of the motor with the heat flows

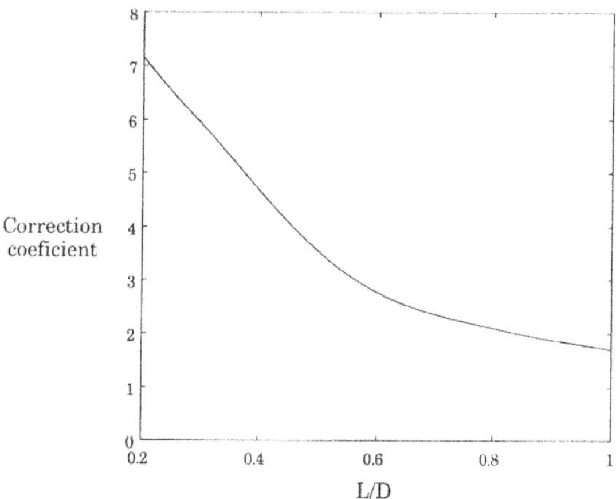

Figure 2. The correction coefficient for the radial model

RESULTS

The identified parameters were used for the thermal calculation of the fan-cooled induction motor operating at different speeds. The temperature field in the motor was measured with 15 thermocouples. Both models yielded good results, since the differences between the measured and calculated temperatures did not exceed 5% (Table 1).

The combination of the radial and axial model provided a reliable information on the temperature field in the motor. For the accurate calculation the air speeds along the ribs have to be predicted well enough.

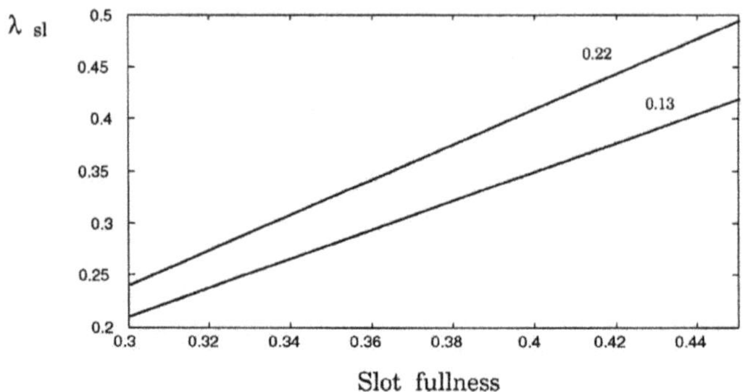

Figure 3. λ_{sl} as a function of slot fullness for different slot insulation conductivities

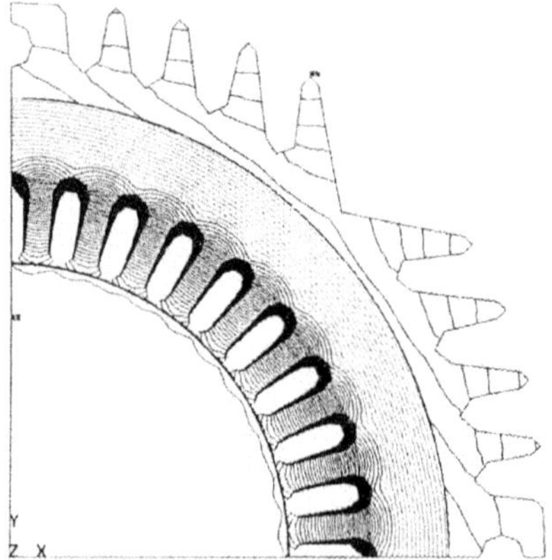

Figure 4. The temperature field in the radial cross section at n = 1389 min^{-1}
Temperatures : rotor = 114 K, slots = 108 K, housing = 79 K

REFERENCES

1. A.Di Gerlando, I.Vistoli : Improved Thermal Modelling of Induction Motors for Design Purposes, Proc. 6th. Int. Conf. on Electrical Machines and Drives, University of Oxford, Sept. 1993
2. V. Hatziathananassiou, J. Xypteras, G. Archontoulakis : Electrical-thermal coupled calculation of an asynchronous machine, Archiv fur Elektrotechnik 77 (1994), 117-122
3. M.Plejić, M.Trlep, A.Hamler, V.Goričan, B.Hribernik : A Thermal Study of the Brushless DC Motor, ICEM' 92, Manchester

THE DESIGN AND ANALYSIS OF RECIPROCATING NON-LINEAR ELECTRO-MECHANICAL SYSTEMS

R.E. Clark, P.H. Mellor, D. Howe

Electronic and Electrical Engineering, University of Sheffield
PO Box 600, Sheffield, S1 4DU. U.K.

INTRODUCTION

Low power reciprocating diaphragm air compressors are commonly used in medical applications where high efficiency and reliability, minimum space envelope and low cost are critical design considerations. The use of reciprocating electromagnetic linear actuators as prime-movers is preferred over other drive mechanisms such as motor/rotating cam systems due to increased reliability and reduced volume envelope. The possible topologies for such linear actuators include moving magnet, moving coil and moving iron devices. Moving magnet configurations have the advantages of improved heat dissipation and reliability, due to the absence of flying leads, than moving coil devices and have an inherent bi-directional action unlike moving iron devices. Existing systems employing moving magnet devices generally utilise open planar configurations, however these have the disadvantages of undesirable forces normal to the intended motion, poor utilisation of materials and low efficiencies. The paper considers two alternative axisymmetric moving magnet actuators, air cored [1] and iron-cored actuator. Figure 1 shows an axisymmetric schematic of the air cored device which consists of a shuttle with an axially magnetised permanent magnet, supported by a non-magnetic shaft. The stator has two opposing current carrying coils housed in a non-magnetic former. The iron cored device shown in figure 2 is a bipolar device and comprises of a soft magnetic shaft carrying two radially magnetised permanent magnet rings and a coaxial soft magnetic shell housing the opposing current carrying coils.
Reciprocating air compressors can have highly non-linear characteristics, predominantly from the stiffening spring characteristic of the pumping membrane (bellows or diaphragm) and as it is desirable to operate the device at the mechanical resonant frequency to achieve optimum efficiency and displacement, the dynamic performance and the frequency response must then be considered in detail. Therefore, in order to design an actuator which meets the specified performance criteria it is necessary to optimise at the system level, i.e. the reciprocating device, the mechanical system and the electrical power source using a methodology which accounts for the system non-linearities. The paper describes such a design procedure and applies it to air cored and iron cored devices.

DESIGN METHODOLOGY

Initial Optimisation Procedure

The design procedure is initially based on a parameter scanning process whereby the leading dimensions of the actuator are scanned between pre-specified limits, however the synthesis process may be readily integrated with more formal constrained optimisation techniques [2]. In order to

Electric and Magnetic Fields, Edited by A. Nicolet
and R. Belmans, Plenum Press, New York, 1995

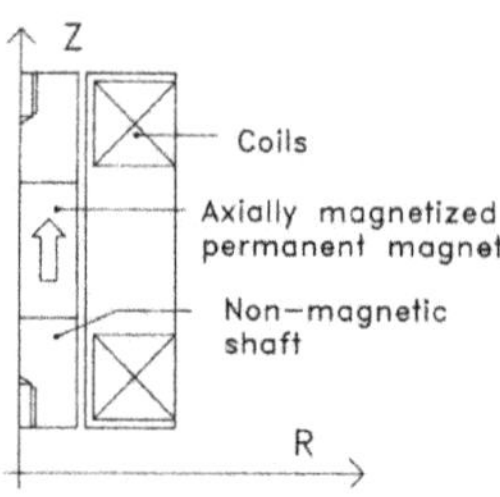

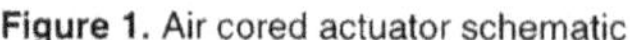

Figure 1. Air cored actuator schematic

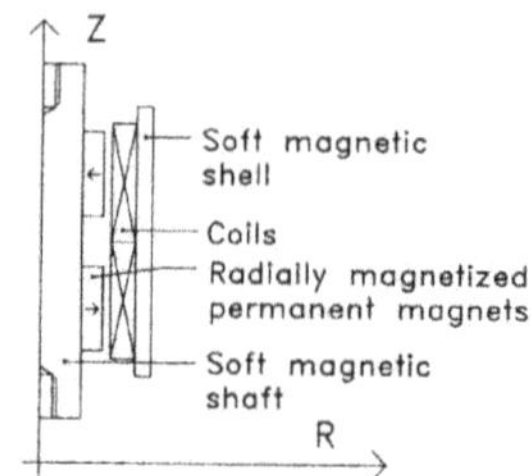

Figure 2. Iron cored actuator schematic

implement such a system the design kernel must employ computationally efficient analytical techniques to predict the static electromagnetic and electrical parameters. In the case of the air cored device, the actuator force is derived from an axisymmetric Biot-Savart calculation of the radial component of flux density [3] due to the coils along a current sheet representation of the magnet shuttle, by integrating over the cross section of the coils. The desired axial component of force is calculated using the Lorentz formula. The self and mutual inductances of the coils are calculated using standard formula for annular conductors. In the case of the iron-cored device the magnetic parameters are established using simple magnetic circuit techniques or a more detailed iterative non-linear lumped parameter model although this has a time penalty on design. To ensure the device is not driven into saturation, the central limb is dimensioned by an iterative procedure.
The static electromagnetic and electrical parameters, and moving mass, are then coupled to equivalent circuit mobility analogues [4], as shown in figure 3, which represent the entire system and provide a first-order approximation of the system performance over the range of possible operating frequencies. The non-linear stiffening spring characteristic of the load, shown in figure 4, is modelled using quasi-linearised approximations where for a given displacement the non-linear spring is modelled as a modified spring constant which ensures continuity in the total energy in the system (i.e. potential energy stored in the spring element and kinetic energy released to the shuttle). The effects of gas compression have been neglected at this stage for simplicity. The viscous damping of the load may be quantified experimentally, however a range of possible values can be used to produce a performance envelope due to the difficulty of accurately assessing this quantity. The circuit is solved iteratively, the displacement being fed back to modify the spring mobility analogue. This technique allows prediction of dynamic steady state performance including displacement, input current, efficiency, etc. at a given frequency or over a specified range. The ability to rapidly produce a large number of designs allows the user to observe trends as well as identifying optimum designs by plotting one parameter against another in the post processing stage. Figure 5 shows an example of post processing outputs for an operating frequency of 50 Hz.

Refined Non-linear System Simulation

Designs which initially appear to meet the performance requirements of the system, are subsequently modelled by non-linear magnetostatic finite element analysis, to give refined calculation of the spatial field distribution and the non-linear force displacement characteristics. From these analyses a plane of coil flux linkage for variations in both shuttle position and current is created which facilitates calculation of the incremental inductance, per unit back-emf and electromagnetic force and is incorporated into a device model for use in the subsequent simulation [5]. The frequency response of the non-linear system is then predicted using a number of discrete state variable time domain simulations, the dynamic performance of the electro-mechanical system being derived from the simultaneous solution of the differential equations which govern its behaviour. Such a technique allows the time consuming field analysis to be effectively decoupled from the system simulation.

VALIDATION OF DESIGN PROCEDURE

The design procedures outlined above have been applied to develop both air and iron cored devices for an application where the target electromagnetic force is 3 Newtons, a stroke of ±3mm

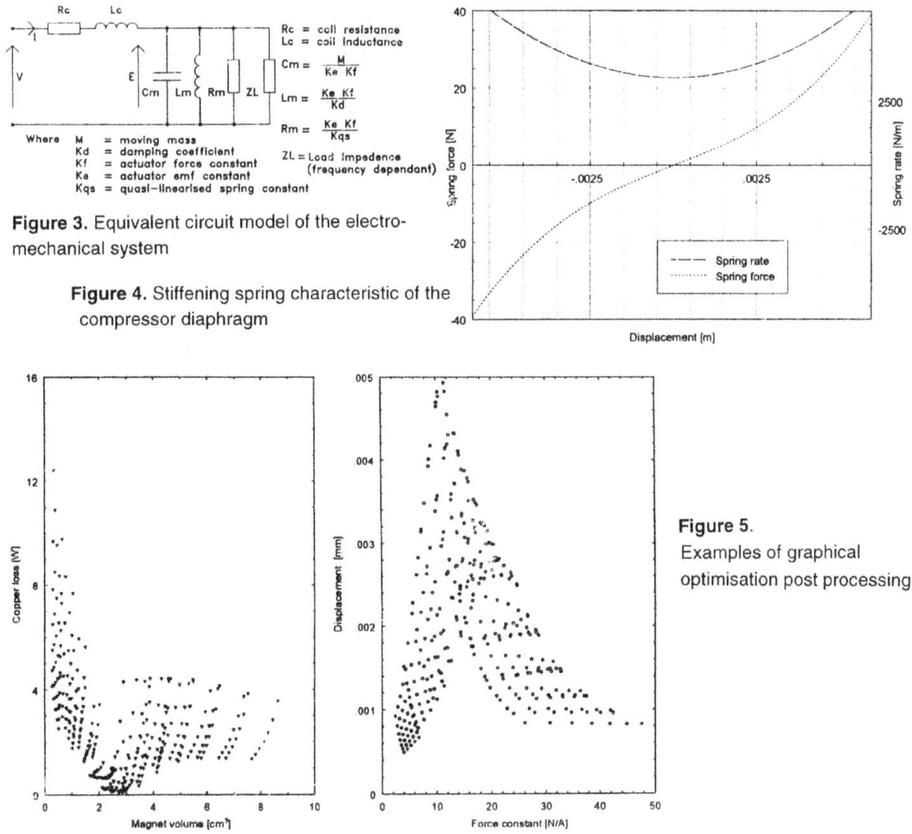

Figure 3. Equivalent circuit model of the electro-mechanical system

Figure 4. Stiffening spring characteristic of the compressor diaphragm

Figure 5. Examples of graphical optimisation post processing

operating over a frequency range from 40 to 60 Hz when supplied from a 12 Volt a.c. power supply in a volume envelope of 40Ø×50mm. Suitable designs were chosen on a basis of predicted efficiency, displacement and minimum magnet volume, the permanent magnet being the dominant cost factor. Table 1 compares the principle electromagnetic parameters assessed at various stages of the design procedure viz. the total device inductance and the peak electromagnetic force constant with the shuttle in the central position.

Figures 6 and 7 compare the frequency responses for displacement and current for the iron cored device, comparing measured results with those predicted at the optimisation stage and refinement stage, whilst figure 8 shows the frequency response for the air cored device, which exhibits an unstable "jump" phenomenon, which is often encountered in stiffening spring non-linear systems. Discrepancies occur between the measured and predicted results for both devices due to joule heating of the windings and consequent increase in winding resistance, whilst for the iron cored topology there remains a discrepancy in the predicted and measured force constant which is most likely due to the radial field pattern approximated by diametrically magnetised magnet segments on the prototype device. Further, the inductive position transducer introduces an additional mechanical damping to the system whilst the performance of such systems is highly sensitive to this parameter variation at or near the resonant frequency. Figure 9 shows results obtained using the equivalent circuit model but using measured parameters, which demonstrates the effectiveness of the model, given improved prediction of parameters.

Table1. Comparison of leading electromagnetic parameters

	Iron cored actuator			Air cored actuator		
Parameter	Analytical	F.E.A	Measured	Analytical	F.E.A	Measured
Inductance (mH)	26.7	35.0	30.1	9.5	9.8	10.8
Peak Force Constant (N/A)	12.4	12.5	9.5	4.16	4.17	4.15

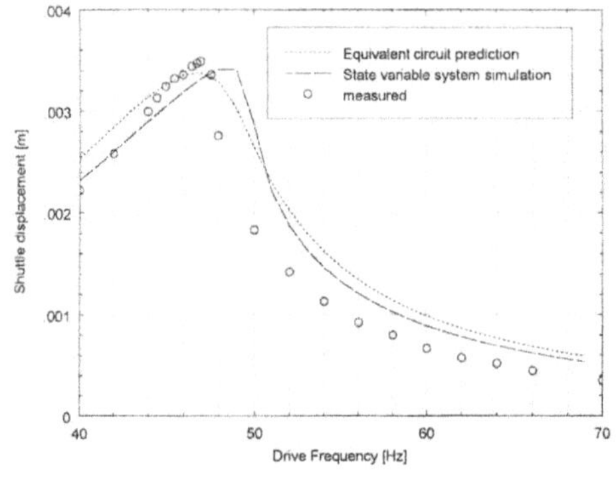

Figure 6. Displacement Frequency responses for iron cored device

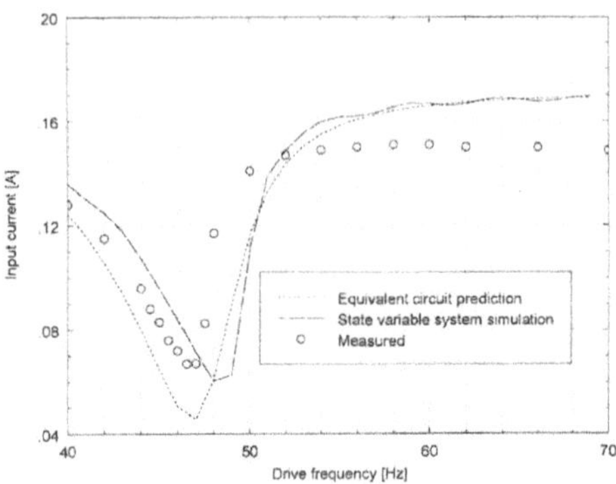

Figure 7. Input current Frequency responses for iron cored device

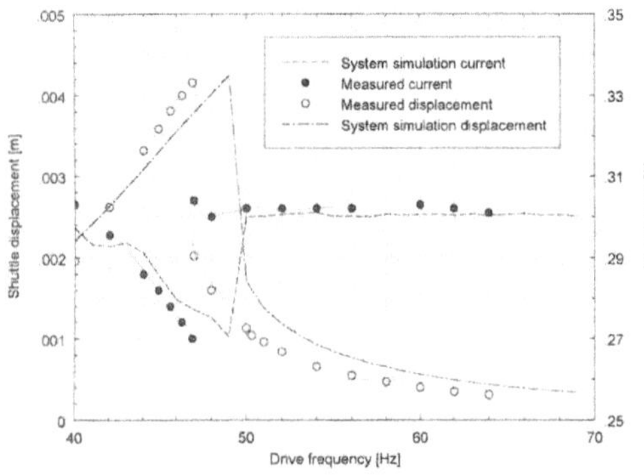

Figure 8. Frequency responses for

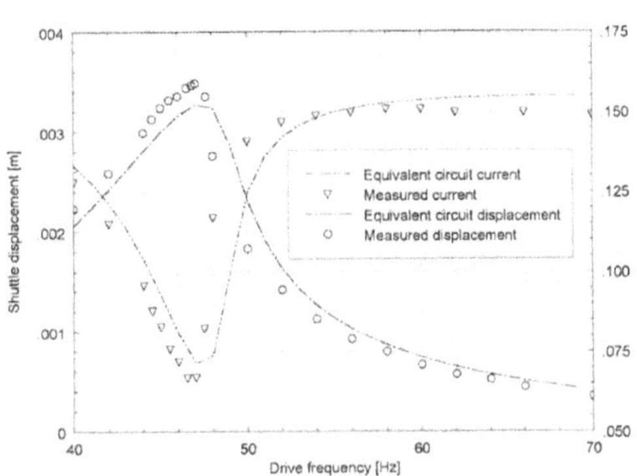

Figure 9. Comparison of equivalent circuit model

CONCLUSIONS

The paper has outlined a suitable design methodology for linear reciprocating actuators where the prediction of the dynamic steady state performance and frequency response of the system is a critical component of the synthesis process. The facility requires further development to account for the coil temperature rise whilst additional losses due to hysteresis and eddy currents may also need to be quantified at the design stage. The addition of the pressure volume work done at each compression cycle will also need to be reliabily incorporated into the overall design procedure.

ACKNOWLEDGEMENTS

The authors wish to acknowledge the support of the U.K. EPSRC and Huntleigh Nesbit Evans Ltd. for the provision of a CASE studentship for R.E. Clark.

REFERENCES

[1] Oliver E. et al., A linear actuator for robotics., Proc. Conf. Power Electronics and Applications 16 - 18 Oct. 1985 Vol 2 Pt 3 pp 197-201

[2] Widdowson G.P.., Design optimisation of permanent magnet actuators, PhD Thesis, University of Sheffield, 1992.

[3] Blewett J. , Magnetic field configurations due to air core coils., J. App. Physics, Vol 18, pp 968-976

[4]Klippel W., Dynamic measurement and interpretation of the nonlinear parameters of electrodynamic loudspeakers., J. Audio Eng Soc., Vol 38, No 12, 1990 pp 944-955

[5] Smith D. S. et al., Simulation of the dynamic performance of an electro-hydraulic solenoid valve., Proc. 1st Int. Workshop on Elec. and Mag Fields pp 59.1-59.6

FINITE DIFFERENCE AND FINITE ELEMENT PREDICTION OF CURRENT DENSITY AND HEAT GENERATION RATE IN AXISYMMETRIC CONTACT ELEMENTS

O. Bottauscio (*), M. Chiampi (§), D. Chiarabaglio (*), G. Crotti (*)

(*) Istituto Elettrotecnico Nazionale "Galileo Ferraris" - Torino, Italy
(§) Dip. Ingegneria Elettrica Industriale - Politecnico di Torino, Italy

INTRODUCTION

The improvement of contact performances requires the knowledge of the temperature distribution in the constriction region, because the temperature increase has a detrimental influence on the electrical and mechanical characteristics of the contact materials. Since the major heating source is the Joule loss, also the current distribution must be known for a correct prediction of the contact behaviour. The experimental investigation is made complex by the small size of the contact elements, the rapidity of the heating and, usually, the lack of reliable and exhaustive information about the material properties. Thus, this phenomenon is often investigated through numerical models of current flow and thermal fields[1,2].

In this paper the contact behaviour is analyzed by a step-by-step procedure based on the separate solution of the electric and thermal problems which are linked through the resistivity variation with temperature and the rate of heat generation. The finite difference (FD) and the finite element (FE) methods are employed in the field solutions. In such a way, two sets of results are always available and the comparative analysis allows the validation of the accuracy of the computed results.

The analysis is performed in an axisymmetric geometry representing a simplified model of an electric contact. Current distribution, constriction resistance and temperature evolution are evaluated varying some geometrical parameters, in order to investigate their influence on the contact performances.

FIELD EQUATIONS AND NUMERICAL METHODS

The temperature evolution inside the domain is faced by a step-by-step procedure, introducing a time sampling, and by a separate solution of the electric and thermal fields. The electric and thermal problems are linked through the variation of resistivity with temperature and through the rate of heat generation.

The electric current problem is described by the equations:

$$curl\boldsymbol{E}=0, \quad div\boldsymbol{J}=0, \quad \boldsymbol{E}=\rho\boldsymbol{J} \tag{1}$$

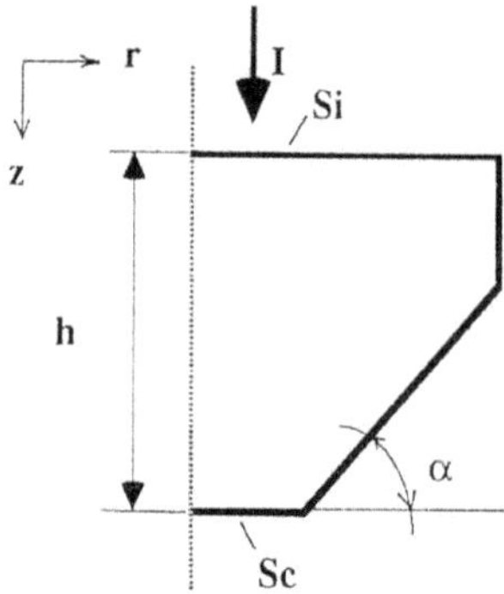

Fig. 1 Studied domain

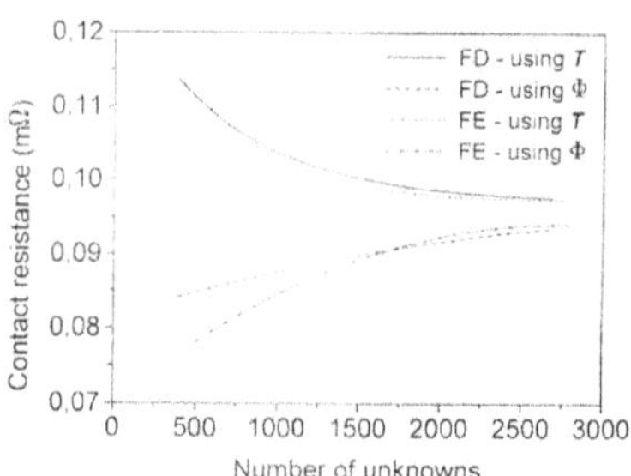

Fig. 2 Resistances vs. number of unknows (α=6°, S_i/S_c=100)

with the constraint: $\int_\Omega J dS = i(t)$, where $\boldsymbol{J}$ is the current density, $\boldsymbol{E}$ the electric field, ρ the electrical resistivity depending on temperature ϑ, Ω is a generic cross section and $i(t)$ is the imposed instantaneous current value. A vector potential $\boldsymbol{T}$ ($\boldsymbol{J}=curl\boldsymbol{T}$) is derived from $div\boldsymbol{J}=0$. Using cylindrical coordinates (r,τ,z), current density $\boldsymbol{J}$ and vector potential $\boldsymbol{T}$ are $\boldsymbol{J}=(J_r(r,z),0,J_z(r,z))$ and $\boldsymbol{T}=(0,T_\tau(r,z),0)$, respectively.

The thermal field in an axisymmetric domain is described by:

$$\frac{\partial^2\vartheta}{\partial r^2}+\frac{1}{r}\frac{\partial\vartheta}{\partial r}+\frac{\partial^2\vartheta}{\partial z^2}+\frac{\rho J^2}{\lambda}=\frac{c}{\lambda}\frac{\partial\vartheta}{\partial t} \tag{2}$$

where c is the heat capacity and λ the thermal conductivity. A stated temperature is imposed to the terminal surfaces; the heat transmission by convection is neglected due to the fast thermal evolution.

The study of the two fields is performed both by FE and FD methods. The FE technique employs triangular elements with piecewise linear shape functions; the FD method approximates the spatial derivatives using Taylor expansion truncated at the second order.

DESCRIPTION OF THE DOMAIN

The device under study is constituted by two axisymmetric copper contact elements; due to their symmetry with respect to the contact surface S_c, only one of them is studied (Fig. 1). In particular, the influence on the current density and temperature distribution of the space angle α and of the contact surface S_c (keeping constant S_i) is analyzed.

The comparison between FE and FD methods is developed during the heat transient state when a constant or variable current is imposed through the contacts. In particular, the short-circuit current behaviour:

$$i(t)=\sqrt{2}I_{rms}\left[\sin(\omega t+\psi-\varphi)-\sin(\psi-\varphi)\exp(-\omega t/\tan\varphi)\right] \tag{3}$$

is considered, where I_{rms} is the root mean square value of the steady-state current, ψ the so-called making angle and φ the phase angle.

Previous investigations[3] have evidenced that the solution accuracy of the problem of current flow in conducting media is strongly influenced by the mesh. Thus, an approach based on complementary variational principles has been used in order to define a grid which ensures satisfactory results without excessively increasing the problem dimensions. The computational procedure is derived considering (1) which can be expressed either by the vector potential $\boldsymbol{T}$ or by a scalar potential ϕ ($\boldsymbol{E}=-grad\phi$). In the first formulation, the vector potential is specified on the insulated boundaries and the power loss is: $P'=\int_\Omega \rho J^2 dS$. Thus, the conductor resistance is: $R'=P'/I^2$. The second approach, which requires that potential ϕ

is imposed on the terminal surfaces, leads to: $P'' = \int_{\Omega} \sigma E^2 dS$. Then, the conductor resistance is evaluated as $R''=V^2/P''$, where V is the voltage drop across the elements. It can be proved[4] that $R''<R<R'$, where R is the true contact resistance; the bound estimates become closer by refining the mesh as shown for example in Fig. 2. This procedure allows the choice of a grid ensuring a good accuracy.

ANALYSIS OF RESULTS

The first comparison between FD and FE methods is developed on the electric current field. To this end, the distribution of the current density on the contact surface S_C is computed for different values of the angle α and the ratio S_i/S_C (Fig. 3). In these diagrams the values of the current density are normalized to the one which corresponds to a uniform distribution. The current is concentrated in the external region of the contact surface; this effect is increased when angle α reduces, according to the results already presented[5]. On the contrary, the ratio S_i/S_C does not modify the current density distribution on the contact area.

The comparison is then developed in the case of coupled electric and thermal fields. The time evolution of the constriction resistance, normalized to the value at 0°C (R_0), due to the flow of a constant current (I = 500 A) is presented in Fig. 4 varying the space angle and the ratio S_i/S_C, respectively. The predictions of both the methods are in good agreement; when the angle α is small the constriction resistance slowly increases with time.

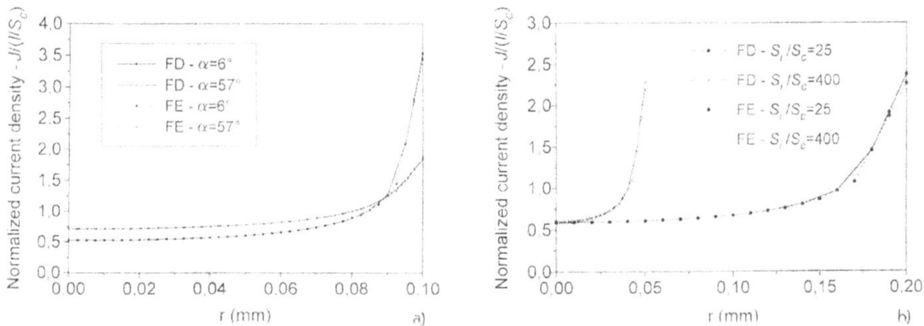

Fig. 3 Normalized current density distribution on the contact surface: (a) S_i/S_C=100, (b) α=29°

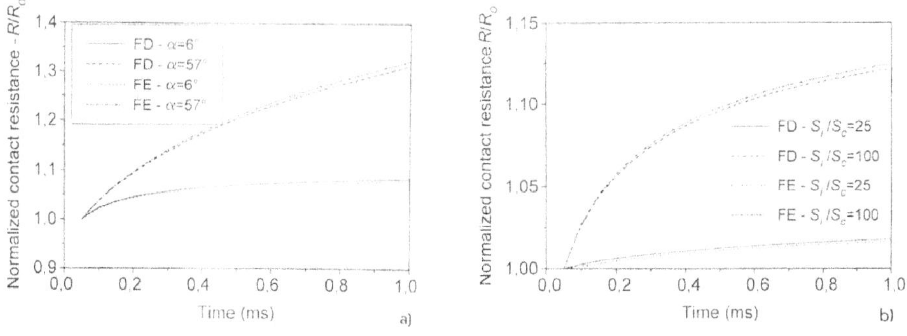

Fig.4 Constriction resistance vs. time: (a) S_i/S_C=100, (b) α=29°

Finally, the case of an imposed variable current is investigated (I_{rms} = 300 A, ψ = 0°, φ = 60°) The evolution is analyzed for a time interval of 10 ms, using a time step of 0.05 ms. Fig. 5 shows the evolution of the temperature on the contact surface computed by the two methods varying the two geometrical parameters. It is worth noting that if the ratio S_i/S_C becomes greater than 100 (i.e. too small contact radius) the temperature reaches values higher than the one corresponding to the softening of the material.

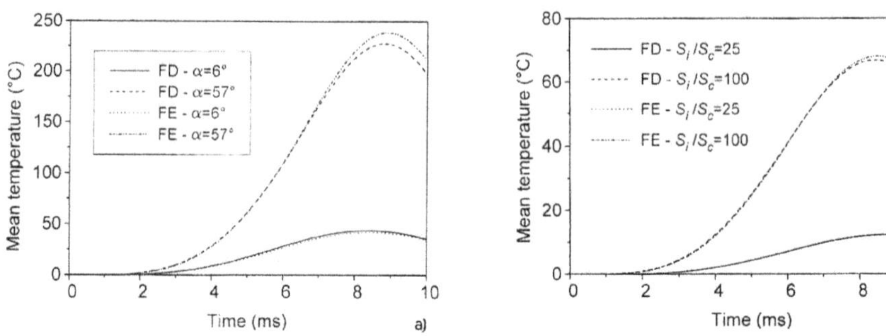

Fig. 5 Mean temperature evolution on the contact surface: (a) S_i/S_C=100, (b) α=29°

CONCLUSIONS

The influence of some geometrical parameters on the current density distribution and Joule heating rate in axisymmetric contact elements is investigated by finite difference and finite element methods. Different time evolutions are taken into consideration. In all the cases the agreement between the two methods is good.

The analysis shows that the ratio S_i/S_C does not appreciably influence the current density distribution on the contact surface; as expected, it affects the constriction resistance and the temperature on the contact surface. On the contrary, the space angle influences all the considered electric and thermal quantities.

Finally, the study evidences that the contact behaviour is worse when the angle is greater in spite that the current density distribution is more uniform on the contact surface.

REFERENCES

1. O. Bottauscio, G. Crotti, G. Farina, Numerical analysis of heating transient of electric contacts under short-circuit conditions, *IEEE Trans. CHMT*, vol.16, n.5, 563:570 (1993).
2. S. R. Robertson, A finite element analysis of the thermal behaviour of contacts, *IEEE Trans. CHMT*, vol.5, n.1, 3:10 (1982).
3. O. Bottauscio, M. Chiampi, D. Chiarabaglio, G. Crotti, Numerical methods for coupled thermal and electric current fields in axisymmetric device, *IMACS-TC1'93 Proc.* 49:54 (1993).
4. J.A.H. Rikabi, Complementary solutions of direct current flow problems, *IEEE Trans. Mag.*, vol.29, n.1, 98:107 (1993).
5. S.W. Park, S.J. Na, Current density and heat generation rate in the neighborhood of a single-line contact with various space angles, *IEEE Trans. CHMT*, vol.12, n.1, 105:108 (1989).

FIELD MODEL OF THE INTERLEAVED TRANSFORMER COIL

Augustin Moraru, Felicia Anghel

University "Politehnica" from Bucharest

INTRODUCTION

Behaviours of transformer windings to withstand voltage surges must take into account internal oscillations of their interleaved disk coils, due to propagation phenomena[1,...,6]. Usually a lumped parameter model is used for the disk coil, but propagation phenomena only by distributed parameters can be described and this leads to field models.

THE FIELD MODEL

Let be a twin disk coil, whose halfs are interleaved as in Fig.1. The twin coil is entered by an external current $\mathbf{i}_1$, imposed by some global conditions of the winding. This current may be decomposed into two orthogonal components:

- a mean current $\mathbf{i}_m$, not varying along the coil turns but bearing the main magnetic field which couples the coils, and

- a supplementary, internal current **i**, varying along the coil turns and being closed only capacitively, between the turns, due to internal oscillations; this current does not cross the median point M of the twin coil.

The magnetic field due to internal currents decreases steeply in axial direction, therefore internal oscillations may be investigated on an insulated interleaved disk coil pair.

In Fig. 2a is represented the straightened conductor of an interleaved coil pair, divided in two halves AM and BM by the mid-connection M; the halves will be named **bars**. The fragments on each bar correspond each to a turn and are numbered like the turns. Full and dashed lines suggest the electric connection, that is the electric field lines between "corresponding" areas on the two faces (external an internal) of the turns.

Because of symmetry, between points A-M, respectively M-B the same voltage V/2 appears, V being the voltage applied to the coil pair (between the terminals A and B).

Taking into account the turns alternance according to Fig.1, the magnetic field given by the internal current **i** will be located mainly in the "flux channel" between the adjacent turns (Fig. 3); its flux can be estimated by means of a leakage line inductance (per unit length), greater for the lateral turns and smaller for the intermediate ones. The magnetic interaction between the two disk coils, due to the internal current **i**, is weak and may be neglected.

The above considerations lead us to the field model of the interleaved coil pair.

Electric and Magnetic Fields, Edited by A. Nicolet
and R. Belmans, Plenum Press, New York, 1995

The established field model is non-homogeneous, both from electric and magnetic points of view. This effect is as stronger as the turn number is smaller. The "exact" field model should lead to very intricated propagation phenomena, at least two kinds being dominant.

SIMPLIFIED, HOMOGENEOUS FIELD MODEL

With allowable approximation, the "exact" model can be transformed successively in a homogeneous one, having the same whole bar-to-bar capacitance and bar inductance. Let 2C be this line capacitance, corresponding to two bar faces, and L be the mean line leakage inductance of a bar.

For instance, if C_t is the capacitance between two adjacent turn faces, l_t - the turn length, and 4n - the number of turns in the twin coil, then $C = C_t * l_t * (n-1)/n$.

The leakage inductance is well estimated by the formula $L = \mu_0/2 * l_t * (d+\delta)/(b+\delta)$, where d is the distance between two adjacent turns, b - the width of the flat conductor, and δ is the skin-depth in the conductor.

The simplified model is composed from two parallel **bars**, having the line capacitance 2C and each bar having a line inductance L. The model may have also an external line magnetic flux Φ_0, the same for each bar (or turn). Between the two bars, at each end, appears the voltage V/2 (Fig. 2b).

MODEL EQUATIONS

Let x be a line coordinate, along a bar $x \in (0, 2*n*l_t)$, with origin at the terminal A on the upper bar, or in the median point M on the lower bar. Let $i_1(x,t)$ and $i_2(x,t)$ be the currents in the two bars, flowing in increasing sense of the coordinate x, and $u_1(x,t)$, $u_2(x,t)$ the voltages of the two bars, assuming the origin in the median point M. For generality and simplicity sake we use the Laplace transformed magnitudes and we denote by F(x,s) the Laplace transform of the function f(x,t).

Without difficulties, neglecting the resistances, the folowing transmission line like equations may be estabished

$$dU_1(x,s)/dx = - s * (\Phi_0(s) + L*I_1(x,s)) \; ; \quad dU_2(x,s)/dx = - s * (\Phi_0(s) + L*I_2(x,s))$$
$$dI_1(x,s)/dx = -dI_2(x,s)/dx = 2*C*s*(U_2(x,s)-U_1(x,s))$$

Now we define new component currents and voltages

$$I_1(x,s) = I_m(x,s) + I(x,s) \; ; \quad I_2(x,s) = I_m(x,s) - I(x,s)$$
$$U_1(x,s) = U_m(x,s) + U(x,s) \; ; \quad U_2(x,s) = U_m(x,s) - U(x,s)$$

With the new variables, the above system splits in two independent ones

$$dU_m(x,s)/dx = - s * (\Phi_0(s) + L*I_m(x,s)) \; ; \quad dI_m(x,s)/dx = 0$$
$$dU(x,s)/dx = - s*L*I(x,s) \; ; \quad dI(x,s)/dx = - 4*s*C*U(x,s)$$

Denoting by $l=2*n*l_t$ the bar length, the boundary values for the voltages are

x	$U_1(x,s)$	$U_2(x,s)$	$U_m(x,s)$	$U(x,s)$
0	V(s)/2	0	V(s)/4	V(s)/4
l	0	-V(s)/2	-V(s)/4	V(s)/4

The components of index 'm' are low frequency ones, which act in the normal state of operation. The current is closed along the conductors and the voltage drop is due only to the inductive effect of the main line flux $\Phi_0(s)$ and leakage line flux $L*I_m(s)$.

Components without index correspond to propagation phenomenon along the coil pair; they are determined only by the bar-to-bar line capacitance and bar line leakage inductance.

For a given external current $I_m(s)$, the same for any x, the mean voltage component results by direct quadrature

$$U_m(x,s) = V(s)/4 - s*(\Phi_0(s)+L*I_m(s))*x$$

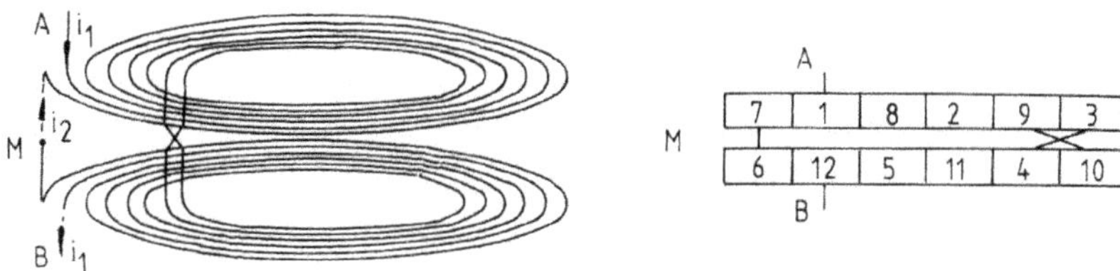

Fig. 1. Interleaved coils

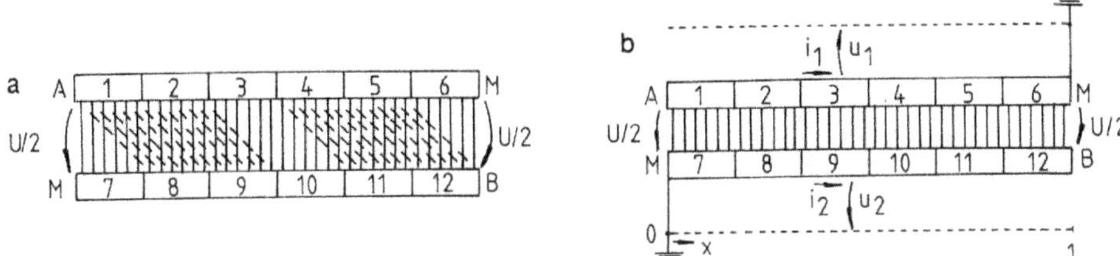

Fig. 2. a) "Exact" and b) homogeneous field model

Fig. 3. Leakage field lines

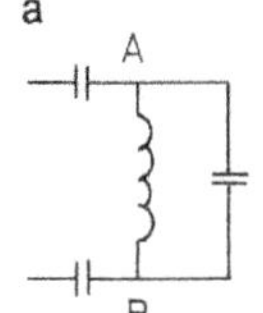

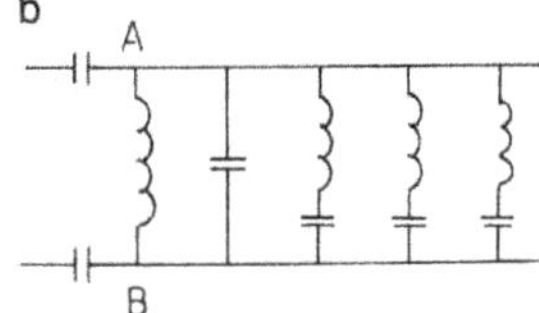

Fig. 4. a) Conventional and b) new equivalent circuit

and from the right end boundary condition it results

$$V(s) = 2*s*(\Phi_o(s)+L*I_m(s))*l$$

Internal components satisfy a second order homogeneous differential equation:

$$d^2U(x,s)/dx^2 - \lambda^2 * U(x,s) = 0 \quad ,\text{with} \quad \lambda^2 = 4 * s^2 * L * C$$

Given the boundary conditions, the solutions are

$$Ux,s) = V(s)/4 * ch(\lambda (x-l/2))/ch(\lambda l/2)$$

$$I(x,s) = V(s)/2 * \sqrt{C/L} * sh(\lambda (x-l/2))/ch(\lambda l/2)$$

EQUIVALENT CIRCUIT

At the terminals A and B they result the folowing voltages and currents

$$U_A(s) = U_1(0,s) = V(s)/2$$

$$U_B(s) = U_2(l,s) = -V(s)/2$$

$$I_A(s) = I_1(0,s) = I_m(s) + V(s)/2 * \sqrt{C/L} * th(\lambda l/2)$$

$$I_B(s) = I_2(l,s) = I_A(s)$$

The entering operational admittance of the twin coil is

$$Y(s) = I_A(s)/(U_A(s)-U_B(s)) = I_m(s)/V(s) + .5 * \sqrt{C/L} * th(\lambda l/2)$$

The first term corresponds to the magnetic coupling and leakage, while the second term - to the propagation phenomenon or self oscillations of the twin coil.

In conventional equivalent circuits (Fig. 4a) the interleaved coil is represented only by its inductance ($\Phi_o(s)/I_m(s)$) and the whole capacitance $2*C*l$ [1,2,4,5,6]. The circuit resulting from the field model in Fig. 4b, replaces the unique capacitor by several parallel branches of series resonant circuits, with the inductances $L*l$ and capacitances $4*C*l/(k\pi)^2$, where k is odd (1, 3, 5, ...). This results from the series development of the hyperbolic tangent. The number of resonant circuits is infinite, but the used field model is approximative, therefore only 1 or 2 resonant branches will be considered, and the remaining capacitance is connected without series inductance. In the equivalent circuit of the interleaved coil are represented also some coupling capacitances with other winding elements.

CONCLUSIONS

The proposed field model allows to calculate the parameters for a new, improved equivalent circuit of the interleaved coil, which takes into account the internal oscillations or propagation phenomena. This model points out the fact that the interleaving do not improve the "very short time initial distribution" as were expected, because the increased longitudinal capacitance obtained by interleaving appears in series with the leakage inductance.

RFERENCES

1. B.M.Dent, E.R.Hartill, J.G.Miles. "A method of analysis of transformer impulse voltage distribution using a digital computer", Proc. IEE, Vol.150A, No. 23, pp.445-459 (1958).
2. Y.Kawaguchi. "Calculation of circuit constants for computing internal oscillating voltage in transformer windings", Electrical Engng. in Japan, Vol.89, No.3, pp.44-53, (1969).
3. A.G.Bunin, L.N.Kontorowitch. "Mathematic model for calculation the internal overvoltages in interleaved coils of high voltage transformers" (in russian). Elektrotchnika (Moskow), No.4, pp.22 -27, (1976).
4. A.Miki, T.Kosoya, K.Okuyama. "A calculation method for impulse voltage distribution and transferred voltage in transformer windings", IEEE Trans. on PAS, Vol.PAS-97, No.2, pp.930-938, (1978).
5. T.Teranishi, M.Ikeda, M.Honda, T.Yahari. "Local voltage oscillation in interleaved transformer windings". IEEE Trans. on PAS, Vol.PAS-100, No.2, pp.873-881, (1981).
6. J.Leohold. "Zur Genauigkeit von Netzwerkmodellen fuer die Berechnung der Resonanzvorgangen in Transformatorwicklungen". ETZ Archiv, Bd.8, H.2, pp.41-50, (1985).

OPTIMIZATION OF THE MEMORY EMPLOYMENT FOR STUDYING ELECTRICAL MACHINES BY FINITE ELEMENT METHOD

G. Cannistrà[1], M. Minenna[1], G. Negro[1], M. Sylos Labini[1]

[1]Politecnico di Bari - Dipartimento di Elettrotecnica ed Elettronica

ABSTRACT

The Authors propose a procedure which allows us to reduce the memory needed to store the terms of the matrix [C] of the coefficients, matrix which is relative to the finite element method. This procedure is based on two fundamental points. The first one consists in subdividing the matrix [C] into four submatrices, each of them can be calculated one at a time. The second one consists in introducing adequate algorithms in the Cholesky method. The proposed procedure has been applied to study the temperature field in a three-phase deep-bars squirrel-cage asynchronous motor.

INTRODUCTION

As is well-known, the finite element method consists in subdividing the system under study into elements linked by nodes, in assuming appropriate "shape functions" and in imposing boundary conditions[1]. In this way, for steady-state analyses, we obtain the following system of linear equations [C] [X] = [K] where [C] is the matrix of the coefficients and [K] is the column vector of the known terms. The solution of this system gives the values taken by the unknown function in the nodes.

The Authors have carried out various finite element calculation programs which calculate, in different ways, the matrix [C] of the coefficients and solve, by means of different methods, the system of linear equations[2,3,4]. In this paper they propose a procedure which allows to reduce the memory needed to store the terms of the matrix [C]. This procedure is based on two fundamental points. The first one consists in subdividing the matrix [C] into four parts, each having the same size, and in introducing an adequate algorithm in the finite element method, so that the terms of each of the four submatrices can be calculated one at a time[2]. The second one consists in introducing adequate algorithms in the Cholesky method, i.e. in the method for solving the system of linear equations, so that the program can operate on two submatrices at a time.

THE SYSTEM SOLUTION

It is well-known that Cholesky method allows the solution of a system made up of N linear equations, each having N unknowns. It is based on a matrix $[C_t]$ (where t stands for total) having N rows and N+1 columns. $[C_t]$ is obtained from the matrix [C] of the coefficients of the linear equations by adding to such matrix a (N+1)th column consisting of the vector of the known terms [K].

Electric and Magnetic Fields, Edited by A. Nicolet
and R. Belmans, Plenum Press, New York, 1995

Cholesky method consists of N calculation steps. In each of these steps, the "column terms", which we will indicate with l_{ij}, and then the "row terms", which we will indicate with u_{ij}, are determined alternately. At the end of the whole calculation process we pass from the original matrix $[C_t]$ to the following final matrix:

$$[C_t'] = \begin{bmatrix} l_{11} & u_{12} & u_{13} & u_{14} & u_{15} & - & u_{1N} & u_{1,N+1} \\ l_{21} & l_{22} & u_{23} & u_{24} & u_{25} & - & u_{2N} & u_{2,N+1} \\ l_{31} & l_{32} & \mathbf{l_{33}} & \mathbf{u_{34}} & \mathbf{u_{35}} & - & \mathbf{u_{3N}} & \mathbf{u_{3,N+1}} \\ l_{41} & l_{42} & \mathbf{l_{43}} & l_{44} & u_{45} & - & u_{4N} & u_{4,N+1} \\ - & - & - & - & - & - & - & - \\ l_{N1} & l_{N2} & \mathbf{l_{N3}} & l_{N4} & l_{N5} & - & l_{NN} & u_{N,N+1} \end{bmatrix}$$

where, for example, the "column and row terms" calculated in the step i=3 are in bold type.

The first column is unchanged and the terms of the first row are determined by dividing each term of the first row of the original matrix $[C_t]$ by C_{11}. Moreover, the formulae to calculate the "column terms" and the "row terms" for the generic step i ($i \neq 1$) are the following.

- "column terms" (that is, terms of the column i):

$$\begin{aligned} &l_{ii}=C_{ii}-(l_{i1}\cdot u_{1i}+l_{i2}\cdot u_{2i}+.......+l_{i,i-1}\cdot u_{i-1,i}) \\ &l_{i+1,i}=C_{i+1,i}-(l_{i+1,1}\cdot u_{1i}+l_{i+1,2}\cdot u_{2i}+.......+l_{i+1,i-1}\cdot u_{i-1,i}) \\ &\cdots\cdots\cdots\cdots\cdots\cdots \\ &l_{Ni}=C_{Ni}-(l_{N1}\cdot u_{1i}+l_{N2}\cdot u_{2i}+.......+l_{N,i-1}\cdot u_{i-1,i}) \end{aligned} \tag{1}$$

- "row terms" (that is, terms of the row i):

$$\begin{aligned} &u_{i,i+1}=[C_{i,i+1}-(l_{i1}\cdot u_{1,i+1}+l_{i2}\cdot u_{2,i+1}+......+l_{i,i-1}\cdot u_{i-1,i+1})]/l_{ii} \\ &u_{i,i+2}=[C_{i,i+2}-(l_{i1}\cdot u_{1,i+2}+l_{i2}\cdot u_{2,i+2}+......+l_{i,i-1}\cdot u_{i-1,i+2})]/l_{ii} \\ &\cdots\cdots\cdots\cdots\cdots\cdots \\ &u_{i,N+1}=[C_{i,N+1}-(l_{i1}\cdot u_{1,N+1}+l_{i2}\cdot u_{2,N+1}+......+l_{i,i-1}\cdot u_{i-1,N+1})]/l_{ii} \end{aligned} \tag{2}$$

After having calculated the whole matrix $[C_t']$, it is possible to determine the unknowns x_i (i=1,2,....N) by a back substitution procedure, using the following formulae:

$$x_N = u_{N,N+1}; \qquad x_i = u_{i,N+1} - \sum_{k=i+1}^{N} u_{i,k}\cdot x_k \quad \text{where } i=1, 2, ... N-1 \tag{3}$$

THE PROPOSED PROCEDURE

The proposed procedure is based on two points. The first one consists in calculating the whole matrix of the coefficients [C] by means of a procedure by rows. The second one consists in introducing adequate algorithms in the Cholesky method, so that the program can operate on two submatrices at a time.

Calculation of the matrix [C]

It is well-known that the matrix [C] is calculated by integrating in [C] the terms of the various "element matrices" $[c]_e$. The process of integration can be carried out by reading row by row the "matrix of the vertices" [V] and by calculating for each row of [V] the corresponding "element matrix" $[c]_e$[2]. In order to determine the matrix [C] in the traditional way, it is necessary to store the whole matrix dimension. Obviously, this limits the use of personal computers, particularly when 3-D problems must be solved. Conversely, the proposed computation procedure allows us to determine each individual row of [C]. It has been possible to obtain this procedure thanks to the algorithm exposed in[2], which calculates each term of [C] without determining in advance any "element matrix" $[c]_e$.

Briefly, to calculate the generic term C(i,j) of the total matrix [C], the program carries out a "thorough reading" of the "matrix of the vertices" [V]. This "thorough reading" consists in running row by row through the whole matrix [V] and in calculating only the terms of the "element matrices" needed for the determination of the required term C(i,j).

How to operate on two submatrices at a time

In the hypothesis that the total number of nodes N is an even number, let's subdivide the original matrix $[C_t]$ into four submatrices A, B, C, D, each having N/2 rows and N/2 columns (submatrices A and C) and (N/2)+1 columns (submatrices B and D). Obviously the submatrices B and D have, as last column, the terms of the vector of the known terms K_1, K_2, ... K_N.

Moreover, let's subdivide the final matrix $[C_t']$ into four submatrices A', B', C', D', each having N/2 rows and N/2 columns (submatrices A' and C') and (N/2)+1 columns (submatrices B' and D').

A thorough examination of the Cholesky method allows us to assert that the calculation procedure of the whole matrix $[C_t']$ can be carried out by determining one by one the various submatrices A', B', C' and D' in an appropriate sequence.

To calculate the terms of A' it is necessary to work only on the terms of A. To calculate the terms of B' (or C') it is necessary to have previously calculated the terms of A'. In other words, to determine the terms of B' (or C') we must work on the terms of A' and B (or on the terms of A' and C). Finally, to calculate the terms of D' it is necessary to have previously calculated the terms of B' and C'. In other words, to determine the terms of D' we must work on the terms of B', C' and D.

It is useful to consider that the calculation of D' can be also done in two consecutive phases, in the following way:
1st phase: calculation of an intermediate submatrix D'_{int} having the same size of D. The generic term $D'_{int}(i,j)$ consists of the product of the ith row of C' and the jth column of B'.
2nd phase: calculation of the final submatrix D'. To determine D', the terms of D'_{int} must work with the terms of D.

According to the previous remarks, the Authors have carried out the following method. Two matrices M_1 and M_2, each consisting of N/2 rows and (N/2)+1 columns, must be dimensioned into the computer. In these matrices the four submatrices A, B, C and D making up the original matrix $[C_t]$, which are supposed to be arranged in four files of the hard disk, must be stored in turns and in an ordered sequence. By applying Cholesky method only to A, it is possible to determine A'. By applying Cholesky method to A' and C and to A' and B it is possible to determine C' and B' respectively. By multiplying the rows of C' by the columns of B' it is possible to determine D'_{int}. Finally, by getting D'_{int} worked with D it is possible to determine D'.

From what it has been said, it is clear that the proposed procedure, which carries out various appropriate operations of reading and printing from and on hard disk respectively, allows us a saving of memory equal to nearly half the memory required for the traditional finite element method.

THERMAL ANALYSIS IN AN ASYNCHRONOUS MOTOR

The proposed finite element procedure has been used to determine the temperature distribution in a three-phase deep-bars squirrel-cage asynchronous motor whose rated data are P=120 kW, V=380 V, f=50 Hz, n_r=1500 r.p.m.. The thermal analysis of this motor has already been carried out in[5] using grids having about a hundred nodes (108 for the stator and 66 for the rotor). In order to study grids having a higher number of nodes, the proposed procedure has not been applied to the matrix [C], but only to its half-band. In other words, each of the four submatrices A,B,C,D contains a quarter of the half-band of [C]. By means of an appropriate numeration of the nodes, this stratagem has made possible to study grids having about 2000 nodes.

The study has been carried out on a longitudunal section of the motor (see Fig.1) and under the same simplified assumptions made in[5]; in particular, the hypothesis of null heat exchange between stator and rotor has been made (and this is a consequence of an adequate ventilation). As a result, the temperatures we have obtained are not much higher than those we found in[5]. To be precise, if we refer to the points which are more stressed from the thermal point of view, that is to say, to the points lying in the middle of the machine, the positive per cent variations of temperature turn out to be equal to 0.7% and 0.8% for stator and rotor respectively. These results confirm that beyond a given subdivision the benefits we get in terms of accuracy are very little; on the contrary, the calculating times become heavy.

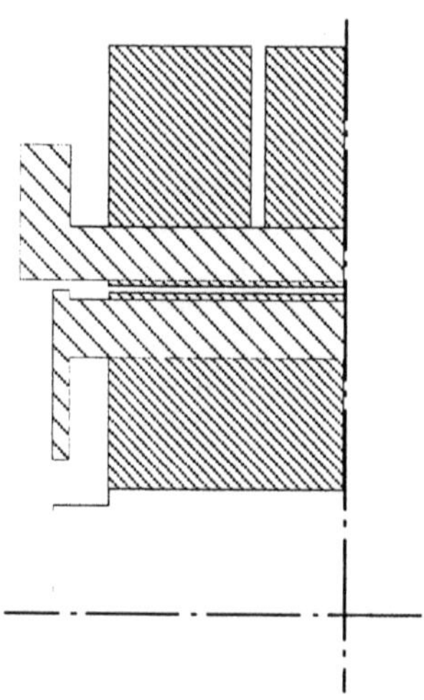

Fig.1 Domain under study relative to a quarter of a longitudinal section of the motor

CONCLUSIONS

It is well-known that the finite element method accuracy increases by increasing the number of the nodes. Obviously, this involves a large memory employment and can cause a break in the personal computer. The procedure proposed by the Authors gives a saving of memory equal to nearly half the memory required for the traditional finite element method, thus allowing to study grids having about two thousand nodes by a personal computer.

REFERENCES

1. O.C. Zienkiewicz, R.L. Taylor: "The finite element method", Fourth edition, Mc Graw-Hill Book Company Europe
2. P. Settanni, M. Sylos Labini, A. Tricarico: "The finite element method in engineering problems: time and memory reduction", International Journal of Modelling and Simulation, vol.9, n.1, 1989
3. M. Carnimeo, A. Covitti, M. Minenna, M. Sylos Labini: "A new approach to reduce memory employment in the field problems studied by the finite element method", Jornadas Hispano-Lusas de Ingenieria Electrica, Vigo, 4-6 July 1990
4. F.Blasi, G. Cannistrà, A. Covitti, M. Sylos Labini: "Analysis of the leakage magnetic field in induction motors by finite element method", Int. Workshop on Electric and Magnetic Field from Numerical Models to Industrial Applications, Liege, Belgium, 28-30 September 1992
5. G.ni Cannistrà, G.pe Cannistrà, M.lo Sylos Labini: "Thermal analysis in an induction machine using thermal network and finite element methods", Fifth IEE Int. Conf. on "Electrical Machines and Drives", London, 11-13 September 1991

A FINITE DIFFERENCE FREQUENCY DOMAIN APPROACH FOR THE ANALYSIS OF THE MAGNETIC DIFFUSION OF THE CURRENT DISCHARGED IN PLATES

M. Angeli and E. Cardelli

Istituto di Energetica Università di Perugia
Strada S. Lucia Canetola
06125 Perugia Italia

ABSTRACT

This work deals with a finite difference frequency domain approach for the numerical computation of the current density distribution into non perfectly conductive plates in which a current is injected. The two-dimensional magnetic scalar formulation proposed allows to limit the discretization only to the current flow region, and the grid used for the finite difference numerical simulation comprises only the flat plate. The numerical model has been tested by comparison with the results obtained by means of analytical expressions, valid only under limiting conditions. Just to have an idea about the possible applications of the model proposed, the values of the internal impedance between two contact points of a rectangular ground plane, against the signal frequency have been computed. Moreover the influence of the shape of the plane, on the equivalent internal ground parameters, has been investigated, and the interfering current values in order to have a given ground overvoltage have been deduced. The results obtained encourage to develop three-dimensional magnetic scalar formulations in order to obtain a more realistic simulation of the diffusion of the current and to avoid errors and possible over-estimations due to the known limits of two-dimensional approaches.

INTRODUCTION

An important subject in the studies about electromagnetic compatibility and interferences between electronic circuits and external signals is the analysis of the electromagnetic current diffusion in flat conductors of finite conductivity, such as shielding, ground connections, etc., both in space and time. The knowledge of the current density distribution contributes to evaluate the magnitude of the electromagnetic field generated around the conductors and the value of the internal ground plane impedance, that can produce significant unwanted voltages which may create bad functioning in the electronic circuits [1,2,3]. In this paper a method to compute the current distribution in a conductive plate, in which a known discharge current has occurred, is presented. The electromagnetic

Electric and Magnetic Fields, Edited by A. Nicolet
and R. Belmans, Plenum Press, New York, 1995

diffusion equation is solved in the frequency domain. By using the H-scalar formulation the boundary conditions at the interface conductor-air are derived and the numerical calculation can be reduced to the only conductive region. The numerical solvers used are finite difference schemes with successive over-relaxation techniques.

CURRENT DIFFUSION IN CONDUCTIVE PLATES

In this paragraph the model used for the computation of the current density distribution in conductive plates is described. The problem to be analysed is pictured in fig. 1.

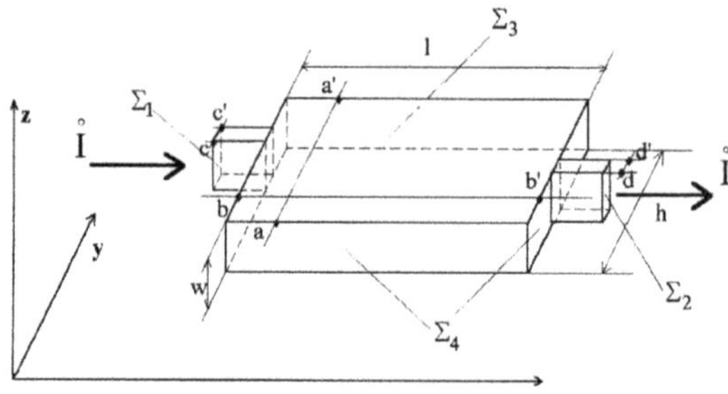

Figure 1. Schematic picture of the analysed problem.

A known discharge current i(t) enters the conductive plate (length l, height h, width w) in the region Σ_1 (length γ_1, width w) and flows toward the ground level throw the region Σ_2 (length γ_2, width w). As known, the electromagnetic current diffusion is described by Maxwell's equations. In the case of good conductor the displacement current in the plane can be neglected over the range of frequencies of practical interest. We treated the problem as two-dimensional, i.e. we neglected the variation of the current desity along z-direction.

Following this simplifying assumptions we can write the electromagnetic current diffusion in the frequency domain as it follows:

$$j\omega\mu\sigma \dot{J}_x = \nabla^2 \dot{J}_x; \qquad j\omega\mu\sigma \dot{J}_y = \nabla^2 \dot{J}_y \tag{1}$$

where ∇^2 is the scalar Laplacian operator, $\dot{J}_x$ and $\dot{J}_y$ are the component of the current density phasor $\dot{\mathbf{J}}$, $\omega = 2\,\pi\,f$ is the electric pulsation, μ and σ are respectively the magnetic permeability and the electric conductivity of the mean. Moreover in our problem we hypothised to know the values of the current density around all the limiting surface Σ of the plate: $\dot{J}_n = 0$ along the edges of the plate in the regions Σ_3 and Σ_4 ; $\dot{J}_n$ known along the remaining regions Σ_1 and Σ_2 . In this case it is convenient to describe the electromagnetic diffusion inside the plate by means of the magnetic field:

$$j\omega\mu\sigma \dot{H}_z = \nabla^2 \dot{H}_z \tag{2}$$

Being $\dot{H}_z$ the only component of $\dot{\mathbf{H}}$ different from 0. Now, taking into account the current continuity condition in integral form, we obtain (see Fig. 1):

$$\frac{\dot{I}}{w} = \int_a^{a'} \dot{J}_x dy = \int_a^{a'} \frac{\partial \dot{H}_z}{\partial y} dy = \dot{H}_z(a') - \dot{H}_z(a) \tag{3}$$

considering a section aa' along y-axis, and

$$0 = \int_b^{b'} \dot{J}_y dx = \int_b^{b'} \frac{\partial \dot{H}_z}{\partial x} dx = \dot{H}_z(b') - \dot{H}_z(b) \tag{4}$$

considering a section bb' along x-axis. Due to the fact that the input and the output sections of the current are narrower than the plate height we can suppose a current density in Σ_1 and Σ_2 :

$$\dot{J}_x = \dot{I} / \Sigma_1 \text{ along } \Sigma_1; \qquad \dot{J}_y = \dot{I} / \Sigma_2 \text{ along } \Sigma_2 \tag{5}$$

and we obtain the boundary conditions for $\dot{H}_z$:

$$\dot{H}_z = \frac{\dot{I}}{2w} \quad \text{along } \Sigma_3 \qquad \dot{H}_z = -\frac{\dot{I}}{2w} \quad \text{along } \Sigma_4$$

$$\dot{H}_z = \frac{\dot{I}}{w}\frac{y - y_c - \frac{\gamma_1}{2}}{\gamma_1} \quad \text{along } \Sigma_1 \qquad \dot{H}_z = \frac{\dot{I}}{w}\frac{y - y_d - \frac{\gamma_2}{2}}{\gamma_2} \quad \text{along } \Sigma_2 \tag{6}$$

It may be noted that the model above described can be extended to configurations that contain different positions of the current input and output, and more than one current. Moreover in the general case of not-constant current density in Σ_1 and Σ_2 the magnetic field boundary conditions can be deduced integrating the relation:

$\frac{\partial \dot{H}_z}{\partial y} = \dot{J}_x(y)$ along Σ_1 and Σ_2, where $\dot{J}_x(y)$ is a known function. Equation (2) has been integrated by a classic central finite difference algorithm:

$$j\omega\mu\sigma \dot{H}_{z(i,j)} = \frac{\dot{H}_{z(i+1,j)} - 2\dot{H}_{z(i,j)} + \dot{H}_{z(i-1,j)}}{\Delta x^2} + \frac{\dot{H}_{z(i,j+1)} - 2\dot{H}_{z(i,j)} + \dot{H}_{z(i,j-1)}}{\Delta y^2} \tag{7}$$

The numerical solver chosen uses successive over-relaxation techniques, and ensures in general a shorter computation time than those necessary with direct inversion methods.

NUMERICAL RESULTS AND VALIDATION

The model described in the previous paragraph has been tested by comparison with the analytical expressions derived by Ney [4], valid under these geometrical conditions:

- The structure has two axis of symmetry x and y;
- The length l is relatively large, so that the striction effect can be neglected at $x = l/2$;
- The contact width is sufficiently small so that the current distribution can be considered uniformly distributed.

The dimensions of the flat plate chosen for the test were: $l = 5$ cm, $h = 1$ cm, $w = 10$ µm, $\gamma_1 = \gamma_2 = 2$ mm. We found that the analytical expressions reach the final value taking into account less than 100 terms and this number remains the same for every frequency in the range examined from 100 Hz to 1 GHz. To avoid numerical instabilities we decided to use constant spatial steps of discretization Δx and Δy. Therefore, the number of the nodes of the discretizing grid used for this finite difference algorithm must be increased with the increasing of the frequency, to simulate correctly the consequent decreasing of the skin depth. This fact causes a strong growth of the time necessary for the simulation. For istance, the computational time necessary at 1 GHz resulted more than 1000 times longer than those necessary at 1 KHz. To avoid this numerical complication we used for the numerical simulations in the field of the higher frequencies (above 10 KHz) a finite difference scheme based on a exponential function interpolation given by:

$$\dot{f}(x) = \dot{A}e^{\alpha x} + \dot{B}e^{-\alpha x} + \dot{C} \tag{8}$$

where $\dot{A}$, $\dot{B}$ and $\dot{C}$ are suitable constant, functions of the values assumed by $\dot{H}$ in the nodes of the grid and $\alpha = (1+j)/\delta$, being δ the value of the one-dimensional skin depth. In this case the finite difference scheme generates the following numerical algorithm:

$$j\omega\mu\sigma \dot{H}_{z(i,j)} = \alpha^2 \left(\frac{\dot{H}_{z(i+1,j)} - 2\dot{H}_{z(i,j)} + \dot{H}_{z(i-1,j)}}{\frac{e^{2\alpha\Delta x} - e^{-2\alpha\Delta x}}{e^{\alpha\Delta x} - e^{-\alpha\Delta x}} - 2} + \frac{\dot{H}_{z(i,j+1)} - 2\dot{H}_{z(i,j)} + \dot{H}_{z(i,j-1)}}{\frac{e^{2\alpha\Delta y} - e^{-2\alpha\Delta y}}{e^{\alpha\Delta y} - e^{-\alpha\Delta y}} - 2} \right) \tag{9}$$

Using this contrivance, because the model is two-dimensional, and following the approach described in the previous section, the analysis can be limited to the only conductive region, the numerical calculation time resulted comparable with those necessary for the analytical evaluation. Moreover it must be outlined that this numerical model does not have the geometrical limitations of the Ney' s formulas. In figure 2 the graphic maps of the current density distribution obtained, for various frequencies of the input current, are reported.

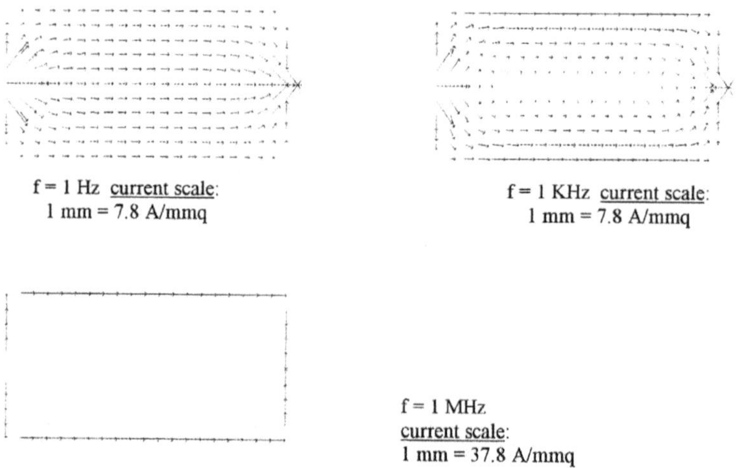

Figure 2. Current density distribution into a rectangular ground plane at t=0 for 1Hz, 1KHz and 1MHz frequencies. $i(t) = 1\cos(\omega t)$ [A].

The percentage difference among the values obtained by the finite difference numerical computations and by the analytical model resulted, limited to 2% either for the low- or for the high-frequencies. After this validation we decided to demonstrate the capability of this model to predict the internal impedance of ground plates, having rectangular shape, by defining its equivalent internal resistance and inductance as follows:

$$R = \frac{\iiint_\Omega \frac{J^2}{\sigma} d\Omega}{I^2}; \qquad L = \frac{\iiint_\Omega \mu B^2 d\Omega + \iint_\Sigma (\mathbf{H} \times \mathbf{A}) \cdot \mathbf{n} d\Sigma}{I^2} \tag{10}$$

where Ω is the plate volume, Σ its surface and $\mathbf{A}$ is the vector potential, given by:

$$\mathbf{A} = \frac{\mu}{4\pi} \iiint_\Omega \frac{\mathbf{J}}{r} d\Omega \tag{11}$$

The expressions (8) and (9) have been integrated numerically. In fig. 3 and 4 the values of the equivalent electrical parameters of the ground plate used for the test, against the input signal frequency, are reported. The value of the internal resistance increases about as the square root of the frequency up to 1MHz, while for higher values of frequency it grows linearly. The value of the internal inductance remains practically constant up to frequency of 100 KHz, then tends to increase as the square root of the frequency. This trend, even if for different frequency limits, has been observed also in rectangular plates with different dimension. We have then calculated the variation of R and L against the length of the plate.

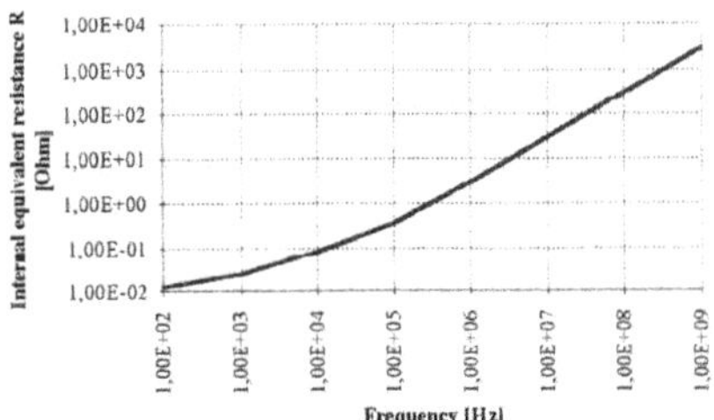

Figure 3. Internal equivalent resistance of a rectangular ground plane versus the input signal frequency. The geometrical parameters of the plate are: length l = 5 cm; height h = 1 cm; width w = 10 μm; input and output in symmetrical position.

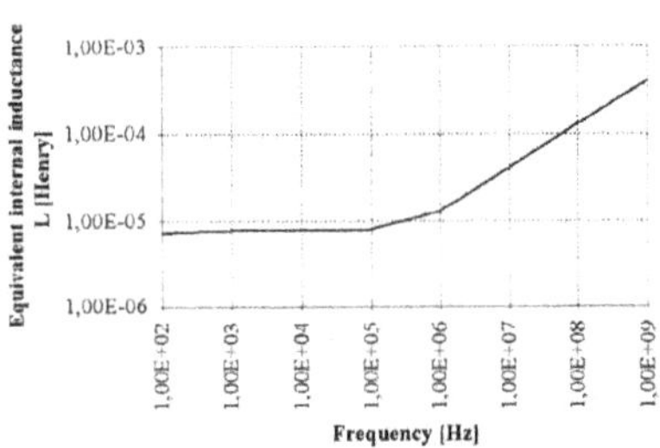

Figure 4. Internal equivalent inductance of a rectangular ground plane versus the input signal frequency.
The geometrical parameters of the plate are the same of Figure 3.

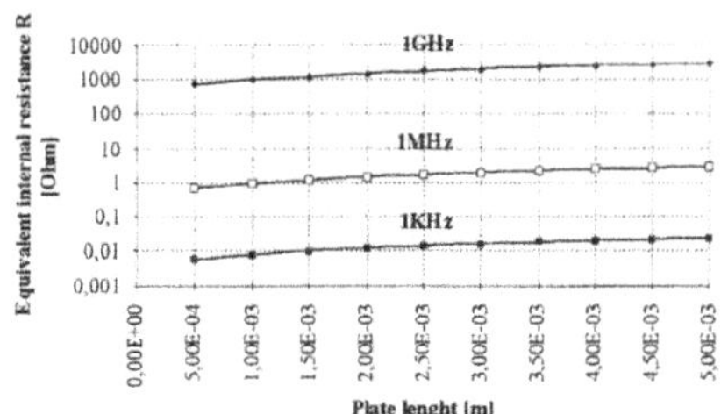

Figure 5. Internal equivalent resistance versus the length l of the ground plate and for various input signal frequencies.

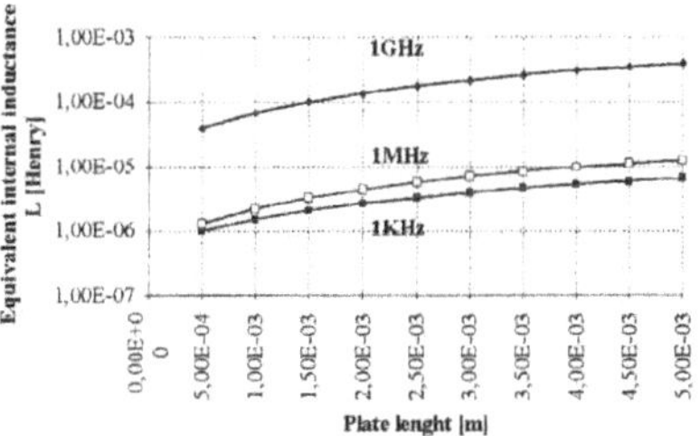

Figure 6. Internal equivalent resistance versus the length l of the ground plate and for various input signal frequencies.

The numerical results, reported in Figures 5 and 6, show that there is a weakly dependence of the internal parameters of the ground plane on the electrical parameters per unit length. Finally, in figure 7 the values of the input current to have a given 5 V overvoltage are pictured, against the frequency.

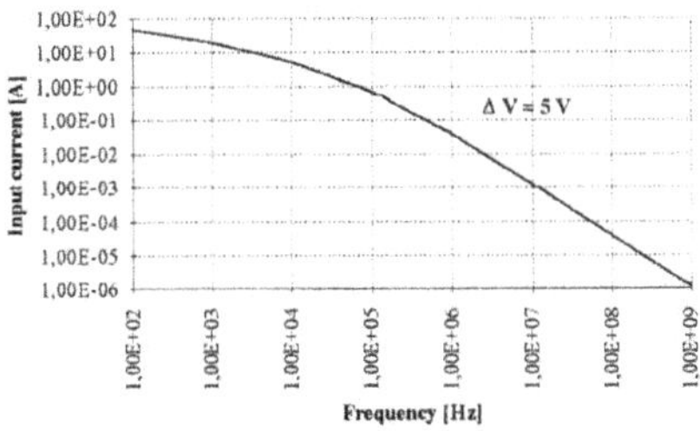

Figure 7. Values of the interfering current versus the frequency of the signal to have a fixed 5V overvoltage.

It may be noted that, while in the range of low frequencies it is not probable that the interfering current may reach values such that possible damages or resets in electronic circuits occur, nevertheless when the frequency of the interfering signal grows up to the order of GHz, few microamperes may be sufficient to create problems.

CONCLUSIONS

The finite difference numerical model presented in this paper has demonstrated effective capability to simulate the magnetic diffusion of current in flat plates. In particular the two-dimensional magnetic scalar formulation allows the limitation of the discretization region to the only conductive plate, and a consequent strong reduction of the nodes of the grid and of the computational time. Therefore, taking into account that two-dimensional computations generally involve an over-estimation of J, or anyway may introduce excessive approximations, it seems important, also in consideration of the encouraging results obtained, to continue the research in this field, trying to develop a three-dimensional model of the magnetic diffusion of the current into non perfectly conducting bodies based on magnetic scalar formulations, limiting the domain of analysis to the conductive body.

REFERENCES

1. H. Ott, Noise Reduction Techniques in Electronic Systems, John Wiley & Sons, New York, 1988.
2. C.R. Paul, Modelling of electromagnetic interference properties of printed circuit boards, IBM J. Res. Dev., vol. 33, N. 1, pp. 33-50. 1989.
3. M. M. Ney, G. J. Costache, Transient striction and skin effects on voltage drop across flat conductors, Can. J. Electr. & Comp. Eng., vol. 18, N. 4, pp. 199-204, 1993.
4. M. M. Ney, Striction and skin effects on the internal impedance value of flat conductors, IEEE trans. on Electromagnetic Compatibility, vol. 33, N. 4, pp. 321-327, November 1991.

AUTOMATIC 2D DISCRETIZATION WITH VARIABLE MESH DENSITY FOR NUMERICAL METHODS

Marko Jesenik, Mladen Trlep, Božidar Hribernik

University of Maribor, Faculty of Technical Sciences
Smetanova ul. 17, P.O. Box 224, 62000 Maribor, R. Slovenia

INTRODUCTION

Most of the time necessary to solve a problem by finite element method is spent for the data generation. Consequently, the best solution for the user is automatic data generation. It is advantageous to generate a mesh with varying density in different areas in dependence on the given problem. An automatic data generation consists of the subdivision into convex subregions, the creation of nodes and elements, and the correction of discretizated areas, which affects the accuracy of calculation [2,3,4].

BASIC REQUESTS

There are some basic requests that have to be satisfied in the discretizated area: elements have to be as similar as possible to equilateral triangles, the number of obtuse inside angles in elements has to be as small as possible, elements have to coincide with the geometry and each element can be only in one material[1].

It is advantageous to generate a mesh with the different density in different areas in dependence on the given problem, so we have to make less adaptive corrections as in case of the uniform mesh density and we can save computer time.

DEFINITION OF THE GEOMETRY

The geometry definiton is directly connected with the definition of the mesh density in different areas. The geometry is specified by means of creating shapes that form areas.

Each area - material is given as a separate shape, so the sides of elements coincide with the boundaries between materials. Each material can be given with more shapes. The mesh density in each area is defined by specifying the length of the element side.

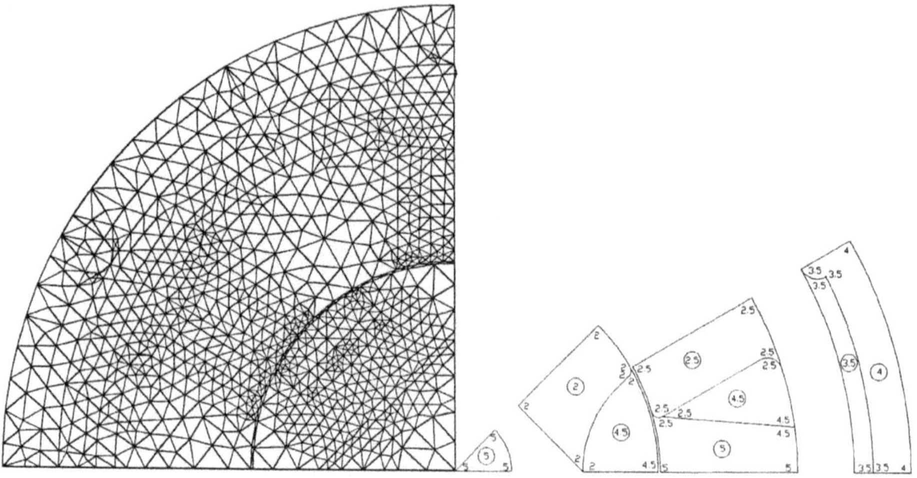

Figure 1. Given shapes with element sides length.

DIVISION OF BOUNDARY SIDES - ARCS

Figure 1 shows some given shapes together with the specified element side length (numbers in circles) and part of final mesh made from these given shapes. With this the mesh density near each of the nodes is defined. If a node is common for more shapes, the mesh density near that node is the highest density of the surrounding shapes.Now that the mesh density around the nodes is known, the connecting lines and arcs between nodes can be divided.The change of the mesh density between the shapes with different given element side lengths has to be gradual. The size of the elements is changing gradualy, so that they are not deformed. In spite of different sizes the elements are still similar to equlateral triangle.

From the node with smaller element side length the distances between nodes increase lineary to the break point and then again lineary to the node with bigger element side length.The break point is defined with two factors: F_1 and F_2. The value of factor F_1, ranging from 1 to 2, defines the position of the break point. The factor F_2 is bigger then 0. With these two factors is possible to define the divisions of the given sides and arcs (Figure 2).

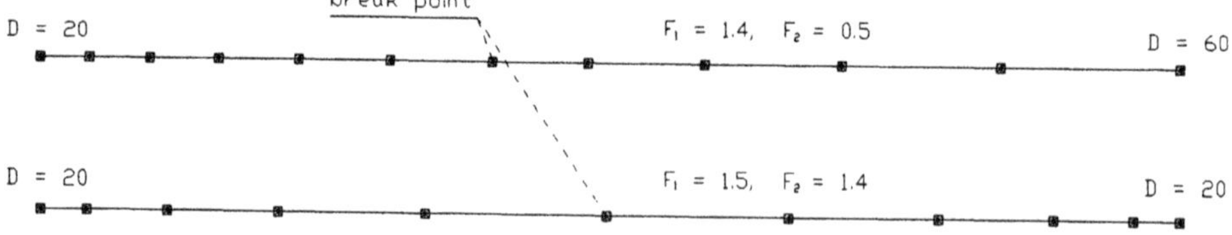

Figure 2. Division of the given sides.

DIVISION OF THE GIVEN SHAPE INTO CONVEX SUBREGIONS

Each given shape should be divided into convex parts, because the algorithm for creating nodes and the algorithm for creating elements can function only in convex shapes, as it is described [1].

The new line inside the nonconvex shape is divided in the same way as the lines that surround the shape.

Element side lengths around the division points T_1 and T_2 (Figure 3) are defined as a mean value of lengths to neighbouring nodes:

$$D_1 = \frac{D_{11} + D_{12}}{2} \quad ; \quad D_2 = \frac{D_{21} + D_{22}}{2} \tag{1}$$

In case, the element side lengths D_1 and D_2 are smaller then the given element side lengths inside the given shape, the factors F_1 and F_2 are set so that the nodes on the connecting line are generated as it is shown in the second example in Figure 2.

ADDITIONAL DIVISION OF THE GIVEN SHAPES

For each shape with divided lines the mean value of the element side lengths on the periphery can be calculated:

$$D_{sr} = \frac{\sum_{i=1}^{n} D_i}{n} \tag{2}$$

In case there are more nodes with the same distances smaller then Dsr, it is good if this shape is additionaly divided (Figure 3):

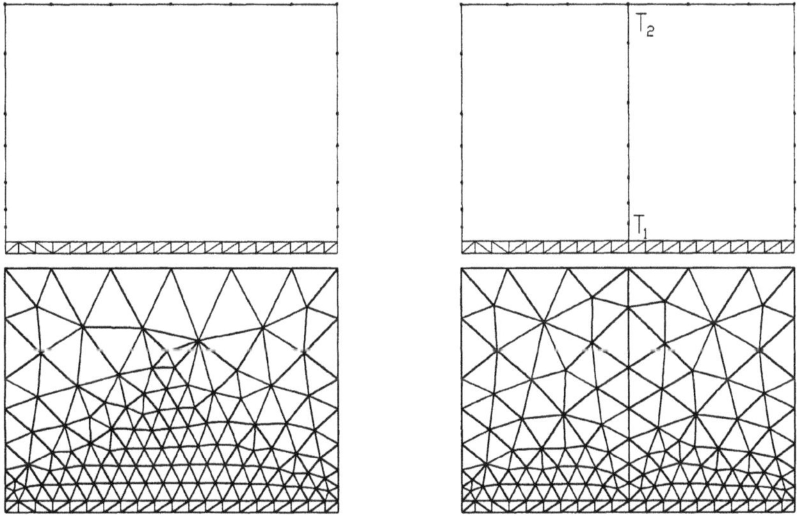

Figure 3. Discretization without and with additional dividing line.

GENERATION OF NODES

Appropriate disposition of the nodes inside the convex shape is a base for the generation of a good mash. Generation is made in hoops as it is described [1]. Three successive nodes are the base for new nodes (Figure 4).

The obtained values of l from the equation (3) are not the real distances from the old nodes to the new one. In case l is bigger than D_{sr} (former chapture), it is multiplied with 0.9, so the grading refinement is reached. In case that the l is smaller than D_{sr} it is multiplied with 1.1 so the mesh becomes courser.

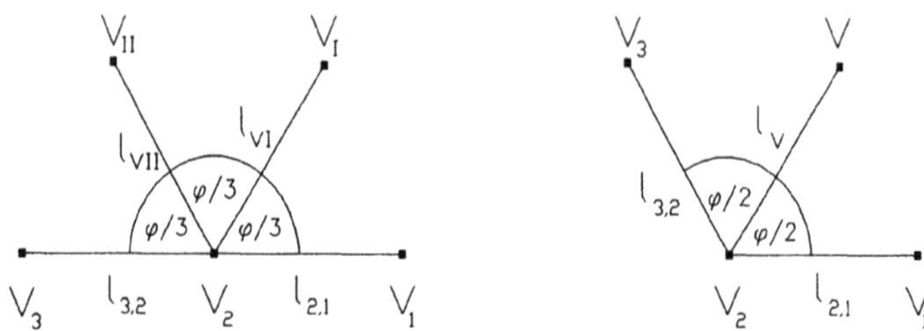

Figure 4. Creating of nodes.

$$l_{VI} = \sqrt[3]{l_{2,1}^2\, l_{3,2}} \quad l_{VII} = \sqrt[3]{l_{3,2}^2\, l_{2,1}} \quad l_V = \sqrt{l_{2,1}\, l_{3,2}} \tag{3}$$

For the generation of each node l had to be calculated and this could be the approximate value of the element side length around the new node. This value is used to erase some new nodes. The nodes that are too close to the nodes created two iterations back are erased, the nodes that are too close to the nodes created one iteration back, are eliminated. Two to three nodes are combined, if they are too near together, and the nodes that are outside the given shape or outside the hoop which generate the nodes are erased.

GENERATION OF ELEMENTS

Generation of elements and the correction of the discretizated area is described [1]. A base for good generation of the elements are well arranged nodes - their density is changing gradualy from the area with one given element side length to the area with another given element side length. Correction of the discretizated area consists of elimination of too large elements, correction of neighbouring element sides and shifting of the node in the middle of the elements[1].

This algorithm was successfully used for many real applications.

REFERENCES

1.Marko Jesenik, Mladen Trlep, Božidar Hribernik; AUTOMATIC 2D DISCRETIZATION FOR NUMERICAL METHODS, 5th International IGTE Symposium on Numerical Field Calculation in Electrical Engineering, Graz, Austria September 28-30 1992

2.EDVARD A. SEDAK; Faculty of Engineering, Cairo University, Egipt (A.R.E.); A SCHEME FOR THE AUTOMATIC GENERATION OF TRIANGULAR FINITE ELEMENTS; INTERNATIONAL JOURNAL FOR NUMERICAL METHODS IN ENGINEERING, VOL. 15, 1813-1822 (1980).

3.J.P. Assumpcao Bastos, R. Carlson; GRUCAD/EEL/UFCS C. P. 467 Floranopolis 88049 Brazil; AN EFFICIENT AUTOMATIC 2D MESH GENERATION METHOD; COMPUMAG 8th CONFERENCE ON THE COMPUTATION OF ELECTROMAGNETIC FIELDS, July 7.11.1991- Sorrento, Italy, VolumeI(83,86)

4.M. Fujita and M. Yamana; Fuji Electric Corporate Research & Development, LTD. TWO-DIMENSIONAL ADAPTIVE FEM MESH GENERATION; MAGNETICS JANUARY 1988 VOLUME 24 NUMBER 1.

PECULARITIES OF BOUNDARY ELEMENT METHOD IN EDDY CURRENT ANALYSIS

Andrzej Krawczyk and Tadeusz Skoczkowski

Department of Fundamental Research in Electrotechnics
Institute of Electrical Engineering
04-703 Warsaw, Poland

INTRODUCTION

The Boundary Element Method (BEM) has for the last dozen of years attracted researchers' interest, especially when applied to eddy current (*i.e.* time-dependent) problems either quasi-stationary (sinusoidal) or transient[1]. That is for many reasons but two of them are of special importance:

1. BEM allows to analyse the skin-effect in a very accurate way since the internal point can be chosen as close to the boundary as an analyst wishes,

2. using BEM in transient problems one avoids numerical differentiation in time which may cause some indeterminate errors.

The above counted items complement the Finite Element Method (FEM), or generally, domain methods, which just fail in these two areas. Indeed, the problems of mesh refinement or stability of solution constitute crucial difficulties while eddy currents are analysed by domain methods. Since BEM is believed to solve the above problems, one has to expect simultanously some new difficulties, as one has to pay for everything, especially for advantages.

This paper intends to show some special, peculiar areas of BEM which are either positive or negative but not occuring, in turn, in the domain methods. Inversely, the discussed below problems are strictly complemented to domain methods, especially to FEM.

It is assumed in the paper that the background of BEM is well-known, thus will not be repeated here.

Electric and Magnetic Fields, Edited by A. Nicolet
and R. Belmans, Plenum Press, New York, 1995

INTEGRATION

In the quasi-stationary case, eddy current problem is described by Helmholtz's equation, fundamental solution (free-space Green function) of which behaves as logarithm in the 2-D case and as r^{-n} in the 3-D case.

$$u_j^*(x,y) = \mathcal{F}(\ln r_{kj}) \qquad \text{for 2-D} \tag{1}$$

or

$$u_j^*(x,y,z) = \mathcal{F}(r_{kj}^{-n}) \qquad \text{for 3-D} \tag{2}$$

where r_j-distance between the field point j and the boundary element k. The normal derivatives of the functions behave in the same way.

Besides the normal difficulties connected with the numerical integration of the above functions a special problem appears which is related to the integration with singularity, *i.e.* when $i = j$ or in other words when r_j tends to zero*.

There are some special approaches to remedy this situation. Among them there are the use of series expansion of the integrand, the approximation of the integrand by a well-integrable function or calculating the integral analytically, if possible. The latter is commonly used in the stationary case where the integrand is expressed in terms of logarithms. In the case of Helmholtz's equation it would be difficult to find either an analytical formula or a series expansion of the integrand. Then it is proposed here to use the technique consisting in adding and subtracting the part of the integrand which produces a singularity. For example, in the two-dimensional case this part will be just a logarithm and one can calculate the integral in the following manner:

$$I = \int_{e_j} (K_0(\alpha r_{jj}) - \ln(r_{jj}))\, dl + \int_{e_j} \ln(r_{jj}) dl \tag{3}$$

The first integral is deprived of singularity while the second integral may find an analytical representation.

Another solution may be adopted to transient problem. Taking again the two-dimensional case, one can show the integral over element e_j as a sum of integrals[2]:

$$I = \int_{e_j - 2\epsilon} E_1 \left(\frac{\alpha r_{jj}^2}{4(t - t_{i-1})} \right) dl + \int_{-\epsilon}^{+\epsilon} E_1 \left(\frac{\alpha r_{jj}^2}{t - t_{i-1}} \right) dl \tag{4}$$

where E_1 is the exponent integral.

Choosing ϵ such that the argument of E_1 is equal to 1 one can find an analytical solution of the second integral expanding function E_1 into series. Finally, one obtains:

$$I = \int_{e_j - 2\epsilon} E_1 \left(\frac{\alpha r_{jj}^2}{4(t - t_{i-1})} \right) dl + 2 \mid \epsilon \mid 0.29024786... \tag{5}$$

In the three-dimensional case, when indirect formulation is employed, the numerical integration fails since the integrands involve the singularity of $1/r^n$ type. Then the special procedure should be adopted and it is presented below. The integrals (only existing if $k = j$) can be transformed to the forms:

$$I_1 \mid_k = \int_{e_j} \boldsymbol{n}_j \times \frac{\boldsymbol{r}_{kj}}{r_{kj}^3} dS = \int_{e_j} \boldsymbol{n}_j \times \text{grad} \frac{1}{r} dS \tag{6}$$

---

* Note that for straight line elements normal derivative vanishes and for curvelinear elements at least the vicinity of the point of singularity and the point itself are excluded from the integration process as the normal derivative of r is very small or even equal to zero.

$$I_2 \mid_k = \int_{e_j} \boldsymbol{n}_j \cdot \boldsymbol{\tau} \times \boldsymbol{r}_{kj} \Psi dS = \int_{e_j} \boldsymbol{n}_j \cdot \mathrm{curl}(\boldsymbol{\tau} K) dS \tag{7}$$

Then one can apply the vector analysis theorems which gives:

$$I_1 \mid_k = \int_{l(e_j)} \frac{1}{r_{kj}} dl \tag{8}$$

$$I_2 \mid_k = \int_{l(e_j)} \boldsymbol{\tau} K dl \tag{9}$$

where K is the time-dependent fundamental solution and Ψ is the kernel of boundary integrals (see Krawczyk and Tegopoulos[3])

In many cases the problem of quasi-singularity exists, *i.e.* the calculation of integrals from the point being placed close to the element of integration. Then the procedure shown in eqn (3) is to be adopted. In such cases the integration of normal derivative is quasi-singular too and the procedures described remain valid.

CALCULATION OF EDDY CURRENTS

The validity of the BEM in transient process analysis reveals itself when eddy currents are to be examined. It is most clearly seen in the 2-D case treated by the magnetic vector potential A. Then the eddy current density results from the time differentiation of A which is normally made numerically in the domain method. It may introduce and really introduces an additional error which may, in fact, be indeterminate, as numerically obtained results of magnetic vector potential are loaded with non-physical oscillations in time[3].

The use of the BEM enables the solution of the problem without the loss of accuracy since the differentiation in time can be made analytically. Indeed, due to the analytical form of integrands (fundamental solution) one can compute eddy currents directly making use of some elements of the convolution theorem. Then the eddy current density is expressed in terms of the boundary values obtained previously and hence any additional error is not introduced. In the two-dimensional case the magnetic vector potential at the point k inside the domain in question is calculated as a sum of convolution integrals:

$$A \mid_k = \sum_{i=1}^{N} \int_0^t f_i(\tau) K_{ik}(t-\tau) d\tau \tag{10}$$

Then the eddy current density is given as:

$$j_s \mid_k = -\gamma \frac{\partial A}{\partial t} \mid_k = \sum_{i=1}^{N} \int_0^t f_i(\tau) \frac{\partial}{\partial t} K_{ik}(t-\tau) d\tau \tag{11}$$

where $\{f_i\}$ is the set of the boundary values (either potential or its derivative) at the point i and K_{ik} - analytically given kernel of the integral. Since the differentiation in time may be done analytically, the eddy current density is computed as accurately as the boundary values were.

Conceptually similar but essentially more difficult is the calculation of eddy current density in the three-dimensional problem that is formulated in $\boldsymbol{H} - \varphi$ terms. After having solved the boundary-integral equation one has the boundary values of fictitious sources τ and σ. Then the value of eddy current density at the point P lying inside the domain in question is:

$$j(P,t) = \int_S \sum_{i=1}^{n} \tau(M,t_i), [K(P,M,t,t_{i-1}) - K(P,M,t,t_i)]dS +$$
$$\int_S \sum_{i=1}^{n} \text{grad}(\tau(M,t_i) \cdot \boldsymbol{r}_{MP} \Psi(P,M,t,t_{i-1},t_i))dS \qquad (12)$$

where K and Ψ are as in eqn (7).

Having the density of eddy currents one can easily compute an eddy current loss, forces and the other integral parameters. It should be underlined that such possibilities are unapproachable by means of the other methods, especially those of the domain type. The semi-analytical way of eddy current calculation is of special importance when the skin effect has to be analysed. Then the analyst can choose the internal points as close to the boundary as he or she wishes and hence the analysis gives more accurate results near the skin than by using other methods which are strongly limited by a regular mesh[2,3]. Note that in this case the procedures of integration, described above, should be employed.

CONCLUSIONS

The paper intended to show some special properties of BEM when applied to eddy current problems. In spite of all the difficulties, the method may be successfully employed in eddy current calculation as it gives some unbeatable merits. One of the main advantages of BEM are its boundary nature (geometrically) and an analytical form (mathematically), thus so the skin effect and the non-uniform distribution of the eddy current density can be analysed more accurately.

References

[1] A. Krawczyk, J. Turowski, Recent development of eddy current analysis, *IEEE Transaction on Magnetics*, 6:3032 (1987).

[2] A. Krawczyk, The analysis of transient eddy currents by means of the boundary element method, *in:* "Boundary Element VIII", C.A. Brebbia, M. Tanaka, eds, Springer Verlag, Berlin (1986).

[3] A. Krawczyk, J.A. Tegopoulos. "Numerical Modelling of Eddy Currents", Clarendon Press, Oxford (1993).

OPTIMIZATION TECHNIQUES IN THE DESIGN OF ELECTROMAGNETIC DEVICES

Vasile Topa, Emil Simion and Calin Munteanu

Technical University of Cluj-Napoca
Electrical Engineering Department
C.Daicoviciu street, 15
Cluj-Napoca, 3400, Romania (RO)

INTRODUCTION

The interest in the study of methods and the development of algorithms of design optimization, particularly in the field of electromagnetic device design, has largely increased in the recent years. Many electromagnetic problems require to evaluate the distribution of an unknown quantity (the source), which produces a specified quantity (the effect) in a given region of the space. Typical specified quantities include local parameters, such as magnetic flux densities or global parameters, such as forces or torques acting on some parts of the device, coil inductance, stored energy, etc. These problems belong to the class of inverse ones and are usually constrained by a number of geometrical and technical aspects, giving rise to the constrained optimization problems, as shown in [1,2,3].

This important class of inverse electromagnetic problems can be formulated and solved using a numerical field calculation, by means of an optimization algorithm included in the computing process. Various techniques have been developed to solve the constrained optimization problems. Unfortunately there does not exist a universal algorithm that works well for all problems, because the convergence and the efficiency of a particular algorithm are dependent on the problem to be solved. For one particular problem, one method may provide a faster convergence that the other, whereas in other cases it may fail to converge. The choice of an adequate optimization method is one of the central questions.

Different higher-order deterministic optimization methods (HODOM) were tested in many magnetostatic design problems to set up a software package for the optimization of electromagnetic devices. The numerical field calculations have involved the Finite Element Method (FEM) in 2D, both linear and non-linear cases.

NON-LINEAR PROGRAMMING METHODS

Mathematically, a simplified model of the magnetostatic non-linear constrained optimization problem can be formulated as follows :

Electric and Magnetic Fields, Edited by A. Nicolet
and R. Belmans, Plenum Press, New York, 1995

$$\begin{aligned} &\text{Minimize}: \; f(p,A)\,, && p \in R^n \\ &\text{Subject to}: \; g_i(p,A) \le 0\,, && i=1,2,..,m.\,, \end{aligned} \tag{1}$$

where the n design variables are collected in the column vector p and could be geometric dimensions, material properties, current distributions, etc. When optimizing electromagnetic devices, the objective function f and the constraints functions g_i represent performance characteristics that are formulated with field quantities. The fields are obtained numerically with the magnetic vector potential A, which itself depends implicitly on the design variables. In FEM the state variable A is computed from a matrix equation of the form:

$$M(p)A = Q(p)\,, \tag{2}$$

where M(p) is the stiffness matrix and Q(p) the column vector of applied loads.

If the design variables are the geometric device dimensions, we are talking about a so called shape optimization problem. In this case, in FEM mesh generation program we used the concept of parametrized triangulation of the domain.

There are two main approaches for solving the constrained optimization problems (1), the optimality criteria (OC) method and the mathematical programming method (MP). On the basis of the extreme principles, the OC method derives the state conditions at the optimum and then updates the current design toward these state conditions. On the other hand, the MP method starts with the current design, and uses the information concerning the current design to find how to reduce the objective function and satisfy the constraints. However, each method offers some advantages and disadvantages. The MP method is more general than the OC one, because the optimality conditions are difficult to be formulated in many practical cases. Therefore the MP method is employed in the current research. To eliminate the constraints functions, different methods of penalty (the linear extended interior function, the quadratic extended interior function and the exterior penalty function) were used.

The numerical MP optimization methods usually start with an initial design p_0,, then update the improved design using the following iteration process:

$$p^k = p^{k-1} + \alpha S^k, \tag{3}$$

where k is the iteration number, S^k is the vector of the search direction and. α is a scalar parameter that specifies the movement of the design variables, determined by the one-dimensional minimization problem. The individual HODOM differ mainly in the way they construct the slope vector S^k. We have carried out tests with the following optimization algorithms presented in detail by Vanderplaats[4]: steepest descent, conjugate directions, the quasi-Newton method by Davidon, Fletcher and Powell (DFP) and the quasi-Newton method by Broyden, Fletcher, Goldfarb and Shanno (BFGS).

SENSITIVITY ANALYSIS

Design sensitivity analysis provides the gradients of the objective function and constraints concerning the design variables:

$$\frac{df}{dp} = \frac{\partial f}{\partial p} + \left(\frac{\partial f}{\partial A}\right)^T \frac{\partial A}{\partial p}\,. \tag{4}$$

By calculating the derivatives of field quantities with respect to the design variables, sensitivity analysis gets the essential information for coupling numerical field analysis with the non-linear programming methods. The capability to obtain the design sensitivity analysis in an accurate and efficient manner is essential for all problems. It is well known that, there are two main approaches to calculate the design sensitivity in FEM: the discretized approach (DA) and the continuum approach (CA), as shown in[5,6]. The DA uses a discretized model to carry out the sensitivity analysis, which includes three methods: finite difference method (FDM), analytical method (ANM) and semi-analytical method (SAM). The ANM differentiates the system equation (2) directly with respect to the design variables :

$$M\frac{\partial A}{\partial p} = \frac{\partial Q}{\partial p} - \frac{\partial M}{\partial p}A \equiv R \tag{5}$$

The adjoint variable method (or state space) starts by defining a vector of adjoint variables λ, that satisfies the equation:

$$M\lambda = \frac{\partial f}{\partial A} \tag{6}$$

Using equations (4), (5) and (6), the exact expression of the total derivative becomes:

$$\frac{df}{dp} = \frac{\partial f}{\partial p} + \lambda^T R \tag{7}$$

This approach has the advantage of being exact, but cannot be generalized. The above overall procedure is summarized as follows: first solve the system equation (2) for the state variable A, second solve the adjoint equation (6) for the adjoint variable and third evaluate the sensitivity df/dp from the previous solutions. In all the test magnetostatic problems, that we have solved, the described sensitivity analysis was used.

APPLICATION

A software package for optimization of electromagnetic devices using the shown principles have been developed. Several studied cases for the design of electromagnetic devices, such as synchronous machine and various kinds of electromagnets were performed using this computer code. Every device structure was analysed with all the proposed HODOM algorithms, and the best result was retain. The efficiency of the numerical optimization methods depends on the type of problem to be solved. The choice of numerical optimization methods mainly depends on the number of design variables and the properties of the objective functions and constraints (i.e. the smoothness and the order of non linearity).

As a numerical result obtained, with this software package, the optimal pole shape of a salient pole synchronous machine is considered, to achieve a sine-distribution of the magnetic flux density in the air gap. The objective function was defined as a least square function that describes the deviation between the target magnetic flux density and the computed values by FEM analysis.

The contour lines of the magnetic potential for the initial geometry of the magnetostatic problem (a) and the final one (b) obtained in the optimization process are given in the figure 1 (only for a half because its symmetry). The best solution was obtained using as optimization algorithms the quasi-Newton method (Davidon, Fletcher and Powell). After 20 iterations the objective function converges as shown in figure 2. The distribution of the magnetic flux density converged to the prescribed sine-distribution flux density with the maximum deviation of 0.6%. In figure 3 the distributions of magnetic flux density for the initial structure, the target value and the final one are presented..

The proposed software package is competitive with other known efficient ones. Our results and conclusions were compared with similar ones obtained by other authors[6,7].

This software package will be developed in the future by including the zero-order stochastic optimization methods (ZOSOM).

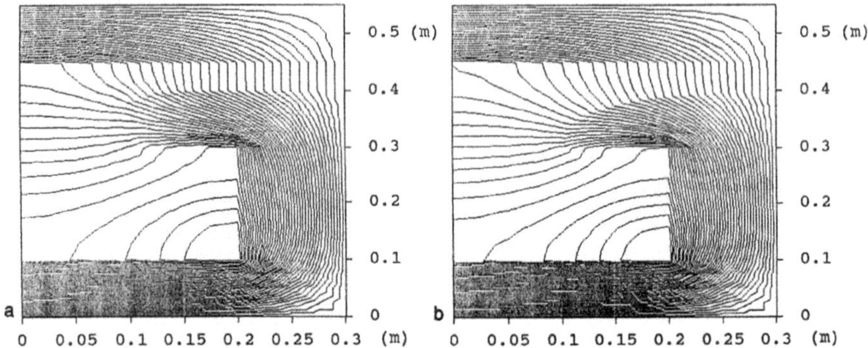

Figure 1. The contour lines of the magnetic potential for the initial (a) and the optimal (b) structure

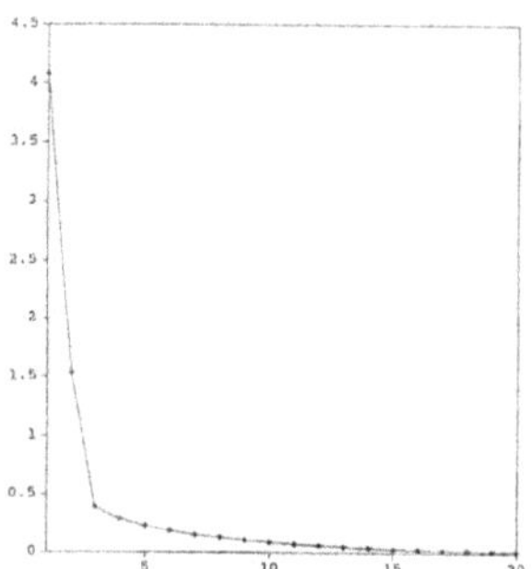

Figure 2. Variation of objective function versus iteration number

Figure 3. Magnetic flux density distribution in the air gap

REFERENCES

1. M. Guarnieri. A. Stella, and, F. Trevisan, A methodological analysis of different formulations for solving inverse electromagnetic problems. *IEEE Trans. on Magnetics.* 622:625, (1990).
2. J. Simkin and C.W. Trowbridge. Optimization problems in electromagnetics. *IEEE Trans. on Magnetics.* 4016:4019. (1991).
3. K. Preis. O. Biro. M. Friedrich. A. Gottvald and. C. Mangele. Comparison of different optimization strategies in the design of electromagnetic devices. *IEEE Trans. on Magnetics,* 4154::4157. (1991).
4. G. Vanderplaats. "Numerical Optimization Techniques for Engineering Design ". McGram-Hill, New-York. 1984.
5. H. M. Adelman. and R.T. Haftka. Sensitivity analysis of discret structural systems. *AIAA Journal,* 823: 832, (1986).
6. I. Park, B. Lee. and S. Hahn. Design sensitivity analysis for non-linear magnetostatic problems using finite element method. *IEEE Trans. on Magnetics,* 1533 :1536, (1992).
7. S.R.H. Hoole. S. Subramanian, and S. Kanaganathan. Two requisite tools in the optimal design of electromagnetic devices, *IEEE Trans. on Magnetics.* 4105: 4109, (1991).

ELECTRIC FIELD OPTIMIZATION PROBLEMS USING THE BOUNDARY ELEMENT METHOD

Calin Munteanu, Emil Simion, and Vasile Topa

Technical University of Cluj-Napoca
Electrical Engineering Department
C. Daicoviciu street, 15
Cluj-Napoca, 3400, Romania (RO)

INTRODUCTION

This paper presents a shape optimization problem based on the boundary element method coupled with optimization techniques. The class of the studied problems is the two-dimensional, homogenous and linear domains, described by Laplace's equation.

The solution of shape optimization problems requires the determination of the location of an initially unknown movable boundary. A beginning position for the free boundary is assumed and its location is adjusted by minimising an objective function in the numerical procedure. More specific, in the present work the problem's formulation aims to get an imposed electric field distribution inside a sub domain D_O by modifying the domain's D boundary divided in a fixed one Γ_1 and a movable one Γ_2.

Due to its easier way of implementation and its accurate solutions, the boundary element method is mostly used in the analysis of electromagnetic field. Unlike the finite element method that requires a mesh for the whole analysed domain, the boundary element method needs only the discretization of the domain boundary, so an important economy of calculus time and hard resources are obtained. These facilities of the proposed method can be extended to shape optimization algorithms where regriding the mesh of the boundary is easier than in domain methods and also leads to accurate solutions .

A numerical example is performed and the results are compared with those obtained by R. Palka[1] using the regularisation method combined with finite element method, and D. Micu[2] using the Monte Carlo method of synthesis in electric fields.

BOUNDARY ELEMENT ANALYSIS

In two-dimensional boundary element analysis, the boundary of the domain is divided into small elements where the potential u and its normal derivative q are approximated, as

Electric and Magnetic Fields, Edited by A. Nicolet
and R. Belmans, Plenum Press, New York, 1995

shown by Brebbia[3]. In order to determine the unknowns' values of u and q at the nodal mesh, the starting matrix equation system can be expressed as:

$$GQ = HU, \tag{1}$$

where the terms g_{ij} of matrix G and h_{ij} of matrix H are connection coefficients between the node i and the segment j. Regrouping the terms in equation (1), the system becomes:

$$G_1X = H_1K, \tag{2}$$

where X, K are the vectors of the unknown, respectively known nodal values of u and q on the boundary.

Solving the equation (2), the method allows to determine potential values of internal points of the domain as follows:

$$U_c = G_iQ - H_iU, \tag{3}$$

where U_c is the vector of the internal point potentials and U, Q vectors that contain the u, q values of the boundary mesh nodes.

DESIGN SENSITIVITY ANALYSIS

Above all, the shape optimization assume to impose design parameters p_m and to define an objective function F. Deterministic methods use the gradient information and require the differentiability of the objective function. By calculating the derivatives of field quantities with respect to the design variables, design sensitivity analysis provides the essential information for coupling boundary element analysis with mathematical optimization. In this paper, the design sensitivity analysis is computed by directly differentiating the boundary element matrices[4,5]. The result of this analysis is a gradient vector, noted R, used in the optimization algorithm.

The design parameters are the nodal coordinates of the movable discretized boundary and they generate the design vector P. The objective function, which has the minimum value for the optimal shape, is defined as a least square function. This function describes the deviation between the target (u_{kc}) and the calculated values (u_{ko}) of electric potential for internal testing points:

$$F = \frac{1}{2}\sum_k (u_{kc} - u_{ko})^2 . \tag{4}$$

The derivative of the objective function (4) with respect to the parameters p_m can be expressed as follows:

$$\frac{\partial F}{\partial p_m} = \sum_k (u_{kc} - u_{ko}) \frac{\partial u_{kc}}{\partial p_m} \tag{5}$$

The values of the potential derivatives from equation (5) can be determined by deriving the equation (3):

$$\frac{\partial U_c}{\partial p_m} = \frac{\partial G_i}{\partial p_m} Q - G_i \frac{\partial Q}{\partial p_m} - \frac{\partial H_i}{\partial p_m} U - H_i \frac{\partial U}{\partial p_m} \tag{6}$$

The derivatives of vectors U and Q from the equation (6) can be determined by deriving the equation (2) as follows:

$$\frac{\partial G_I}{\partial p_m} X + G_I \frac{\partial X}{\partial p_m} = \frac{\partial H_I}{\partial p_m} K \tag{7}$$

and regrouping the terms of the X vector derivative.

It is mentioned that the derivatives of the metric coefficients from equations (6), (7) are analytically calculated starting from expressions resulted in the boundary element analysis.

OPTIMIZATION

The efficiency of numerical algorithms depends on the type of the problem to be solved. In this paper, a Fletcher-Reeves algorithm has been used[6] due to its rapid convergence and small number of iterations.

If $P^{(i)}$ is the vector of the design parameters at iteration i, the new vector $P^{(i+1)}$ is determined from a linear equation:

$$P^{(i+1)} = P^{(i)} + \alpha S^{(i+1)} \tag{8}$$

In equation (8) $S^{(i+1)}$ is the new searching direction determined in the current iteration from:

$$S^{(i+1)} = -R^{(i+1)} + \beta^{(i)} S^{(i)}, \tag{9}$$

where:

$$\beta^{(i)} = \frac{\left|R^{(i+1)}\right|^2}{\left|R^{(i)}\right|^2} \tag{10}$$

It is mentioned that $S^{(i)}$, $R^{(i)}$ from equations (9), (10) are respectively the searching direction vector and the gradient vector determined in the previous iteration.

The movement of the length step α along the search direction is determined by the one-dimensional minimisation method[6]. In this way the objective function F will have the minimum value at the end of each iteration.

NUMERICAL EXAMPLE AND RESULTS

The proposed optimization algorithm is applied to the domain presented in figure 1. The movable boundary is defined as the shape of two electrodes of constant potentials V_0 and $-V_0$. Because of the problem symmetry, only a quarter part of the given domain, more specifically the region placed on the first front was taken into account. The testing points are situated in the area inscribed by D_O and the demanded conditions inside this region are the values of the potential in order to obtain an uniform distribution of the electric field $E(x,y)=E_0 y$.

The imposed boundary conditions are: $u(x,0)=0$; $u(a,y)=0$; $u'_x(0,y)=0$;

The numerical values used are[1] (figure 1): $a=b=1$; $c=d=0.4$; $V_0=1$; $E_0=1$.

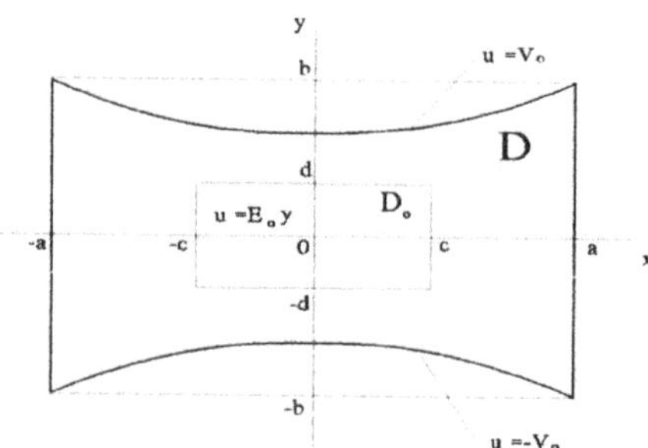

Figure 1. Domain under investigation in the numerical example.

The coordinates of 10 points located on the surface of the electrode are taken as the design variables and allowed to move in y-direction and 20 testing points were selected inside the domain D_0.

The optimized shape is obtained after 9 iterations. The maximal deviation between the prescribed and imposed potential values is about 0.2%. Initial and optimized shapes of the electrode are presented in figure 2, compared with those obtained by R. Palka[1] and D. Micu[2].

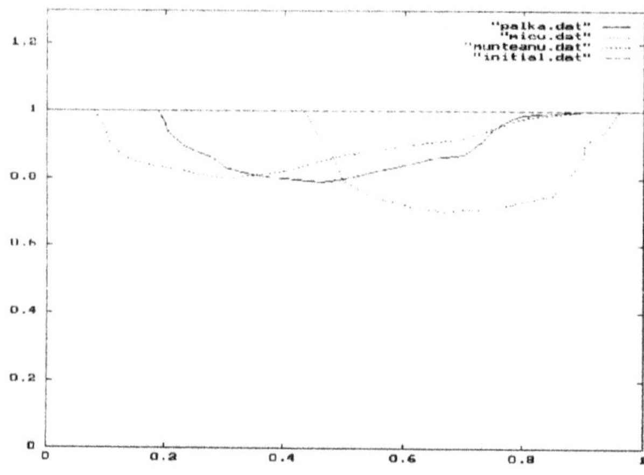

Figure 2. Optimized boundaries.

REFERENCES

1. R. Palka. "Synthesis of electrical fields by optimization of the shape of the region boundaries", *IEEE Proceedings*, Vol.132 Pt. A. No. 1 (1985).
2. D. Micu. Doctoral dissertation. Cluj-Napoca (1993).
3. C.A. Brebbia. "The Boundary Element Method for Engineers". Pentech Press. London (1980).
4. Z. Zhao. "Shape Design Sensitivity Analysis and Optimization Using the Boundary Element Method". Springer-Verlag Berlin. Heidelberg (1991).
5. C.S. Koh and S.Y. Hahn. "A sensitivity analysis using boundary element method for shape optimization of electromagnetic devices". *IEEE Transaction on Magnetics*, Vol. 28. No. 2 (1992).
6. I. Dancea. "Metode de Optimizare". Editura Dacia. Cluj-Napoca (1973).

A VECTORIAL FINITE ELEMENT PROCEDURE FOR SOLVING THREE-DIMENSIONAL FIELD PROBLEMS

G. Cannistrà[1], M. Minenna[1], G. Negro[1], M. Sylos Labini[1]

[1]Politecnico di Bari - Dipartimento di Elettrotecnica ed Elettronica

ABSTRACT

All the steps underlying the finite element method, i.e. calculation of the matrix of the coefficients [C], and solution of the linear system, are analysed; consequently, some algorithms are carried out to speed up the whole finite element computation process. These algorithms have been implemented on a package previously carried out at the Polytechnic of Bari. The new package, which is useful for studying both electrostatic and current field problems, has been applied to the study of the 3-D current field generated by leakage grounding grids.

INTRODUCTION

The finite element method (FEM) is based on some steps such as for example calculation of the matrix of the coefficients [C], and solution of the linear system [C][X]=[K]. In this paper such steps are analysed and adequate algorithms are carried out to speed up the whole finite element computation process.

The algorithms are focused on: calculation of the terms of a single row of the matrix [C]; vectorization of the only non-null terms of this row; solution of the system of equations by means of the pre-conditioned conjugate gradient method.

It has been possible to calculate the only non-null terms of a row of [C] thanks to the algorithm exposed in[1], which calculates each row of [C] without determining in advance the various "element matrices" $[c]_e$.

As for the vectorization of the non-null terms, an algorithm derived from[2,3,4] has been employed. This algorithm turns out to be very useful for speeding up the matrix-vector products.

Finally, the well-known pre-conditioned conjugate gradient method has been employed. It consists of many matrix-vector products, so it is very quick to solve large sparse systems providing that only the non-null terms of the matrix [C] are vectorized.

The proposed procedure has been applied to the study of the 3-D current fields generated by leakage grounding grids. The results, in terms of memory employed and time required, compared with those obtained from the previous traditional finite element package[5], are good.

THE PROPOSED VECTORIAL FINITE ELEMENT PROCEDURE

Let's refer to a 2-D problem and let N and N_e be the total number of nodes contained in the grid and the total number of triangular elements respectively. The dimension of the

Electric and Magnetic Fields, Edited by A. Nicolet
and R. Belmans, Plenum Press, New York, 1995

global matrix [C] is NxN. It is well-known that [C] is obtained by properly storing in [C] the terms of the "element matrices" $[c]_e$ (e=1, 2, ...N_e) and by summing up in the same matrix [C] the terms of the various matrices $[c]_e$, terms which occupy the same place in [C]. Each "element matrix" $[c]_e$ whose dimension is 3x3, is a symmetric matrix; thus, the global matrix [C] is symmetric too.

The transfer procedure and the following process of summing up are carried out by means of the "matrix of the vertices" [V], as follows[1]. For each triangular element eth (e=1, 2, ...N_e) we must define the 1st, 2nd, and 3rd vertices of this element. To each vertex corresponds a number of node n (where n=1,2,...N). Each value of n, which will be called "node index", according to the vertex which it refers to (that is, to the 1st, 2nd, or 3rd vertex), must be stored in a matrix [V] having N_e rows and three columns; [V] will be called "matrix of the vertices".

Briefly, the traditional procedure for calculating the global matrix [C] is based on reading row by row all the "matrix of the vertices" [V] and on the calculation of one "element matrix" $[c]_e$ for each row of [V]; on the appropriate transfer and on the following accumulation of the terms of this "element matrix" $[c]_e$ in the global matrix [C].

The proposed procedure

It is evident that the above-mentioned traditional method requires a memory employment equal to NxN. In other words, all the memory required for the whole matrix [C] must be stored in the computer; this can obviously be a limit to the use of personal computers. In[1] the Authors have got round this drawback by defining a method which is able to calculate the global matrix [C] by means of a procedure by rows, without determining in advance all the nine terms of each "element matrix" $[c]_e$.

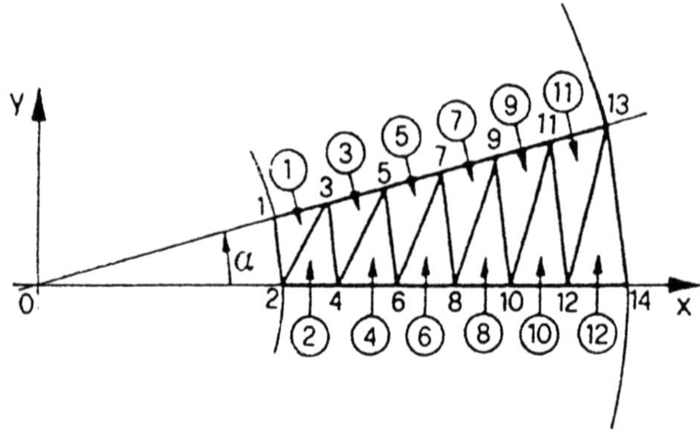

Fig.1 A simple plane domain to be studied by finite element method

Let's refer, for example, to the grid of Fig.1. As regards the triangular element e=2, we must refer to the row e=2 of [V], row in which there are the "node indices" 2, 3, 4. At this point if we concentrate our attention on the place each of the three "node indices" has in the row under examination (that is, 1st, 2nd or 3rd place), we can calculate all the nine terms $c_2(r,c)$ of the "element matrix" $[c]_e$=2, by means of the formulae given in[1]. By applying the traditional method, each of such terms can be properly transferred in the global matrix [C] according to what Table I shows.

Table I

$c_2(1,1) \rightarrow C_2(2,2)$;	$c_2(1,2) \rightarrow C_2(2,3)$;	$c_2(1,3) \rightarrow C_2(2,4)$
$c_2(2,1) \rightarrow C_2(3,2)$;	$c_2(2,2) \rightarrow C_2(3,3)$;	$c_2(2,3) \rightarrow C_2(3,4)$
$c_2(3,1) \rightarrow C_2(4,2)$;	$c_2(3,2) \rightarrow C_2(4,3)$;	$c_2(3,3) \rightarrow C_2(4,4)$

From such Table, however, we can see that the contributions given by the triangle e=2 (that is, by the row e=2 of [V]), for example, to the third row of the global matrix [C] are only $C_2(3,2)$, $C_2(3,3)$, $C_2(3,4)$. In other words, the eth triangle gives a contribution to the rth row of [C] only if in the eth row of [V] the index r appears; in this case the triangle e gives contributions to the row r of the global matrix [C] and these contributions are always three.

In conclusion, to calculate the generic rth row of the global matrix [C], the computer has to carry out a "thorough reading" of the "matrix of the vertices" [V]. This "thorough reading" consists in running the whole matrix [V] and in calculating only the terms of the matrices $[c]_e$ needed for the determination of the required row r of [C].

During the calculation of each row of [C], the computer stores in a vector only the non-null terms of such row, thanks to an algorithm derived from[2,3,4]. Briefly, it consists in storing in a sequential way:

- all the non-null terms of the matrix [C] in a vector MAT;
- all the column indices of the non-null terms of [C] in a vector COL;
- only the position which the first non-null term appearing in each row of [C] has in the vector MAT; this position is stored in a vector RIGA.

This algorithm of vectorization turns out to be very useful for speeding up the matrix-vector products. For, in the row-column product, all the products between the null terms of the row under examination and the corresponding terms of the column vector are automatically avoided.

The well-known pre-conditioned conjugate gradient method, which is suitable for solving large sparse systems, consists of many matrix-vector products, so it is very quick providing that only the non-null terms of the matrix [C] of the coefficients are vectorized.

The pre-conditioned conjugate gradient method has been implemented by the Authors in programming language C.

THE FINITE ELEMENT METHOD TO STUDY THREE-DIMENSIONAL FIELD PROBLEMS

The algorithms described in the previous paragraph have been implemented on a package carried out at the Polytechnic of Bari; this package allowed us to study both electrostastic and current field problems[5]. The new package has been applied to the study of the 3-D current fields generated by leakage grounding grids.

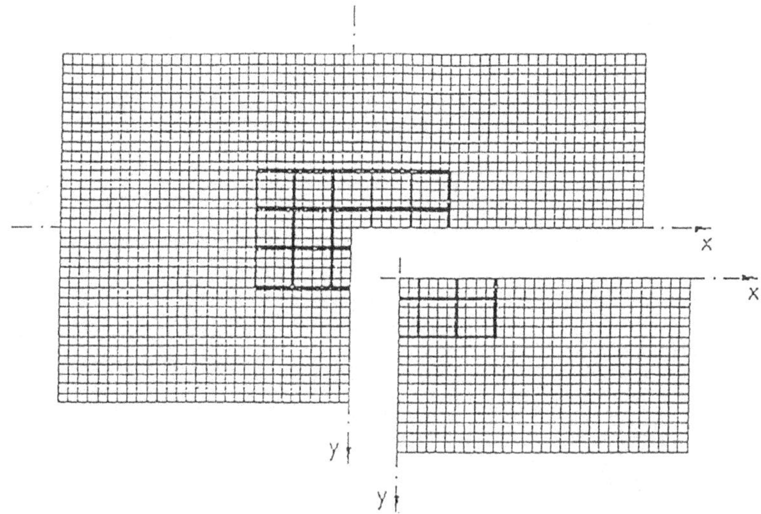

Fig.2 Grounding grid system to be studied by finite element method

Let's assume for example that the grounding system to be studied is the one in Fig.2. This system is made up of a rectangular grid having 15 square meshes inside. The program requires the following input variables:
- grid size along x- and y-axes;
- buring depth of the grid;
- length of the side of each mesh;
- number of the divisions one wishes to do in each side of the grid (see Fig.2);
- number of the divisions one wishes to do in the area between the soil surface and the grid.

According to these data, the program recognizes the symmetry existing with respect to x- and y- axes and decides to operate only in a quarter of the grid, so as to reduce the employment of required memory. That being done, the program automatically computes the size of the domain (width, length and depth), the total number of nodes and elements, and the cartesian co-ordinates of the nodes. Finally, the program proceeds to divide the domain in tetrahedrons as follows. Each prismatic element having rectangular (or square) base is divided in two prismatic elements each having rectangular (or square) base; each of these elements is divided in its turn into three tetrahedral elements.

RESULTS AND CONCLUSIONS

Since the computation program works only on the non-null terms, it allows us to solve quickly, by personal computer, problems having several thousands of nodes.

The results, in terms of memory employed and time required, compared with those obtained from the previous traditional finite element package[5], are very good. For, the previous package allowed us to study, by personal computer, domains having at most 800 nodes in a time equal to about 50 seconds; on the contrary, the new package can study, by the same personal computer, domains having 6000 nodes in a time equal to about 4 minutes.

REFERENCES

1. P. Settanni, M. Sylos Labini, A. Tricarico: "The finite element method in engineering problems: time and memory reduction", International Journal of Modelling and Simulation, vol.9, n.1, 1989
2. H. Magnin, J.L. Coulomb: "A parallel and vectorial implementation of basic linear algebra subroutines in iterative solving of large sparse linear systems of equations" IEEE Transactions on Magnetics, vol.25, n.4, July 1989
3. J. Flower, A. Kolawa, T. Liang, V. Weingarten: "Finite-Element Analysis on a PC", IEEE Software Magazine, September 1991
4. H. Magnin, J.L. Coulomb, R. Perrin-Bit: "Parallel and Vectorial Solving of Finite Element Problems on a Shared-Memory Multiprocessor", IEEE Transactions on Magnetics, vol.28, n.2, March 1992
5. M. Sylos Labini: "Una procedura per lo studio del campo di corrente generato da un impianto di terra mediante il metodo degli elementi finiti", L'Energia Elettrica, n.12, 1991

DIFFERENT FORMULATIONS IN AXISYMMETRIC MAGNETOSTATIC PROBLEMS

U. Pahner, R. Belmans, K.Brandiski
Dept. E.E. - Electrical Energy
K.U. Leuven
Kardinaal Mercierlaan 94
B-3001 Heverlee - Leuven
Belgium

J. Webb, D. Lowther
Infolytica Corp.
1140 De Maisonneuve
Suite 1160
Montreal H3A 1M8
Canada

F. Henrotte, W.Legros
Dept. EE-Inst. Montefiore
University of Liège
Sart Tilman Bât. B28
B-4000 Liège
Belgium

ABSTRACT

In the paper a number of formulations for axisymmetric finite element analysis are compared. Different types of auxiliary potentials and the accuracy of the results with these potentials are discussed. Special attention will be paid to a new and very simple way of enhancing the local accuracy of the post-processing results near the axis. A model containing high permeability near the axis is taken as example for demonstrating the benefits of the new mixed potential as well as the new post-processing approach.

1. INTRODUCTION

Finite element programmes capable of handling two-dimensional, axisymmetric problems, often use different formulations for the auxiliary vector potential required for solution. Furthermore, the order of the approximation may be increased, improving the accuracy, at the expense of calculation time. In general two methods are found. The first is based on the auxiliary potential:

$$V = \frac{A_\phi}{r} \tag{1}$$

In the second method, the auxiliary potential is defined as:

$$W = rA_\phi \tag{2}$$

It may be shown that both approaches are inaccurate in some cases. Especially in systems, having a highly permeable material on the axis, problems may occur. This may be in the V auxiliary potential, when the distance to the axis increases, or due to the singularities in the W auxiliary potential at the axis. Therefore, a mixed approach was used [3], combining the benefits of both approaches. Consider the auxiliary potential

Electric and Magnetic Fields, Edited by A. Nicolet
and R. Belmans, Plenum Press, New York, 1995

$$U = A_\phi r + f\frac{A_\phi}{r} \tag{3}$$

where f is a continuous function, a first-order polynomial in each triangle, taking the value f_0 on the z-axis and zero at every node of the mesh. As the z-axis is approached:

$$U \rightarrow f_0 \frac{A_\phi}{r} \tag{4}$$

the potential tends to potential (1). Also on the z-axis:

$$B_z = \frac{2U}{f_0} \tag{5}$$

so that having computed U, no singularity arises in deriving the flux density from it. This benefit of potential V is preserved by the mixed approach. On the other hand, in triangles having no vertices on the axis, f is identically zero and U is exactly rA_Φ, potential W. This potential generally gives higher accuracy than potential V. Especially the flux density in the iron core at the axis is not correctly represented by the V- and W-formulations.

Using auxiliary potential W it can be shown, that sufficiently good results can be achieved by using a special post-processing algorithm. In [1,2], an approach is discussed, dealing with a coordinate transformation and the introduction of quadrilateral elements. The very simple approach shown here, can be implemented in the post-processor only.

2. TEST EXAMPLE

A practical example is shown on Fig. 1. The problem is solved as linear one, specifying the iron permeability as a constant (μ_r=1000). The mesh has been built using a Delaunay based mesh generator. The mesh for solution with the auxiliary potential V is slightly different from the one shown in Fig. 1 due to the usage of another program. All programs use second order elements. The local values of the magnitude of B are computed along a specified contour from the axis through the airgap. There is no significant difference in the equipotential plots of using potential W and U, as the effect of using the mixed potential applies only to the elements with nodes at the axis.

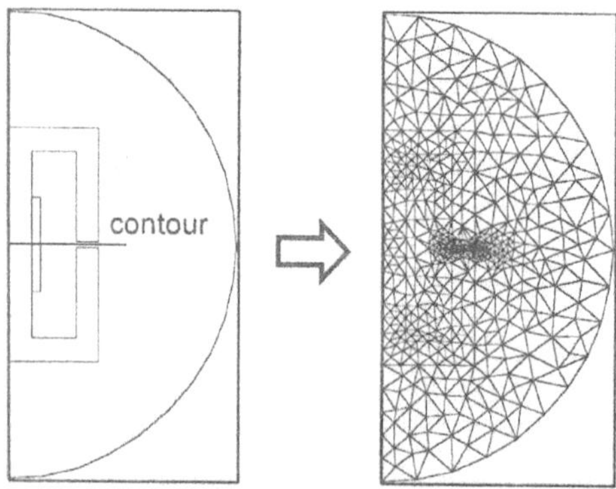

Fig. 1 Geometry and mesh

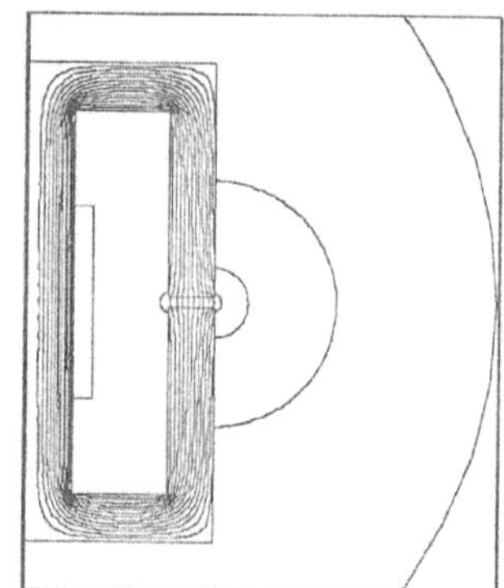
Fig. 2 Flux plot using mixed potential

In Fig. 3a the flux density along the contour is given for the solution with auxiliary potential W. The results using the auxiliary potential V are unacceptable, as the flux density distribution is not only far from homogeneous, there is also a significant error of the flux density in the core (Fig. 3b). Best results are obtained by the mixed formulation, but there is still a variation of B inside the first element (Fig.3c).

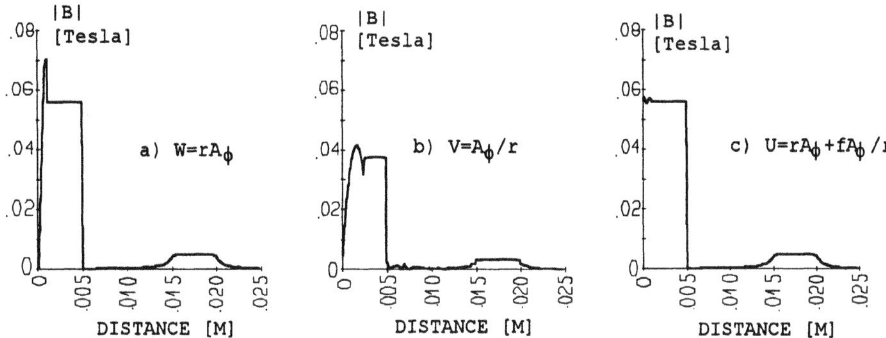

Fig. 3 Bmag along the specified contour through the airgap

3. POST-PROCESSING NEAR Z-AXIS

A coordinate transformation together with the introduction of quadrilateral elements near the axis is discussed in [1]. Using the auxiliary potential W together with bilinear shape functions the misbehaviour of B at the axis is be removed.

A simple weighting algorithm in post-processing will be discussed, which leads to similar results. For this approach, the auxiliary potential W is used. The mixed potential could also be used, but the results prove that the W-potential is fully sufficient.

Consider elements as shown in Fig. 4. $B_{z\phi}$ are nodal values. The radial component of the flux density at the axis is zero. A distance-weighted averaging of the two nearest nodal B_z-values is applied to calculate the axial component of B on the axis (eqn.(6)). This calculation is done in each element separately to allow material discontinuities from one element to the next. It is shown that this simple algorithm can be easily implemented for higher order elements.

$$B_{ax} = \frac{B_{z1}d_2 + B_{z2}d_1}{d_1 + d_2} \tag{6}$$

The main difference to the approach discussed in [1] is that there is no attempt made to compute the nodal values of B at the axis directly. Together with adaptive mesh refinement, the accuracy of this approach is sufficient even for non-linear problems.

Again the magnitude of B is calculated along the contour through the airgap (Fig. 5). The averaging affects only the first element, so that the good accuracy of potential W is maintained for the rest of the model.

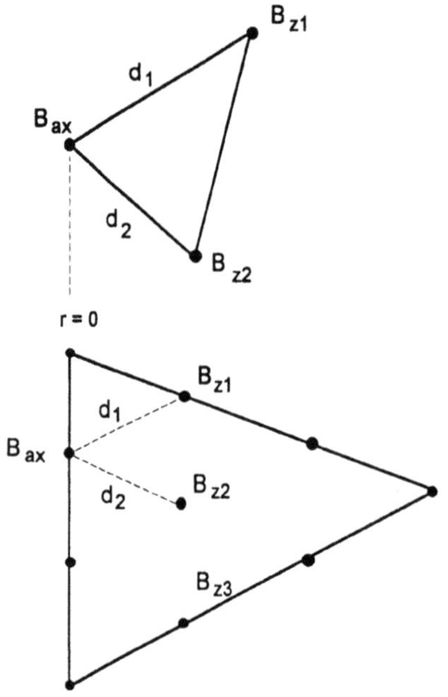

Fig. 4 Distance weighted averaging near the axis

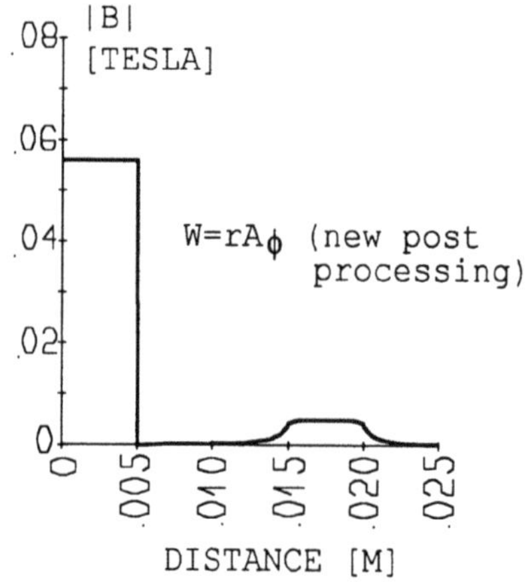

Fig. 5 Bmag along contour using auxiliary potential W with special post-processing

4. CONCLUSION

The modelling of axisymmetrical problems with the auxiliary potential V=A/r and first order elements has shown its limitations. Two methods were presented in this paper which aim to allow a better representation of the physical quantities derived of the calculation itself (especially the flux density and magnetic field intentsity). In particular it has been shown that the mixed potential approach gives a better solution than the potential W approach. A very simple distance averaging has been introduced which is well suited for reallife problems as it allows to be implemented in the post-processor stage only. Test calculations have proven that this post-processing approach allows the usage of the simple potential formulation W instead of the mixed potential U.

5. ACKNOWLEDGEMENTS

The authors are grateful to the Belgian Nationaal Fonds voor Wetenschappelijk Onderzoek for its financial support of this work and the Belfian Ministry of Scientific Research for granting the IUAP No.51 on Magnetic Fields

6. REFERENCES

1. F.Henrotte,H.Hedia,N.Bamps,A.Genon,A.Nicolet,W.Legros, "A comparison between different approaches for finite element solution of axisymmetrical problems", Workshop Liège 1992, pp29.1-29.5
2. J.B.M. Melissen, J. Simkin, "A new coordinate transform for the finite element solution of axisymmetrical problems in magnetostatics", IEEE vol. MAG-26, n°2, March 1990.
3. Infolytica Corp.: "MagNet 5 Users Manual", Montreal Canada, 1992

FIELD COMPUTATION OF PERMANENT MAGNET SYSTEMS WITH CONSIDERATION OF THE KNEE POINT

A. Miraoui, L. Kong, J.M. Kauffmann

Insitut de Génie Energétique de Belfort
2, Avenue Jean Moulin
Belfort, 90000 (France)

INTRODUCTION

Modern high-coercivity permanent magnets are widely used in various types of electric machines and devices and the demagnetization characteristics are generally, supposed to be a straight line. But, for permanent magnets which present a knee point such as for Neodymium-Iron-Boron (Nd-Fe-B) in high temperature and ferrites in low temperature, the whole characteristic needs to be taken into account. Indeed, the knee point location moves considerably when temperature changes . So, neglecting this effect in these conditions will lead to unacceptable errors in computed evaluations.

In field computation softwares, it is generally considered, that the magnets have a linear characteristic. But it is evident that this imposes restrictions on the use of these softwares when it is question of non linear characteristics what happens frequently for Nd-Fe-B magnets in thermal limit working. Some models have been studied in the literature. Among the most known we may evoke the approximation with two straight lines .Most of the models are valid, at a given temperature. For another one, it would be necessary to use another characteristic. And if the temperature changes continuously, the use of these models is very difficult. So, in this study, the authors propose a thermal dynamic model of Nd-Fe-B magnets. It is expressed by a double series of Tchebycheff polynomials.

DESCRIPTION OF THE THERMAL DYNAMIC MODEL

The flux density B should be considered as a function of two variables: the temperature θ and the magnetization field H. It can expressed by Tchebycheff polynomials of two variables. For a given temperature:

Electric and Magnetic Fields, Edited by A. Nicolet
and R. Belmans, Plenum Press, New York, 1995

$$B = B(\theta, H) = \sum_{i=0}^{n} a_i T_i(H) \quad (1)$$

B: Flux density (Tesla)
H: Intensity of magnetic field (A/m)
θ : Temperature (°C)
a_i : i^{th} coefficient
T_i : i^{th} Tchbycheff polynomial

If we suppose that both variables θ and H may be separated, the induction $B(\theta, H)$ may be expressed by a double series of Tchebycheff polynomials:

$$B(\theta, H) = \sum_{i=0}^{n} \sum_{j=0}^{m} b_{ij} T_j(\theta) T_i(H) \quad (2)$$

In equation (2) b_{ij} are constant coefficients for a material and it is the main interest of the method. For example, for a Nd-Fe-B magnet of type NEIBON 35, the behaviour described by this model is shown in figure 1.The relative error does not exceed 3 % in a large range of temperature. The use of an adequate rotation of the axis allowed to limit the order of the polynomials to 7.

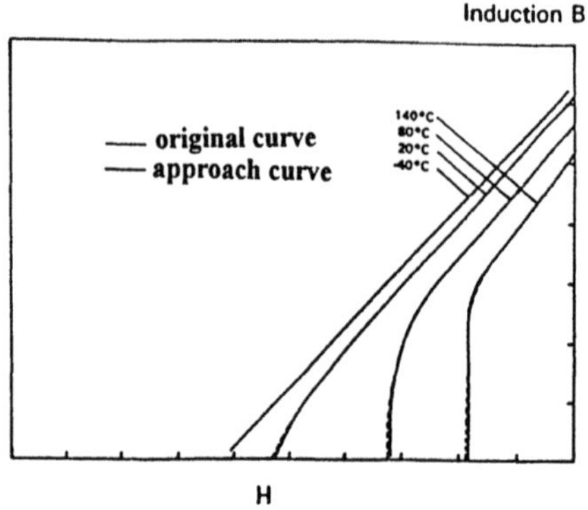

Figure 1. Behaviour of Nd-Fe-B (NEIBON35) - Temperature changes from -40 °C to 140 °C

The use of FluxExpert, which is a partial derivative equation solving software by the finite element method has permitted to realize the magnetostatic modulus.The previous thermal dynamic model is integrated in this modulus.The governing field, in a two dimensional analysis and cartesian co-ordinates equation when taking into account the magnets non linearity is written:

$$\vec{\nabla}(\upsilon_r \vec{\nabla} A) = \mu_0 J + \mu_0 \vec{\nabla}(\upsilon_r \vec{M}_r) \quad (3)$$

After discetization and all intermediate computations, we may write with a matrix notation:

$$[M][A] = [K] \quad (4)$$

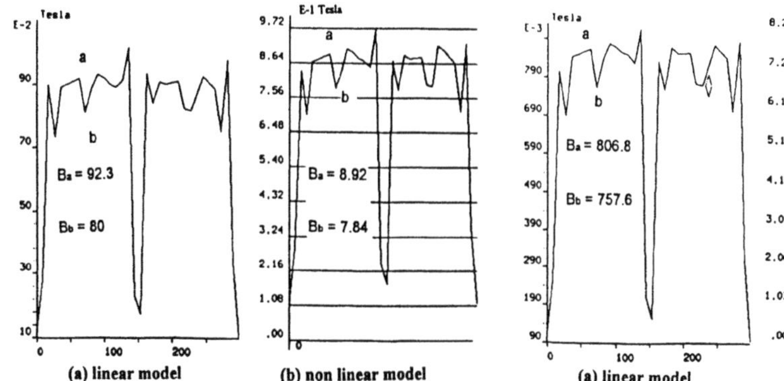

Figure 2. Comparison of both models (θ=80, $I=I_n$)

Figure 3. Comparison of both models (θ=140, $I=I_n$)

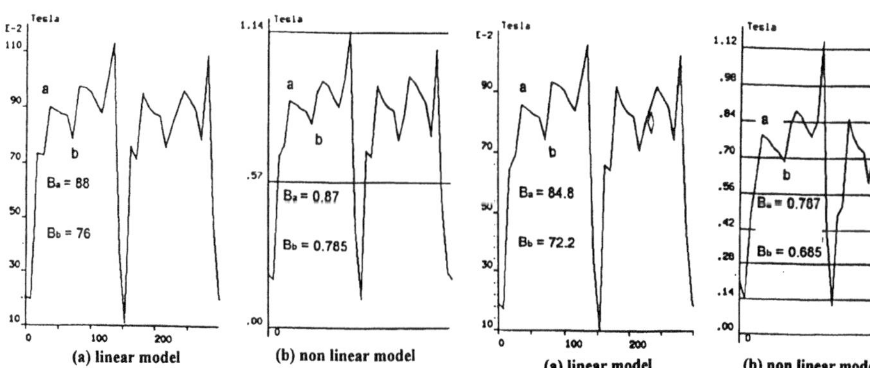

Figure 4. Comparison of both models (θ=80, $I=5I_n$)

Figure 5. Comparison of both models (θ=140, $I=5I_n$)

where

$$[M]=\iint_{\Omega}\vec{\nabla}\alpha_i\vec{\nabla}\alpha_j\upsilon_r dxdy \qquad (5)$$

$$[K]=\iint_{\Omega}\alpha_{ij}J\mu_0 dxdy-\int_{\ell}\vec{n}\wedge(\upsilon_r\vec{M}_r)\alpha_i\mu_0 dl+\iint_{\Omega}\vec{\nabla}\alpha_i\wedge(\upsilon_r\vec{M}_r)\mu_0 dxdy \qquad (6)$$

COMPARISON OF BOTH MODELS: LINEAR AND NON LINEAR

To compare both models, the computation for the linear model is effected by Flux2d software and for the non linear model with the FluxExpert software. All computations are effected for two temperatures 80 and 120 °C.

Figures 2, 3, 4 and 5 represent the air gap flux density. Two marks, a and b have been introduced in these figures to value the differences between both models with more precision.

It can be established, that at a temperture of 80 °C, the relative difference between both models remains very low (from 1.1 to 3.4 %). Indeed, it is evident that the magnet operating point is in the linear part of the demagnetization curve. This 3.4 % is due, on one side, to the fact that Flux2d operates in double precision when Fluxexpert (in this case) operates in single precision, on the other side the B(H) curve approximation of the ferromagnetic materials are not the same. Indeed, we use the Tchebycheff polynomials when Flux2d uses the Spline functions.

At a temperature of 140 °C, the differences are quite important and reach 10.5 %. This shows that the magnet operating point is located in the non-linear zone. The error is too important to be neglected.

CONCLUSION

A continuous thermal dynamic model integrated to a magnetostatique modulus has been developed by double series of Tchebycheff polynomials. The precision is sufficient to the engineering calculating needs. The method can be applied to all types of permanent magnets but also to describe ferromagnetic materials.

Through their thermal sensitivity, the Nd-Fe-B magnets impose to take into account the non-linearity of their magnetization characteristics. Indeed, this model permits to solve the problems of reversible and irreversible properties; it means when the working point is located over or under the knee point.

REFERENCES

1. S. Chen, K.J. Binns, Z. Liu and D.W. Shimmin, "Finite Element Analysis of the Magnetic Field in Rare-Earth Permanent Magnetic Systems with Consideration of Temperature Dependence" IEEE Trans. on Magnetics, Vol. 28, No.2, March (1992).
2. C. Qishan and C. Lin "Field computation with knee point" IEE proceeding, Vol.136, No.6 November (1989)
3. L. Kong "Contribution à la conception d'un moteur à aimnats permanents du type jante et à son alimentation" Thése de Doctorat de l'Institut National Polytechnique de Lorraine, (1994)

TREATMENT OF NON-HOMOGENEOUS REGIONS IN CHARGE ITERATION

G. Aiello, S. Alfonzetti, S. Coco and N. Salerno

Dipartimento Elettrico Elettronico e Sistemistico
Facolta' di Ingegneria - Universita' di Catania
Viale A. Doria, 6 - 95125 Catania, Italy

INTRODUCTION

Charge iteration is an iterative finite-element procedure to deal with unbounded electrical field problems [1,2,3]. In the procedure a fictitious boundary B_F, enclosing all the conductors, defines a bounded domain. By assuming an initial guess for the potential on B_F a standard FEM solution can be carried out. Starting from this first-step solution one can compute the charge density lying on the conductor surfaces and from it a new estimate (closer to the true one) of the potential on B_F, by means of an appropriate free-space Green's function. The procedure is iterated until a convergence test is satisfied.

If the dielectric medium is non-homogeneous, the applicability of the procedure depends on the availability of the appropriate Green's function for the non-homogeneous unbounded problem, which may be difficult to find.

In this paper a generalized way to perform the potential evaluation on the fictitious boundary is given which allows systems with localized non-homogeneities to be dealt with.

POTENTIAL COMPUTATION IN THE PRESENCE OF NON-HOMOGENEITIES

To overcome the difficulties outlined above a closed surface B_M is introduced, internal to the fictitious boundary, but enclosing all the conductors and all the dielectric non-homogeneities (possibly at minimum). The only aim of this surface is to allow computation of the potential on B_F, without directly considering the charge on the conductors, by means of the following formula:

$$v(P_F) = -\int_{B_M}\left(G(r)\frac{\partial v(P)}{\partial n} - v(P)\frac{\partial G(r)}{\partial n}\right)d\sigma \tag{1}$$

where P_F and P are points on B_F and B_M respectively, r is the distance between them, n is the unit vector normal to the surface B_M, oriented outwards from the surface (towards B_F)

Electric and Magnetic Fields, Edited by A. Nicolet
and R. Belmans, Plenum Press, New York, 1995

and G is the free-space Green's function given by: $G(r)=(1/2\pi)\ln(1/r)$ for space dimensionality N=2 or $G(r)=1/4\pi r$ for N=3.

By selecting B_M in such a way that it is constituted by a set S_M of element sides, the potential on B_F is written as

$$v(P_F) = \sum_{S\in S_M} \sum_{n=1}^{N_E} h_n V_n \tag{2}$$

where S is the generic element side in set S_M, N_E is the number of nodes of the element to which the side S belongs, V_n are the nodal values of the potential in the same element and where the adimensional coefficients h_n are given by:

$$h_n = -\int_S \left(G(r)\frac{\partial \alpha_n(P)}{\partial n} - \alpha_n(P)\frac{\partial G(r)}{\partial n} \right) dS \tag{3}$$

where α_n are the shape functions. A simple Gauss quadrature technique can be used to compute the integral in (3). Note that no singularities arise in these integrals since the surfaces B_F and B_M are not intersecting. For simplex Lagrangian elements with straight sides this computation can be also performed in terms of universal matrices as:

$$h_n = -\frac{(N-1)!\,S_i}{NE} \sum_{j=1}^{N+1} \sum_{m=1}^{N_E} M_{mn}^{(i,j)} G(r_m) S_j \cos\theta_{ij} \tag{4}$$

where: N is the space dimensionality (N=2 or 3); E is the element measure; S_i (coincident with S) and S_j are the i-th and j-th local sides of the element ($1\le i,j \le N+1$); θ_{ij} is the angle between the vectors normal to S_i and S_j; r_m is the distance between P_F and the m-th local node P_m of the element; M_{mn} is the universal value:

$$M_{mn}^{(i,j)} = \int_{S_i'} \left(\alpha_m \frac{\partial \alpha_n}{\partial \xi_j} - \alpha_n \frac{\partial \alpha_m}{\partial \xi_j} \right) dS_i' \tag{5}$$

where S_i' is the i-th side of the standard simplex finite element.

In the case of axisymmetric problems (with respect to the z-axis), the coefficients h_n are computed as:

$$h_n = -\int_S \left(xU_1(P)\frac{\partial \alpha_n(P)}{\partial n} - U_2(P)\alpha_n(P) \right) dS \tag{6}$$

where x is the abscissa of point P=(x,0,z), and

$$U_1(P) = \frac{1}{4\pi}\int_0^{2\pi} \frac{1}{\rho} d\varphi \qquad\qquad U_2(P) = \frac{1}{4\pi}\int_0^{2\pi} \frac{\partial}{\partial n_\varphi}\frac{1}{\rho} d\varphi \tag{7}$$

with: $\rho = \left[(x_F - x\cos\varphi)^2 + x^2\sin^2\varphi + (z_F - z)^2 \right]^{\frac{1}{2}}$ and $\bar{n}_\theta = n_x \cos\varphi\, \bar{i} + n_x \sin\varphi\, \bar{j} + n_z \bar{k}$ (n_x and n_z are the components of the unit vector n normal to the triangle side S).

For Lagrangian triangles with straight sides:

$$h_n = -\frac{S_i}{2E}\sum_{k=1}^{3}\sum_{j=1}^{3}\sum_{m=1}^{N_E} x_k L_{mn}^{(i,j,k)} U_1(P_m) S_j \cos\theta_{ij} + S_i \sum_{k=1}^{3}\sum_{m=1}^{N_E} x_k R_{mn}^{(i,k)} U_2(P_m) \tag{8}$$

where x_k is the abscissa of the k-th vertex of the triangle, L_{mn} and R_{mn} are the universal values:

$$L_{mn}^{(i,j,k)} = \int_{S_i'} \xi_k \alpha_n \frac{\partial \alpha_m}{\partial \xi_j} dS_i' \qquad R_{mn}^{(i,k)} = \int_{S_i'} \xi_k \alpha_n \alpha_m dS_i' \tag{9}$$

AN EXAMPLE OF APPLICATION

Consider a condenser constituted by two parallel circular plates (of radius R and distance 2h), with a dielectric medium (with relative permittivity ε_r>1) between them alone, the whole embedded in air. The plates are assumed to be thin and voltaged with opposite potentials V=±0.5 V. A reference frame is set having the z-axis coincident with the axisymmetrical one and the origin in the middle of the two plates. For symmetry reasons, the analysis is conveniently restricted to one quarter of the x-z plane: homogeneous Dirichlet and Neumann conditions hold on the x- and z-axis, respectively. In Fig. 1 the selected fictitious boundary B_F and closed surface B_M are reported. Several analyses were performed by means of the ELFIN code[4] for different values of the ratio h/R and of ε_r. The results are summarized in Tab. 1. The second column shows the values of the normalized capacitance in vacuo

$$\gamma = \frac{2hC_0}{\varepsilon_0 \pi R^2} \tag{10}$$

The other columns give the filling factor η, defined as[5]

$$\eta = \frac{C}{\varepsilon_r C_0} \tag{11}$$

for several values of ε_r. The actual capacitance C is then obtained as $\eta\gamma\varepsilon_r\varepsilon_0\pi R^2/2h$.

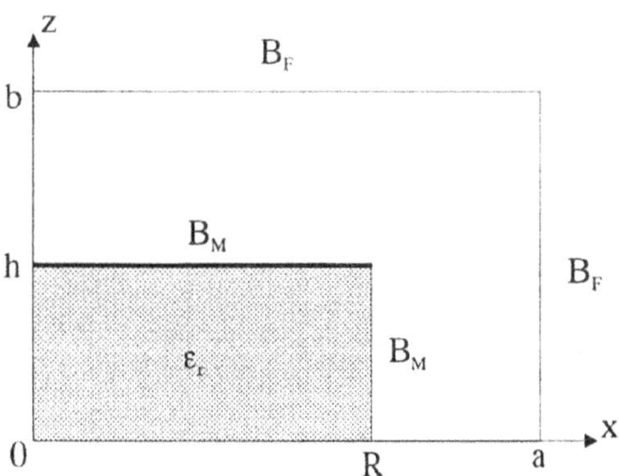

Fig.1 - System of the example

Table 1. Normalized capacitance in vacuo and filling factors.

h/R	γ	η			
		ε_r=2	ε_r=4	ε_r=8	ε_r=16
0.25	1.55	0.832	0.742	0.690	0.674
0.50	2.08	0.758	0.631	0.563	0.522
0.75	2.61	0.721	0.559	0.476	0.425
1.00	3.13	0.692	0.517	0.420	0.373

Fig. 2 reports the contours of the potential relative to the values v_k=k 50 mV, k=1-9, in the case of h=0.5 R, a=R+h, b=2h, ε_r=8. A mesh was employed, constituted of 240 2nd-order triangles and 525 nodes. Starting from a homogeneous Dirichlet condition on B_F, after 5 iterations the mean absolute relative error on the boundary was less than 0.1%.

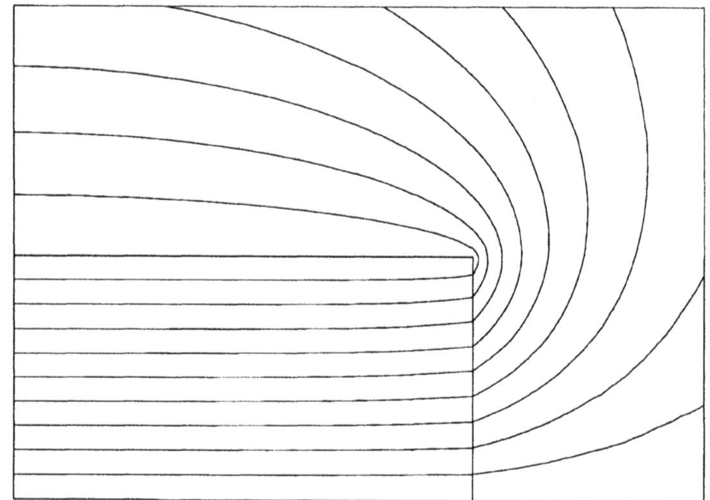

Fig. 2 - Contours of the potential

REFERENCES

1. G. Aiello, S. Alfonzetti, S. Coco: "Charge Iteration for N-Dimensional Unbounded Electrical Field Computations". *IEEE Trans. Magn.*, Vol. 28, No. 2, March 1992.
2. G. Aiello, S. Alfonzetti, S. Coco,N. Salerno: "Axisymmetric Unbounded Electrical Field Computation by Charge Iteration". *IEEE Trans. Magn.*, Vol. 29, No. 2, March 1993.
3. G. Aiello, S. Alfonzetti, S. Coco, N. Salerno: "Convergence analysis of the Charge-Iteration Procedure for Unbounded Electrical Fields". *Compumag Conf.*, Miami, Oct. 1993.
4. S. Alfonzetti, S. Coco:"ELFIN: an N-Dimensional Finite-Element Code for the Computation of Electromagnetic Fields". *IEEE Trans. Magn.*, Vol.24, No. 1, Jan. 1988.
5. P. Benedek, P. Silvester: "Capacitance of Parallel Rectangular Plates Separated by a Dielectric Sheet." *IEEE Trans. Microwave Th. Tech.*, Vol. 20, No. 8, Aug. 1972.

SOLVING 3D STATIC FIELD PROBLEMS BY DUAL FORMULATIONS USING POTENTIAL VARIABLES

Zhuoxiang Ren

Laboratoire de Génie Electrique de Paris
CNRS URA 127, Universités Paris VI & XI, ESE
Plateau du Moulon, F-91192 Gif sur Yvette cedex, France

INTRODUCTION

Recent study in electromagnetic field computation shows the interest of dual formulations: They provide complementary energy bounds, which permits to obtain global parameters such as impedance with minimum computation costs[1-3]. The calculation of electromagnetic field by dual formulations ensures "strongly" Maxwell equations, the numerical errors are on the constitutive laws describing physical properties, which leads to an efficient indicator to auto-adaptive mesh refinement[4,5]. Solving a same problem by dual formulations gives also an indication of numerical errors.

An electromagnetic field can be described by two field variables: the field intensity (**E** or **H**) and the flux density (**D** or **B**). The two variables are related by the constitutive law, and governed by the curl equation and divergence equation, respectively.

Two dual finite element formulations can be derived by working respectively with the field intensity and the flux density. Since the curl equation associates **E** or **H** with the integration of curves, and divergence equation associates **D** or **B** with the integration of faces, the elements to be used are respectively the edge element and facet element[6,7], where the degrees of freedom are associated with the edges and facets of finite element mesh. However, in a static field with zero curl or zero divergence, additional conditions should be imposed .

In the case of a static field, it is more suitable of working with scalar or vector potential variables[8].The nodal (or edge) element provides a natural discretisation of the scalar (or vector) potential, and leads to a diminution of the number of degrees of freedom.

This paper gives a synthetic description of dual formulations in terms of potential variables for static field problems. The numerical implementation is realised by Whitney's elements[7]. In the case of non-zero curl or non-zero divergence field, the calculation of the source field is discussed. Numerical examples of magnetostatic and electrostatic problems are given. The convergence of the matrix system is examined, the complementary energy bounds are also illustrated.

SCALAR POTENTIAL FORMULATIONS

In the case of an electrostatic field, curl $\mathbf{E} = 0$ implies a gradient field of an electric scalar potential V such that $\mathbf{E} = -\text{grad } V$. Solving weakly div $\mathbf{D} = \rho$ by applying a test function V', we get the following variational formulation:

$$\int_\Omega \varepsilon \text{ grad } V' \cdot \text{grad } V \, d\Omega + \int_\Omega V' \rho \, d\Omega - \int_\Gamma V' \, \mathbf{D}_0 \cdot \mathbf{n} \, d\Gamma = 0, \tag{1}$$

where Ω is the study domain bounded by Γ and $\mathbf{n}$ the outside normal of Γ. This is a very simple formulation. The numerical implementation can be easily realised using nodal elements. The uniqueness of V is easily ensured by setting V=0 in ∞ or in a symmetric plane. The continuity of V ensures the tangential continuity of **E**, because the difference of V gives the circulation of **E**.

In a magnetostatic field, curl **H** is different from zero in a current carrying region. Nevertheless, a scalar potential formulation can be derived by writing: $\mathbf{H} = \mathbf{T} - \mathrm{grad}\,\phi$, where ϕ is a magnetic scalar potential and **T** is an any source field providing curl $\mathbf{T} = \mathbf{j}$, which is calculated from a given **j**. The variational formulation has the form:

$$\int_{\Omega} \mu\, \mathrm{grad}\,\phi' \cdot \mathrm{grad}\,\phi \; d\Omega - \int_{\Omega} \mu\, \mathrm{grad}\,\phi' \cdot \mathbf{T}\, d\Omega + \int_{\Gamma} \phi' \; \mathbf{B}_0 \cdot \mathbf{n}\, d\Gamma = 0. \tag{2}$$

It is a weak equation of div $\mathbf{B} = 0$. Where ϕ is discretised by nodal element and **T** by edge element.

Concerning the calculation of **T**, there are different possibilities. An immediate choice is taking **T** as a source field $\mathbf{H}_0$ calculated from Biot-Savart law. In this case, ϕ is known as a reduced potential. In order to avoid the cancellation error in iron, $\mathbf{H}_0$ can be calculated in assuming the magnetic permeability of iron infinite[9]. Some authors calculate **T** in current carrying conductors[10,11], where **T** is named as an electric vector potential. To solve the multivalued problem of ϕ for a multiply connected conductor, **T** is set up also in the hole of the conductor. For a conductor of regular geometry and with a constant current distribution, **T** can be obtained in a analytical way. Otherwise, a numerical field solution of curl $\mathbf{T} = \mathbf{j}$ with boundary condition $\mathbf{n} \times \mathbf{T} = 0$ is necessary.

VECTOR POTENTIAL FORMULATIONS

A vector potential can be introduced when the flux density is divergence free. In the case of a magnetostatic field, a magnetic vector potential **A** such that $\mathbf{B} = \mathrm{curl}\,\mathbf{A}$, is introduced. A variational formulation can be established by solving weakly the curl $\mathbf{H} = \mathbf{j}$:

$$\int_{\Omega} \frac{1}{\mu} \mathrm{rot}\,\mathbf{A}' \cdot \mathrm{rot}\,\mathbf{A}\, d\Omega - \int_{\Omega} \mathbf{A}' \cdot \mathbf{j}\, d\Omega + \int_{\Gamma} \mathbf{A}' \cdot \mathbf{n} \times \mathbf{H}_0 \; d\Gamma = 0. \tag{3}$$

where **A'** is a vector test function. The edge element is used for the discretisation of **A**, which ensures normal continuity of the flux density **B**. Whereas the current density **j** is discretised by the facet element.

To ensure the uniqueness of **A**, in the case of edge element, a gauge condition $\mathbf{A} \cdot \mathbf{w}=0$ can be used, where **w** is an any vector field defined by a tree constructed by a set of edges[12]. According to author's experiences, the numerical accuracy depends on the choice of a tree. The best choice is setting **w** closed to the direction of **B** (orthogonal to A), but it is difficult of predetermining **B** direction for a sophisticated geometry without the field solution. However, numerical experiences show that without the gauge condition, the system converges to an any solution of A, while the curl of **A** , i.e. **B**, is unique. This point will be discussed later through a numerical example.

For an electrostatic field problem, div $\mathbf{D} \neq 0$ in a region contaning electric charges. A vector potential formulation can be acheived in expressing: $\mathbf{D} = \mathbf{S} + \mathrm{curl}\,\mathbf{U}$, where **U** is an electric vector potential and **S** is an any source field preserving div $\mathbf{S} = \rho$. Solving weakly curl $\mathbf{E} = 0$, we have[13]

$$\int_{\Omega} \frac{1}{\varepsilon} \mathrm{curl}\,\mathbf{U}' \cdot \mathrm{curl}\,\mathbf{U}\, d\Omega + \int_{\Omega} \frac{1}{\varepsilon} \mathrm{curl}\,\mathbf{U}' \cdot \mathbf{S}\, d\Omega + \int_{\Gamma_1} \mathbf{U}' \cdot \mathbf{n} \times \mathbf{E}_0 \; d\Gamma = 0. \tag{4}$$

The discretisation is ensured by using edge element for **U**, and facet element for **S**. The uniqueness of **U** can be ensured by introducing a tree[13]. However, as the case of the magnetostatic field, a unique solution of **D** can be obtained without gauge condition.

To preserve div $\mathbf{S} = \rho$, the source field **S** can be set up only in electric charge carrying domains. In charge free region, $\mathbf{D}=\mathrm{curl}\,\mathbf{U}$. It can be shown that in this case, the electric potential **U** is multivalued. This multivalued problem can be solved by setting up the source field **S** in a region bringing the domain simply connected, i.e. in a region connecting two domains having charges of opposite signs[13]. Otherwise, in the case of a voltage boundary condition, the source field **S** is unknown. An additional equation in forcing curl $\mathbf{U} + \mathbf{S} = \mathbf{D}$ should be introduced for solving the whole system[13].

Compared with the scalar potential, the number of unknowns of the vector potential formulation, related to the number of edges, is much more important.

COMPLEMENTARY ENERGY BOUNDS

The scalar and vector potential variational formulations (1) and (4) (or (2) and (3)), for a static field problem, correspond to two energy functionals: Φ_S and Ψ_V, respectively. It can be shown that, for a numerical solution of field, the functional Φ_S provides the upper energy bound and Ψ_V the lower energy bound[1-4]. The complementary bounds property is helpful in numerical modelling: Solving the same problem by two dual formulations with a same mesh, the true energy value is bounded by two numerical ones. The difference of two results gives an indication of numerical accuracy of the problem. In addition, taking the average of numerical values provided by two formulations, a better result, near to exact solution, is obtained even for a rather coarse mesh. This leads to a gain in CPU time.

NUMERICAL EXAMPLES

The first example concerns a magnetic circuit with an air-gap and a non-linear B(H) characteristic (Fig.1). One fourth of the domain is meshed by tetrahedral elements. The problem is solved by the two dual formulations with different meshes. The non-linear system is solved by Newton-Raphson method. The linear algebraic equation is solved by the conjugate gradient method with diagonal precondition.

Figure 2 shows the magnetic energy obtained by the two formulations with different size of meshes. The phenomenon of complementary energy bounds is clearly observed.

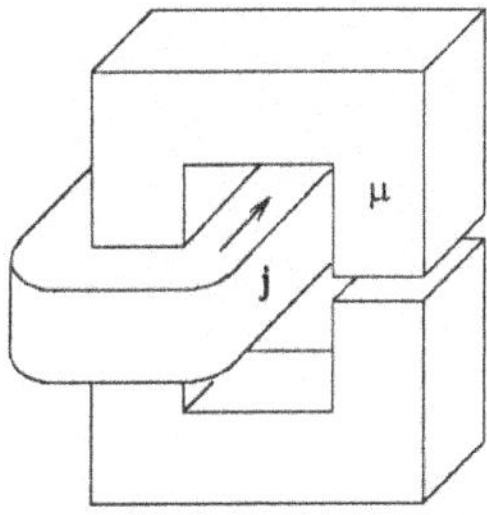

Fig.1. Mesh of a magnetic circuit

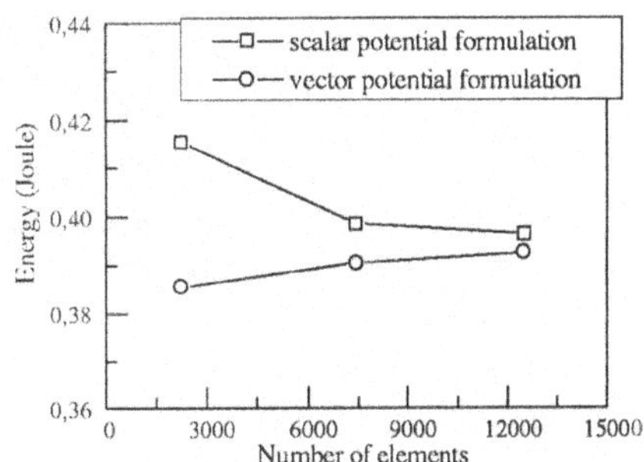

Fig.2. Magnetic energy versus mesh fineness

The convergence of the linear system in the case of **A**-formulation without gauge condition is examined. The residues of A and of **B** (curl **A**) after each iteration are shown in figure 3. It is found that the residue of A diminishes slowly and becomes constant as the number of iterations increases. Whereas which of **B** decreases quickly. That means the method provides a good convergence of **B**, which is physically what we need. When the residue of A becomes constant, the further iteration adds a gradient to the solution of A, which has no influence on the solution of **B**. This property suggests a new idea to test the convergence of linear equation: when the residue of A converges to a fix value, the iteration can be stopped.

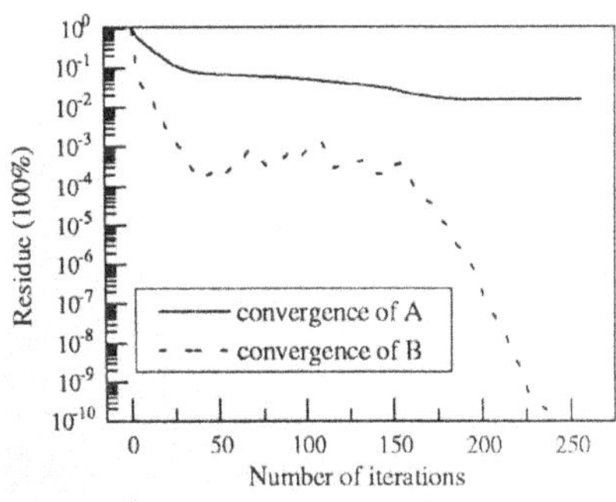

Fig.3 Convergence of A and **B**

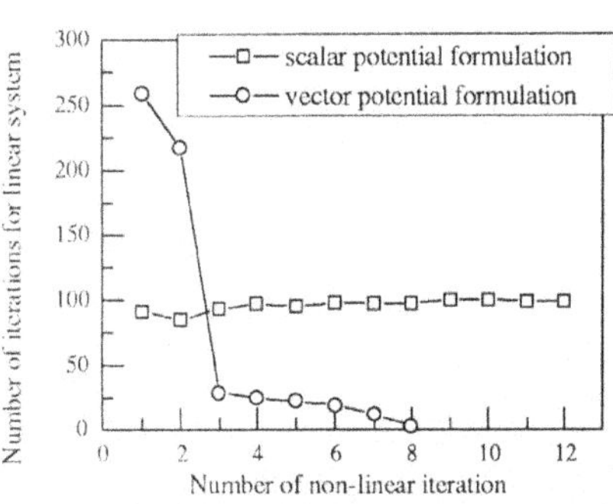

Fig.4. Number of iterations of linear system versus

Another interest phenomenon is observed when solving a non-linear system with Newton Raphson method. Figure 4 gives, for each non-linear iteration, the number of iterations needed for solving the linear equation. With the scalar potential formulation, the number of iterations for linear system is almost the same for each non-linear iteration. However, with the vector potential formulation, this number decreases very quickly when the non-linear system converges. Globally, the total number of iterations is not important.

The second example is a micro-motor based on electrostatic driving principle. Figure 5 shows a mesh of the one sixth of the machine. Two dual formulations are used for solving this problem[14]. The convergence of the capacitance for a given rotor position with the mesh refinement is plotted in figure 6. A complementary bounds of the capacitance value is observed, as predicted by the theory.

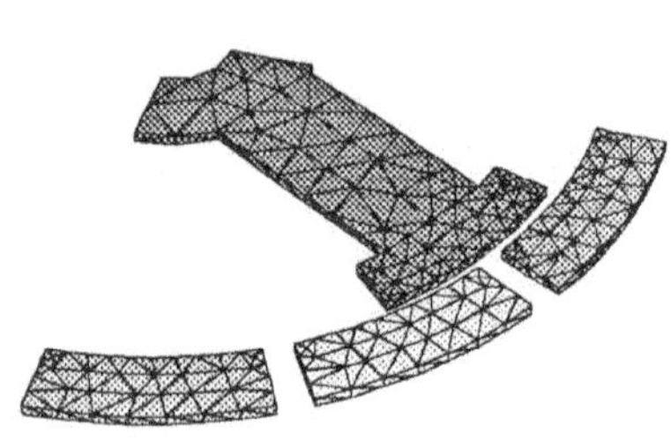

Fig.5 Modelling of a micromotor.

—□— scalar potential formulation
—○— vector potential formulation
Capacitance (femto farad)
Number of elements

Fig.6. Capacitance values versus mesh fineness.

Remark: Results of figures 2 and 6 show that the scalar potential formulation depends more on the mesh fineness while as the vector potential formulation is more stable. This observation worth a further theoretic study.

REFERENCES

1. P.Hammond and J.Penman, 'Calculation of inductance and capacitance by means of dual energy principles', IEE Proc., Vol.123, No.6, 1976, pp.554-559
2. P.Hammond and T.D.Tsiboukis, 'Dual finite element calculations for static electric and magnetic fields', IEE Proc. A, Vol.130, No.3, 1983, pp.105-111
3. J.Penman and J.R.Fraser, 'Complementary and dual energy finite element principles in magnetostatics', IEEE Trans. Mag. Vol.18, No.2, 1982, pp.319-323
4. J.Rikabi, C.F.Bryant and E.F.Freeman, 'Error based derivation of complementary formulations for eddy current problems', IEE Proc. A, 135(4), 1988, pp.208-216
5. C.Li, Z.Ren and A.Razek, 'An approach to adaptive mesh refinement for 3-D eddy current computations', To appear in IEEE Trans.Mag. 1994.
6. J.C.Nedelec, 'A new family of mixed finite elements in R3', Numer. Math, 1986, Vol.50, pp.57-81
7. A.Bossavit, 'Mixed finite elements and the complex of Whitney forms', Conference MAFELAP, London, UK, 1987
8. P.Hammond, 'Use of potentials in calculation of electromagnetic fields', IEE Proc. A, Vol.129, No.2, 1982, pp.106-112
9. I.D.Mayergoyz, M.V.K.Chari and J.D'Angelo, 'A new scalar potential formulation for three-dimensional magnetostatic problems', IEEE Trans. Mag., Vol.23, No.6, 1987, pp.3889-3894
10. T.Nakata, N.Takahashi, K.Fujiwara and Y.Okada, 'Improvements of the T-Ω method for 3-D eddy current analysis', IEEE Trans. Mag., Vol.24, No.1, 1988, pp.94-97
11. A.G.Kladas and J.A.Tegopoulos, 'A new scalar potential formulation for 3-D magnetostatics necessiting no source field calculation', IEEE Trans. Mag., Vol.28, No.2, 1992, pp.1103-1106
12. R.Albanese and G.Rubinacci, 'Magnetostatic field computations in terms of two component vector potentials', Int.J. for numerical Methods in Engineering, Vol.29, 1990, pp.515-532
13. Z.Ren, 'A 3D vector potential formulation using edge element for electrostatic field computation', International Conference CEFC, Aix les Bains, France, July, 1994.
14. Z.Ren and A.Razek, 'Calculation of 3D electrostatic field and force in micromachines by dual formulations', International Conference ICEM-94, Paris, September, 1994

COMPARISON OF VARIOUS 2-D MESHING TECHNIQUES FOR FINITE ELEMENT SOLUTIONS.

J.-Fr. Remacle, M. Umé, A. Nicolet, A. Genon, W. Legros

University of Liège - Dept of Electrical Engineering
Institut Montefiore - Sart Tilman Bât. B28 - B-4000 Liège (Belgium)

INTRODUCTION

A wide range of two and three-dimensional meshing techniques are available and the choice of one of them is quite difficult[1]. It is well known that the quality of a mesh has a big influence on the quality of a finite element (F.E.) solution. Many people who use industrial mesh generators do not know much about meshing techniques and then use their mesh generator like a black box. They hope that the final result is convenient for their modelization.

In this paper, some meshing techniques such as the Delaunay triangulation or the advancing front method are presented. Meshes based on the same characteristic lenght field have not the same efficiency if generated with different techniques.

An algorithm that merges couples of triangles and creates an hybrid triangle-quadrangle mesh is also presented.

MESHING TECHNIQUES

The perfect mesh generator should have some required properties. It should be automatic, no human interaction is needed, should generates elements with prescribed sizes, good shapes and it should be robust enough to mesh complex geometries.

Note that offset and transfinite techniques are not studied here because they cannot be considered as automatic methods.

Techniques based on the Bowyer-Watson algorithm

The principle of the method is to create a Delaunay triangulation with the nodes on the boundary of the structure and then to add nodes inside the structure until all the triangles have the size prescribed [1]. The procedure to add a new vertex in a Delaunay triangulation is called the Bowyer-Watson algorithm. It results in a new Delaunay tr

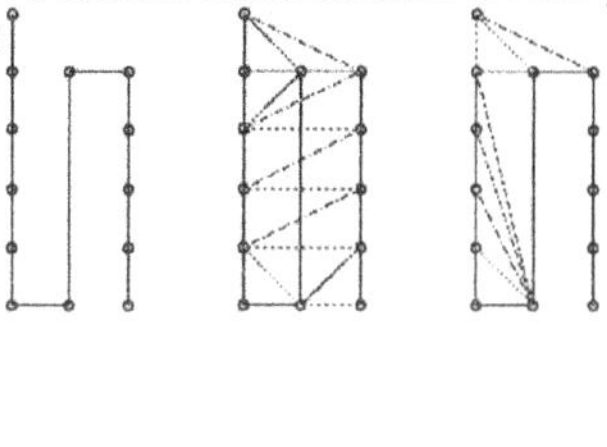

Figure 1. The left figure shows the boundary discretisation, the center figure shows the Delaunay triangulation that does not respect the boundary conformity and the right figure shows a mesh that respects the mesh conformity but not the Delaunay criterion.

This text presents research results of the Belgian programme on Inter university Poles of Attraction initiated by the Belgian State, Prime Minister's Office, Science Policy Programming. The Scientific responsibility is assumed by its authors.

Electric and Magnetic Fields, Edited by A. Nicolet
and R. Belmans, Plenum Press, New York, 1995

A major issue in the Delaunay triangulation approach is the one related to the boundary integrity. Given a set of points connected by the Delaunay criterion, there is no guarantee that the resulting construction will ensure that the edges of the elements will conform to the prespecified boundary (figure 1). We have then to construct a mesh that does not satisfy the Delaunay triangulation next to the boundary.

A large amount of Bowyer techniques that differ on the location of the new inserted point are available [1,3]. These iterative techniques are automatic and a background mesh[2,3,4,6] is used to have a mesh density prescribed

The Advancing Front method

Starting with a set of points on the boundary (initial front), the principle is to progress in the domain and create simultaneousely the point and its connections [4]. Using this technique allows a good control on element shapes and sizes and it is therfore easy to create anisotropic meshes. The technique has been extended to automatic quadrangulation [5]. Another main advantage of this method is that the boundary conformity is certain.

Note that anisotropic meshes or purely quadrangular meshes are not compared in this paper but they are available in our mesh generator.

The Frontal-Delaunay method

This technique developped by Rebay et al.[3] is a combination of frontal and Delaunay methods. This technique has the advantage to be easily implemented starting with a usual Delaunay code. The control on shapes of elements is also excellent.

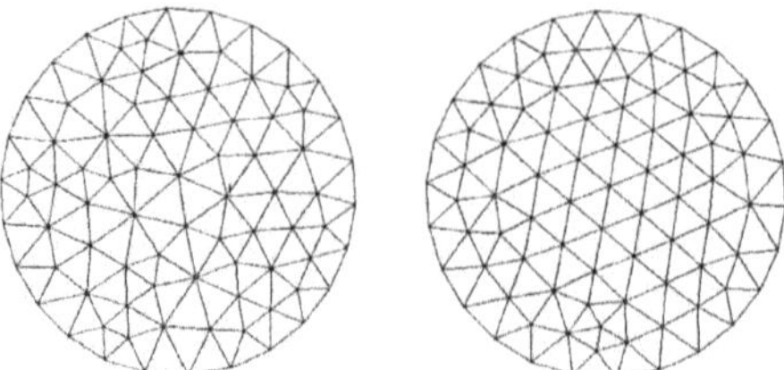

Figure 2. The left mesh is constructed with the Delaunay criterion and the right one with the advancing front method.

GRID QUALITY

The evaluationt of grid quality is performed by a statistical approach ; two statistical distributions are computed, one about mesh topology, the other one about geometry of cells. The example considered is a motor (figure 5).

The first statistical variable is the number of elements surrounding a point. The optimum connectivity is 6 for triangles and 4 for quadrangles. Figure 3 shows that frontal techniques give optimal connectivities. The peek on the value 3 is dued to boundary nodes for which optimal connectivity is 3. The bandwidth of the system of F.E. equations is directely related to this connectivity.

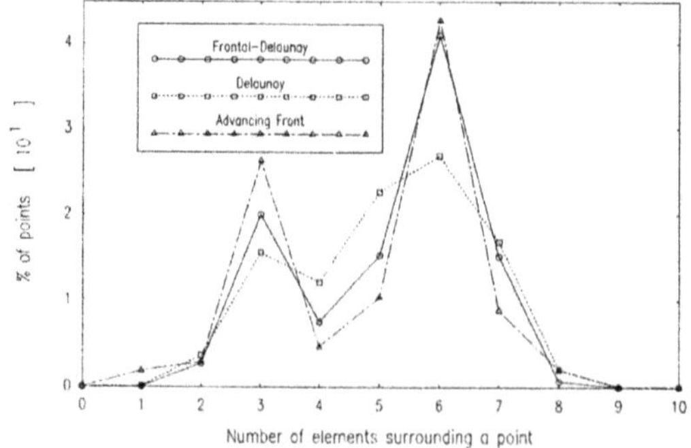

Figure 3. Connectivity of points

The second variable is computed for an element. It is the measure of the maximum difference between an optimum angle (60 deg. for triangles and 90 deg. for quadrangles) and the angles of the element. It is obvious by seeing the figure 4 that frontal techniques give better element shapes. This good behaviour will have some influence on the quality of the solution.

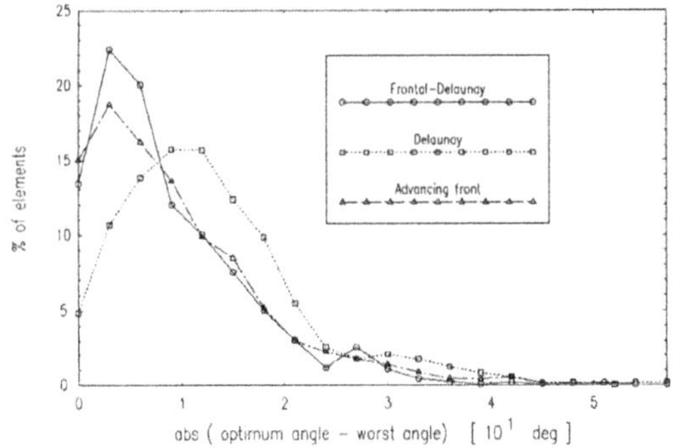

Figure 4. element qualities

RECOMBINATION ALGORITHM

Starting with a triangulation, a certain amount of quadrangles can be easily created by recombining triangles. The difficulty is to find the best set of triangle couples in order to obtain a mesh with a maximal number of well shaped quadrangles [2,4]. The following algorithm is used :

- Create a list with all the couples of neighbour triangles.
- Associate a parameter to each couple representing the quality of the quadrangle formed by recombining the two triangles. The quality criterin of a quadrilateral element can be the diagonal ratio or the value of its most obtuse angle.
- Sort the couples setting with respect to the quality factors.
- Recombine all the acceptable couples. When a couple of triangles is recombined, the two triangles are marked as already recombined.

This method gives quite good results starting with a Delaunay triangulation. The elements formed are always well shaped even in very stretched zones. It is thus a way to avoid ill conditioned meshes coming to a Delaunay triangulation.

NUMERICAL EXAMPLE

As an example let us consider the permanent magnet motor represented on figure 5. The electro-magnetic field has been calculated in the device with different discretizations and the energy of the system has been evaluated in each case.

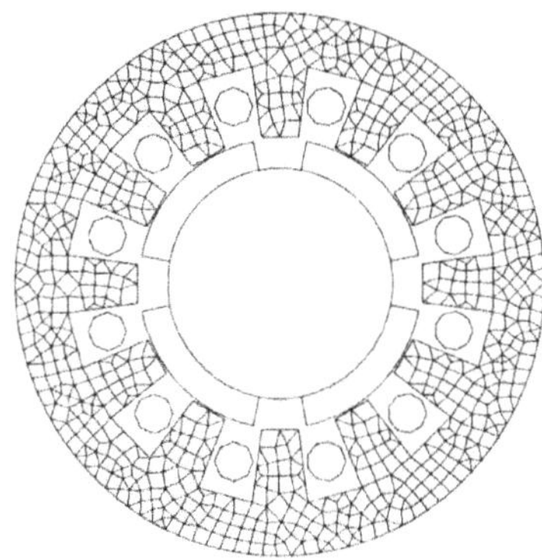

Figure 5. Recombined Delaunay mesh for the stator

For the same kind of discretization, the Delaunay technique is really the worst choice in all cases (figure 6). Frontal techniques are better because the filling of the space is homogeneous : the final mesh contains a large amount of optimum hexagonal cells. Recombination is better because shape functions of quadrangles span a larger space than shape functions of triangles.

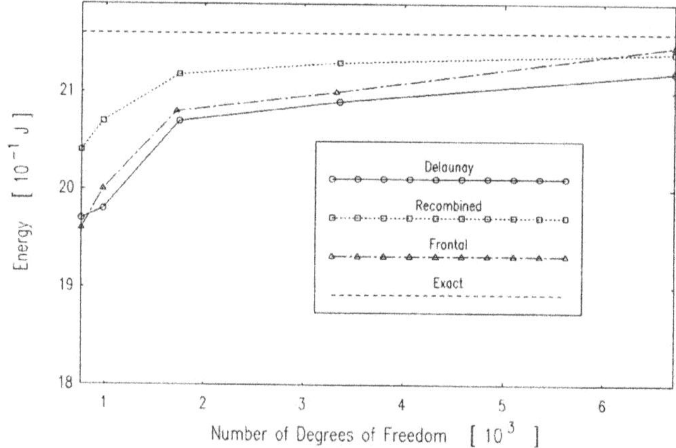

CONCLUSIONS

The choice of one technique of mesh generation is really important for the quality of a F.E. model. Delaunay techniques are often used but the mesh issued from it is not optimal neither on the point of view of the topology nor on the one of the shape of the cells. Frontal and Delaunay-Frontal mesh generators give quite better results on all aspects considered.

An algorithm of recombination has been developped. This technique really increases the quality of a finite element solution with neglectible additional computation.

REFERENCES

1. N.P. Weatherhill, O. Hassan, D. L. Marcum, M. J. Marchant, "Grid generation by the Delaunay triangulation", grid generation short course, von Karman Institute for Fluid Dynamics, January 24-28, 1994.

2. Y. Du Terrail Couvat, J. Morandini, "A General Algorithm for Automatic Hybrid Mesh Generation", recorded on the 9th COMPUMAG Conference, Miami Florida , 1993.

3. S. Rebay, "Efficient Unstructured Mesh Generation by Means of Delaunay Triangulation and Bowyer-Watson Algorithm", Journal of computational physics 106, pp 125-138, 1993.

4. P. Beckers, H. G. Zong, "Revue des estimateurs d'erreur a posteriori et expériences numériques sur l'adaptation des maillages élément finis", Rapport interne du L.T.A.S, Liège, 1992.

5. J. Peraire, M. Vahdati, K. Morgan, O. C. Zienkiewicz, "Adaptive remeshing for compressible flow computation", Journal of computational physics, 72, pp 449-466, 1987.

6. J. Z. Zhu, O. C. Zienkiewicz, E. Hinton, J. Whu, "A new approach to automatic mesh generation", International Journal of computational physics, 77, pp 849-866, 1990.

A HYBRID FE-BE METHOD FOR ACCURATE FIELD AND TORQUE CALCULATION IN ELECTRICAL MACHINES

A. Nysveen[1], R. Nilssen[1] and G. Sande[2]

[1]University of Trondheim, Norwegian Institute of Technology (NTH), Norway
[2]ABB Corporate Research Norway, Norway

INTRODUCTION

The Finite Element Method (FEM) has emerged to be the standard numerical method for analysis of fields in electrical machines. The success of the method is due to its ability to handle problems with complex geometry, source distributions and material properties. For some specific field problems (e.g. open boundary problems), integral techniques as the Boundary Element Method (BEM) are more suitable.[1]

This paper presents a hybrid FE-BE method suitable for field analysis of electrical machines. In a magnetic analysis of the machine the finite element method is used on solid parts (iron, windings) and the boundary element method on homogeneous, linear parts (air).

The motivation for developing a hybrid FE-BE program is to be able to perform non-linear time domain analysis of problems involving moving geometry.

When the method is applied to a rotating electrical machine, the magnetic field in the air gap can be computed by the boundary element method. In the post-processing stage, the magnetic flux density can then be computed without performing a numerical derivation of the potential solution. The cost is that (nearly) Cauchy and hyper-singular integrals must be evaluated with high accuracy. Numerical integration methods, recently published in numerical literature, are applied to minimize the numerical errors. It should therefore be possible to compute the electromagnetic torque in the machine accurately without applying a dense finite element mesh in the air gap.

HYBRID FORMULATION

The magnetic fields is assumed to be either static or quasi-static at power frequencies, meaning that the displacement current is negligible. A simple problem domain consisting of two sub-domains Ω_B and Ω_F is presented in fig. 1. In the "BEM-domain" Ω_B, no eddy currents are present and the magnetic permeability is constant. In the "FEM-domain" Ω_F the field can be either static or quasi-static.

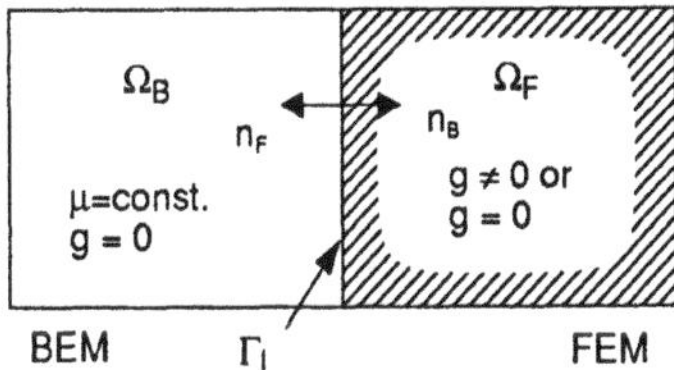

Figure 1 Hybrid FE-BE problem domain

Based on Galerkin's method, the finite element formulation gives rise the following algebraic equations:

$$\sum_i \int_{\Omega_F} \nabla N_j \frac{1}{\mu} \nabla N_i d\Omega A_i - \sum_i \int_{\Gamma_I} N_j \frac{1}{\mu} N_i d\Gamma q_i = \int_{\Omega_F} N_j J d\Omega \tag{1}$$

A_i denote the nodal values of the potential A and q_i the nodal values of $\frac{\partial A}{\partial n}$ on the interface boundary Γ_I.

In the quasi-static case the same approach yields a system of differential equations:

$$\sum_i \int_{\Omega_F} N_j g N_i d\Gamma \Omega \dot{A}_i + \sum_i \int_{\Omega_F} \nabla N_j \frac{1}{\mu} \nabla N_i d\Omega A_i - \sum_i \int_{\Gamma_I} N_j \frac{1}{\mu} N_i d\Gamma q_i = \int_{\Omega_F} N_j J d\Omega \tag{2}$$

The boundary element equations can be derived following different approaches. Here, a weighted residual scheme is used taking the Green function as weighting function with the singular points on the boundary nodes.

$$\frac{1}{\mu}\alpha A_j + \sum_i \{ \int_{\partial\Omega_B} \frac{1}{\mu} N_i \frac{\partial G_j}{\partial n} d\Gamma \} A_i - \sum_i \{ \int_{\partial\Omega_B} \frac{1}{\mu} N_i G_j d\Gamma \} q_i = \int_{\Omega_B} J G_j d\Omega \tag{3}$$

G_j denotes the Green function with the singular point on node j.

At the interface Γ_I the potential must be continuous, otherwise the flux density becomes infinite. According to Maxwell's equations, the tagential component of the magnetic field strength **H** is continuous, i.e.:

$$\frac{1}{\mu} q_i |_{FEM} = -\frac{1}{\mu} q_i |_{BEM} \tag{4}$$

The unknown normal derivatives q_i in the FEM-domain are now eliminated by substituting eq. 4 into eq. 2.

RESOLVING NUMERICAL PROBLEMS

At a sharp corner the normal derivative is discontinuous, i.e. $q_L \neq q_R$. To obtain a satisfactory result, two values of the normal derivative q are associated with the node on the sharp corner. Now, depending on the boundary conditions, the following cases must be considered: 1) q_L and q_R known, 2) q_L or q_R known, 3) q_L and q_R unknown. Case 1 and 2 do not cause any specific problems since there is no need for introducing a new equation.

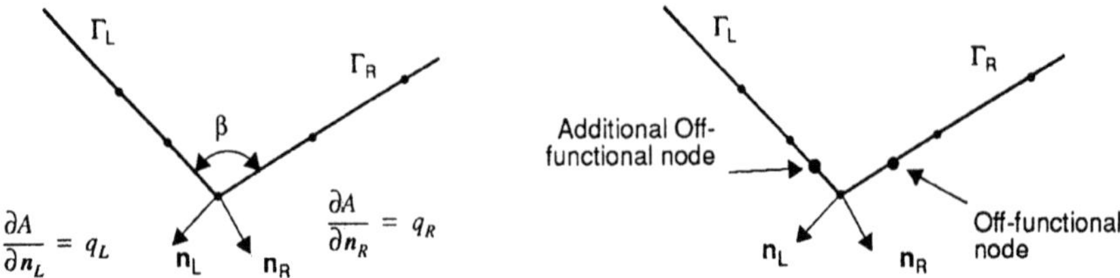

Figure 2 a) Multiple values of q at a sharp corner b) Location for the "off-functional nodes"

Various methods have been proposed to solve case no. 3. A method was suggested by S.P. Walker and R.T. Fenner[2] in 1989 in which the geometry near the corner were investigated to obtain a relationship between the normal and tangential derivatives of A and the angle β. The method is easy to implement, but cause ill-conditioned equation systems for large angles ($\beta \approx 180°$). Better results were obtained by A.K. Mitra and M.S. Ingber generating a new boundary equation.[3] The question is where to place the singular point of the new Green function. In their first proposal, the singular point was located outside the BEM-domain. This method does not generate a new unknown variable since the factor α in eq. 3 is zero outside the BEM-domain. However it leads to ill-conditioned matrices. They presented an improved method in 1993 where the singular point is placed on the boundary on an "off-functional node".[4] The "off-functional node" is placed between two original nodes. A new unknown variable is not necessary since the potential on the "off-functional node" is a linear combination of the nodal potential values.

In this paper the method is slightly modified, i.e. an additional "off-functional node" is introduced by moving the singular point of the Green function associated with the sharp corner as shown in fig. 2. This gives a more symmetric treatment of the corner.

The singularity of the kernel in the integrals in eq. (3) can cause inaccuracies since the integrals are difficult to evaluate in geometric configurations that frequently appear in electrical machines.

The singular integrals occur when the singular point is located at the boundary element under consideration. The integrand in the second integral in eq. (3) is singular, but the integral does have a definite value. A gaussian quadrature rule with logarithmic weight is needed for accurate evaluation. When higher order iso-parametric boundary elements are used, the elements can have a curvature and this must be taken into account during the implementation. An accurate method is presented by M.H. Lean and A. Wexler.[5]

The nearly singular integrals are more difficult to evaluate since the special quadrature rule developed for singular integrals cannot be applied. Various techniques have been proposed to handle these integrals. A well known method is the element sub-division method. The original element is sub-divided into smaller non-overlapping sub-elements until a sufficient accuracy is reached. A higher accuracy can be archived with the same number of quadrature points if a transformation which is lumping the quadrature points toward the singularity is applied.[6]

In a standard reference element the nearly singular integral takes the form:

$$I = \int_{-1}^{1} f(\xi)\, d\xi \qquad (5)$$

The transformation of the quadrature coordinates is cubic and it is defined by:

$$\xi = a\gamma^3 + b\gamma^2 + c\gamma + d \qquad (6)$$

The transformation is set to fulfil the following constraints (denoting the projection of the singular point down to the ξ-plane by $\bar{\xi}$):

$$\xi(1) = 1 , \quad \xi(-1) = -1 , \quad \left.\frac{\partial\xi}{\partial\gamma}\right|_{\bar{\xi}} = r \quad 0 < r \le 1 , \quad \left.\frac{\partial^2\xi}{\partial\gamma}\right|_{\bar{\xi}} = 0 \qquad (7)$$

These four conditions make it possible to calculate the coefficients a, b, c and d when the parameter r is known. The derived formulas are listed in the references.[6] The projection of the singular point may be outside the considered element, i.e. $|\bar{\xi}| > 1$.

FIELD AND TORQUE CALCULATIONS

After the solution is computed, the field inside the BEM-domain is given by (compare with eq. (3)):

$$A(p) = -\sum_i \int_{\partial\Omega_B} N_i \nabla_r G_p \mathbf{n} dr A_i + \sum_i \int_{\partial\Omega_B} N_i G_p dr q_i + \int_{\Omega_B} J G_p d\Omega \qquad (8)$$

Space differentiating eq. (8) with respect to the field point p, the field strength (flux density) is given by:

$$\nabla_P A(p) = -\sum_i \int_{\partial\Omega_B} \{N_i \nabla_P (\nabla_r G_p \mathbf{n})\, dr\} A_i + \sum_i \int_{\partial\Omega_B} \{N_i \nabla_P G_p dr\} q_i + \int_{\Omega_B} J \nabla_P G_p d\Omega \qquad (9)$$

By applying this formula, the flux density is computed without performing a derivation of the basis functions N_i. The cost is that (nearly) hyper-singular and Cauchy-singular boundary integrals must be evaluated. The non-linear coordinate transformation described above must be applied.

The force acting on a part, is computed by integrating the Maxwell stress tensor along a surface Γ enclosing the part. The electromagnetic torque in an electrical machine can be computed along a circle with radius r_0 enclosing the rotor as:

$$T_e = \oint_\Gamma \frac{r_0}{\mu_0} B_r B_\varphi d\Gamma \qquad (10)$$

B_r and B_ϕ denote the radial and tangential components of the magnetic flux density in the air gap, respectively. They are computed from eq. (9).

SAMPLE CALCULATIONS

The computed values of the implemented boundary element method have successfully been compared to analytical solutions on models presented by Walker et al.[2]

A problem by using eq. (10) in conjugation with the finite element method is that the computed value can be highly sensitive to the radius of the integration path and the mesh in the air gap. Usually, a rather dense mesh is need in the air gap to obtain satisfactory results.

The presented method have been tested on two different machines (a DC-machine and a reluctance machine), and the values are compared with results from a commercial FEM program named ACE supplied by ABB Corporate Research, Sweden.[7] ACE provides the possibility to use adaptive mesh refinement, which are applied in these test calculations. Linear elements are used.

The sensitivity of the computed torque has been tested by taking the integral along paths with different radii. Two hybrid FE-BE meshes are generated for each machine. Bi-quadratic finite elements are used in conjugation with quadratic boundary elements. The computed torque values are presented in table 2.

Table 1 Mesh data for DC-machine and the reluctance machine

	DC machine			Reluctance machine		
	Mesh # 1	Mesh # 2	ACE	Mesh # 1	Mesh # 2	ACE
BE on rotor	57	76	-	52	112	-
BE on stator	36	78	-	76	148	-
Nodes in whole mesh	1985	7540	11777	1610	4856	12588 *

* Mesh generated in the whole cross section of the machine

Rotor diam.: 216 mm
Stator bore: 220 mm
Air gap: 2.0mm

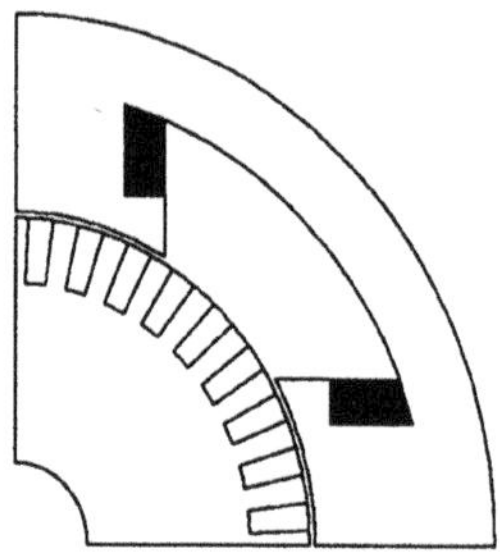

Rotor diam.: 158 mm
Stator bore: 160 mm
Air gap: 1.0mm

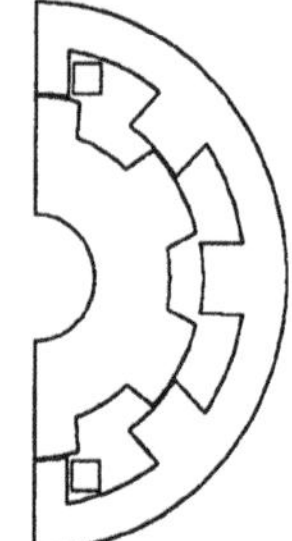

Figure 3 a) DC-machine b) Reluctance machine (zero position)

Table 2 Computed torque values for different radii.

DC machine							Reluctance machine, rotor position 15 degrees						
	Mesh # 1		Mesh # 2		ACE			Mesh # 1		Mesh # 2		ACE	
Radius r_0 [mm]	Torque [Nm]	Var. [%]	Torque [Nm]	Var. [%]	Torque [Nm]	Var. [%]	Radius r_0 [mm]	Torque [Nm]	Var. [%]	Torque [Nm]	Var. [%]	Torque [Nm]	Var. [%]
108.4	112.7	0.0	116.7	0.1	117.6	0.6	79.3	3.40	0.9	3.32	0.0	3.23	0.6
108.7	112.7	0.0	116.8	0.0	118.0	0.9	79.4	3.39	0.6	3.32	0.0	3.22	0.3
109.0	**112.7**	**0.0**	**116.8**	**0.0**	**116.9**	**0.0**	**79.5**	**3.37**	**0.0**	**3.32**	**0.0**	**3.21**	**0.0**
109.3	112.7	0.0	116.8	0.0	117.1	0.2	79.6	3.35	-0.6	3.32	0.0	3.20	-0.3
109.6	112.1	0.5	116.8	0.0	117.2	0.3	79.7	3.33	-1.2	3.32	0.0	3.20	-0.3

CONCLUSIONS

The presented results show that the torque computed by the presented hybrid FE-BE method is almost insensitive to the integration path. The quality of the calculation depends on the finite element mesh in the solid parts of the machine. We therefore conclude that the presented method can be applied in field analysis of electrical machines.

REFERENCES

1. D'Angelo, J., "Hybrid finite element/boundary element analysis of electromagnetic fields". *Topics in Boundary Element Research*, vol. 6, Brebbia, C.A. (ed), Springer-Verlag, Berlin (1989).
2. Walker, S. P. & Fenner, R. T., "Treatment of corners in BIE analysis of potential problems". *International Journal for Numerical Methods in Engineering*, vol. 28, pp 2569-2581 (1989).
3. Mitra, A. K. & Ingber, M. S., "Resolving Difficulties in the BIEM Caused by Geometric corners and Discontinuous Boundary Conditions". *Boundary Elements*, vol. IX (1987).
4. Mitra, A. K. & Ingber, M. S., "A multiple-node method to resolve the difficulties in the boundary integral equation method caused by corners and discontinuous boundary conditions". *International Journal for Numerical Methods in Engineering*, vol. 36, pp 1735-1746 (1993).
5. Lean, M. H. & Wexler, A., "Accurate numerical integration of singular boundary element kernels over boundaries with curvature". *International Journal for Numerical Methods in Engineering*, vol. 21, pp 211-228 (1985).
6. Telles, J. C. F., "A self-adaptive co-ordinate transformation for efficient numerical evaluation of general boundary element integrals". *International Journal for Numerical Methods in Engineering*, vol. 24, pp 959-973 (1987).
7. ACE 2.2 Users Manual, ABB Corporate Research, Västerås, Sweden (1993).

NUMERICAL ASPECTS OF THE CALCULATION OF RADAR CROSS SECTIONS FROM 2D FINITE ELEMENT FREQUENCY-DOMAIN NEAR FIELDS

A. Kedadra, A. Nicolas, L. Nicolas, J.L. Yao-bi

Centre de Génie Electrique de Lyon
URA CNRS 829 - Ecole Centrale de Lyon
BP 163 - 69131 Ecully cedex - France

INTRODUCTION

Numerical methods, such as finite difference (FD), boundary element (BE) or finite element (FE), are efficiently used to compute near scattering fields: by comparison with other analytical methods, advantages are that scattering by any geometrical shape may be computed, and the physical nature of the scatterer may be complex (perfect electric conductor coated with an inhomogeneous dielectric for example). In many applications, the near field solution has to be transformed into the far field in order to get the radar cross section (RCS) for instance. The purpose of this paper is to present two different methods to perform this transformation (boundary element method and harmonic expansion), and to describe some numerical aspects attached to their implementation in a FE package.

THE 2D FE FORMULATION

We have developed[1] a general 2D FE formulation for the modeling of scattering of plane waves by arbitrarly shaped objects. Main features of this formulation are:

- the fields are modeled in the frequency domain
- the formulation is directly written in terms of vector fields
- nodal or mixed elements are used
- open boundary is taken into account either by absorbing boundary condition or by boundary element method.

For the purpose presented here, we used nodal elements. Because of the 2D approximation, only one component of the main near field is computed. The outside of the FE domain is taken into account by a second order Engquist-Majda absorbing boundary condition (ABC). The general form may be written as:

Electric and Magnetic Fields, Edited by A. Nicolet
and R. Belmans, Plenum Press, New York, 1995

$$\iint_S \left[\nabla W . \frac{1}{j\omega\mu} \nabla E + j\omega\varepsilon W . E \right] ds - \frac{k}{\omega\mu} \int_\Gamma W . E d\Gamma + \frac{1}{2k\omega\mu} \int_\Gamma \frac{\partial W}{\partial \tau} . \frac{\partial E}{\partial \tau} d\Gamma =$$

$$\frac{1}{j\omega\mu} \int_\Gamma W . \frac{\partial E^i}{\partial n} d\Gamma - \frac{k}{\omega\mu} \int_\Gamma W . E^i d\Gamma + \frac{1}{2k\omega\mu} \int_\Gamma \frac{\partial W}{\partial \tau} . \frac{\partial E^i}{\partial \tau} d\Gamma \quad (1)$$

where E is the total electric field,
E^i is the incident field,
W is the weighting function,
Γ is the boundary of the FE domain,
S is the FE domain itself,
k, ω, ε and μ are the classical physical constants.

FIRST METHOD: BOUNDARY ELEMENT (BE) METHOD

From the Green's identity, it follows that[2] the value of the electric field E_{far} at any point outside the FE domain S may be expressed as a surface integral relation of the magnetic near field E_{fe} and its normal derivative over the boundary Γ of the surface S:

$$E_{far} = \int_\Gamma \left[G \frac{\partial E_{fe}}{\partial n} - \frac{\partial G}{\partial n} E_{fe} \right] d\Gamma, \quad \text{with } G = \frac{j}{4} H_0^{(2)}(kr) \text{ and } \frac{\partial G}{\partial n} = \frac{jk}{4} H_1^{(2)}(kr) \quad (2)$$

where $H_0^{(2)}$ and $H_1^{(2)}$ are Hankel functions of the second kind,
r is the distance between the calculation point and Γ.

By extension of this method, the boundary Γ may be anything surrounding the scatterer. The sources of the scattered field are surface currents located on the surface of the scatterer, and sufficient condition to apply (2) is that the sources have to be inside the boundary Γ_{fe} used for the calculation of the far field.

RCS is then obtained by:

$$\sigma(\phi) = 2\pi \lim_{r \to \infty} \left(r \frac{|E^s|^2}{|E^i|^2} \right) \quad (3)$$

where E^s and E^i are the scattered and the incident fields.

SECOND METHOD: HARMONIC SERIES (HE) EXPANSION

Using cylindrical coordinates system (r,ϕ), the field is expanded into the serie[3]:

$$E(r,\phi) = \sum_{n=-\infty}^{+\infty} A_n(kr) e^{jn\phi} \quad \text{with } A_n(kr) = j^{-n} a_n H_n^{(2)}(kr) \quad (4)$$

where $H_n^{(2)}$ is the Hankel function of the second kind and order n,
and a_n is a constant coefficient.

The coefficients A_n are obtained from the near field E_{fe} on a circle of radius r by:

$$An(kr) = \frac{1}{2\pi}\int_0^{2\pi} E_{fe}(kr,\phi)\, e^{jn\phi} d\phi \tag{5}$$

On a far point (r_2,ϕ_2), the same relation (4) may be written, with the same coefficients a_n, and the far field E_{far} is then obtained by:

$$E_{far}(r_2,\phi_2) = \sum_{n=-\infty}^{+\infty} A_n(kr).\frac{H_n^{(2)}(kr_2)}{H_n^{(2)}(kr)} e^{jn\phi_2} \tag{6}$$

From a practical point of view, the Hankel functions are calculated analyticaly by using recurrence relations, and n is equal to 240.

Compared to the BE method, the HE method does not need the derivation of the main field, which introduces some numerical inaccuracy. Both methods work with the FE near field located on a closed boundary Γ_{fe} surrounding the scatterer (fig.1). Table 1 compares CPU times for the computation of the far field for both methods, depending on the number of points of this boundary.

Table 1. CPU times for HE and BE methods depending on the number of points of Γ_{fe} - computation of 180 far points.

number of points	60	88	116	168	224
HE method	1.8	3.3	5.3	9.8	16.6
BE method	1.4	2.2	3.2	5.4	7.8

NUMERICAL COMPARISON

Scattering of a plane wave by a perfect electric conducting cylinder is considered as test case. Numerical conditions of this test are given in figure 1. For both methods, we have studied the effect of the density of the FE mesh, and the effect of the distance between the boundary Γ_{fe} and the scatterer. Results are summarized in table 2. A global error may be quantified by comparison with the analytical solution of the field[4]:

$$E_{ana}(r,\phi) = E_0 \sum_{n=-\infty}^{+\infty} j^{-n}\left[J_n(kr) - \frac{J_n(ka)}{H_n^{(2)}(ka)} H_n^{(2)}(kr)\right] e^{jn\phi} \tag{7}$$

where J_n is the Bessel function of the first kind,
a is the radius of the cylinder.

The global error is calculated as:

$$\varepsilon = \int_0^{2\pi} \frac{|E_{far} - E_{ana}|}{|\max(E_{ana})|} \partial\phi \tag{8}$$

where E_{far} and E_{ana} are computed on a circle of radius 10m.

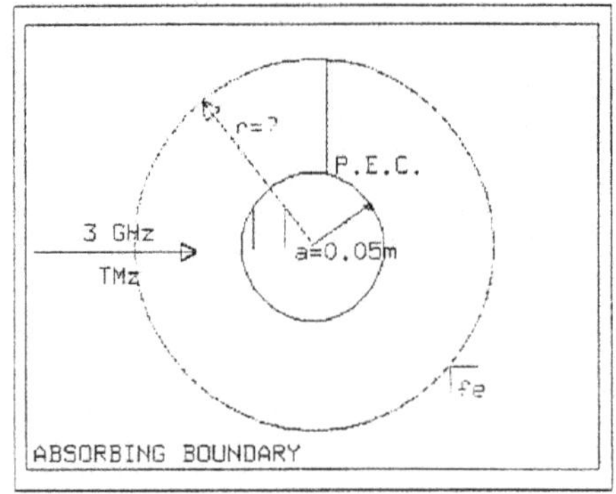

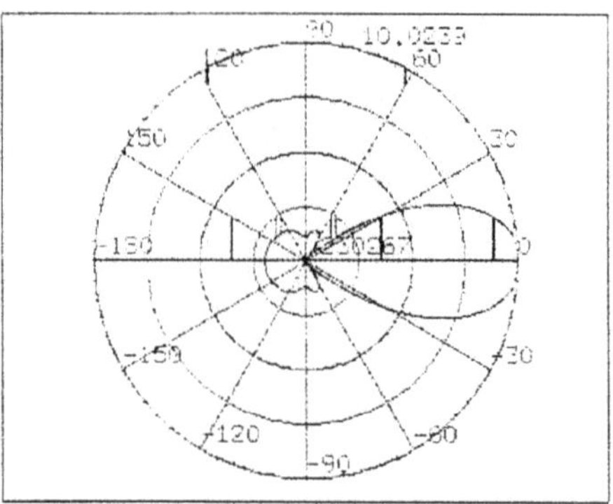

Figure 1. Scattering by a perfect electric conducting cylinder of radius 0.05 m: 3 GHz, TMz polarization, 0° incident angle. The external boundary is a rectangular box (0.4 m x 0.3 m). Left: definition of the boundary Γ_{fe} for the computation of the far field. Right: bistatic RCS.

Table 2. Global error (in %) for the far field, depending on the FE mesh density and on the distance beetwen the scatterer and the boundary Γ_{fe}. Mesh density is defined as the mean number of nodes per wavelength.

Boundary Element Method

density	5.6	7.6	9.9	14.2	18.7
r=0.051	6.4	4.9	3.7	1.9	1.7
r=0.075	9.5	4.6	3.6	1.9	1.8
r=0.1	11.7	5.3	3.6	2.3	2.0
r=0.125	15.7	6.1	4.0	2.5	2.2
r=0.149	16.4	8.7	5.5	3.2	2.5

Harmonic Expansion Method

density	5.6	7.6	9.9	14.2	18.7
r=0.051	1.4	0.4	0.2	0.1	0.1
r=0.075	8.1	3.6	2.4	1.1	1.17
r=0.1	12.7	3.8	2.3	1.7	1.8
r=0.125	11.0	3.4	2.5	2.1	2.0
r=0.149	13.2	6.2	4.2	2.7	2.7

CONCLUSION

From table 2, it can be conclude that:

- in most cases, HE method is more accurate than BE method. But, on an other hand, it is more CPU time consuming.
- when the boundary Γ_{fe} is close to the boundary of the PEC, the HE method gives almost exact result. And this is not sensitive to the mesh density. But, on an other hand, the boundary Γ_{fe} has to be a circle, while it can be anything when using BE method.
- when the boundary Γ_{fe} is close to the external boundary, the error becomes really higher: this is due to spurious reflexions produced by the ABC.

REFERENCES

1. L. Nicolas, K.A. Connor, S.J. Salon, B.G. Ruth, L.F. Libelo, "Modélisation 2D par éléments finis de phénomènes micro-ondes en milieu ouvert." *J. Phys. III*, France 2 (1992), pp.2101:2114.
2. J.A. Stratton, "Electromagnetic Theory," McGraw-Hill, New York (1941).
3. R.F. Harrington, "Time-Harmonic Electromagnetic Fields," McGraw-Hill, New York (1961)
4. C.A. Balanis, "Advanced Engineering Electromagnetics," Wiley (1989)

A GENERAL ELEMENT STRUCTURE FOR FINITE ELEMENT PROGRAMMES

Fr. Henrotte, J.-Fr. Remacle, A. Nicolet, A. Genon, W. Legros

University of Liège - Dept of Electrical Engineering
Institut Montefiore - Sart Tilman Bât. B28 - B-4000 Liège (Belgium)

INTRODUCTION

The power for a finite element (F.E.) programme to model a large range of physical phenomena depends mainly on its capability to allow the definition of various kinds of finite elements.

From a general point of view, a finite element is a small piece (which volume is V_e) of a modelled system (which volume is V_s; $V_e << V_s$). Depending on the particular media the element is made of, on the physical problem and on the kind of modelling adopted, a set of N degrees of freedom (DoF or connectors) is selected to represent the finite element. Those connectors are related to one another by a set of M relations which are discretized expressions for the physical laws the finite element obeys. In a matrix form, one can write

$$M_{kl}\, x_k = 0 \qquad (l=1,2,\ldots,M) \qquad (1)$$

where $\{x_k\}$ is the set of N connectors of the finite element, M_{kl} is called the Elementary Matrix of the finite element and where the implicit summation on repeated indices is assumed as it will always be in the rest of the paper.

Whereas old fashioned computer codes were entirely organised around the notion of node, modern finite element approaches, following the present tendency to geometrisation in physics, have introduced new types of connectors besides the classical nodal values (and so doing new types of shape functions), i.e. edge circulations, facet flows and volume densities[1].

The specificity of each kind of finite element is defined by a collection of combinative characteristics which are for instance :

— type of geometry (3D, 2D, axisymmetrical, ...)
— linearity or isotropy of the physical media
— physical type of the connectors (magnetical, thermal, ...)
— geometrical nature of the connectors (nodal value, edge circulation,...)
— shape of the element (triangle, quadrangle, tetrahedron,...)
— degree of the shape and trial functions
— ...

This text presents research results of the Belgian programme on Interuniversity Poles of attraction initiated by the Belgian State, Prime Minister's Office, Science Policy Programming. The scientific responsability is assumed by the authors.

Electric and Magnetic Fields, Edited by A. Nicolet
and R. Belmans, Plenum Press, New York, 1995

Despite that diversity, the treatment each finite element undergoes during calculation remains basically the same. It consists in evaluating the coefficients M_{kl} of its associated elementary matrices and in assembling them in the general matrix of the system, in the lines and rows corresponding to the connectors of the element.

So, it is important from the viewpoint of a computer science to find the most general formal structure of the Elementary Matrix Coefficient expressions in order to show that they have some interesting properties. One will then demonstrate how naturally that formal structure leads to a concise and efficient computer implementation using the capabilities of the standard C language.

NUMERICAL INTEGRATION

Numerical integrations on a given domain are usually performed by evaluating the function to be integrated at N particular points p_i called 'Gauss points'. Each Gauss point is associated with a weight w_i and the approximate result of the integration is :

$$\int_{V_0} f\, dV_0 \equiv \sum_N w_i\, f(p_i). \tag{2}$$

The theory of integration gives the number of Gauss points, their location and the associated weights so as to have an exact integration of a polynomial n-order function f on a given reference domain V_0 (line, triangle, quadrangle, tetrahedron, ...). In order to take advantage of those tabulated data, a change of coordinates is performed in each element to reduce it to the normalised reference element V_0 corresponding to its own shape. In the programme, it is then possible to write a general integration subroutine

$$I[\, f, \{N, p_i, w_i\}\,] \tag{3}$$

whose arguments are :

— a pointer towards the function f to be integrated taking the change of coordinates into account,

— a Gauss point table $\{N, p_i, w_i\}$, i.e. a C language structure[2] containing the number of Gauss points, their location and their associated weights.

The choice of such a table for a given finite element depends on its shape (line, triangle, quadrangle, tetrahedron, ...) and on the degree of the associated shape functions (higher order shape functions need more Gauss points). It is clear that the Gauss point tables are shared by all elements that have the same type; they are initialized only once at the beginning of the programme.

GENERAL STRUCTURE FOR ELEMENTARY MATRIX COEFFICIENTS

Each elementary matrix coefficient M_{kl} is obtained by integration of a density called m_{kl} on the reference volume V_0 of the finite element :

$$M_{kl} = \int_{V_0} m_{kl}(u^1,\dots,u^N)\, du^1 \dots du^N. \tag{4}$$

The density m_{kl} is an expression combining a physical characteristic function λ and the 'product' of either a shape function ω_k or its derivative (curl ω_k or grad ω_k) and either a trial function or its derivative (ϕ_l, curl ϕ_l or grad ϕ_l). For the sake of generality, the expression is written in order to make appear explicitly the Jacobian matrix J_{ij} of the change of coordinates from the 3D Euclidean coordinates $\{X^i\}$ (in which the physical laws are expressed) towards the reference element coordinates $\{u^i\}$ (in which the integration (4) is performed).

Consider as an example the following elementary matrix (5) which appears in 2D magnetostatics (with λ representing the magnetic reluctivity ν of the media) and its associated density (6) :

$$M_{kl} = \int_V \lambda \, \mathrm{grad}\omega_k . \mathrm{grad}\phi_l \, dV \tag{5}$$

$$m_{kl}(u^1,\ldots,u^N) = \lambda \frac{\partial \omega_k}{\partial u^i}(J^{-T}J^{-1})_{ij} \frac{\partial \phi_l}{\partial u^j} \det(J). \tag{6}$$

Multiple Coordinate Transformations

Change of coordinates is a major tool in finite element modelling. A lot of very important kinds of finite elements are based upon one or several successive changes of coordinates, e.g. 2D axisymmetrical problems in cylindrical coordinates, anisotropic media modelling in a well chosen local coordinate system in which it becomes isotropic, stretched elements[3], open boundary finite elements based on a conformal transformation of coordinates[4], ...

Let be

$$\{x_1^i\},\ldots,\{x_N^i\}$$

N intermediate coordinate systems between the reference element coordinates $\{u^i\}$ and the Euclidean coordinates $\{X^i\}$. We have

$$\begin{aligned} J_{ij}(u^1,\ldots,u^N) &= \frac{\partial X^i}{\partial u^j} = \frac{\partial X^i}{\partial x_N^k} \frac{\partial x_N^k}{\partial x_{N-1}^l} \cdots \frac{\partial x_2^p}{\partial x_1^q} \frac{\partial x_1^q}{\partial u^j} \\ &= (J_{X,N} J_{N,N-1} \cdots J_{2,1} J_{1,u})_{ij}. \end{aligned} \tag{7}$$

The Jacobian Matrix which appears in (6) is then simply the matrix product (ordered) of the individual Jacobian matrices of the successive coordinate transformations; its determinant is the product of the individual determinants :

$$\det(J) = \det(J_{X,N}) \det(J_{N,N-1}) \ldots \det(J_{2,1}) \det(J_{1,u}). \tag{8}$$

Standard Density Functions

Formally, the density (5) do not depend on the degree and the shape of the finite element. Moreover it is neither dedicated to one kind of elements nor to 2D magnetostatics. For instance, it also appears in thermal problems (with λ representing the thermal conductivity k) and in 3D magnetostatics with scalar potential (where λ is the magnetic permeability μ).

In the programme, the idea is to define a "standard density function" (it is a function in the computer sense) which has the same level of generality as the mathematical expression (6). One standard density function is associated with each density m_{kl} and we shall call it m_{kl} as well . It has six arguments which are pointers towards respectively : the shape functions of the finite element and their derivatives, the trial functions and their derivatives, the Jacobian matrix (which can be the product of several simple Jacobian matrices) and the physical characteristic function of the element :

$$m_{kl} = m_{kl}[\omega_k, \partial\omega_k/\partial u^i, \phi_l, \partial\phi_l/\partial u^j, J_{ij}, \lambda]. \tag{9}$$

Generation of The Elementary Matrix Coefficients

With the "general integration subroutine" (3) and the "standard density functions" (9), we are now able to write the general structure of the Elementary Matrix Coefficients expressions :

$$M_{kl} = I[\ m_{kl}[\omega_k, \partial\omega_k/\partial u^i, \phi_l, \partial\phi_l/\partial u^j, J_{ij}, \lambda]\ ,\ \{N, p_i, w_i\}\ ]. \tag{10}$$

The evaluation speed of those coefficients can be hugely improved if the contribution of one Gauss point is evaluated for *all* the Elementary Matrix Coefficients associated with the finite element before passing to the next Gauss point. For that purpose, tables containing the values at the Gauss points of the element of all the arguments of the m_{kl} are filled beforehand and are transferred rather than transferring pointers towards the functions permitting to value them within each density function m_{kl}. Those "Gauss Point Tables" are represented with brackets in (11).

$$M_{kl} = I[\ m_{kl}[\{\omega_k, \partial\omega_k/\partial u^j\}, \{\phi_l, \partial\phi_l/\partial u^j\}, \{J_{ij}\}, \lambda], \{N, p_i, w_i\}\ ]. \tag{11}$$

The number of standard density functions is relatively small; a whole 2D magnetic programme (including non linear materials and eddy currents) works with only three of them and nevertheless deals with various degrees and shapes of elements :

$$m^1_{kl} = \lambda \frac{\partial\omega_k}{\partial u^i}(J^{-T}J^{-1})_{ij}\frac{\partial\phi_l}{\partial u^j}\det(J),$$

$$m^2_{kl} = \lambda\omega_k\phi_l \det(J),$$

$$m^3_{kl} = \lambda x_p x_q \left(\frac{\partial\omega_k}{\partial u^i}(J^{-T}J^{-1})_{ij}\frac{\partial\omega_p}{\partial u^j}\right)\left(\frac{\partial\omega_q}{\partial u^i}(J^{-T}J^{-1})_{ij}\frac{\partial\phi_l}{\partial u^j}\right)\det(J).$$

FINITE ELEMENT LIBRARY

In the programme, the C language structures (11) and (12) are defined; they contain pointers towards the tables and the functions which are the 'data' associated with each type of finite element :

$$\{\ \text{label}, \{N, p_i, w_i\}, J_{ij}, \text{EMS}\}, \tag{12}$$

$$\{\ \{x_k\}, m_{kl}, \lambda, \omega_k, \phi_l\ \}. \tag{13}$$

In the structure "Finite_Element_Type" (12), the label is a number or a character string which allows to make reference to that kind of finite element, EMS is a table of "Elementary_Matrix" structures (13). The other elements keep the same signification as before except that J_{ij}, ω_k and ϕ_l are the subroutines that fill the Gauss Points Tables rather than the functions themselves.

Finally, the finite element library is a table of "Finite_Element_Type" structures. The operation of defining a new type of finite element consists now simply in filling one new line in the finite element library with the 'data' that endows it with the desired features.

CONCLUSION

A general definition of the "finite element" notion has been made. Together with a philosophy close to an "object oriented" one, it leads to a concise and efficient computer implementation. The finite element is associated with a set of data (connectors, Gauss Point Tables, standard density functions, ...) in such a way that it can be handled by general operators for integration and assembling in the global system matrix. Because this implementation requires the useful notions of "structure" and "pointer", it has been done with the *Standard C language*; the use of a non standard object oriented language as C++ is not required.

References

1. A. Bossavit. "Whitney forms: a class of finite elements for three-dimensional computations in electromagnetism" IEE Proceedings, Vol. 135, Pt. A, No. 8, NOVEMBER 1988
2. B.W. Kernighan, D.M. Ritchie. "Le Langage C", manuels informatiques Masson (1988).
3. A. Nicolet, J.Fr. Remacle, B. Meys, A. Genon, W. Legros, "Transformation methods in computational electromagnetism", J. Appl. Phys. 75(8), 15 May 1994.
4. D.A. Lowther, E.M. Freeman, B. Forghani "A sparse matrix open boundary method for finite element analysis", IEEE Transactions on Magnetics, VOL. 25, NO. 4, JULY 1989.

UP-WIND FEM ITERATIVE SOLUTION OF UNBOUNDED TRAVELLING MAGNETIC FIELD PROBLEMS

G. Aiello, S. Alfonzetti and S. Coco

Dipartimento Elettrico Elettronico e Sistemistico
Facolta' di Ingegneria - Universita' di Catania
Viale A. Doria, 6 - 95125 Catania, Italy

INTRODUCTION

The finite element method (FEM) solving of electromagnetic field problems in unbounded domains generally represents a very onerous task and requires specialized techniques such as ballooning, infinite elements, hybrid FEM/BEM and use of transformations[1]. In the last few years the authors have successfully developed an iterative procedure to cope with unbounded electrical field problems[2]. The main advantage of the procedure lies in its simplicity of implementation in a standard FE code for bounded problems and in its robustness.

In this paper a similar iterative procedure is proposed to deal with 2D motional eddy current problems in unbounded domains governed by a diffusion-like equation in terms of the magnetic vector potential. The boundlessness is treated by means of a fictitious boundary enclosing the system and defining a bounded domain where the problem can be solved by using a classical up-winding FEM. In each step of the iterative procedure, the magnetic vector potential on the fictitious boundary is updated according to the source currents and the value and the normal derivative of the magnetic vector potential on the boundary of the non-homogeneous and/or moving regions. This procedure is applicable both to transient and steady-state problems; however, in the following only the transient form is addressed in view of a solution to problems involving travelling magnetic fields[3,4].

STATEMENT OF THE PROBLEM

Consider a bidimensional system consisting of a conductor moving with constant velocity **v** through a magnetic field generated by assigned time-varying balanced source currents flowing in the orthogonal direction (z-axis) with respect to the system plane. The moving conductor is assumed to be infinitely extended in the direction of motion. An inductor core may be present in order to provide the required magnetic field configuration in the moving conductor. The whole system is embedded in a homogeneous unbounded dielectric medium.

Electric and Magnetic Fields, Edited by A. Nicolet
and R. Belmans, Plenum Press, New York, 1995

The electromagnetic phenomena are governed by a diffusion-like equation in terms of the magnetic vector potential A (directed along the z-axis) including both time-dependent and motional terms:

$$\nabla \cdot \left(\frac{1}{\mu} \nabla A \right) - \sigma \frac{\partial A}{\partial t} - \sigma \bar{v} \cdot \nabla A + J(t) = 0 \tag{1}$$

All the materials are assumed to be linear. Moreover in the coils ($\mu=\mu_0$, $\sigma = 0$) no skin effect is considered to occur and a uniform distribution of source currents J(t) is assumed to have been assigned; in the moving conductor ($\mu \geq \mu_0$, $\sigma \neq 0$) no source currents are present, but eddy currents are induced due to both motional effects and time-varying coil sources; in the inductor core ($\mu > \mu_0$, $\sigma = 0$), if any, no source currents are present.

In literature the FEM solution of equation (1) is obtained by using truncation to deal with the boundlessness of the problem: homogeneous Dirichlet and/or Neumann conditions are imposed on a boundary placed at a suitably great distance from the "centre" of the system, so that negligible effects are assumed to occur on the solution[3,4].

A specific treatment of boundlessness can be included in the FEM solution process by introducing a fictitious boundary B_F, which allows us to obtain an improvement in the accuracy of the results. This boundary encloses all the regions where (source or eddy) currents flow, the only exception being the moving conductor which is cut by B_F at such a distance that the eddy currents are practically zero. In this way the region outside B_F can be considered to be Laplacian. In the bounded domain D delimited by B_F, equation (1) is discretized in space by applying the Galerkin FEM:

$$\mathbf{T} \frac{d}{dt} \mathbf{A}_D(t) + \mathbf{S} \mathbf{A}_D(t) = \mathbf{K}(t) - \mathbf{T'} \frac{d}{dt} \mathbf{A}_F(t) - \mathbf{S'} \mathbf{A}_F(t) \tag{2}$$

where $\mathbf{A}_D$ and $\mathbf{A}_F$ are the column vectors of the instantaneous magnetic vector potential values at the internal and B_F nodes respectively; vector **K** and matrices **T** and **S** are obtained by assembling the following element arrays:

$$K_i^{(k)}(t) = J_k(t) \int_{E_k} \alpha_i \, dxdy \qquad T_{ij}^{(k)} = \sigma_k \int_{E_k} \alpha_i \alpha_j \, dxdy$$

$$S_{ij}^{(k)} = E_k \sigma_k \tilde{\alpha}_i \bar{v} \cdot \nabla \tilde{\alpha}_j + \frac{1}{\mu_k} \int_{E_k} \nabla \alpha_i \cdot \nabla \alpha_j \, dxdy \tag{3}$$

where E_k is the k-th finite element, α_i is the shape function of the i-th local node in E_k and the symbol ~ denotes the evaluation at some point in the element depending on the degree of upwinding[5]. Similar expressions hold for **T'** and **S'**.

Discretizing equation (2) in time by a fixed time-step Δt and denoting by $t_n = n\Delta t$ the generic discrete time, one can write:

$$\begin{aligned} &\frac{1}{\Delta t} \mathbf{T}[\mathbf{A}_D(t_n) - \mathbf{A}_D(t_{n-1})] + \frac{1}{2} \mathbf{S}[\mathbf{A}_D(t_n) + \mathbf{A}_D(t_{n-1})] = \\ &= \frac{1}{2}[\mathbf{K}(t_n) + \mathbf{K}(t_{n-1})] - \frac{1}{\Delta t} \mathbf{T'}[\mathbf{A}_F(t_n) - \mathbf{A}_F(t_{n-1})] - \frac{1}{2} \mathbf{S'}[\mathbf{A}_F(t_n) + \mathbf{A}_F(t_{n-1})] \end{aligned} \tag{4}$$

This equation relates $\mathbf{A}_D$ to $\mathbf{A}_F$ at time t_n provided that the same vectors are known at time t_{n-1}. Another equation needs to be derived. This can be accomplished by computing the

potential A on B_F due to the various regions of the system as explained in the following.
a) *Coil regions*: for simple coil shapes analytical formulas can be utilized; alternatively, all the finite element contributions can be added, each one being expressed as:

$$A(P_F) = -\frac{\mu_0}{2\pi} I_k \ln r_k \tag{5}$$

where I_k is the total current flowing in the k-th element given by $I_k = J(t_n)\,\text{meas}\{E_k\}$ and r_k is the distance between P_F and the barycentre of the element.
b) *Moving conductor region:* this region is taken into account through its surface; if S_h denotes the generic element side lying on this surface, its contribution to $A(P_F)$ is given by:

$$A(P_F) = -\frac{1}{2\pi}\sum_{j=1}^{N_E} A_j \int_{S_h}\left[\ln\left(\frac{1}{r}\right)\frac{\partial \alpha_j(P)}{\partial n} - \alpha_j(P)\frac{\partial}{\partial n}\ln\left(\frac{1}{r}\right)\right] dS \tag{6}$$

where r is the distance between P_F and the generic point P of the element to which S_h belongs and n is the outward normal unit vector to the side S_h.
c) *Inductor core regions*: the effect of these regions is taken into account as in b).

All the above contributions can be assembled together in the following matrix relation:

$$\mathbf{A}_F = \mathbf{R}\,\mathbf{A}_D + \mathbf{R}_0 \tag{7}$$

where $\mathbf{R}_0$ is due to the source currents and $\mathbf{R}$ is a rectangular matrix of geometrical coefficients where null columns appear for the nodes not 'near' to the above surfaces.

Equations (4) and (7) allow solution at time t_n. This solution can be obtained in an iterative way. Selecting an arbitrary initial guess $\mathbf{A}^0{}_F(t_n)$ for the magnetic vector potential on B_F, equation (4) gives the corresponding $\mathbf{A}^0{}_D(t_n)$ for the internal nodes. A new estimate $\mathbf{A}^1{}_F(t_n)$ is then computed by means of (7), by using $\mathbf{A}^0{}_D(t_n)$. The procedure is iterated until a convergence test is satisfied. Convergence to the true solution takes place in a few steps if the initial guess is reasonably chosen, for example by selecting the distribution of A on B_F at the previous time t_{n-1}. The procedure starts at time $t_0=0$ with a static unbounded analysis, whose equation is obtained by dropping the time derivatives in (2) and (4).

AN EXAMPLE

The iterative procedure to deal with boundlessness is performed at each time step; however for the sake of simplicity the results presented hereafter refer to the static unbounded analysis occurring at the beginning of the time analysis. The system considered consists of a conducting plate moving with a constant velocity v=10 m/s in the magnetic field created by two U-shaped inductor cores with two sets of coils each. The system is embedded in an unbounded dielectric medium (vacuum). Fig. 1 shows half of the system geometry ($a=7.5\cdot10^{-2}$ m, $b=6.5\cdot10^{-2}$ m) and the position chosen for B_F. A homogeneous Neumann condition hold on the bottom. All the materials are assumed to be linear: for the inductor core $\mu_r = 500$, whereas for the conducting plate $\sigma=10^7$ S/m. The source current flowing in the coils is I=1 kA. The problem domain was discretized by using 625 first-order finite elements: quadrilaterals were used for the moving-conductor whereas triangles were used in all the other regions. Starting from a homogeneous Dirichlet condition on B_F, four iteration steps were needed to obtain convergence (an end-iteration tolerance of 0.1% was selected for two consecutive solutions). The resulting solution is shown in Fig. 2 where equal magnetic vector potential contour lines are plotted.

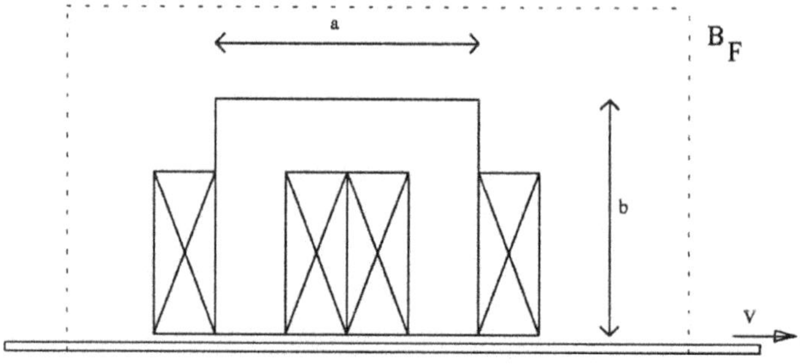

Fig. 1 - An half of the system of the example.

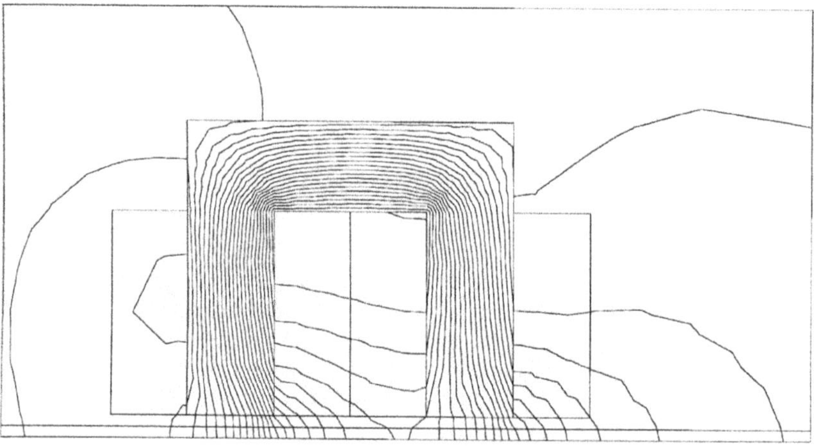

Fig. 2 - Magnetic vector potential contour lines.

REFERENCES

1. C.R.I. Emson: "Methods for the Solution of Open Boundary Electromagnetic Field Problems". *Proc. IEE*, vol. 135, Pt. A, No. 3, March 1988.
2. G. Aiello, S. Alfonzetti, S. Coco: "Charge Iteration for N-Dimensional Unbounded Electrical Field Computations". *IEEE Trans. on Magn*, Vol. 28, No. 2, March 1992.
3. M. Ito, T. Takahashi, M. Odamura: "Up-Wind Finite Element Solution of Travelling Magnetic Field Problems". *IEEE Trans. on Magn.*, Vol. 28, No. 2, March 1992.
4. M. Odamura, M. Ito: "Up-Wind Finite Element Solution of Saturated Travelling Magnetic Field Problems". *4th Int. Symp. on FEM in Flow Problems, July 1982.*
5. T.J.R. Hughes: "A Simple Scheme for Developping 'Upwind' Finite Elements". *Int. J. Num. Meth. Engng*, Vol. 12, p. 1359-1365, 1978.

ON "HYBRID" ELECTRIC-MAGNETIC METHODS

A. Bossavit

Électricité de France, 1 Av. du Gal de Gaulle, 92141 Clamart, France

INTRODUCTION

"Hybrid", "mixed", etc., are dangerous words: you never know what people mean. Here, we deal with hybrid methods in the following sense. Due to a well-known duality inherent in Maxwell equations, and to the impossibility to enforce all continuity conditions with finite-elements, basic methods can be classified as "e-oriented" or "h-oriented" (or as "electric" versus "magnetic") according to whether they enforce the tangential continuity of e or of h (hence the normal continuity of b or of j, respectively) at element interfaces. Now, in "hybrid" approaches, one makes a *patchwork* of methods, using h-oriented ones in some regions of space and e-oriented ones elsewhere. Hybrids we specifically study here are h-oriented in the [illegible] in the conductor.

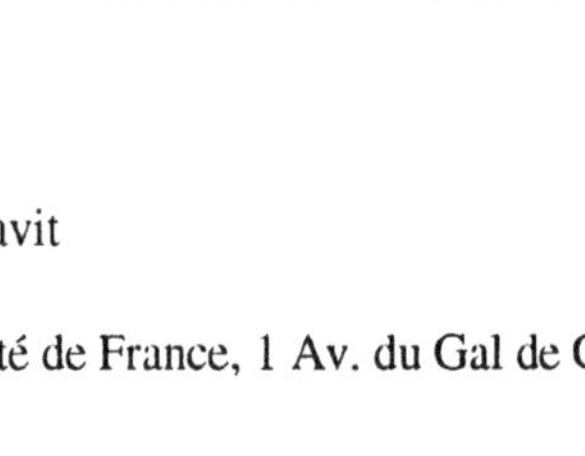

Figure 1. Situation and notation. The computational domain, D, contains a coil with given current density j^g (time-dependent), a conductor C with non-zero conductivity σ, and possibly some "magnetic" regions, where $\sigma = 0$ and $\mu \neq \mu_0$. The boundary B is supposed to be an "electric" one, on which $n \times e = 0$ (and hence $n \cdot b = 0$). The air-conductor interface is S. We assume topological triviality (no current loops, no inside holes). Note the orientation of the unit normal field (outward with respect to D – C).

We use $L^2(R)$ to denote the space of square-integrable functions over some region R, (and $\mathbb{L}^2(R)$ for fields). Then we set $L^2_{grad}(R) = \{\varphi \in L^2(R) : \text{grad } \varphi \in \mathbb{L}^2(R)\}$, and $\mathbb{L}^2_{rot}(R) = \{u \in \mathbb{L}^2(R) : \text{rot } u \in \mathbb{L}^2(R)\}$. All derivatives are denoted with a ∂. Constructs like φ_S, h_S, $\text{grad}_S\varphi$, etc., should respectively be understood as the restriction of φ to S, the tangential part of h on S, the surface gradient of φ on S, etc.

Electric and Magnetic Fields, Edited by A. Nicolet and R. Belmans, Plenum Press, New York, 1995

FORMULATION

We denote $\mathbb{L}^2_{rot}(C)$ by U, and $L^2_{grad}(D-C)$ by Φ. Let us introduce the field $a^g(t, x) = (4\pi)^{-1} \int |x-y|^{-1} j^g(t, y)\, dy$, where dy is the volume element, and define the "source-field" h^g as $h^g = \text{rot}\, a^g$. There exists, in the case of Fig. 1, a region containing C in which $h^g = \text{grad}\, \varphi^g$, for some suitable potential φ^g (not defined, of course, over all D). This way, we may look for the magnetic field h in the air (the region $D-C$) in the form $h = h^g + \text{grad}\, \varphi$, where the magnetic potential φ lies in Φ.

Elements of U will be "modified vector potentials" of sorts: more precisely, calling e(t) the electric field at time t, and assuming that all fields, including the given current density j^g, were null up to time $t = 0$, we set $u(t) = -\int_0^t e(s)\, ds$. This way, $e = -\partial_t u$, and $h = \mu^{-1}\, \text{rot}\, u$.

Note that knowing u in the conductor and φ in the air is enough to fully know the electromagnetic situation[1], *provided the required continuity conditions* at the interface S *are enforced.* These so-called "transmission conditions" are

$$n \cdot \text{rot}\, u = n \cdot [\mu(h^g + \text{grad}\, \varphi)], \quad n \times (h^g + \text{grad}\, \varphi) = n \times [\mu^{-1}\, \text{rot}\, u]. \tag{1}$$

Unfortunately, this doesn't make much mathematical sense if taken *a priori*[2], so we cannot simply say "search for a pair $\{u, \varphi\}$ that satisfy (1) and ...", where the dots would stand for some weak formulation that should also be enforced. This difficulty is typical of *hybrid* methods.[3]

There is a way around it. Let's provisionally pretend that both h_S and $n \cdot b$ (that is, $\mu\, n \cdot h$) are known on S. Then, starting from Ohm's law in C, $\sigma \partial_t u + \text{rot}\, h = 0$, taking the scalar product with a test-field u', and integrating by parts, we get

$$\int_C \sigma \partial_t u \cdot u' + \int_C \mu^{-1}\, \text{rot}\, u \cdot \text{rot}\, u' = \int_S n \times h \cdot u' \tag{2}$$

as a necessary condition to be satisfied by u, for all u' in U. On the other hand, and because div b has to be zero in $D-C$, and $n \cdot b \equiv \mu\, n \cdot h$ has to vanish on B, φ should satisfy

$$\int_{D-C} \mu\, (h^g + \text{grad}\, \varphi) \cdot \text{grad}\, \varphi' = \int_S \mu\, n \cdot h\, \varphi' \tag{3}$$

for all φ' in Φ. Since eventually $n \cdot b$ has to equal $n \cdot \text{rot}\, u$, and $n \times h$ has to equal $n \times (h^g + \text{grad}\, \varphi)$, we see that the pair $\{u, \varphi\}$ must satisfy the coupled equations:

$$\int_C \sigma \partial_t u \cdot u' + \int_C \mu^{-1}\, \text{rot}\, u \cdot \text{rot}\, u' - \int_S n \times \text{grad}\, \varphi \cdot u' = \int_S n \times h^g \cdot u' \quad \forall\, u' \in U,$$

$$\int_{D-C} \mu\, (h^g + \text{grad}\, \varphi) \cdot \text{grad}\, \varphi' - \int_S n \cdot \text{rot}\, u\ \varphi' = 0 \quad \forall\, \varphi' \in \Phi,$$

where now all integrals make perfect sense (and have, as we see in a moment, unambiguously defined discrete counterparts). Thanks to the identity $\text{div}_S(n \times u) = -(n \cdot \text{rot}\, u)_S$, we

---

[1] The outside e is rarely needed, and if so, may be recovered by solving a standard electrostatic problem in the air, once h is known.

[2] These relations link "Dirichlet-like" data (on the left-hand side) to "Neumann-like" ones (on the right), and whereas the former (called "traces") depend on u or φ in a continuous way, the latter don't: fields that merely belong to $\mathbb{L}^2$, like rot u or grad φ, have no trace, be it tangential or normal. No mathematical pedantry here, but a very fundamental difficulty: a priori constraints on the solution that, like (1) above, involve other than *continuous* mappings, simply *cannot* be enforced (as simple algebraic constraints) at the discretized level of finite elements.

[3] Contrast the situation with that of a magnetic-magnetic method (h inside, φ outside): same space Φ, but h roams in $H \equiv \mathbb{L}^2_{rot}(C)$. The transmission condition would then be $h_S = \text{grad}_S(\varphi + \varphi^g)$, which defines a *closed* affine subspace in the Cartesian product $H \times \Phi$, because both traces $h \to h_S$ and $\varphi \to \varphi_S$ are continuous in a suitable sense.

may give a more symmetrical appearance to the previous equations, and finally specify the problem as *find* $\{u, \varphi\}$ *in* $U \times \Phi$ *such that*

$$\int_C \sigma \partial_t u \cdot u' + \int_C \mu^{-1} \operatorname{rot} u \cdot \operatorname{rot} u' + \int_S \varphi \; n \cdot \operatorname{rot} u' = \int_S n \times h^g \cdot u', \tag{4}$$

$$\int_S n \cdot \operatorname{rot} u \; \varphi' - \int_{D-C} \mu \operatorname{grad} \varphi \cdot \operatorname{grad} \varphi' = \int_{D-C} \mu \, h^g \cdot \operatorname{grad} \varphi' \tag{5}$$

for all test-pairs $\{u', \varphi'\}$ *in* $U \times \Phi$.

DISCRETIZATION

Of course, this was a one-way derivation, and we still have to verify that problem (4)(5) is well-posed, that is, has a unique solution that continuously depends on the data h^g. (If so, it will yield the right electromagnetic field, since conditions (1) result, *a posteriori*, from (4) and (5), as one may check by doing the integrations by parts.) But first, let us make sure that it discretizes in a straightforward way. Consider a mesh of D. Call **G**, **R**, **D**, the incidence matrices of this mesh (e.g., $\mathbf{G}_{en} = 1$ if edge e abuts on node n, etc.; cf. Ref. 1). They are the discrete counterparts to grad, rot, div. Let λ_n be the barycentric function for node n, and $w_e = \lambda_n \operatorname{grad} \lambda_m - \lambda_m \operatorname{grad} \lambda_n$, the edge-element for edge e = {n, m}. For a region R and a scalar function α, define $\mathbf{M}_R(\alpha)$ as the matrix of entries $\int_R \alpha \, w_e \cdot w_{e'}$ (where e and e' are edges). Finally, for a node m and an edge e *both in* S, set $\mathbf{N}_{m,e} = \int_S n \cdot \operatorname{rot} w_e \; \lambda_m$, and let **N** be the matrix obtained by filling-in with zeros for other node- and edge-indices. Now, by the standard Galerkin treatment, (4)(5) becomes a system of ODE's: *find* time-dependent *vectors* of degrees of freedom **u** *and* $\boldsymbol{\varphi}$ *such that*

$$\partial_t [\mathbf{M}_C(\sigma)\, \mathbf{u}] + \mathbf{R}^t \mathbf{M}_C(\mu^{-1})\, \mathbf{R}\, \mathbf{u} \qquad + \mathbf{N}^t \boldsymbol{\varphi} = - \mathbf{N}^t \boldsymbol{\varphi}^g, \tag{6}$$

$$\mathbf{N}\mathbf{u} - \mathbf{G}^t \mathbf{M}_{D-C}(\mu)\, \mathbf{G}\, \boldsymbol{\varphi} = \mathbf{G}^t \mathbf{M}_{D-C}(\mu)\, \mathbf{h}^g \tag{7}$$

(where t stands for "transpose"). Data $\boldsymbol{\varphi}^g$ (node-values of φ^g) and $\mathbf{h}^g$ (edge-circulations of h^g) are easily computed from j^g (using now standard "spanning tree" techniques if needed). Note the remarkable structure of the overall matrix of this system (globally symmetric, but *not* positive definite). It is characteristic of what numerical analysts call "mixed", or "two-field" approaches. Remark also that if $\mu = \mu_0$ outside C, one may replace the term $\int_{D-C} \mu \, h^g \cdot \operatorname{grad} \varphi' \equiv \int_S \mu \, n \cdot h^g \; \varphi'$ by $\mu_0 \int_S n \cdot \operatorname{rot} a^g \; \varphi'$, and thus replace the right-hand side in (7) by $\mu_0 \mathbf{N} \mathbf{a}^g$.

WELL-POSEDNESS

Is (4)(5) well-posed? To study this, let us introduce the "Neumann-to-Dirichlet map" Q, as follows. Let v (a scalar function) be given on S. There is a unique $\varphi \in \Phi$ such that $\int_{D-C} \mu \operatorname{grad} \varphi \cdot \operatorname{grad} \varphi' = \int_S v \, \varphi' = 0 \;\; \forall \varphi' \in \Phi$. Then $\varphi_S = Qv$ defines Q. Let us also define ϕ^g as the unique solution to

$$\int_{D-C} \mu \operatorname{grad} \phi \cdot \operatorname{grad} \varphi' = \int_{D-C} \mu \, h^g \cdot \operatorname{grad} \varphi' \quad \forall \varphi' \in \Phi,$$

that is, eq. (5) with $u = 0$ and a change of sign. (Note that ϕ^g_S is *not* equal to φ^g_S.)

Now, eq. (5) means that $\varphi_S = Q(n \cdot \operatorname{rot} u) - \phi^g_S$. Substituting in (4), we get

$$\int_C \sigma \partial_t u \cdot u' + \int_C \mu^{-1} \operatorname{rot} u \cdot \operatorname{rot} u' + \int_S Q(n \cdot \operatorname{rot} u)\, n \cdot \operatorname{rot} u'$$

$$= \int_S (\phi^g - \varphi^g)\, n \cdot \operatorname{rot} u' \quad \forall u' \in U, \tag{8}$$

a perfectly well-posed problem in space U, since Q is positive definite. (More precisely, Q is the "Riesz isomorphism" associated with the space of traces of Φ on S, usually denoted as $H^{1/2}(S)$: Q maps the dual $H^{-1/2}(S)$ of $H^{1/2}(S)$ onto the latter. It is well known, on the other hand, that the mapping $u \rightarrow n \cdot \text{rot } u$ sends U into $H^{-1/2}(S)$, continuously.)

Eq. (8), by the way, forms the basis of another "hybrid" method, that uses edge-elements for u, or equivalently for e, inside the conductor, and a boundary element method to precompute Q. See Ref. 2. Such hybrid methods, which shun finite elements in the air, are better adapted to the case of open (infinite) air regions.

VARIANTS

From this point on, one can make many variations. One, that seems to be popular these days (cf., e.g., Ref. 3), consists in expressing u as the sum $a + \text{grad } \psi$, where a is the standard vector potential and $\psi(t) = \int_0^t v(s)\, ds$, a suitable primitive of the electric potential v. Thus one deals with *triples* $\{a, \psi, \varphi\}$. Fixing the gauge by requiring (for instance) $\text{div } a = 0$ in C and $n \cdot a = 0$ on S makes the solution unique. But one may as well penalize the constraint $\text{div } a = 0$, which is simply done by adding in (4) the term $\int_C \mu^{-1} \text{div } a \text{ div } a'$. This means that a may be looked for in a subspace of U made of more regular fields, namely $A = \{a \in U : \text{div } a \in L^2(C)\}$, equipped with the scalar product $(a, a') = \int_C a \cdot a' + \int_C \text{rot } a \cdot \text{rot } a' + \int_C \text{div } a \cdot \text{div } a'$. Denoting $L^2_{grad}(C)$ by Ψ, we arrive at the following formulation: *find* $\{a, \psi, \varphi\}$ *in* $A \times \Psi \times \Phi$ *such that*

$$\int_C \sigma\, \partial_t(a + \text{grad } \psi) \cdot (a' + \text{grad } \psi') + \int_C \mu^{-1} \text{rot } a \cdot \text{rot } a' + \int_C \mu^{-1} \text{div } a \text{ div } a'$$

$$+ \int_S \varphi\, n \cdot \text{rot } a' = \int_S n \times h^g \cdot a', \tag{9}$$

$$\int_S n \cdot \text{rot } a\ \varphi' - \int_{D-C} \mu \text{ grad } \varphi \cdot \text{grad } \varphi' = \int_{D-C} \mu\, h^g \cdot \text{grad } \varphi' \tag{5}$$

for all test-triples $\{u', \varphi'\}$ *in* $A \times \Psi \times \Phi$, hence

$$\sigma(\partial_t a + \text{grad } v) + \text{rot}(\mu^{-1} \text{rot } a) - \text{grad}(\mu^{-1} \text{div } a) = - j^s \tag{10}$$

inside the conductor.

We *know* there are solutions: starting from the u that solves (4)(5), take ψ such that $\int_C (u - \text{grad } \psi) \cdot \text{grad } \psi' = \int_C w\, \psi'\ \ \forall\, \psi' \in \Psi$, then set $a = u - \text{grad } \psi$. Here, w is an arbitrary element of $H^{-1/2}(S)$, which shows to what extent the solution of (9)(5) is non-unique. Uniqueness can be enforced by narrowing either A or Ψ (or both), which is done by specifying either $n \cdot a$ on S or ψ on S (or both, on complementary parts of S). Quite independently of these "gauge fixing" procedures, however, $\mu^{-1} \text{div } a = 0$ if (9) holds, which shows that the right problem is solved, in spite of the strange look of eq. (10). (To check this assertion, start from $f' \in L^2(C)$, take ψ' such that $\psi' = 0$ on S and $-\Delta \psi' = f'$ in C, and set $a' = 0$. Then (10) reduces to $\int_C \mu^{-1} \text{div } a\ f' = 0\ \ \forall\, f' \in L^2(C)$, hence $\mu^{-1} \text{div } a = 0$.)

Standard nodal elements (vector-valued) can then be used for a, which seems to be perceived as an advantage by some.

REFERENCES

1. A. Bossavit. "Électromagnétisme, en vue de la modélisation," Springer-Verlag, Paris (1993).
2. Z. Ren, F. Bouillaut, A. Razek, A. Bossavit, J.C Vérité. A New Hybrid Model Using Electric Field Formulation for 3-D Eddy Current Problems, *IEEE Trans., MAG-26*, 2, pp. 470-73 (1990).
3. O. Biro, K. Preis. On the use of the magnetic vector potential in the Finite Element Analysis of three-dimensional eddy currents, *IEEE Trans., MAG-25*, 4, pp. 3145-59 (1989).

NON-ABELIAN SYMMETRY IN THE 3-D EDDY-CURRENT ANALYSIS

Jacques Lobry and Christian Broche

Faculté Polytechnique de Mons
Bd. Dolez, 31 7000 Mons, Belgium

INTRODUCTION

Geometrical symmetry is found in many electrical engineering problems. It is often the case in computational electromagnetics with finite element type methods. Symmetry can be taken into account in a rational way by using the Group Representation Theory that is scarcely used in this context. Briefly, it consists in reducing an original problem to a set of sub-problems, or "modes", defined on a symmetry cell of the domain under study; the global solution is obtained from superposition of the partial ones. Recent experiences[1,2] exist and show that there is much to be gained by way of economy in computation. Symmetry groups are of two types according as the composition of the isometries happens to be commutative or not so that they can be either abelian or non-abelian. This property is of great importance because the modes described above involve tedious coupled problems in the latter case. Further to a previous paper[3] that only considered the abelian case, we propose the implementation of the 3 D eddy current analysis in the general non-abelian case[4]. A mixed FEM-BEM method is used, the degrees of freedom are the H-circulations along the inner edges of the conducting regions (edge elements) and the nodal values of a scalar magnetic potential on their boundaries (boundary elements). An example points out that it is worth managing the non-abelian case in spite of a more difficult implementation since some substantial improvement is brought in the performances.

SYMMETRY IN FINITE ELEMENTS

Geometrical symmetry often occurs in computational electromagnetics. Taking advantage of it is easy when the excitation fields share some part of the symmetry like, for instance, an even or odd parity with respect to a reflection plane. But intuition fails with complex geometries and general excitations. Nevertheless, it is possible to take full advantage of symmetry in such cases thanks to the Group Representation Theory that we briefly describe below (see e.g. Hamermesh[5] for more details).

Let Ω be a spatial domain that presents some symmetry we describe by the isometries g (rotations and reflections) that leave Ω globally invariant. Those transformations form

Electric and Magnetic Fields, Edited by A. Nicolet
and R. Belmans, Plenum Press, New York, 1995

a finite group G = {e, f, g,...} called *symmetry group*. The number of elements in the group is the *order of the group* n_G. If the composition of isometries is commutative, the group G is said *abelian*. In this paper, the general non-abelian case is considered. Figure 1 discussed in the next section illustrates this definition with the non-abelian group D_{3h}. It clearly shows that the subdomain C (called *symmetry cell*) regenerates Ω from the symmetry operations.

Linear operators O_g acting on complex functions of a vector space V are defined, they are such that any function ϕ is shifted by O_g as a point x is shifted by g. These operators form a group G' isomorphic to G and the mapping O between the groups G and G' is called a *linear representation of the group G in the representation space V.* If we consider the group of matrices D(g) related to the operators O_g and a basis of V, we get a *matrix representation of G in V.* A full theory about finite groups shows that any linear representation of a finite group can be decomposed in some sub-representations called *irreducible representations*. Their number k and degree d_ν are well defined for a given group and finding all of them is an important point in group theory. We now present the *fundamental decomposition theorem*.

Let V be a representation vector space of a finite group G, of order n_G, some operators $P_{ij}^{(\nu)}$ $(i,j=1,\ldots,d_\nu)$ called *projectors* we do not explicit here are defined for each irreducible representation ν (from 1 to k). Then we have the following properties :
1°) the projectors $P_{ij}^{(\nu)}$ decompose any function ϕ of V into functions $\phi_{jj}^{(\nu)}$ so that :

$$\phi = \sum_{\nu=1}^{k} \sum_{j=1}^{d_\nu} \phi_{jj}^{(\nu)} \qquad where \qquad \phi_{jj}^{(\nu)} = P_{jj}^{(\nu)} \phi \tag{1}$$

2°) the symmetry properties of these components are related to $(d_\nu\text{-}1)$ "partner functions" $\phi_{ij}^{(\nu)}$ that are built from the $(d_\nu\text{-}1)$ projectors $P_{ij}^{(\nu)}$ $(i \neq j)$:

$$\phi_{ij}^{(\nu)} = P_{ij}^{(\nu)} \phi$$

the functions $\phi_{ij}^{(\nu)}$ form a *ν-symmetric* set and verify the *ν-symmetric* conditions:

$$O_g \, \phi_{ij}^{(\nu)} = \sum_{l=1}^{d_\nu} D_{li}^{(\nu)}(g) \; \phi_{lj}^{(\nu)} \qquad (i = 1, \ldots, d_\nu) \tag{2}$$

This last relationship shows that the only restriction of the $\phi_{ij}^{(\nu)}$ functions on the symmetry cell C is sufficient to regenerate all of them on the whole domain Ω. In particular, this is the symmetry property of the components $\phi_{jj}^{(\nu)}$.

Hence, starting from an original problem described by a set of equations defined on a domain with symmetry, we have to determine a symmetry cell and to solve on it for each of the source fields and boundary conditions derived from the irreducible representations. The overall solution is finally obtained from superposition and symmetry operations. Since the degrees d_ν may be greater than one, coupled problems are expected.

SYMMETRY IN THE 3-D EDDY-CURRENT PROBLEM

Consider a 3-D bounded symmetrical conducting domain Ω (Figure 1) in the euclidian space E (air). A coil Ω_s with a known sinusoidal current density $\mathbf{J}_s$ is the source field. The materials are assumed linear and homogeneous. Without any symmetry consideration, the general 3-D eddy-current problem may be modelled with the following so-called H-formulation[6] in which the unknowns are the H-field in Ω and a harmonic reduced scalar potential Φ in the outer space Ω_c :

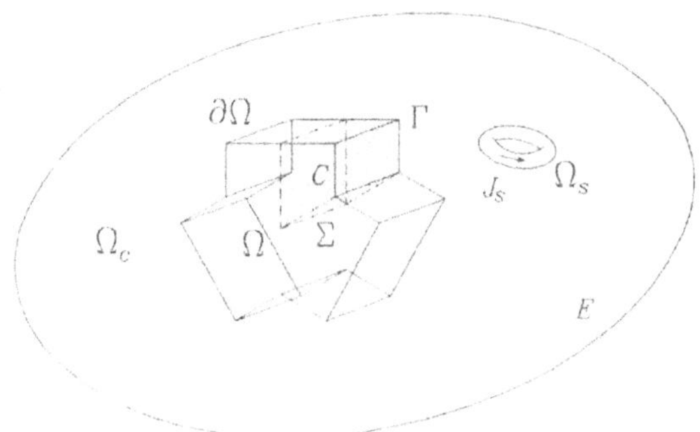

Figure 1. General configuration of the 3-D eddy-current problem.

find H *and* $\phi = \Phi$ *on* $\partial\Omega$ *such that :*

$$H = H_s + grad\Phi \quad in\ \Omega_c$$

$$\int_{\Omega} (\rho\ curl\boldsymbol{H}\ curl\boldsymbol{H}' + j\omega\mu\boldsymbol{H}\boldsymbol{H}')\,d\Omega + j\omega\mu_0 \oint_{\partial\Omega} (H_{sn} + \Re\phi)\,\phi'\,d\Gamma = 0 \quad (3)$$

$$\forall\ \boldsymbol{H}', \Phi' \ :\ \boldsymbol{H}' = grad\Phi' \ in\ \Omega_c,\ \phi' = \Phi' \ on\ \partial\Omega$$

where $\mathbf{H}_s$ is the excitation field that would exist if the coil were alone in space, $\Re$ is an operator relating ϕ to $\partial\Phi/\partial n$. Our numerical scheme for problem (3) is a mixed FEM-BEM method[6]. The degrees of freedom are the circulations h_e of $\mathbf{H}$ along the inner edges (edge elements) of a tessellation of Ω and the nodal values ϕ_n of ϕ on the boundary $\partial\Omega$ (boundary elements).

Let G be the symmetry group of Ω and C and Γ the symmetry cells of Ω and $\partial\Omega$ respectively. The decomposition (1) can be applied to the H-formulation (3). This leads to the following equivalent set of sub-problems[4] :

$\forall\ \nu = 1,\ldots,k,\ j = 1,\ldots,d_\nu$, *find the* ν*-symmetric sets* $\{(\boldsymbol{H}_{lj}^{(\nu)}, \phi_{lj}^{(\nu)})\ ,\ l=1,\ldots,d_\nu\}$ *such that :*

$$\sum_{l=1}^{d_\nu}\left(\int_C (\rho\ curl\boldsymbol{H}_{lj}^{(\nu)}\ curl\boldsymbol{H}'^{(\nu)*}_{lj} + j\omega\mu\ \boldsymbol{H}_{lj}^{(\nu)}\boldsymbol{H}'^{(\nu)*}_{lj})\,d\Omega + j\omega\mu_0 \oint_{\Gamma} (H_{s_n lj}^{(\nu)} + \Re_j^{(\nu)}\phi_{lj}^{(\nu)})\,\phi'^{(\nu)*}_{lj}\,d\Gamma\right) = 0$$

with the following reciprocal constraints on the interior boundary Σ *of* C *:*

$$O_g\,\boldsymbol{H}_{lj}^{(\nu)} = \sum_{i=1}^{d_\nu} D_{il}^{(\nu)}(g)\,\boldsymbol{H}_{ij}^{(\nu)} \quad and \quad O_g\,\phi_{lj}^{(\nu)} = \sum_{i=1}^{d_\nu} D_{il}^{(\nu)}(g)\,\phi_{ij}^{(\nu)} \quad \forall\ g \in G$$

where the test-functions $\mathbf{H}'^{(\nu)}_{lj}$ and $\phi'^{(\nu)}_{lj}$ and the source term $\mathbf{H}_{s,lj}^{(\nu)}$ are derived from group actions on $\mathbf{H}'$ and ϕ' and $\mathbf{H}_s$ respectively. The global $\mathbf{H}$ and ϕ are finally calculated by superposing the partial ones.

EXAMPLE

Consider the problem of a conducting cube made of copper in the vicinity of which a infinite filamentary sinusoidal current is flowing. The excitation shares no symmetry with the cube (Figure 2a). The cube configuration exhibits many isometries so that several symmetry groups may be generated. In order to compare the numerical performances between abelian and non-abelian cases, the dihedral groups D_{2h} (abelian) and D_{4h} (non-abelian) are considered. They are illustrated in figures 2b and 2c respectively.

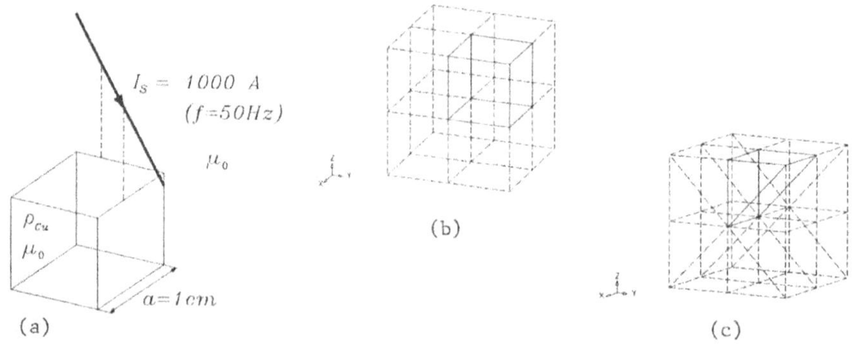

Figure 2. A 3-D eddy-current problem with symmetry. Full geometry (a), groups D_{2h} (b) and D_{4h} (c).

Times measurements for the solution of the problem with no symmetry consideration and with exploitation of the two above symmetry groups give the following conclusions. For a moderate size of the problem (4000 tetrahedra), the abelian treatment gives a speed-up of about 50 whereas the non-abelian case allows a value of 120. Although the gain with the abelian case is high, it can be deduced that the non-abelian algorithm is still about 2.4 times faster than the abelian one. Moreover, the size of the problem is decreased due to a further reduced symmetry cell. These conclusions are general[4].

ACKNOWLEDGMENT

The authors thank the Power Systems Department of F.P.Ms for its financial support.

REFERENCES

1. M. Bonnet, On the use of geometrical symmetry in the boundary element methods for 3D Elasticity', *in:* "Boundary Element Technology VI", Comp. Mech. Publ.. 185:200 (1991).
2. P. Chelius, B. Davat and M. Lajoie-Mazenc, Group theory and electromagnetic field calculation in permanent-magnet synchronous motor, Proc. of the IMACS-TC1'90 Conf., 215:220 (1990).
3. J. Lobry and C. Broche, Edge elements for a 3D induction machine with solid iron rotor - Exploitation of symmetry (the abelian case), Electric and Magnetic Fields Proc., Liège, 36.1:36.6 (1992).
4. J. Lobry. "Symétries et éléments de Whitney en magnétodynamique tridimensionnelle - Application à l'étude d'un moteur asynchrone à rotor massif rainuré," Ph.D. thesis, FPMs (1993).
5. M. Hamermesh. "Group Theory," Addison-Wesley, London (1964).
6. A. Bossavit and J.C. Vérité, A mixed FEM-BIEM Method to solve 3-D eddy-current problems, *IEEE Trans. on Mag.*, Vol. 18, No.2, 431:435 (1982).

REGULARIZATION NEURAL NETWORKS FOR INVERSE PROBLEMS IN NON-DESTRUCTIVE TESTING

Francesco Carlo Morabito and Maurizio Campolo

Dipartimento Ingegneria Elettronica e Matematica Applicata
Università degli Studi di Reggio Calabria
Via E.Cuzzocrea, 48
I-89127 Reggio Calabria (Italy)

INTRODUCTION

Electric and magnetic Non-Destructive Testing (NDT) techniques allow to detect defects in materials by inspecting with suitable sensors the vicinity of the defect. The theoretical analysis of the relation between the defect properties and the perturbation of the fields is generally equivalent to the treatment of an inverse problem. This paper deals with an electrostatic test problem that reminds standard electromagnetic testing. Due to the general lack of closed-form analytic solutions, numerical methods are required. In recent years Neural Networks (NNs) have emerged as powerful numerical tools in a number of multi-sensors problems. Also NDT problems have already been proposed and solved with NNs[1,2,3]. The fast processing of data carried out by NNs in the after-training mode (reference mode) make them match the need of modern NDT applications. Unfortunately, because of a number of reasons, spanning from difficult design and training problems to insufficient number of available examples, the generalization properties of the network can be unsatisfactory. In these cases performances of the network in training and reference modes are not comparable. To cope with such problems one can borrow results from the approximation and regularization theory. This leads to a network implementation known in the literature as Radial Basis Function neural network (RBF NN)[4].

FORMULATION OF THE NDT PROBLEM

The NDT problem here analyzed can be formulated as follows: a point charge q is placed over an earthed metallic plate that can have a defect of hemispherical shape of radius a at a distance $b>a$ from it. Let P be a point exterior to the metallic plate. The electric potential v(P) can be determined by using the method of images and the inversion transformation as follows:

$$v(P) = (q/r_1 - q'/r_2 + q'/r_3 - q/r_4) / 4\pi\varepsilon_0 \qquad (1)$$

where $q'=qa/b^*$, $b'=a^2/b^*$, and b^* is the distance from the inducing point charge to the centre of the boss, i.e. the centre of the inversion. The meaning of the other symbols is clarified in fig.1. A proper selected pattern of potential and/or field measurements allows in principle to characterize the defect in terms of both size and location. Anyway, the problem is an inverse one and it can then be ill-posed and/or ill-conditioned. The problem under study can be viewed as a pattern recognition task. Indeed, by knowing a pattern of measurements which represent a proper sampling of a continuous function, we do not aim to reconstruct the entire function, rather to infer some geometrical quantities related to the object that generate the distribution. By considering the map of the electric potential generated by the point source in presence of the metallic plate, the existence of a defect is pointed out by the perturbation of this pattern (see fig.2). The extent of the perturbation is, in turn, related to size and location of the defect. To set up the database required by the learning process, exact synthetic data are calculated using (1). The radius of the boss, a, can vary between 1.5 and 3.5 mm. The region over which the boss could be present is a rectangular grid of size

Electric and Magnetic Fields, Edited by A. Nicolet
and R. Belmans, Plenum Press, New York, 1995

10 cm along *x* and *5 cm* along *y*. All of the sensors are placed along this rectangular contour at a fixed height. A suitable choice of the locations' measurements within the inspected window is a prerequisite to good detection. After careful investigations, we decide for a configuration of sensors of the type reported in fig. 1.

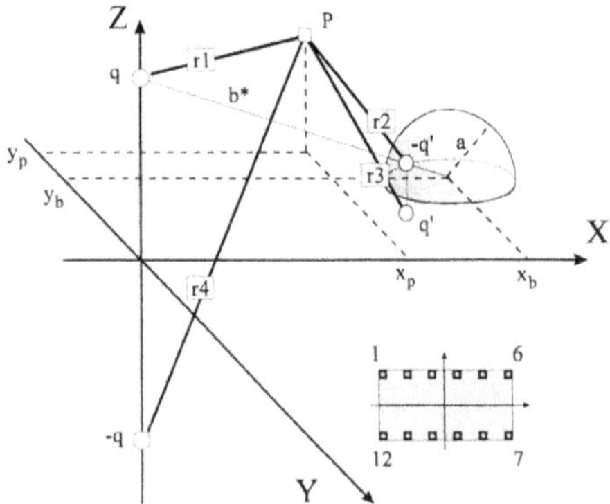

Figure.1 - A pictorial representation of the NDT problem

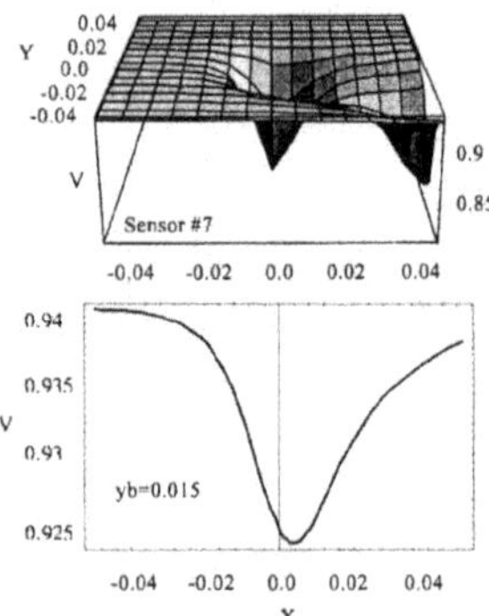

Figure 2 - The electric potential measured by sensor #7 (a=0.5mm).

REGULARIZATION THEORY AND NEURAL NETWORKS

In a recent paper it has been shown that a network representation of the problem of approximating a function from a set of N examples is easily obtained from regularization theory [4]. In particular, the regularized solution of the aforesaid problem lies in an N-dimensional subspace of the space of smooth functions. A basis for this subspace is given by the N Green's functions centred on the data points $\underline{x}_i$, i=1, N. When the Green's function is a radial function, which happens when the complexity term of the cost functional hold some properties, the regularized solution is given by an expansion in terms of radial functions.

The major difference between a basis function and the standard activation function used in the framework of NN theory is that the former produces a localized response to an input pattern whereas the activation function refers to a special feature possibly present in the same pattern. In spite of their similar topology, RBF NNs coming from regularization theory and multilayer feedforward neural networks (MFF NNs) based on activation functions like the sigmoids solve an approximation problem by means of different strategies. While MFF NNs are able to extract rules from the training dataset then combining the detected features in a proper way with suitable weights, RBF NNs implicitly subdivide the multidimensional input space in regions of proper size. In this case, a pattern is then recognized and associated with the desired output on the basis of a measure of distance (or similarity). This means that once defined a metric in the input space a basis function of the hidden layer produces a significant nonzero response only for inputs within a small localized region of the input space. On the contrary, the distance between two patterns is not determinant for the activation of a hidden unit. Fig.3 shows the topology of the regularization network. The input vector is applied to each neuron of the hidden layer that computes its output by means of a Gaussian kernel function of the following form:

$$h_j = \exp\left[-\frac{\left(\underline{x}-\underline{w}_j\right)^T\left(\underline{x}-\underline{w}_j\right)}{2\sigma_j^2}\right] \qquad j = 1,\ N_h \tag{2}$$

where $\underline{x}$ is the input pattern, $\underline{w}_j$ is the weight vector for the jth node in the hidden layer which coincide with the center of the Gaussian function for node *j*, and h_j is the output of the same node. Actually, we compute the distance between the input vector and the centers of each receptive field in a suitable metric. The closer the vector $\underline{x}$ is to the center of the Gaussian, the larger is the output of hidden node, h_j. Each node produces an identical output for inputs that lie a fixed radial distance from the center of the kernel $\underline{w}_j$. Fig.4 shows a typical Gaussian RBF. The outputs of the network can be computed by a linear combination of the radial basis functions:

$$y_k = \underline{v}_k^T \underline{h}\ , \qquad k=1,\ N_{out} \tag{3}$$

Constant and linear terms represented respectively by a connection to a dummy input fixed to 1 and by direct linear connections between the inputs and each output can be added to the output expression. These kinds of corrections may be related to the selected stabilizer which is added to the standard cost function to be minimized in the learning phase. In short, the overall network performs a nonlinear transformation from the space of the inputs to the space of the outputs by forming a linear combination of the nonlinear basis functions. There are no general rules to train an RBF NN. The learning of RBF NNs is usually splitted in more steps: first of all a clustering procedure has to be carried out to organize the dataset in groups. This step allows to roughly determine the number of hidden neurons (at least one for each cluster center). Once the clusters of the input vectors are identified, the vector representative of each cluster is found by averaging all vectors in the clusters. In a successive time of the learning phase we can carry out a gradient descent procedure to tune the location of the centers. Of course, the method is not guaranteed to converge but some versions of either conjugate gradient or simulated annealing techniques can be used to cope with this problem. The second stage of the learning procedure is the determination of the diameters of the receptive regions, i.e. σ_j. This stage can have a profound impact upon the accuracy of the global procedure. The objective is to cover the input space with Gaussian functions as uniformly as possible. If the previous step yields a not uniform coverage due to different spacing between centers, we have to select different values of σ_j for each hidden neuron. The training of the output layer weight matrix, $\underline{\underline{v}}$, can be carried out by using a supervised technique.

Finally, a fine tuning of the model can be simply executed by the backpropagation algorithm used to train MFF NNs. In our software procedure we foresee a normalization step of the data. This is desirable with the aim to give to each dimension the same variance so avoiding to bias the mapping. The clustering technique we have found working are k-nearest neighbour algorithm and Kohonen SOM: both algorithms can be considered as types of unsupervised learning. One strong drawback of both RBF NNs and regularization approaches is that the function minimizing the cost functional must be specified by N coefficients, one for each pattern in the dataset of examples. This means that to reduce the computational complexity of the solution we need a strategy of reduction of the number of centers, and then of the units of the hidden layer. As we have already said the standard way to cope with this problem is the generalization of RBF NNs by means of movable centers functions. The centers of the representation have not necessarily to coincide with patterns of the dataset although this is permitted. In general the optimal location of the centers can be obtained by a clustering procedure. This is a technique of reduction of data in a certain number of groups (clusters) where the elements of each group should be as similar as possible as well as dissimilar from those of other groups. This implies the introduction of a distance measure between the patterns to be grouped. The standard distance is the Euclidean one which give rise to radially symmetric basis functions. It is a common practice to use a different metric in the Gaussian kernel which is known as Mahalanobis distance:

$$(\underline{x}-\underline{w}_j)^T \underline{\underline{K}}^{-1} (\underline{x}-\underline{w}_j)$$

where $\underline{\underline{K}}$ is the covariance matrix of the input vectors. By using this distance we obtain contours of constant distance which are hyperellipsoids and the basis functions are no longer radially symmetric.

Another common drawback of the RBF NNs is that they are generally slower than MFF NNs, and then seem less appealing for on-line inspection. One of the major advantages of the RBF NNs is that learning is much faster than in the MMF NNs. The reason why this happen can be easily understood bearing in mind that the learning process in RBF NNs can be broken in more stages, and each one of them can be carried out very efficiently. Moreover, there is another motivation to use RBF NNs which refers to their better performance in the case of availability of a low number of examples. The generalisation capabilities of a NN are usually tested by means of a separate dataset of previously unseen examples. In the case in which our dataset of examples is not sufficiently wide to allow for a splitting in a test and a training dataset, the standard method of cross validation cannot be used. As this is what we expect in a practical NDT problem, we have to obtain an alternative technique to test our model. In this paper, we use a special measure commonly known as predicted squared error (PSE) expressed in terms of mean squared error (MSE) on the training dataset which has been recently proposed[5]:

$$PSE = MSE + f(N_{eff}, N_p, \sigma^2), \quad (4)$$

where f is a suitable function of the effective number of degrees of freedom of the network, which is actually much less than the actual number of weights and biases because of the use of a regularizing term in the cost function as well as of the number of training patterns and of the variance of the expected noise corrupting the data.

RESULTS AND CONCLUSIONS

The reason why we have examined the possibility of using a different network topology to solve a problem we have already solved elsewhere[3] is related to the complexity of learning in MFF NNs particularly with respect to the generalization issue.

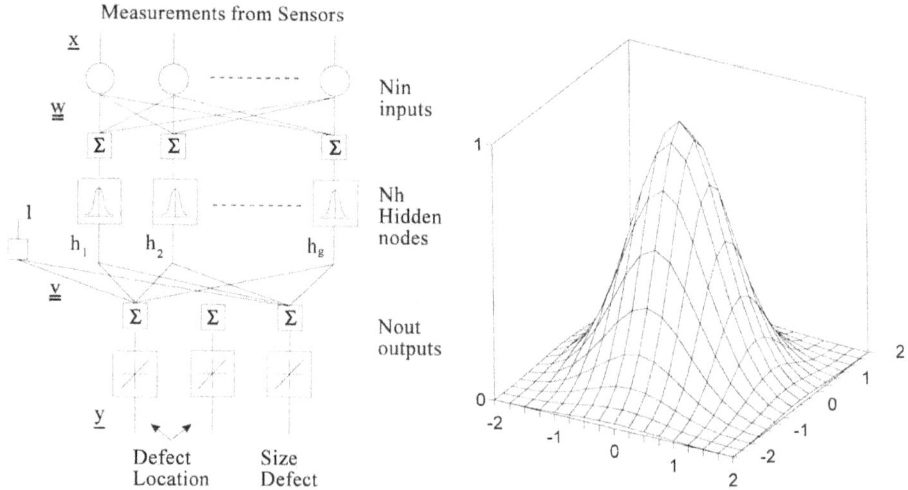

Figure 3 - Gaussian RBF NN topology.

Figure 4 - An example of Gaussian RBF.

Having no availability of sufficient number of examples, the choice of the size of the hidden layer in MFF NNs is very uneasy. Moreover, the learning procedure converges generally very slowly to a not satisfactory local minimum. Indeed, we have introduced a very complex task decomposition model to deal with these issues[3]. RBF NNs yield a computational paradigm solidly grounded on regularization theory which is in addition simply and rapidly trainable. The results obtained by using movable centers and a step of fine tuning via backpropagation are more satisfactory with respect to a MFF NNs when stopping the training to a comparable number of steps, as it is shown in Tab.1. In Tab.1 for modified NN we mean a NN with direct connections between the inputs and the outputs. In this paper we have not treated real measurements. This means that we cannot argue about the robustness of the models to additive noise. Anyway, it can be said that when some assumptions hold the injection of noise in the training dataset can enhance the generalisation capabilities of the NN.

Table 1 - Performance of three different MFF NNs and three RFB NNs in terms of full scale root mean square error. The training set contains 100 cases.

NN Topology	Fully Sigmoidal	Linear Output	Modified	RBF NNs Fixed centers	RBF NNs Movable centers	RBF NNs modified
#input sensors	12	12	12	12	12	12
#hidden nodes	8	10	6	10	10	6
#conn.+biases	131	163	99	153	153	93
%xb MSE/PSE	14 / 16	14 / 16	10 / 13	12 / 12	10 / 11	8 / 9
%yb MSE/PSE	14 / 16	12 / 16	10 / 13	11 / 12	10 / 11	8 / 9
%a MSE/PSE	18 / 22	16 / 22	11 / 15	14 / 15	12 / 13	10 / 11

REFERENCES

1. L.Udpa and S.S.Udpa, "Neural Networks for the Classification of Nondestructive evaluation signals", IEE Proceedings, 138,1:41, (1991).
2. E.Coccorese, R.Martone and F.C.Morabito, 1993, "A Neural Network Approach for the Solution of Electric and Magnetic Inverse Problems", paper accepted for publ. on IEEE Trans. on Magn. (1993).
3. F.C.Morabito, M.Campolo, :"Advanced In Neural Network Non-Destructive Testing", 2nd Intl. Conf. on Computation in Electromagnetics, IEE, Nottingham 1:76, (1994).
4. T.Poggio and F.Girosi, "A theory of Networks for approximating and learning", Artif. Intell. Lab. Memo, 1140, MIT, (1989).
5. M.T.Musavi et al., "On the training of Radial Basis Function Classifiers", N. N. 5:595, (1992).
6. J.E.Moody and C.J.Darken, "Fast Learning in Networks of Locally-Tuned Processing Units", Neural Comp., 1:281, (1989).

ALGORITHMS FOR THE ANALYSIS OF MAGNETIC FIELDS IN 3-D CONDUCTOR SYSTEMS

B. Azzerboni[1] and E. Cardelli[2]

1 Dipartimento Elettrico, Elettronico e Sistemistico, Universita' di Catania,
Viale A. Doria, 6 - 95125 Catania, Italia
2 Istituto di Energetica, Universita' di Perugia,
St. S. Lucia Canetola - 06125 Perugia, Italia

ABSTRACT

This paper deals with a series of algorithms that can be used as basic tools in software packages for the electromagnetic analysis of iron-free media.

The system under analysis is decomposed into a collection of rectangular busbars and annular arc shape conductors, carrying constant current density and arbitrarily oriented in space.

The magnetic field, forces, energy and the self- and mutual coefficients are evaluated using analytic expressions derived by the complete three-dimensional integration of the Biot-Savart's law in closed form. Finally, some examples of application of this algorithms, that have been checked against known results found in the literature, are reported.

INTRODUCTION

The design of electromagnetic components and systems for several devices and applications working with high magnetic induction levels, such as fusion and acceleration machines, high current pulse generators, switches etc., requires a three-dimensional analysis of iron-free media with massive and moving conductors. Studies of this kind involve the use of numerical codes with a long series of input data that can operate only on large size computers.

The need for software tools that allow both reduction in complexity of the input data and computation time, avoiding remeshing and the imposition of conditions in open boundaries, has suggested the development of integral formulations of the problem.

In this paper we present a series of algorithms that utilise exact analytic expressions and allow the computation of magnetic field and vector potential in iron-free media due to a collection of rectangular busbars and annular arc shape conductors, carrying current distributed with constant density and arbitrarily oriented in space. These algorithms have been developed by our research group during a scientific programme on the design of electromagnetic launchers and have been used in numerical codes FEMAN (Forces and

Electric and Magnetic Fields, Edited by A. Nicolet
and R. Belmans, Plenum Press, New York, 1995

Eddy-currents Magnetic Analysis) for the analysis of magnetic fields, forces and eddy-currents in three-dimensional problems.

The software products of the scientific programme are not commercial codes, but they have been conceived and realised as scientific supports for our own usage. Therefore the development of specific graphics pre- and post-processor routines, and in general of any user-friendly routines has been neglected. The numerical codes are open development environments, structured in a series of modular routines, in order to add other expressions, related to other elementary conductors, that are currently under investigation.

BASIC RELATIONS USED IN FEMAN ALGORITHMS

As is well known, it is possible to express the magnetic vector potential due to a conductor in which a current flows, by means of the Biot-Savart's law (see Fig. 1), as follows:

$$A(P) = \frac{\mu_0}{4\pi}\int_{\Gamma}\frac{J(Q)}{r_{PQ}}d\Gamma \tag{1}$$

where μ_0 is the magnetic permeability in the vacuum, Γ is the volume of the conductor, J is the vector current density, and r_{PQ} the distance between the point P in which the vector potential is evaluated and the variable point Q in the integration domain.

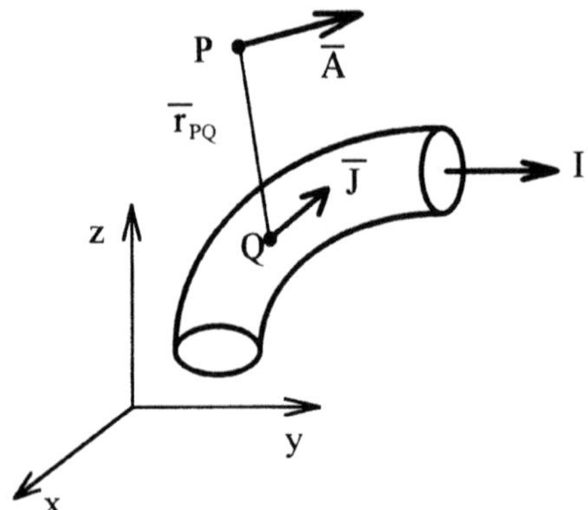

Fig. 1 - Magnetic vector potential due to a current flow.

Under the hypothesis of constant current density the integral (1) can be solved in closed form; in the case of a conductor in slab shape this form of analytical gives results of the type:

$$A_{slab} = \Sigma_i(POL_i \cdot ATAN_i \cdot LOG_i) \tag{2}$$

while in the case of conductor in shape of annular sector we have:

$$A_{ring} = \Sigma_i(POL_i \cdot ATAN_i \cdot LOG_i \cdot E_i \cdot F_i \cdot \pi_i) \tag{3}$$

where POL, ATAN and LOG are respectively suitably polynomial, inverse trigonometric and logarithmic functions of the coordinates of the point considered and of the dimensions of the conductor, and E, F and π are Legendre's Elliptic integrals of the first, second and third kind, having the same independent variables. The complete expressions of (2) and (3) in explicit form and with the main integration passages are reported in the references [1,2] and we not repeated here.

From the knowledge of A we derive the magnetic induction B:

$$B = \text{curl } A \tag{4}$$

Moreover the magnetic energy W_m and force F_m can be evaluated by:

$$W_m = \frac{1}{2}\int_\Gamma A \cdot J d\Gamma \quad (5)$$

and by:

$$F_m = \text{grad} W_m \quad (6)$$

We found that in the case of a conductor in slab shape, these integrals can be solved also analytically in closed form, giving results that can be summarised in the following synthetic form:

$$F_{mslab} = \Sigma_i (PPOL_i \cdot AATAN_i \cdot LLOG_i) \quad (7)$$

where PPOL, AATAN, and LLOG are suitable functions of similar shape (2).

For a more detailed descriptions of the analytic expressions of the magnetic energy and forces see[3].

MAGNETIC FORCES AND EDDY-CURRENTS ANALYSIS

A series of numeric routines have been developed, based on the algorithms presented in the previous section.

The system under analysis is suitably divided into a series of elementary conductors (see Fig. 2), in each of which a known constant current density is supposed. Each elementary conductor volume is specified by the position of its local coordinates system, by its dimensions, given in the local coordinates frame, and by the value and the orientation of its current density.

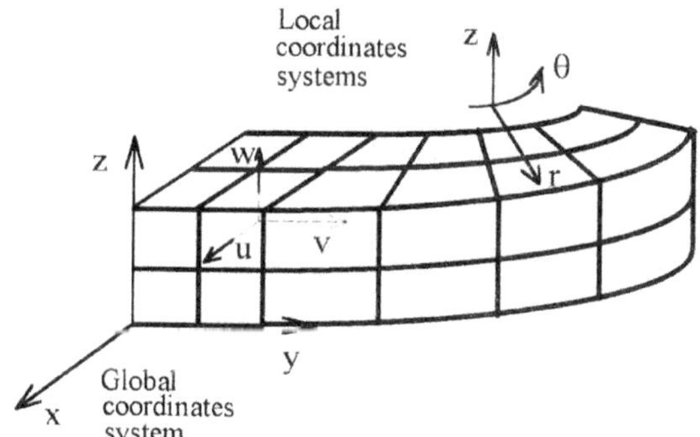

Fig. 2 - A simple example of three-dimensional modelling of a conductor for a FEMAN analysis.

The discretization of the system is determined by the current dispositions assigned and results, after a minimum of training, generally easy and quick still in three-dimensional problems, because only the conductor geometries need be specified, without any other boundary condition apart from the current values.

When the current distribution in the system is unknown and must be determined, the conductor regions are discretized also in a series of elementary conductors, each having suitable size, shape and current direction determined by the type of the problem considered.

The eddy-currents evaluation is made solving a suitable equivalent electrical network[4,5,6]. The determination of the self- and mutual coefficients of the network is given by the integration of:

$$M_{ij} = \frac{\int_{\Gamma_i} A_{ij} J_i d\Gamma_i}{I_i I_j} \quad (8)$$

where A_{ij} is the magnetic vector potential in the i-th conductor due to the current I_j that flows in the j-th conductor, Γ_i is the i-th conductor volume and I_i is the current that flows in the i-th conductor. The use of slabs in the discretization allows generally a remarkable reduction in the time necessary for the computation because, in this case, any numerical integration can be avoided. The algorithms developed have been tested with several benchmark, and with the comparison of numerical calculation results available in literature [7,8,9,10].

In the next paragraph some applications of the algorithms described above are shown.

EXAMPLES OF APPLICATION AND DISCUSSION

In this section two examples of the application of the algorithms described in this paper are summarised, just to give an idea about the possibilities of this type of magnetic analysis.

The first example chosen and shown here deals with the electromagnetic analysis of the augmented railgun represented in Fig. 3.

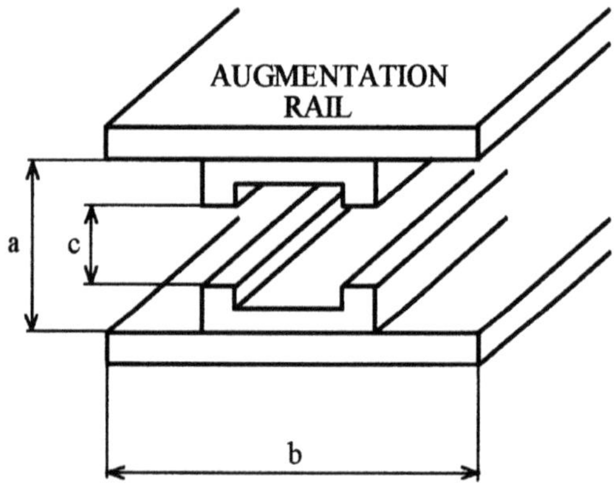

Fig. 3 - The augmented railgun with primary and secondary arc analysed by FEMAN.

Augmentation of railguns by adding one or more rail turns was proposed in order to reduce, for a given accelerating force, the input current and consequently the power dissipation in the plasma armature and in the rails. In this modelling the arcs were assumed to have the same cross section as the bore size, and different lengths.

The forces were evaluated for the opposite cases of complete penetration of the current in the rail and in the arcs, and in the high frequency limit, with the current distributed only around the surfaces of the conductors [11]. So the effective forces acting in the system are expected to be in the range limited between the values obtained for the two different current distributions considered.

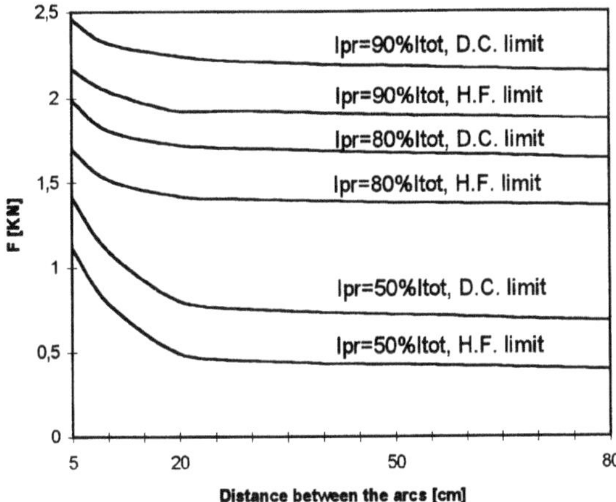

Fig. 4 - Propulsive forces in the primary arc of the augmented railgun vs. the relative position of a secondary arc.
I = 1 MA, a = 5 cm, c = 4.3 cm.

In Fig. 4 the forces exerted on the primary arc, supposing the presence of a secondary parasitic arc born back to the primary one, for different distances and current rates between the arcs, are reported. By means of a FEMAN approach, performed during a dedicated research programme, we deduced that the reduction in accelerating force is mainly caused by the current shunt in the secondary arc and that the decrease in launch efficiency is practically independent of the distance from the primary arc to the secondary one.

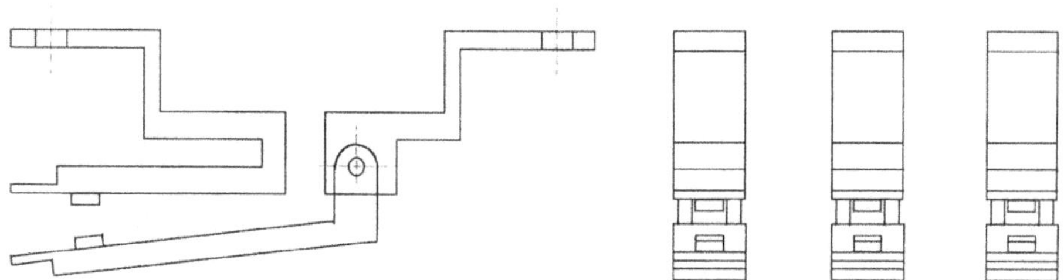

Fig. 5 - Schematics of a low voltage three-phase current limiting circuit-breaker.

Another example reported is the evaluation of the release currents in a low voltage three-phase current-limiting circuit breaker. In this case the algorithms above described have been used to evaluate the Lorentz forces acting on the moving contacts to cause their detachment from the fixed contacts (see Fig. 5), to diminish the tripping time due only to the magnetic release, during the fault current.

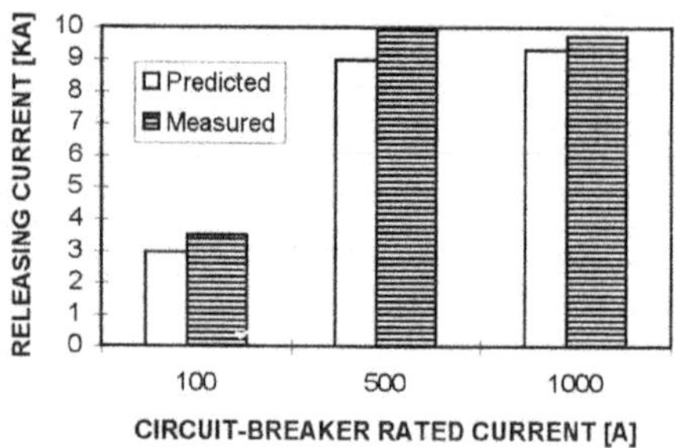

Fig. 6 - Calculated and measured releasing currents during a three-phase short-circuit for various sizes of current limiting circuit-breakers.

In Fig. 6 and 7 the values of the estimated and measured release currents in the case of a three-phase short-circuit and in the case of a earth-fault, for different circuit-breaker sizes, are reported.

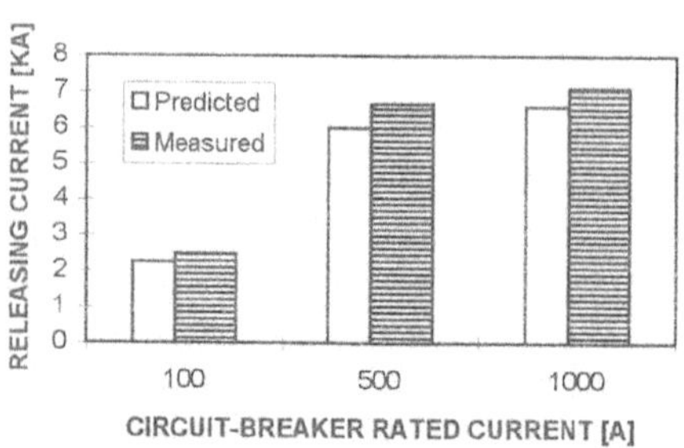

Fig. 7 - Calculated and measured releasing currents in presence of an earth-fault current for various sizes of current limiting circuit-breakers.

Results are in good agreement with the experimental data.

CONCLUSIONS

The algorithms developed for research usage by our group, are useful tools in the analysis and in the design of electromagnetic devices in iron-free media and can be applied to analyse several applications in the field of switching, power plant conductors, nuclear fusion machines, electromagnetic acceleration machines, etc. etc., that are in our field of interest.

Further forecast developments of these algorithms are the introduction of conductors in the shape of disk sectors, and the possibility to the considering current densities distributed with polynomial and exponential laws.

REFERENCES

1. B. Azzerboni, E. Cardelli , and A. Tellini, Computation of the magnetic field in massive conductor systems, IEEE Trans. on Mag., vol. MAG-25, no. 6, Nov. 1989.
2. B. Azzerboni, E. Cardelli, M. Raugi, A. Tellini, and G. Tina, Analytic expressions for magnetic field from finite curved conductors, IEEE Trans. on Mag., vol. MAG-27, no. 2, March 1991.

3. B. Azzerboni, E. Cardelli, A. Tellini, and G. Tina, Analytical evaluation of the energy and stresses in 3-D conductors system, To be published in IEEE trans. on Mag.
4. F. M. Grover, "Induction Calculation," New York: Dover Publications Inc., 1973.
5. W. H. Hayt, Engineering electromagnetics, Mc Graw Hill, 1989.
6. L. K. Urankar, Vector potential and magnetic field of current carrying finite arc segment in analytical form, part III: Exact computation for rectangular cross section, IEEE Trans. on Mag., vol. MAG-18, no. 6, Nov. 1982.
7. R. T. Honjo, and R. M. Del Vecchio, A program to compute magnetic fields, forces and inductances due to solid rectangular conductors arbitrarily positioned in space, IEEE Trans. on Mag., vol. MAG-22, no. 6, Nov. 1986.
8. A. Y. Wu, K. S. Sun, Formulation and implementation of the current filament method for the analysis of current diffusion and heating in conductors in railgun and homopolar generators, IEEE Trans. on Mag., Vol. 25, n. 1, pp. 610-615, January 1989.
9. B. Azzerboni, E. Cardelli, M. Raugi, A. Tellini, Current distribution in rail launchers via an equivalent network simulation approach , IEEE Trans. on Mag., vol. 28, no. 2, March 1992.
10. B. Azzerboni, E. Cardelli, M. Raugi, A. Tellini, Three-dimensional thermal and electromagnetic coupled analysis of railguns, 6th Symposium on Electromagnetic Launch Technology, Austin, April 1992.
11. J. F. Kerrisk, Electrical and Thermal Modelling of Railguns, IEEE Trans. on Mag., Vol. MAG.20, NO. 2, pp. 399-402, March 1984.

FINITE ELEMENT COMPUTATION OF THE ELECTROMAGNETIC FIELDS PRODUCED IN THE BODY BY MAGNETIC RESONANCE IMAGING SURFACE COILS

Olivier Le Dour°, Markus Vester*, Peter Henninger* and Wolfgang Renz*

°E.N.S.P.S., Université Louis Pasteur, Strasbourg (France),
[current address: European Commission, Life Sciences and Technologies, Medical Research, DG XII-E4, Rue de la Loi 200, SDME 2/50, B-1049 Brussels]
*SIEMENS AG Research Laboratories, ZFE BT EP 42, Paul-Gossen-Strasse 100 - Postfach 3220 D-91052 Erlangen

INTRODUCTION

The magnetic resonance imaging (MRI) Radiofrequency (RF) surface coil is a circular or rectangular conducting loop closed by a capacitor. In M.R. imaging, it plays the role of a magnifying glass, providing high resolution images with a high signal-to-noise ratio (SNR) on small regions of interest (ROI). Its main drawback is the non-uniformity of the magnetic field whichit produces, which leads to a loss of sensitivity, increasing rapidly with depth: this restricts the use of surface coils to the imaging of areas relatively close to the surface.

The purpose of our study was to evaluate the uniformity of the magnetic RF field, the amplitude of E.M. fields, currents, and power losses in the body for different models of surface coils, in order to determinate the best coil design and the most likely improvements most likely to minimize the currents induced in the body by the coil-patient coupling. Such a coupling constitutes a menace to the patient's safety and comfort; it reduces the signal-to-noise ratio, the quality factor Q and the sensitivity S and shifts the resonance frequency. All these factors cause a loss in the image quality. The electrical fields and currents involved in coil-body coupling are of two kinds [1,2]: the eddy currents created by the coil excitation and the currents created by the capacitive coupling, mainly localised close to the capacitors (i.e. where the highest voltage appear on the coil itself). Both types of current increase with the frequency.

The access to the quantities to be evaluated (fields, power loss densitiy, etc.) is difficult since they are to be determined in the human body. Hence we used the finite element calculation method to compute numerically these quantities for the coil-body system [3], using the MacNeil-Schwendler EMAS package, and compared the results with experiments.

Electric and Magnetic Fields, Edited by A. Nicolet
and R. Belmans, Plenum Press, New York, 1995

FIRST SIMULATIONS

We used a simple surface coil model[4]: a copper conducting loop with 10 cm diameter, 1 cm circular section and closed by a capacitor with a value chosen so that the coil resonates at 40 MHz, at a current of 1 A amplitude. This capacitor is a cylinder of 8 mm height and 1 cm diameter, containing about 120 cells. The body is placed at 5 mm from the coil. It is homogeneous and represents the characteristics of the muscle at 40 MHz (r = 80, = 0.8 S/m)[5]. Conductor zones which include a bend or angle are typically critical zones for finite element method, and the locus of hot spots. One of the areas of greatest interest in our model corresponds precisely to this description: the intersection between conductor and capacitor.

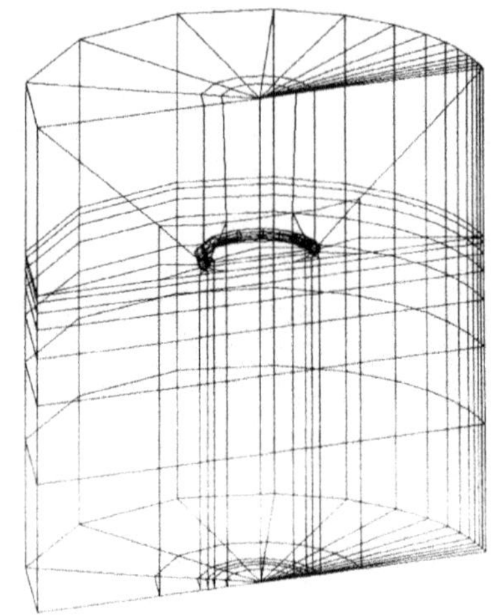

Figure 1. Aspect of a first model (with about 1500 gridpoints). Only the coil and external faces of the model are shown. For obvious symmetry reasons, only half of the space was simulated.

Only by using a very fine mesh in this area and by choosing the best possible excitation method were we able to reduce these hot spots sufficiently to prevent their influence from being noticeable outside the coil. The coil excitation was obtained by imposing an arbitrary current entering the coil in a gridpoint located on this intersection between conductor and capacitor. More and more detailed models were generated until we reached a stable result (i.e. a stage where a higher number of gridpoints and cells did not sensibly modify the results any more).

With a final model of ca. 4500 knots we obtained the following results: Q is 750 for the empty coil, and 26 for the loaded coil. The maximum power loss density, located on the surface under the capacitor, exceeds by 4% the value of the power loss density under the coil, on the opposite side to the capacitor. The maximum azimuthal component of the electric field along the loop is twelve times stronger than the maximum vertical component (under the capacitor extremities). This means that the capacitive coupling is very small compared to the eddy current.

We have also observed the sensitivity reduction with depth (29 % for 2 cm), which is typical for the sensitivity of this kind of coil. We did not observe any notable difference in sensitivity or uniformity between the empty and loaded coil. A change in the frequency (from 40 to 63.6 MHz) logically reduces Q and causes the power loss density and its local maximum to reach values which are more than twice the former ones.

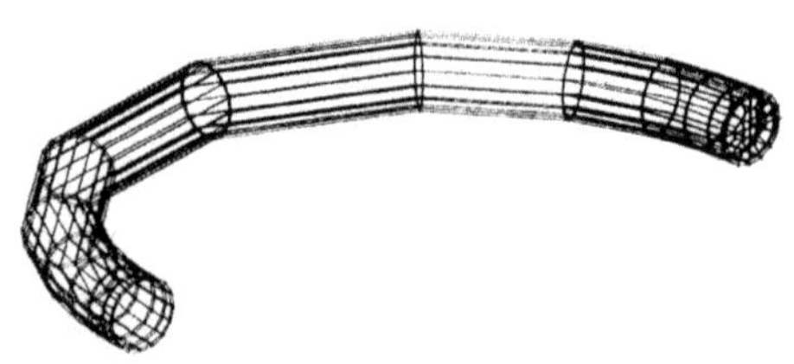

Figure 2. The final structure of the coil model (10 cm diameter, 1 cm section).

The capacitive interaction increases more strongly with the frequency than the inductive interaction. The stored energy, virtually independent of the frequency, remains stable. The skin depth reduction causes S and U (uniformity) to decrease.

Experimental validation was then necessary before these first results, and more generally our simulation method, could be deemed reliable.

EXPERIMENTAL VALIDATION AND FURTHER SIMULATIONS

In order to measure the electrical field produced by a coil identical to the model, resonating at 40 MHz, placed at 5 mm from the surface of a salt water solution - presenting the same impedance as the simulated body - we used the principle of reciprocity, which states that the field configuration created by the coil as a transmitter provide an exact indicator of the behaviour of the coil as a receiver. A diode (with a ca. 1 cm volume), fed by an optical fibre created the electrical field to be detected by the coil. The diode was fixed to a rigid rod bound to a displacement table capable of moving the diode into the three space directions inside a plexiglas water-filled container, 5 mm under which the receiver coil was fixed. The system [coil + container] was placed on a rotating table, with the same axis of rotation as our circular coil. A sheet of copper surrounded the container to protect it from the noise and from any parasitic signal.

Since we did not measure the field created by the coil in the water, but a quantity proportionnal to it, we did not try to obtain directly quantitative values but to confirm the aspect of the electric field observed during the simulation. We detected a difference of about 5 % between the maximum field (under the capacitor) and the field on the surface, on the opposite side to the capacitor (we obtained 3% with the simulation). The general aspect of the amplitude and phase of the measured field corresponds to the field configuration of our simulation. From this we concluded that our simulation method is successful.

We achieved a closer simulation of real experimental conditions by adding a skin layer to the body (a single 2mm thick row of elements on the body surface) and observed the effects of changes in the frequency (from 40 to 63.6 MHz), in the coil diameter (from 10 to 15 cm) and the coil design (simulation of a coil closer to the thin layer model). Possible improvements to reduce the capacitive coupling (use of two capacitors, in order to reduce the voltage on the coil by a factor 2, addition of a guard ring) were also tested.

CONCLUSION

The capacitive coupling affects the MRI image quality very little: decoupling measures are very unlikely to improve the sensitivity. At 63.6 MHz, as far as dissipated energy is concerned, the capacitive coupling created by the surface coil plays a leading role in the skin and can be neglected in the muscle. As far as safety is concerned, the division of capacitors is very effective at this frequency. The addition of a guard ring such as the one we simulated, although also effective in reducing hot spots, is very likely to affect the coil performances in regard to what already constitutes its main drawback: namely its sensitivity. Some already well-known facts (skin depth reduction when the frequency increases, deeper sensitive area when the coil diameter is greater) have also been confirmed.

Generally speaking, these conclusions show that the F.E. method is very likely to bring useful information concerning MRI coil performances.

REFERENCES

[1] Gadian & al. Radiofrequency Losses in NMR Experiments; J. Magn. Res. 34 pp. 449-455 (1979)

[2] Redpath et al. Estimating Patient Dielectric Losses in NMR Imagers. J. Magn. Res. Im. 2 pp. 295-300 (1984)

[3] Zientkewicz et al. F.E. in the Solution of Field Problems, The Engineer, 42 (Sept. 1965)

[4] Ackerman et al. Mapping of metabolites in Whole Animals by 31P NMR, using Surface Coils. Nature (1980)

[5] Stoy et al. Dielectric properties of Mammalian Tissues from 0,1 to 100 MHz. Phys. Med. Biol. Vol. 4 pp. 501-513 (1982)

MODELLING AND CHARACTERISATION OF PULSED EDDY CURRENTS - APPLICATION TO NON DESTRUCTIVE TESTING IN RIVETED ASSEMBLIES USED IN AERONAUTICS

F. Thollon, N. Burais

CEGELY - URA CNRS 829
Département d'Electrotechnique - Ecole Centrale de Lyon (ECL)
BP163 - 69131 Ecully, Cedex, France

INTRODUCTION

This study is devoted to the numerical simulation of pulsed eddy currents. A new method is proposed to characterise these currents. New parameters describing the induced current distribution are also defined.

These parameters allow us to describe eddy currents in a more complete and more general way as compared to a simple knowledge of the standard depth of penetration. For this, the software package FISSURE[1] developed in CEGELY has been used and improved. The latter is designed to resolve both 2D and 3D axisymetric electromagnetic problems. It is well adapted to the study of the electromagnetic methods used in non destructive testing and in non destructive evaluation, more specifically eddy currents method. Based on the resolution of Maxwell's equations with finite element method, this package is particularly optimised for the automatic study of geometrical and physical parameters influence, for sinusoidal and pulsed current. The obtained results have been used to optimise a probe designed to detect defects in riveted assemblies used in aeronautics. Assuming that the response of a defect is maximum when maximum energy is injected at the presumed location of the defect, excitation coil dimensions and pulse duration are calculated in this aim for one assembly structure and a single pulse shape.

GENERAL MATHEMATICAL FORMULATION OF THE PROBLEM.

Physical phenomena on the set probe - assembly under control are modelled with Maxwell's equations.Using the magnetic potential vector **A** defined by $\mathbf{B} = \nabla \times \mathbf{A}$, only one equation characterises the behaviour of the system. For materials with linear physical properties and for 2D and 3D axisymetric structures, the equation is expressed as [2]:

$$\frac{1}{\mu}\Delta A - \sigma\frac{\partial A}{\partial t} = -J_{ex} \tag{1}$$

Electric and Magnetic Fields, Edited by A. Nicolet
and R. Belmans, Plenum Press, New York, 1995

This is the classical equation of magnetodynamic where J_{ex} is the current density in the coil, σ and μ are the conductivity and the permeability of the different materials. If the supply is pulsed, the equation can be discretized as a first order temporal derivative. A backward difference approximation is used which maintains a stable solution but for which the precision depends on the value of time step Δt. The equation (1) can be expressed as:

$$\frac{\Delta A_t}{\mu} - \sigma\frac{A_t}{\Delta t} = -\sigma\frac{A(t-\Delta t)}{\Delta t} - J_{ex}(t) \quad (2)$$

It is therefore necessary to solve the equation with the finite element method for each time step. The number of calculations is linked to the value of each time step Δt and the duration of the studied phenomena. For pulsed excitations, that implies a large number of resolutions (short pulses: short time step Δt, deep defects: large examination duration). Therefore, different values of Δt can be used for one calculation in order to optimise calculation time.

The flux density B is obtained using the spatial derivative of A, the voltage induced in a measure coil, by the integration of A.

DESCRIPTION AND CHARACTERISATION OF EDDY CURRENT

Despite the linearity of materials concerned in this study, the different signals do not have the same temporal shapes. More especially, the shape of eddy current depends on the depth in the material. As a pulse is a finite energy signal, we propose to characterise eddy currents with their energy (in mathematical meaning). The software FISSURE allows to calculate:

$$w_{cf} = \int_0^{t_{max}} j^2(t)dt \quad (3)$$

where w_{cf} is energy of the eddy currents at a given point and $t_{95\%}$ the time interval during which $0.95w_{cf}$ is dissipated. w_{cf} is an image of the energy dissipated in materials under control. Indeed, the volumic density of energy dissipated by joule effect can be expressed as $W = 1/\sigma\, w_{cf}$, where σ is the material conductivity.

Eddy currents are created by the quantity $n\, i = \sigma\, j_{inj}$ where n is the number of turns of excitation coil , i(t) the intensity of the injected currents, s the cross-section of the coil, j_{inj} the density of injected currents. To compare different excitations (coil and signal), the excitation signal energy is normalised. The following condition is then imposed:

$$W_{exc} = \int_0^{t_{max}} ni^2(t)dt = \int_0^{t_{max}} s\, j_{inj}^2 dt = 1 \quad (4)$$

A dimensions equation shows that W_{exc} and w_{cf} are linked with the square of a surface. Therefore, another parameter to take into account is the mean section of the coil (or the mean radius).

This method is first used to describe eddy current distributions in a semi-infinite plate of aluminium excited by a cylindrical coil supplied with a pulsed half-sinusoidal current (figure 1). Eddy current energy has been calculated for a coil mean radius varying between 4.5mm and 16.5mm and a normalised pulse duration varying between 100µs and 2200µs.

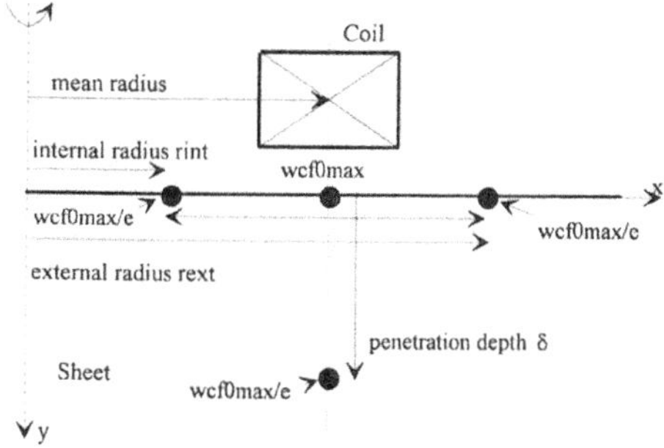

Figure 1. studied structure

Results have allowed us to identify the density by two equations:

$$w_{cfx} = w_{cfx}(0) \exp(-y/\delta(x)) \qquad w_{cfx}(0) = w_{cf0max} \exp(-b(x - r_{moy})^d) \tag{5}$$

where w_{cfx} is the energy over a vertical line situated at a distance x of the axis of revolution, $w_{cfx}(0)$ is density at the surface of the metal, and b and d parameters depending on r_{mean} and the duration of the pulse.w_{cf0max} represents the maximal density. w_{cfx} takes this value for $x = r_{mean}$. This identification is accurate (except in the neighbourhood of the revolution axis) and adapted to the studied situation[3].

The distribution of eddy current can be characterised by 4 parameters:

-) w_{cf0max}: maximum value of density ($x = r_{mean}$, y=0)

-) the penetration depth δ defined by $w_{cfx}(\delta) = w_{cf0max} / e$ for $x = r_{mean}$. It allows us to define variations of w_{cfx} with depth.

-) the internal radius and the external radius defined by $w_{cfx}(y=0)$ = wcf0max/e. Values obtained are designated r_{int} and r_{ext}. The spread is defined by $r_{ext}-r_{int}$. These quantities allow us to define w_{cfx} variations on the surface.

An analysis of the results leads to the following conclusions:

-) w_{cf0max} increases with the radius and decreases with the duration of the pulse.

-) the spread increases with r_{mean} and pulse duration.

-) the penetration depth increases with the duration of the pulse and the radius.

To optimise the value of energy injected at a given point, equivalues w_{cfx} are traced. Figure 2 represents these lines for a 12.5mm mean radius coil according to the duration of the pulse. We can see, for example, that, to have a maximum of energy at 10mm of the axis, it is necessary to have a pulse of duration 200μs for a depth of 2.7mm, and a pulse of 400μs for a depth of 5mm.

APPLICATION TO NON DESTRUCTIVE TESTING OF RIVETED ASSEMBLIES

This method is then applied on the control of riveted assemblies used in aeronautics (figure 3). Assuming that the response of a defect is maximum when maximum energy is injected at the presumed location of the defect, optimal mean radius and pulsed duration are searched. Figure 4 represents injected energy at 10mm depth and at 5mm from the axis (along the rivet) for different coil diameters and pulse durations. It clearly shows that to inject a maximum of energy at this spot, it is necessary to use a coil of mean radius 11mm and a pulse of duration 0.6ms.

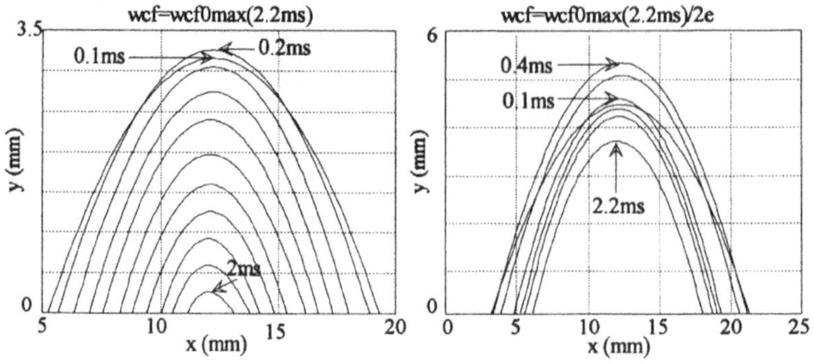

Figure 2. w_{cf} equivalues

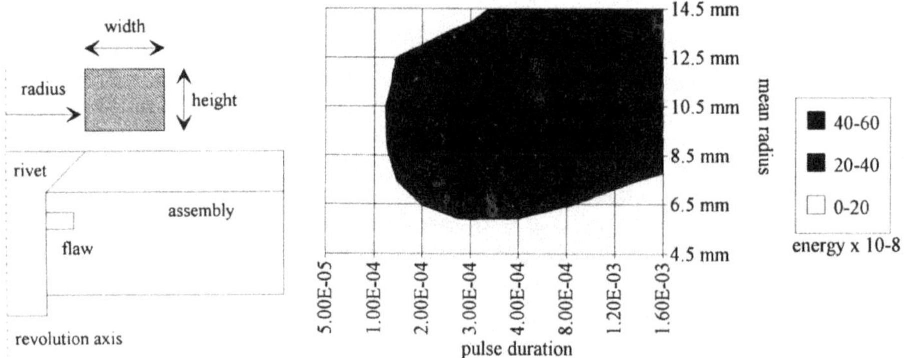

Figure 3. studied assembly

Figure 4. excitation optimisation

CONCLUSION.

In this study, we have presented a new method for the characterisation of pulsed eddy currents using a notion of energy. Introduction of new parameters allows us to define the distribution of these currents in a more complete manner. When applied to the control of riveted assemblies used in aeronautics, the method allowed us to detect defects at a depth of 20mm in non magnetic assemblies.

ACKNOWLEDGMENTS

This study is supported by AIA of Clermond-Ferrand and STPA, in collaboration with DASSAULT AVIATION and CEAT of Toulouse.

REFERENCES

1. N.Burais, A.Nicolas, L.Nicolas, "Développement de la modélisation en CND et END par méthodes électromagnétiques", 1er Congrès COFREND, NICE FRANCE 1990, pages 41-46.
2. B.L.Allen, W.Lord, "Finite element modelling of pulsed eddy current phenomena", Proceedings Congres Review of Progress in Quantitative NDE", vol 3, Aug.83, Santa-Cruz, pages 561-568.
3. Z. Mottl, " The quantitative relations between true and standart depth of penetration for air-cored probe coils in eddy current testing.", NDT International, Vol 23, n° 1, feb 90, pp 11-18.

DESIGN SHAPE OPTIMISATION OF ELECTROMAGNETIC DEVICES USING THE FINITE ELEMENT METHOD

Jaime A. Ramirez, C. Chat-uthai, Ernie M. Freeman

Imperial College of Science Technology & Medicine
Electrical & Electronic Engineering Department
Exhibition Road, London SW7 2BT, UK

INTRODUCTION

The design of electromagnetic devices, such as electrical machines, transformers, electromagnets, etc., has always been a challenge for electrical engineers. Amongst the factors that defines such designs are the device's dimensions, its material properties and field sources. Also, the determination of the magnetic flux density B can be regarded as crucial.

In the past, such designs and the assessment of B were an extremely difficult task based very much on the engineer's experience and intuition. The know-how of the engineer was used to change some parameters and to predict the behaviour induced in the device but this practice was limited due to the inherent complexity of the design. Some analytical methods such as separation of variables were used for the field analysis but its applications, in most of the cases, were limited to simple two dimensional problems because these calculations were tedious and time consuming.

After the advent of the computer the whole process of design changed. The minimisation of functions of many variables, say a hundred or more, became possible. Powerful numerical methods such as the Broyden-Fletcher-Goldfarb-Shanno (BFGS) algorithm were developed. On the field analysis side the calculation of B became more accurate, less difficult and less time consuming.

In recent years there has begun to emerge a methodology for formalising this optimisation process based on the finite element method, see for instance Gitosusastro et al (1989), Weeber and Hoole (1992).

This paper describes how the finite element method can be used in conjunction with classical deterministic methods for the optimum design of electromagnetic devices. The field analysis is made using the finite element method which defines a set of equations in terms of a generic design parameter p that can be the dimensions, material properties or fields sources of the electromagnetic device. The optimisation process is carried out using an iterative approach based on the quadratic extended penalty function. In this process the direction of the minimum is found by direct differentiation of the finite element matrices. The results of the shape optimisation of the poles of an electromagnet are shown as an example of application of the theory presented. The work is part of a wider study of optimisation of electromagnetic devices, particularly coupled problems involving electromagnetic and thermal effects.

THE FIELD ANALYSIS

Starting from Maxwell's equations and using the finite element method together with the variational approach it is possible to get the following equation for 2D magnetostatic problems

$$[K(x,y,\mu)]\{A\} = \{R(x,y,J)\} \tag{1}$$

where $[K]$ is the stiffness matrix which represents a differentiable function of the dimensions (x and y) and the material properties that constitute the device (μ). The vector $\{A\}$ represents the unknown magnetic potentials at each node of the mesh, and the vector $\{R\}$ is a differentiable function of the dimensions (x and y) and sources (J) that characterise the device. These features of (1) are exploited in the next section.

Electric and Magnetic Fields, Edited by A. Nicolet
and R. Belmans, Plenum Press, New York, 1995

THE OPTIMISATION PROCESS

Mathematically, non-linear shape optimisation problems can be stated as

Minimise
$$F = F(\{p\}, A(\{p\})) \tag{2}$$

Subject to
$$g_j(\{p\}, A(\{p\})) \le 0 \quad j = 1,\ldots,m \tag{3}$$

$$h_k(\{p\}, A(\{p\})) = 0 \quad k = 1,\ldots,l \tag{4}$$

$$p_i^l \le p_i \le p_i^u \quad i = 1,n \tag{5}$$

where F represents the objective function, g the inequality constraints, h the equality constraints and p stands for the design parameters. It has been assumed that the objective function and both inequality and equality constraint are dependent functions of p and A (A itself being a function of p).

The method used to solve the system of equations given by (2) to (5) is known as a transformation or penalty method which is defined by

$$\Phi(\{p\}, r_p) = F(\{p\}) + r_p P(\{p\}) \tag{6}$$

where F is the original objective function (2), r_p is a scalar multiplier and P represents an imposed penalty function. Following Haftka and Starnes (1976), the Quadratic Extended Penalty Method (QEPM) was used in this work which has the form

$$P(\{p\}) = \sum_{j=1}^{m} \tilde{g}_j(\{p\}) \tag{7}$$

where

$$\tilde{g}_j(\{p\}) = \frac{-1}{\varepsilon}\left\{\left[\frac{g_j(\{p\})}{\varepsilon}\right]^2 - 3\left[\frac{g_j(\{p\})}{\varepsilon}\right] + 3\right\} \quad \text{if} \quad g_j(\{p\}) > \varepsilon \tag{8}$$

or $\tilde{g}_j(\{p\}) = -1/g_j(\{p\})$ if $g_j(\{p\}) \le \varepsilon$ which is the original interior penalty function. The transition parameter ε is a small negative number given by $\varepsilon = -C(r_p)^a$, where $1/2 \le a \le 1/3$ and C is a constant.

After the transformation, the pseudo-objective function Φ is minimised using any classical unconstrained technique. Those techniques are characterised by an iterative algorithm where the design variables are updated according to $\{p\}^q = \{p\}^{q-1} + \alpha S^q$ where q is the iteration number and S is a vector search direction in the design space. The scalar α is the step size which defines the distance to be moved in the direction S. During this process S is updated using $\{S\}^q = -\nabla F + \beta^q \{S\}^{q-1}$ which clearly shows the need to calculate the gradient of the objective function. This discussion is normally referred to as sensitivity analysis. The calculation of terms α and β is addressed by Vanderplaats (1984).

Sensitivity Analysis

The total design dependence of the objective function is determined by the total derivatives of F with respect to each parameter p which is

$$\frac{dF(\{p\}, A(\{p\}))}{dp_i} = \frac{\partial F}{\partial p_i} + \left\{\frac{\partial F}{\partial A}\right\}^T \left\{\frac{\partial A}{\partial p_i}\right\}, \quad i = 1 \cdots n \tag{9}$$

The first term of (9) is easily computed from the known form of the objective function. The second term involves the calculation of $\{\partial A/\partial p_i\}$ which is based on the differentiation of both sides of (1), that is

$$\left\{\frac{\partial A}{\partial p_i}\right\} = [K(x,y,\mu)]^{-1}\left\{\left\{\frac{\partial R(x,y,J)}{\partial p_i}\right\} - \left[\frac{\partial K(x,y,\mu)}{\partial p_i}\right]\{A\}\right\}, \quad i = 1 \cdots n \tag{10}$$

where the partial derivatives of $[K]$ and $\{R\}$ with respect to any physical parameter are expressed by direct differentiation of the finite element matrices, see Ramirez (1994). Having calculated $\{\partial A/\partial p_i\}$ the sensitivity equation of the magnitude $|B|$ follow directly as

$$\frac{\partial|B|}{\partial p_i}=\frac{1}{2|B|}\left(\{A^e\}^T\left[\frac{\partial K^e}{\partial p_i}\right]\{A^e\}+2\{A^e\}^T\left[K^e\right]\left\{\frac{\partial A^e}{\partial p_i}\right\}\right) \tag{11}$$

where the superscript e denote element. It is this sensitivity information that provides the essential guideline for the search direction S in which the design variables are modified in the optimisation process.

RESULTS

The problem investigated consists in determining the shape of the poles of an electromagnet in order to keep the magnetic flux density B constant in its air gap. This problem has also been discussed in Marrocco and Pironneau (1978), Haslinger and Neittaanmäki (1988). This is shown in fig. 1 which represents half of the original magnet and the two dimensional model.

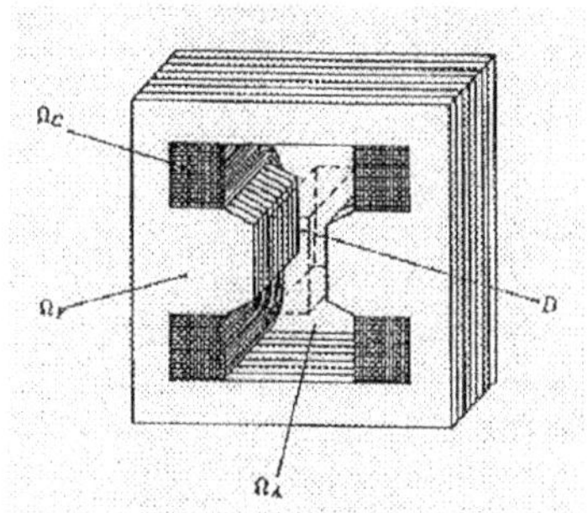

a) Half of the original magnet

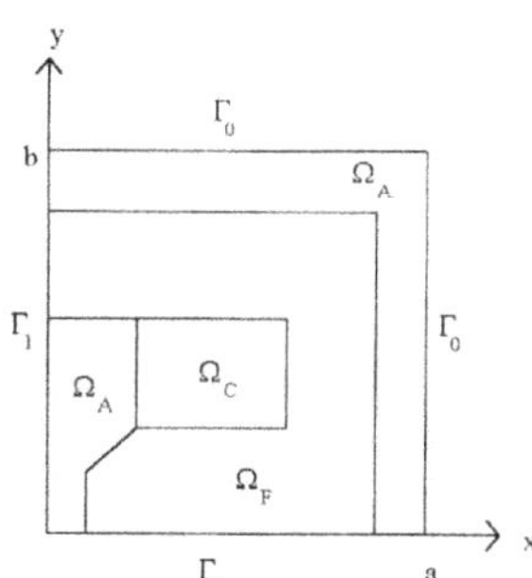

b) Two dimensional model

Figure 1

The domain is represented by $\Omega=\Omega_F+\Omega_C+\Omega_A$ in which Ω_F represents the ferromagnetic region, Ω_C represents the coil region and Ω_A represents the air region. Dirichlet and Neumann boundary conditions are used on Γ_0 and on Γ_1 respectively. The mesh used in the simulation consisting of 156 nodes and 263 elements is shown in fig. 2a. Fig.2b shows the pole region highlighting the moving nodes.

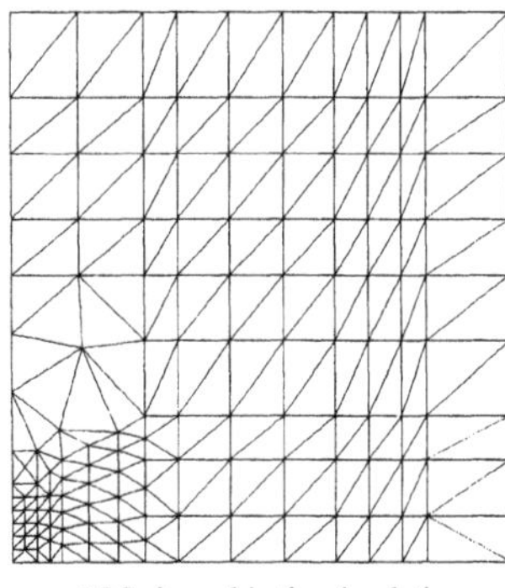

a) Mesh used in the simulation

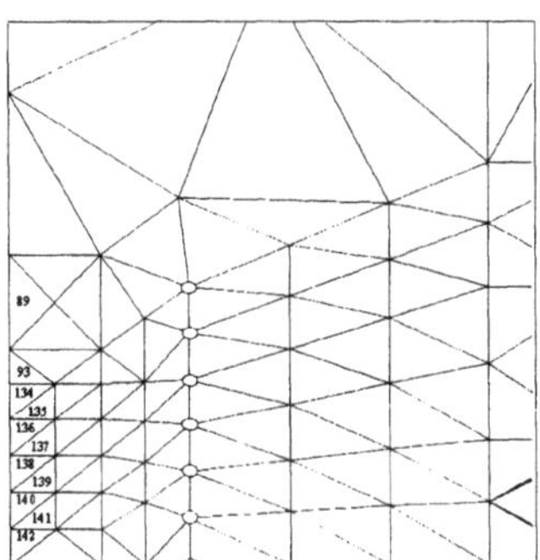

b) Zoom showing the moving nodes

Figure 2

Mathematically, the problem is expressed as the minimisation of the objective function given by

$$\min\ F(\{p\},A(\{p\}))=\sum_{i=1}^{n}\left(\left|B_{c_i}-B_{d_i}\right|^2\right) \tag{12}$$

where the moving nodes are subject to $(0.010 \le x \le 0.018)$. The summation is carried out over the elements numbered in fig.2b. B_c and B_d are respectively the calculated and desired magnitude of the magnetic flux density. The data used in the simulation were: $\Omega_T = (0,a)\times(0,b) = [(0,0.12)\times(0,0.10)]m^2$; $J = 1000\,A/cm^2$ in Ω_C and $\mu = 1000\mu_0$ in Ω_F. Table 1 gives the final value of $|B|$ in the elements considered and fig.3 shows the final shape achieved for the pole. The simulation took 1:00:00 (h:m:s) to reach a tolerance of 10^{-5} in a PC 486/66Mhz, 8Mb of extended memory

Table 1. Final result for $B_d = 0.35$T

Element	B_x	B_y	$\lvert B\rvert$ (T)
89	-0.3292	0.0252	0.3301
93	-0.3533	0.0000	0.3533
134	-0.3504	0.0000	0.3504
135	-0.3532	0.0000	0.3532
136	-0.3516	-0.0010	0.3516
137	-0.3545	-0.0010	0.3545
138	-0.3506	-0.0007	0.3506
139	-0.3512	-0.0007	0.3512
140	-0.3503	-0.0002	0.3503
141	-0.3499	-0.0002	0.3499
142	-0.3500	0.0000	0.3500

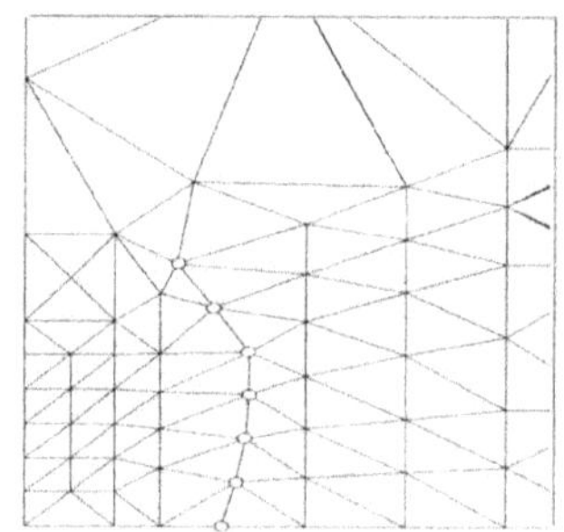

Figure 3. Detail of the pole for $B_d = 0.35$T

CONCLUSION

It has been demonstrated how the finite element method can be used in conjunction with the QEPM for the optimum design of electromagnetic devices. Particular attention was paid to show that the search direction vector, which is a essential guideline in the optimisation process, can be found by direct differentiation of the finite element matrices. This fact confers not only generality to the finite element method but also represent a substantial saving in computational time if the finite difference is applied for the sensitivity analysis. The QEPM proved to behave particularly well in this application.

The authors have also applied the method presented to eddy-current problems and are now on the course to extend it to 3D problems. The results of this work, covering also the Augmented Lagrange Multiplier Method and the Quadratic Programming Method, will be published elsewhere in the near future.

ACKNOWLEDGEMENTS

The authors would like to acknowledge the Brazilian Government (Capes) and the British Council for the financial support given to Mr. J. Ramirez and Mr. C. Chat-uthai respectively.

REFERENCES

Gitosusastro, S., Coulomb, J.L., and Sabonnadiere, J.C., 1989, Performance Derivative Calculations and Optimization Process, *IEEE Trans. on Mag., vol.MAG 25(4)*, pp.2834-2839.

Haftka, R.T., and Starnes Jr., J.H., 1976, Applications of a Quadratic Extended Interior Penalty Function for Structural Optimization, *AIAA Journal, vol. 14(6)*, pp.718-724.

Haslinger, J., and Neittaanmäki, P., 1988, "Finite Element Approximation for Optimal Shape Design", John Wiley & Sons Ltd., Chichester, U.K.

Marrocco, A., and Pironneau, O., 1978, Optimum Design with Lagrangian Finite Elements: Design of an Electromagnet, *Comp. Meth. in App. Mech. and Eng., vol.15*, pp.277-308.

Ramírez, J.A., 1994, "Sensitivity Analysis and Shape Optimisation of Power Frequency Electromagnetic Devices", Ph.D. Thesis (in preparation), Imperial College, Elec. & Electc. Eng. Dept., London, UK.

Vanderplaats, G.N., 1984, "Numerical Optimization Techniques for Engineering Design: With Applications", McGraw-Hill series in mechanical engineering, New York.

Weeber, K., and Hoole, S.R.H., 1992, Geometric Parametrization and Constrained Optimization Techniques in the Design of Salient Pole Synchronous Machines, *IEEE Trans. on Mag., vol.MAG 28(4)*, pp.1948-1959.

ANALYSIS OF THE MAIN PARAMETERS OF AN E-SHAPED ELECTRO-MAGNET WITH SHADING RINGS BY USING FINITE ELEMENTS

Witold Tarczynski,[1] Antonios G. Kladas,[2] and John A. Tegopoulos[2]

[1]Institute of Electrical Apparatus
Technical University of Lodz
B.Stefanowskiego 18/22
90-924 Lodz, Poland

[2]Electric Power Division
Department of Electrical Engineering
National Technical University
42, 28th October street
10682 Athens, Greece

INTRODUCTION

E-shaped electromagnets constitute one of the most common configurations favoured by designers of a.c. contactors and actuators. The main parts of such an electromagnet (Figure 1) are the core, the armature mounted on springs, the exciting coil and the shading rings. A current flow in the coil produces a force pulling the armature towards the core. This force has to be greater than resistance forces in a real device, tending to pull the armature away. In addition, at the instant of zero crossing of the sinusoidal exciting current, the resistance force may be greater than the minimum attractive force value, developed due to the shading ring currents. In this case chattering noise is produced. Therefore it is important for designers to investigate phenomena of the electromagnet in order to obtain the best correlation between characteristics of the resistance force and the time varying magnetic force.

Previous work found in the literature concerning such devices[1,2,3,4] is based on an equivalent magnetic circuit representation and involves mainly zero air-gap length. The present paper uses a 2D finite element model checked through measurements for an existing contactor case. This model has been implemented to provide a systematic analysis for E-shaped electromagnet parameters varying with the air-gap length.

NUMERICAL MODEL

The electromagnet has been modelled by using the finite element method in the two dimensional Cartesian configuration shown in Figure 1. In such a case the magnetic field

Electric and Magnetic Fields, Edited by A. Nicolet
and R. Belmans, Plenum Press, New York, 1995

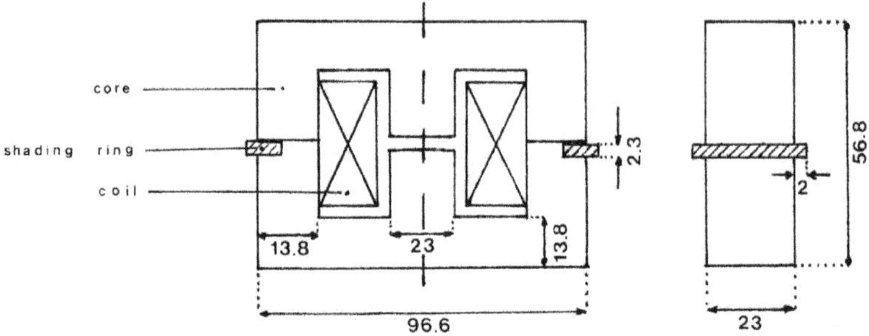

Figure 1. Geometry of contactor's magnetic circuit and main dimensions

distribution is determined by solving the diffusion equation expressed in terms of the magnetic vector potential as follows:

$$\nabla \wedge \left(\frac{1}{\mu} \nabla \wedge A\right) + \sigma \frac{\partial A}{\partial t} = J_s \tag{1}$$

where A is the only existing component along z direction of the magnetic vector potential, μ is the magnetic permeability, σ is the electric conductivity and Js is the component along z direction of the source current density. The second term of (1) represents the eddy current density induced in the ring.

Such a simulation neglects the leakage flux developed at the lateral end regions of the device and a detailed representation necessitates, in general, a three dimensional analysis. However, the main parameters and the design of the apparatus are related to configurations involving very small air-gap lengths and in such cases the approximate two dimensional analysis provides sufficient accuracy. As the excitation varies sinusoidally with time, an important saving in computer time can be obtained by adopting complex variables to consider time variations. In such a model the iron saturation is taken into account in a somewhat approximate way,[5] but this simplification is quite realistic in the above class of problems. The mesh used for the problem analysis comprises 580 nodes and 1117 triangular elements of first order. Due to the problem symmetry only one half of the device is represented.

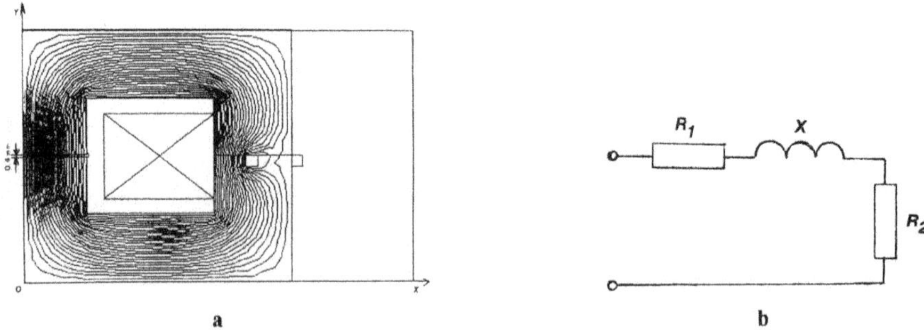

Figure 2 **a.** Field distribution for d=0 at the instant of maximum excitation current.
b. Equivalent electric circuit for the electromagnet.

RESULTS AND DISCUSSION

The magnetic field in the electromagnet has been calculated for different air gap lengths varying from d=0 to d=10 mm. The field distribution for d=0 at the instant of peak excitation current is shown in Figure 2a. In this figure the effects of the shading ring may be noticed, tending to push the flux outside the shaded part of the core cross-section.

In order to investigate the global quantities variations, the equivalent electric circuit shown in Figure 2b has been considered. In this circuit R_1 denotes the coil resistance, X the equivalent total reactance and R_2 the apparent shading rings resistance corresponding to the power loss P dissipated in the shading rings. These quantities can be easily deduced from the field distribution by applying the flux cutting rule and by using Joule loss formula. The variations of the above mentioned parameters with the air gap length are illustrated in Figures 3a and 3b. Figure 3a shows that the power loss P in the shading ring increases with the air-gap length up to 30 W and then decreases slightly. In the same figure it may be noticed that the shading ring resistance is maximum for d = 0.5 mm. Figure 3b shows the reactance decrease as well as the reactive power variation Q with the air-gap length.

Figure 4a compares the variations with the air gap length of the measured time average force Ft to the maximum and minimum values of the calculated force in time (Fm

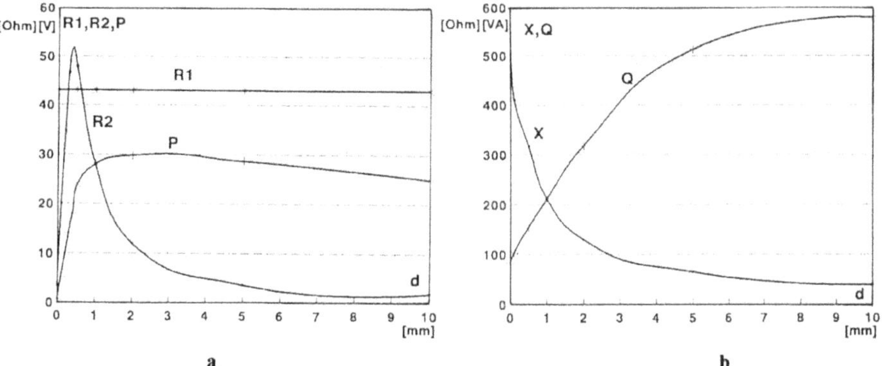

Figure 3 **a.** Resistances R_1, R_2, and shading ring Joule loss variations with the air-gap length d.
b. Reactance X and reactive power Q variations with the air-gap length d.

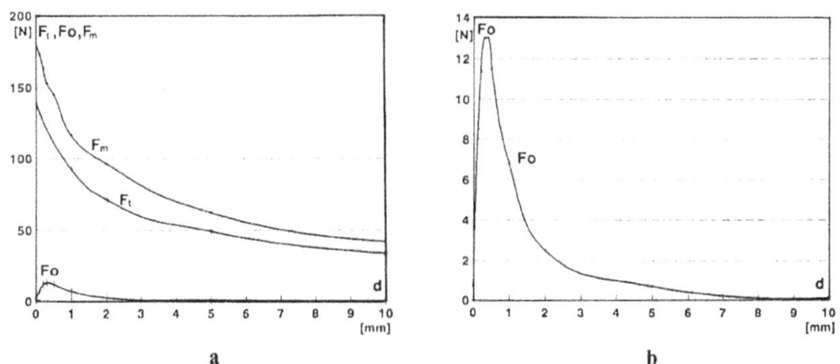

Figure 4 **a.** Force max (Fm) and min (Fo) calculated values and measured time average value (Ft).
b. Minimum value Fo of the calculated electromagnetic force in time.

Table 1. Influence of the shape of the shading ring on the main contactor parameters

d=0	F_m	F_o	P	Q	R_2	X	I_c
	1	1	1	1	1	1	1
	1.41	2.2	1.4	1.04	1.27	0.98	1.02
	0.95	0.53	1.1	1.06	1.0	1.0	1.0
	1.16	1.6	1.28	1.05	1.18	1.0	1.01
	1.06	1.3	1.18	1.05	1.09	0.99	1.01
	1.34	2.2	1.26	1.05	1.15	0.99	1.01
	1.28	1.5	1.4	1.04	1.29	0.98	1.02

and F_o respectively). This figure illustrates that the measured and calculated forces are in quite good agreement. The variation of F_o with the air gap is given with a greater scale in Figure 4b.

The influence of the shape of the shading ring on the main parameters of the device has also been investigated and the results are given in Table 1. The provided values have been normalised to those of the existing contactor. In this table the values of the current absorbed by the coil I_c indicating the impedance variations of the device have been included. The analysis was made for shaded core portion values in the vicinity of 2/3 providing maximum force.

REFERENCES

1. H.J. Kubiak and L.H. Matthias, Shading coil magnet of new design, *AIEE Transactions.* 346:352 (1962).
2. H.J. Kubiak, Analysis of the shading coil magnet, *AIEE Transactions.* 352:357 (1962).
3. P. Chaudhuri, Single and double shading rings in single phase magnets: a general analysis, *IEEE Transactions on Power Apparatus and Systems*, Vol. PAS-89, No 4. 663:669 (1970).
4. C.J. Noh and I.R. Smith, Chattering conditions in electromagnetic contactors, *IEE Proceedings*, Vol. 127, Pt. B, No 5. 324:328 (1980).
5. S. Williamson and J.W. Ralph, Finite element analysis for nonlinear magnetic field problems with complex current sources, *IEE Proceedings*, Vol. 129, Pt. A. 391:395 (1985).

OPTIMIZATION OF INSULATORS USING A GENETIC ALGORITHM

João A. Vasconcelos[2], Laurent Krähenbühl[1] and Alain Nicolas[1]

[1]Centre de Génie Electrique de Lyon, URA CNRS 829 - ECL - France
[2]Electrical Eng. Department - UFMG - Brazil

INTRODUCTION

In the last years, several deterministic algorithms have been used in design optimization in electromagnetics[1-3]. The principal advantage of using such methods is that the number of function evaluations is small as compared to stochastic ones. Usually, the global solution is difficult to find in some cases and its application might require fastidious sensitivity analysis. The stochastic methods are global optimization techniques that require no gradient calculations. The genetic algorithm, denoted **GA**, is one of such methods although its application to design optimization in electromagnetics has been very recent[4].

This paper presents the optimization of insulator contours according to either the total or the tangential electric field strengths using a **GA** coupled with a **BEM** code. Results for a post insulator are presented and compared with those published by Däumling and Singer[5].

GENETIC ALGORITHMS

The genetic algorithm, developed by John Holland[6], is a powerful technique that is both easy to understand and to implement. It is based on mechanisms of natural selection and natural genetics, such as reproduction, crossover and mutation. It acts on a population of individuals in such a way that the new individuals created perform better than their predecessors. This population is simply a set of design configurations, where each design is a possible solution and it must be adequately represented to permit the execution of operations such as crossover, mutation, and inversion. Binary representation is the most frequently used.

The consequence of these operations are that new points to be tested on the optimization space are generated. From the generation **t** to **t+1**, individuals that have performed well have more possibility to be selected to participate in the new generation. This process is commonly known as reproduction. The criterion of selection is based on individual performance evaluated by fitness function.

In a simple **GA** code, an initial population of bit strings is first generated. The bit strings are decoded, the fitness function for each individual is evaluated and individuals are selected to participate in the next operation. After this, the natural operations are performed. The bit strings are decoded again, the fitness functions evaluated and the process repeated until some stop criterion is satisfied.

Reproduction

In this step, individuals are selected to participate in the next generation based on their fitness function values. The probability for an individual to be selected is proportional to the

Electric and Magnetic Fields, Edited by A. Nicolet
and R. Belmans, Plenum Press, New York, 1995

ratio of its fitness function value and the sum of this value for all other individuals in the population. A process based on the roulette wheel principle, where each individual probability represents an area, can be used to determine the strings that participate in the next process[7].

Crossover

The crossover process is feasible with a probability $\mathbf{p_{cros}}$ and the decision for its execution is taken by using a random number generator to generate a number uniformly between **0** and **1**. If the number generated is less than $\mathbf{p_{cros}}$, the operation is carried out. A second number, say **k**, between **1** and ***l*-1** is necessary to determine the crossover site, where *l* is the number of characters **0** and **1** in a bit string. Finally, the **0**'s and **1**'s on the right hand side of the crossover site are exchanged between themselves.

The crossover process is schematically illustrated in Fig. 1.a for two strings **A** an **B**, with **l=10** and the crossover site **k=4**.

Mutation

It is simply a random perturbation of the value of a bit in a string with a probability of mutation $\mathbf{p_{mut}}$. This process safeguards the search against the loss of valuable genetic characteristics during the reproduction and crossover process by introducing new genetic material and allowing for new configurations to be tested.

This process is carried out bit by bit. It is illustrated below for **k=5** where the decision for its execution is taken in a similar way as in the crossover operation.

Inversion

The process of inversion is simply defined as an inversion interior to a string, which is randomly selected with a probability $\mathbf{p_{inv}}$ and two points $\mathbf{k_1}$ and $\mathbf{k_2}$, where $\mathbf{1 \leq k_1 < k_2 \leq} l$, are determined by using a uniform random generator. This operation is illustrated below for a string **A** and for two points $\mathbf{k_1=5}$ and $\mathbf{k_2=7}$. Holland pointed out that inversion acts to enlarge the linkage between characters of a string.

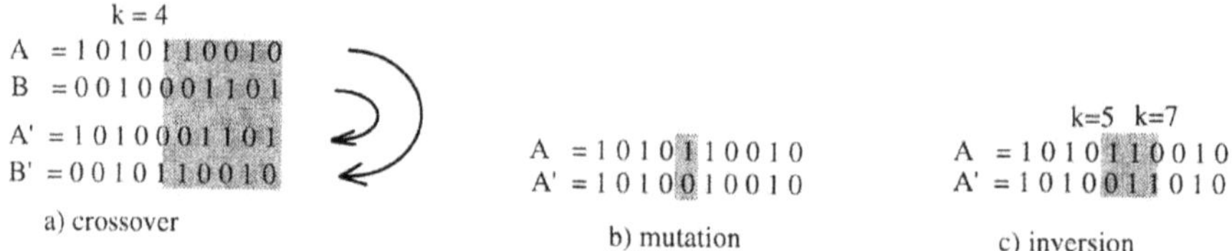

Figure 1. Natural operations.

COUPLING A GA WITH THE BEM

The boundary element method can be coupled with a genetic algorithm via a fitness function. Since the genetic algorithm is naturally stated in maximization form, with a non negative function in all domains of the problem, it is necessary to map out the natural objective function to a fitness function form through an adequate mapping. Using a transformation as in penalty methods, the original problem is easily rewritten as an unconstrained minimization problem, which can easily be transformed to a maximization form adequate to the genetic algorithm.

OPTIMIZATION OF A POST INSULATOR

Optimization of insulator contours has been effected using one of the following criteria: optimization with regards to the total or the tangential electric field strengths. Däumling and Singer[5] pointed out that the first criterion better improves the breakdown voltage than the second one.

The insulator[5] shown in Fig. (2) is optimized according to either the total (Etot-criterion) and tangential (Etan-criterion) electric field strengths. The objective function f(**x**) was determined in the following way,

$$\min \; f(\mathbf{x}) = \sum_{k=1}^{NT} (E_k^2 - E_{0k}^2)^2 \qquad (1)$$

$$\text{with } \; x^i_{min} \le x^i \le x^i_{max} \qquad i = 1,\dots,4$$

where **x** is the vector of design variables (Fig. 2). The E_k and E_{ok} *(tangential if Etan-criterion or vice-versa)* are respectively the calculated and specified electric field strengths on the **k** test point on the movable boundary.

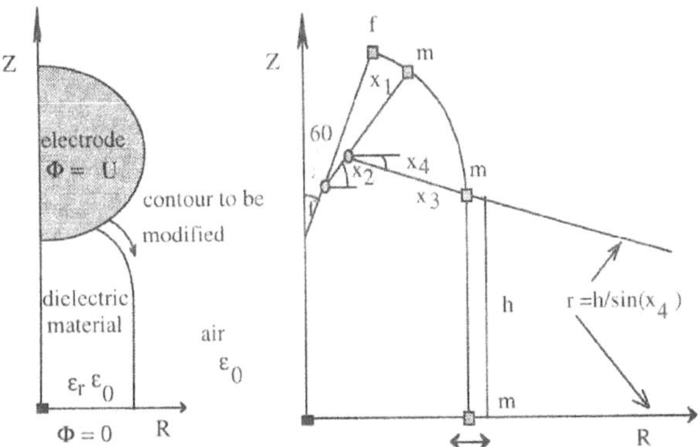

Figure 2. Post insulator and design parameters ($x_1,...,x_4$), where f denotes a fixed and m a movable point (right).

The genetic parameter values **NBPOP** *(number of bit strings)*, **NBGEN** *(number of generations)*, p_{cros}, p_{mut} and p_{inv} used to solve this problem were **10**, **10**, **0.9**, **0.025** and **0.025** respectively. Three optimizations using different criteria were performed. The first and second optimization used the uniform Etan and Etot-criterion and in the third one the maximum Etot-criterion was used, in which case the number **NT** and E_{ok} in (1) were made equals to **1** and **0**, respectively. Table 1 shows the results obtained and Fig. 3.a-c presents the curves for the total, tangential and normal electric fields along the movable boundary for the initial and the optimized geometries.

In all cases, the number of field calculations was kept constant and equal to 100. For the first case, Etan-criterion, the result obtained represents a reduction of 39 % and 15% on the maximum tangential and maximum total electric fields respectively. For the Etot-criterion case, these gains were 20 %. Finally, for the max.-Etot-criterion, Fig. 3.c, these gains were 25 and 22% respectively.

Table 1. Initial and final values for design variables and maximum electric field strengths *[( )[5] denotes the result obtained by Däumling and Singer for the PE insulator].*

	x_1 (mm)	Angle x_2	x_3 (mm)	Angle x_4	Etan. Max. (V/m)	Etot. Max. (V/m)
Inf. limit	0.01	36	0.01	0.8		
Sup. limit	0.09	54	0.09	60		
Initial	0.06	45	0.06	1	23.35 (24.9)[5]	23.35 (24.9)[5]
Etan-un. crit.	0.0100	36.51	0.0392	1.24	14.25 (13.7)[5]	19.92 (22.8)[5]
Etot-un. crit.	0.0313	40.27	0.0396	22.87	18.78 (18.5)[5]	18.78 (19.1)[5]
Etot-max. crit.	0.0139	45	0.0464	3.45	17.42	18.30

These results are in concordance with the results presented by Däumling and Singer[5] except in the Etan-criterion where the maximum total and normal electric fields obtained were inferior. This can be explained by the small number of parameters that were used to describe the geometry and the termination of the optimization process with a minimum value for the first parameter.

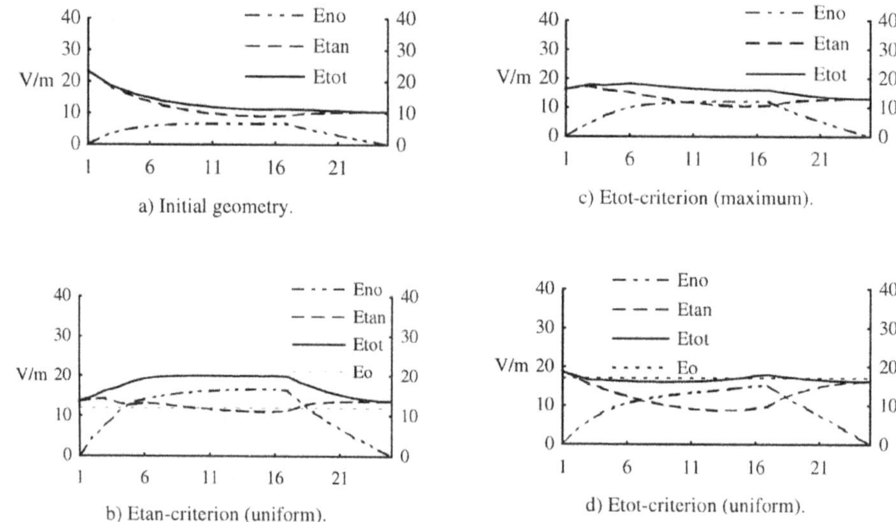

Figure 3. Electric field strengths along the insulator contour for the initial and the optimized ones.

CONCLUSION

The dielectric contour optimization results obtained using a **GA** coupled with a code based on **BEM** show that **GA** is a reliable tool for shape optimization in electromagnetics. Although only four parameters are used to describe the dielectric contour, the results are very close to those published by Däumling and Singer, who used charge simulation techniques with a strategy of boundary changes with continuity to achieve the optimum boundary. Also, shape optimization with different criteria, such as the normal, tangential or total electric field strengths *[uniform or maximum value]*, can easily be effected using a **GA** coupled with the **BEM**.

REFERENCE

[1] M. Defourny - "Optimization in electrostatics coupled with the boundary element method", 5rd ISHVE, paper No. 31.06, Braunschweig (1987).

[2] J. A. Vasconcelos, L. Krähenbühl, L. Nicolas, A. Nicolas- "Design optimization in electrostatic field analysis using the BEM and the augmented Lagrangian method", Compumag'93, Miami (1993).

[3] R. R. Saldanha, J. L. Coulomb, A. Foggia, J. C. Sabonnadiere - "A dual method for constrained optimization design in magnetostatic problems", IEEE Transactions on Magnetics, vol. 27, n° 5, pag. 4136-4141 (1991).

[4] J. A. Vasconcelos, L. Krähenbühl, L. Nicolas, A. Nicolas. - "Design optimisation using the BEM coupled with Genetic Algorithm", IEE-Proceedings CEM'94, Nottingham, UK (1994).

[5] H. H. Däumling, H. Singer - "Investigations on field optimization of insulator geometry", IEEE Trans. Power Delivery, Vol. 4, No. 1 (1989).

[6] J. H. Holland, "Adaptation in natural and artificial systems", MIT Press (1992).

[7] D. E. Goldberg, "Genetic Algorithms in search, optimization and machine learning", Addison-Wesley (1989).

SHAPE OPTIMIZATION OF AN HV CONNECTOR IN A GIS

J.A. Vasconcelos, L. Nicolas, F. Buret, A. Nicolas

Centre de Génie Electrique de Lyon - URA CNRS 829
Ecole Centrale de Lyon
BP163 - 69131 Ecully cedex - France

INTRODUCTION

The basic tools for the design of high voltage apparatus with regard to insulation are:

- the knowledge of the maximum rating value of the electric field, which depends on the nature of the dielectric medium.
- the methods to calculate electric field distribution.

The first point is one of the manufacturer's expertise gained by experimental work. The second point has been greatly improved during the recent years by the development of software packages which are able to compute precisely the E-field in 3D structures. Even with such tools, designer's work is still not easy. He has to find the shapes of the stressed parts so as to obtain an E-field lower than the maximum rating value with respect to the other constraints. In the critical region (the region where the field is maximum), electric field is very sensitive to minor variations in the geometrical parameters. Therefore this approach can require a lot of computations. In the final analysis, it is not certain that the exhibited solution is the best or that it cannot be improved.

This paper gives an example of automatic shape optimization processed by a 2D field software package using boundary integral method (BEM)[1].

HV CONFIGURATION

Description

The system studied in this work is a part of a three phases 145 kV GIS (gas insulated system). In this apparatus the three bars are in the same grounded enclosure filled with a gas under pressure (generally the gas is SF_6 or a mixture at 0.3...0.6 MPa).

More precisely, we are interested in the part of the system which connects the different sections of the buses (figure 1). This connector must have good electrical contacts, permit mechanical expansion due to thermal effects and prevent deformations due to electrodynamic forces during short circuit. Theses properties induce some constraints on the shape and the dimensions which must be taken in account during optimization process.

Electric and Magnetic Fields, Edited by A. Nicolet
and R. Belmans, Plenum Press, New York, 1995

Electric field

The original shape of the connector is shown in figure 2. This profile has been obtained by successive geometric modifications based on the experience of the designer.

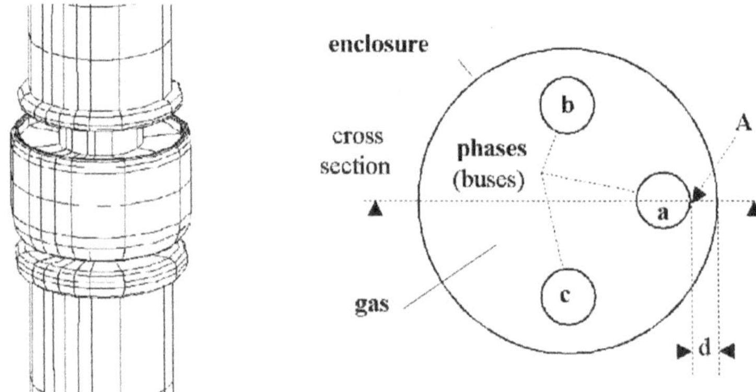

Figure 1. Structure of the GIS and 3D view of the connector.

The E-field is computed assuming that only one bar (a) is energized while the two others (b and c) are grounded to the enclosure. This configuration has to be tested according to the Standard Tests. Three computations have been performed :

- the actual configuration, modelled using a 3D software package using BEM (BEM3D)[2]. The connectors of the grounded phases (b) and (c) have however been replaced by cylinders with the same diameter.

- the axisymmetrical configuration, also modelled using BEM3D. This structure consists of the connector of the phase (a) and a cylindrical enclosure with same axis, the diameter of which permits keeping the same distance d (figure 1).

- the same axisymmetrical configuration, modelled using BEM2D[3].

The maximum value of E appears on connector (a) (figure1 and 2; point A) in front of the enclosure.

Table 1. Comparison of maximum E-field values. Potential of phase (a) = 1 volt.

BEM3D	BEM3D	BEM2D
actual 3D	axisymmetrical configuration	
46.0 V/m	48.3 V/m	49.0 V/m

Table 1 gives the results of the computations for point A. It is clear that this problem can be considered, with reasonable accuracy, as a 2D axisymmetrical configuration with regard to the maximum E-field. Note that in 2D, the external diameter is smaller than the real one in the 3D configuration, which is why the field values are higher in the axisymmetric computation.

The maximum computed value of the E-field corresponds to a withstand voltage of almost 675 kV (under lightning impulse with a SF_6 pressure of 0.4 MPa) which is above the demanded level (650 kV).

Nevertheless, it is interesting to optimize the shape and reduce the maximum value of the electric field, in order to raise the security margin of the insulation level or to try to minimize the size of the entire structure.

OPTIMIZATION

Constraints

The various functions that the connector must produce, induce the following constraints (figure 2):

- R, the outer radius of the connector is minimal in the original profile. We assume that its value cannot be changed and that no part of the connector can have a larger radius.
- distance d_1 is kept invariable, d_2 can be modified.
- distances d_3 and d_4 must be greater than 0.5 cm.

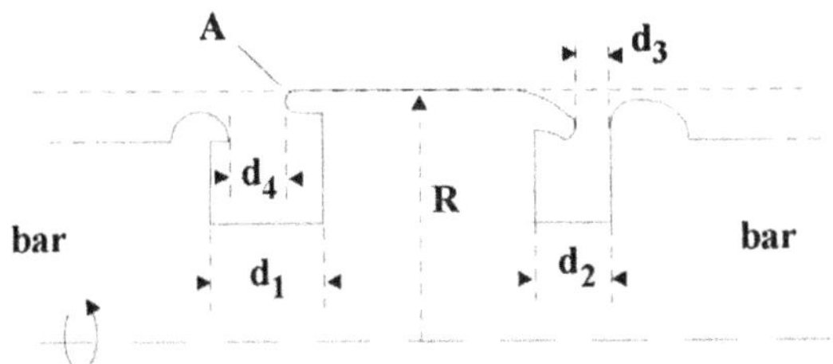

Figure 2. Initial profile of the connector (cross section defined in figure 1)

Optimisation

Optimization with E_{max} criterion: The objective we want to reach is an E-field value lower than 37 V/m (potential of (a) equal to 1 volt) everywhere on the connector.

Full optimization: In this case we attempt to minimize as much as possible the value of E with the same dimensional constraints as previously.

Results

Figure 3 shows the evolution of the profile of the connector for the first optimization (E_{max} criterion). This optimization is performed by seven field analyses and table 2 gives the values of the maximum E-field in the axisymmetrical configuration and in the actual 3D configuration.

Table 2. Maximum values of the electrostatic field. Potential of phase (a) = 1 volt.

BEM2D	BEM3D	BEM3D
axisymmetrical configuration		
36.9 V/m	36.8 V/m	34.9 V/m

The profile corresponding to full optimization is very close to this one and E-field is reduced to 35.5 V/m (figure 4) or 33.8 V/m in the 3D configuration. This last value corresponds to a withstand voltage of 920 kV.

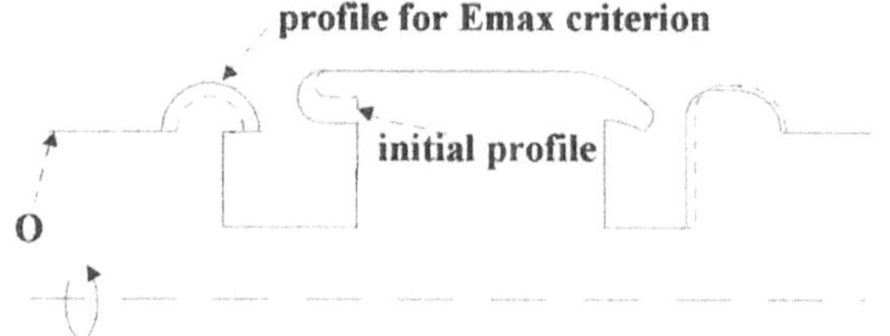

Figure 3. Modification of the profile of the connector.

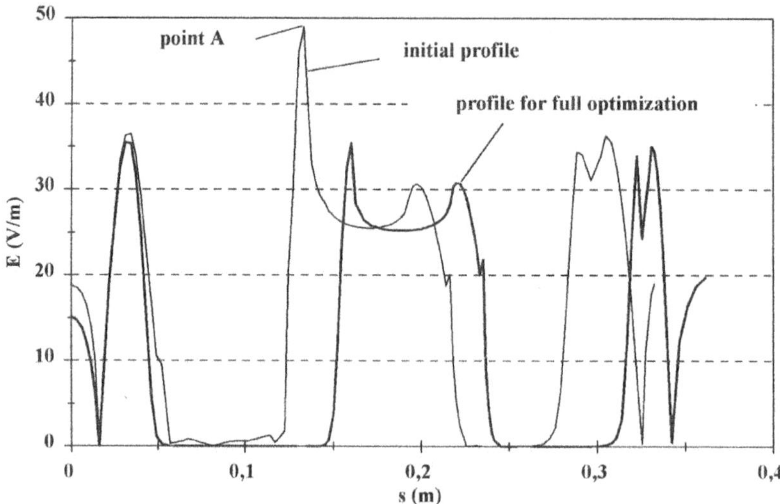

Figure 4. Comparison of the variations of the electrostatic field along the profile of the connector (computed with BEM2D). Point O (figure 3) is the origin for s.

CONCLUSION

This exemple shows clearly the advantages of the BEM2D package. However, it must be noted that, even when the constraints are established for the insulation design, the designer has to define how the profile can move. In this package the points which create the profile can be defined by parameters. The choice of the free parameters which can be optimized is quite free.

REFERENCES

1 J.A. Vasconcelos, L. Krähenbühl, L. Nicolas, A. Nicolas. Design optimization in electrostatic field analysis using the BEM and the augmented Lagrangian method, COMPUMAG, pb 1-15, Miami (1993).

2 L. Krähenbühl, A. Nicolas, L. Nicolas. The C.A.D. package PHI3D, for the computation of electric or magnetic fields in 3D devices - its validation, 3Dmag 89, Okayama, Japan - Compel, International Journal for Computation and Mathematics in Electric and Electronic Engineering - Vol 9, sup. A, pp. 185,189 (1990).

3 J.A. Vasconcelos, L. Krähenbühl, L. Nicolas, A. Nicolas. Potential and electric field computation at point interior to a domain by using the boundary element method, International Workshop on Electric and Magnetic fields, Liege (Belgium), paper 23.1 (1992).

THE DESIGN OF A MAGNETIC SHIELD IN A STRONG EXTERNAL FIELD BY FINITE ELEMENT ANALYSIS

Piergiorgio Sonato and Giuseppe Zollino

Gruppo di Padova per Ricerche sulla Fusione
Associazioni EURATOM - ENEA - CNR - Università di Padova
Corso Stati Uniti, 4 - 35020 Padova, Italy

ABSTRACT

The design of a double layer magnetic shield for the control unit of a mass spectrometer is presented in the paper. A F.E. analysis has been carried out, in order to take into account the non linear B-H curve of the shield soft magnetic material and to minimize the shield weight. The F.E. formulations adopted in both 2-D and 3-D analyses, the optimization algorithm, the main results of the analyses and the design choices are presented. A prototype shield has been built, its performances have been experimentally tested and the results are compared with those coming from the F.E. analyses.

INTRODUCTION

RFX[1] is a fusion device where the plasma current, up to 2 MA, is inductively produced and sustained through an air core transformer. The plasma represents the secondary winding of the transformer. The absence of an iron magnetic circuit makes the magnetic field all around the torus non negligible[2]. In the torus hall a magnetic field that can reach 100 mT, with a gradient in the range of 0.1 T/m, is present, so many equipments require an effective magnetostatic shielding. The shielding produces a perturbation of the magnetic field that must be as little as possible in order to reduce the effects on the plasma equilibrium and stability. So that the shield must be minimized in terms of volume and equivalent magnetic reluctance.

In the paper the design, the optimization and the tests of the shield for a control unit (CU) of a mass spectrometer are presented; the magnetic field peak value in the CU position is approximately 25 mT. Experimental tests showed that the CU is sensitive to magnetic field over 0.5 mT, so a very effective magnetic shield, with a shielding factor above 50, is required. The optimization of the shape and of the wall thicknesses is made with a 2-D analysis, while the final design of the shield including the cover plate is made with a 3-D analysis. In the 3-D design the criteria and the consequences of the different solutions for the design are presented. In the design a uniform external magnetic field has been imposed corresponding to the maximum induction on the shield position.

TWO-DIMENSIONAL ANALYSES AND OPTIMIZATION

A number of 2-D finite element analyses were carried out in order to decide whether the shield had to be designed with two or three layers; in fact it is well known in the theory of magnetic shielding[3] that a better efficiency can be achieved by using a system consisting of multiple thin layers than of a single thick layer. The same analyses allowed to choose the magnetic materials for the shield. At first guess values were chosen for the dimensions of the shield and the thickness of each layer, referring to simplified formulas which hold for cylindrical multiple shields[3,4]. Since the shield is perpendicular to the external field the cylindrical shield which was referred to had the cross section with the same perimeter as the parallelepiped shield. Afterwards a design optimization was executed.

Electric and Magnetic Fields, Edited by A. Nicolet
and R. Belmans, Plenum Press, New York, 1995

Three different arrangements were considered: three layer shield, with the outer and intermediate layers made of soft iron and the inner layer made of very high permeability alloy (mu-metal); two layer shield like the previous one but without the intermediate layer; two layer shield completely made of soft iron.

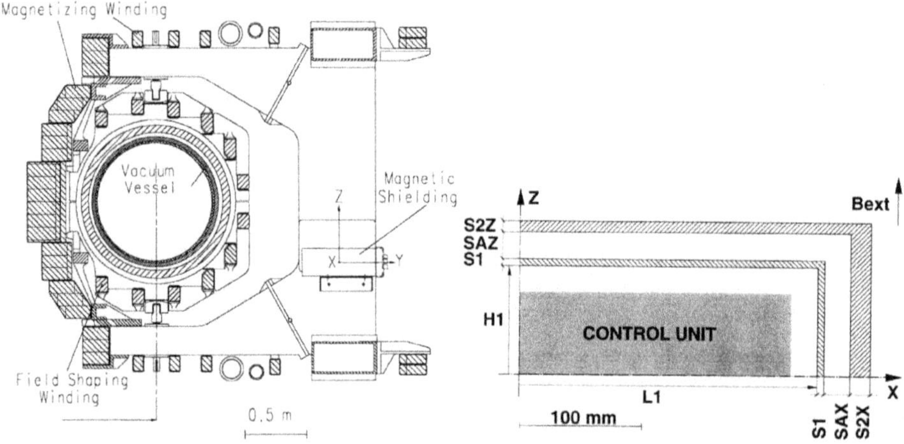

Figure 1. Cross section of the RFX fusion device showing the position of the magnetic shield (left) and 2-D model of the shield showing the design variables used in the optimization.

Finite element model

The 2-D magnetostatic option of the Ansys code[5] is based on the vector potential (out of plane component A_Z only) formulation; four-node and eight-node quadrilateral elements are available. A mesh representing a magnetic shield usually includes oblong elements in the iron region, so higher-order element types should be preferred to improve the accuracy of the results. The mesh of the 2-D model contains about 830 eight-node elements and 2600 nodes for the two layer shield and about 900 elements and 2850 nodes for the three layer shield. Symmetry allowed to limit the model to only 1/4 of the shield. The external magnetic field, which was assumed directed along the y-axis, was taken into account by proper boundary conditions on the potential of the nodes lying on the symmetry planes and on the far boundary planes. The number of degrees of freedom of the problem was 2490 for the two layer shield and 2730 for the three layer shield).

A non linear analysis was carried out by taking into account the first magnetization curve of the ferro-magnetic materials. Such curves were deduced from data sheets given by the manufacturer: special care was taken in linearizing the curves at low m.m.f. (no point of inflection should be left in the curve in order to achieve a proper convergence). The full Newton-Raphson procedure was chosen and the problem was solved in two steps. Convergence was achieved after a number of iteration which depended on the thicknesses of the ferro-magnetic layers, i.e. on the working point of the magnetization curve (the number varied from 6 to 13 during the optimization). In all cases the convergence was based on A_Z and the convergence criterion was set to 10^{-6}. The solution (based on the frontal equation solver of the code[5]) took about 20 s for each iteration on a VAX-station 4000/90, for the two layer cases and 22 s, for the three layer case.

Design optimization

The above mentioned model was defined parametrically to allow a design optimization. The optimization procedure of the Ansys code was used to minimize the shielding cross section area and thus the perturbation produced by the shield on the external field. Details on the procedure can be found in reference 4: however attention should be drawn to the fact that at each optimization loop (OL) the value of the i-th design variable (DV) to be used in the next loop, $X_{i,\ell+1}$ is computed as

$$X_{i,\ell+1} = X_{i,best} + A\,(X_{i,\ell} - X_{i,best})$$

where $X_{i,best}$ is the value of the i-th DV used in the best design up to the current design loop, $X_{i,\ell}$ is the value of the i-th DV computed by minimizing the ℓ-th approximate problem and

$$A = 1 - C_{best} - C_r C_i$$

(C_{best} is the best design contribution fraction, and it can be chosen in the range $0< C_{best}<0.9$; C_r is the random contribution fraction, and it can be chosen in the range $0 < C_r < 1-C_{best}$; C_i is a random number, in the range -0.5, 0.5). By varying C_{best}, $X_{i,\ell+1}$ can be biased towards $X_{i,best}$ or $X_{i,\ell}$; by varying C_r the amount of randomness in the stepping can be changed, making it easier to avoid a local minimum. For these analyses C_{best} was set to 0.25 and C_r to 0.3.

A design optimization was performed for both two layer shields. Seven design variables were considered (see Figure 1): the thickness of the inner layer (S1), the thickness of both the vertical and the horizontal side of the outer layer (S2X, S2Y), the distance between the two layers in the x and y direction (SAX,SAY), the width (L1) and the height (H1) of the inner layer. As state variables the values of the induction inside both layers and in the region of the CU were chosen and proper constraints were introduced in order to keep the ferro-magnetic materials below saturation and the field on the CU below 0.5 mT. The cross section area of the shield was the objective function. The optimal design was achieved (see reference 4 for the convergence conditions) after 25 OLs, for the shield with the mu-metal layer and after 21 OLs, for the soft iron shield. A further optimization, starting from the optimized values of the DVs, confirmed that the achieved constrained minimum was not a local one. For both optimized shields the area was reduced by approximately 25% with respect to the first design. The optimal values for the width and the height of the shield resulted 264 and 108 mm respectively. The perturbation on the external field was almost the same in both cases (as a measure of the perturbation the integral of the x-component of the field outside the shield for y=0 was considered) and was reduced after the optimization by approximately 15%.

Main results

Table 1 reports the results of the first analyses for three shielding arrangements and the results after the optimization for two of them. The three layer shield, in spite of higher complexity and cost, does not achieve much better results than the two layer shields. For this reason the last two options were preferred and optimized. In the same table, as a comparison, a single layer soft iron shield, with the same weight as the two layer soft iron one was considered. It resulted less effective than the equivalent two layer shield, in accordance with the analytical results which hold for a cylindrical or spherical shielding system[3]. At last the two layer soft iron shield, which is the simplest and most economic solution among those fulfilling the constraints, was chosen for the 3-D analyses and the detailed design.

Table 1. Main results of the two-dimensional analyses.

Shielding arrangement	Shield cross section area	Induction on the CU
3 layer shield (first design)	52.4 cm^2	8 10^{-6} T
2 layer shield: outer soft iron, inner mu-metal		
(first design)	47.5 cm^2	1.8 10^{-5} T
(optimized)	35.2 cm^2	3.9 10^{-5} T
2 layer soft iron shield		
(first design)	50.5 cm^2	2.5 10^{-4} T
(optimized)	38.5 cm^2	4.5 10^{-4} T
1 layer soft iron shield	38.5 cm^2	9.8 10^{-4} T

THREE-DIMENSIONAL ANALYSES

For the actual design of the shield the thicknesses coming from the optimization were changed into the closest commercially available ones. Low carbon iron (ARMCO) annealed at 950 °C was chosen (and its magnetization characteristic used in the analyses). The following changes had to be done: S1 from 2.6 to 3 mm, S2X from 8.5 to 8 mm, S2Y from 5.2 to 5 mm. The depth the shield should have in the z-direction depends on whether it is closed at both ends or not: in fact edge effects at both ends have to be taken into account. The 3-D analyses showed that to keep the field on the CU region below the design value an open shield should have a depth not compatible with the overall dimensions of the RFX torus assembly: so a closed shield was designed.

The 3-D magnetostatic formulation of the Ansys code based on the total scalar magnetic potential was used: to make it possible to employ the total potential the external field was represented by means of proper boundary conditions instead of including in the model the actual current sources. Thanks to simmetry only 1/8 of the shield and the surrounding air were represented; the whole mesh was made up of 6520 8-node hexahedron elements and 7440 nodes (6790 degrees of freedom). The solution took 290 s for each iteration on the same CPU mentioned above: convergence was reached after 7 iterations for the shield with open ends and

5 iterations for the shield with closed ends (as far as the solution is concerned the same choices than in the 2D analyses were done, but the problem was solved in only one step and the convergence was based on the flux[5]).

For the open shield the maximum value of induction on the outer layer was higher than for the infinitely long equivalent shield (1.5 T against 1.3 T for the 2-D model), i.e. the magnetic material works closer to saturation. In fact flux lines deflect towards the shield also in the y-z plane. On the countrary for the closed shield the same value was lower than in the 2-D model (1 T). In the latter case the shield section perpendicular to the field direction is higher than in the former, but the increment in the total flux which flows through the shield is less than the section increment. The magnetic material works far from saturation and the permeability is higher. Fig. 2 shows the magnetic induction on the CU region for the closed shield: the maximum value is 0.27 mT, whereas for the open shield the maximum value was 1.9 mT.

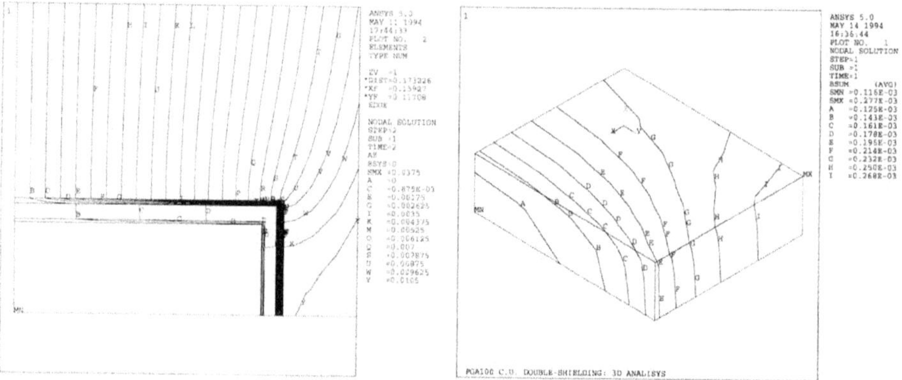

Figure 2. Magnetic flux lines for the two layer soft iron shield after the optimization (left) and magnetic induction on the CU region for the closed shield.

DETAILED DESIGN

The shield must allow access to the CU, so to close the front end of the shield a removable cover was designed. The vertical component of the magnetic field in the region where the shield will be mounted reaches 25 mT only at the side opposite to the RFX main axis (see Figure 1) and reduces by a factor five at the other side: for this reason the actual shield was designed with no cover at that side. The cover is provided with holes for forced ventilation: a detailed design was carried out for the front end cover and for the shields of the holes. For the last detail analytical formulae were used since the shields consist in single layer cylinders. For the first one again 2-D analyses were performed in order to choose the best arrangement. A simple plane cover screwed on the outer layer at the end of the shield and a cover which overlaps the outer layer (see Figure 3) were considered. In both cases the unavoidable air gap between the cover and the shield was taken into account: it was assumed an average air gap thickness of 1 mm for the overlapped cover and 1/2 mm for the plane cover. Annealing at 950 °C in fact usually prevents from achieving high precision in spite of accurate machining. In the case of the overlapped cover a larger air gap was foreseen to allow the cover to be more easily removed. In Figure 3 the effectiveness of the overlapping in reducing the magnetic flux inside the shield due to the reluctance of the air gap can be observed. It was found that an overlapping of 50 mm reduced the field in the region of the C.U. below 0.5 mT, likewise a closed shield. This result strictly depends on the amplitude of the air gap and on the permeability of the ferro-magnetic material: if the permeability could be assumed as a constant, the air gap reluctance for the plane cover is higher by approximately a factor five than the reluctance for the overlapped cover. This means that in the last case a higher shielding effect can be expected, but in order to compute the magnetic flux inside the shield in both cases the reluctance of the whole flux paths must be known. A F.E. analysis was so necessary for an accurate design of the cover.

EXPERIMENTAL TESTS

A full size prototype shield was built and the standard procedure for annealing at 950 °C was followed after machining. Experimental tests were carried out during RFX plasma pulses at a magnetizing current of 21 kA (the full performance value will be 50 kA, and the shield was designed referring to this current value). The prototype shield was positioned at the equatorial plane of the RFX device at a proper distance from the machine axis, with the symmetry plane y-z (see Figure 1) coincident with a RFX poloidal plane, so that the

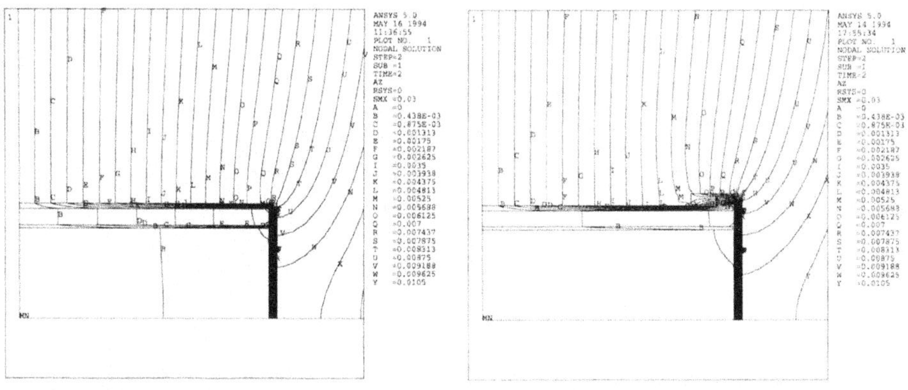

Figure 3. Magnetic flux line for the cover without overlapping (left) and with overlapping.

average induction value on the shield was about 22 mT and it could be assumed as vertical. In order to reproduce the conditions of the numerical analyses, the tests were referred to the open shield (the conditions of the closed shield were not reproducible with only one cover). A three axis pick-up coil was used to measure the magnetic field components inside and outside the shield, by integrating the coil signals by a digital oscilloscope. The value of the induction was measured at 3 points inside the shield on the y-axis (see Figure 1): due to the symmetry of the field and the shield only two field components (vertical and radial) could be measured. The field immediately above at the centre of the shield was measured as well. Each measure was repeated at least three times. Table 2 reports the vertical field component measured during the experimental tests and the corresponding computed value. A good agreement can be observed between measured and computed values, but the prototype shield seems to be more effective than it was expected from calculations. This is partially due to the actual field distribution all around the shield: along the y-axis the actual field is not uniform as in the computations, but it decreases at both side of the shield; the second reason which justifies the higher measured shielding effect is the actual permeability of soft iron which is higher than in the F.E. analyses. The magnetization characteristic was in fact linearized at low field so reducing the value of permeability: the working point of most of the magnetic material is just on the linearized piece.

Table 2. Comparison between analytical results and experimental measurements.

Position	Measured	Computed
Inside the shield on the y-axis (at y=0)	0.73 10^{-3} T	0.92 10^{-3} T
(at y= 20 cm)	0.87 10^{-3} T	1.1 10^{-3} T
(at y= 25 cm))	1. 10^{-3} T	1.15 10^{-3} T
Above the shield at the centre	27.3 10^{-3}	26.1 10^{-3} T

REFERENCES

1. P.Kusstatscher et al., Proc. 17th Symposium on Fusion Technology, Roma, Italy, 1992, pp. 717-721.
2. M.Fauri. P.Sonato, Finite Elements in Analysis and Design, Elsevier Publishers, vol.10, 1991, pp.27-39.
3. T.Rikitake, "Magnetic and Electromagnetic Shielding", Terra Scientific Publishimg Company, Tokyo, 1987.
4. S.P.K. Tavernier, F.Van Den Bogaer, L.Van Lancker, Nuc. Instr. Meth., 167 (1979) 391.
5. P.Kohnke, ANSYS User's Manual - Theory rev.5.0, Swanson Analysis System, 1992.

CALCULATION OF THE ELECTRIC AND THE MAGNETIC FIELD GENERATED BY BUSBAR SYSTEMS

R. Mertens, R. Belmans

Dept. E.E. - Electrical Energy
K.U. Leuven
Kardinaal Mercierlaan 94
B-3001 Heverlee - Leuven
Belgium

INTRODUCTION

More and more people ask questions on eventual influence of electric and magnetic fields. Therefore, accurate numerical prediction methods for geometrically difficult configurations are important to compare with measured values. In this way standards can be checked during design. For magnetic fields the International Radiation Protection Association (IRPA) proposes the limiting values as shown in table 1.

Table 1. Limiting values for magnetic fields.

Workers	• 25 mT for limbs • 5 mT for a maximum exposure of 2 hours a week • 0.5 mT for exposure during a whole day
General Public	• 1 mT for a maximum exposure of a few hours a week • 0.1 mT for exposure during a whole day

The paper describes how the magnetic field generated by busbar systems and other electrical installations can be calculated using the finite element method. Since no symmetry is found in general, the use of the finite element method in three dimensions is necessary.

Electric and Magnetic Fields, Edited by A. Nicolet
and R. Belmans, Plenum Press, New York, 1995

CALCULATION OF THE MAGNETIC FIELD STRENGTH

The magnetic field generated by currents sinusoidally varying in time, at a given point in space is represented by

$$\begin{aligned}\vec{b}(x,y,z,t) = {} & \hat{B}_x(x,y,z)\cos(\omega t+\gamma_x(x,y,z))\vec{e}_x \\ & +\hat{B}_y(x,y,z)\cos(\omega t+\gamma_y(x,y,z))\vec{e}_y \\ & +\hat{B}_z(x,y,z)\cos(\omega t+\gamma_z(x,y,z))\vec{e}_z \end{aligned} \tag{1}$$

All materials are assumed to be linear. If the values of the magnetic field are low enough there are no important saturation effects. (1) is also the description of the magnetic field ellipsoid. Calculating the instantaneous values at two instants in time and using the working scheme of figure 1 is sufficient to know the RMS-values of the three space components (λ can be either x, y or z).

$$\left.\begin{aligned} t_0 = 0 & \rightarrow b_{\lambda,0}(x,y,z) \\ t_1 = \frac{T}{4} & \rightarrow b_{\lambda,1}(x,y,z) \end{aligned}\right\} \rightarrow \gamma_{\lambda(x,y,z)} \rightarrow \hat{B}_\lambda(x,y,z) \rightarrow B_\lambda(x,y,z)$$

Figure 1. Calculation of the RMS-value of the three space components x,y and z.

The calculation of the instantaneous values is done by the finite element method. The instantaneous values of the currents are defined in the problem definition for the model.

The maximum value of the magnetic field strength, the RMS-value of the semimajor axis magnitude of the magnetic field ellipsoid, can be calculated from the RMS-values of the three space components.

CHOOSING A BUSBAR SYSTEM

As an example a busbar system of an existing electrical installation is chosen: a 36 kV distribution system of a large chemical factory. The function is to deliver the electric energy to individual local consuming units. Therefore the system is split up in cells. To achieve the desired level of reliability each unit can be connected to one of both feeders by means of the necessary switchgear. Modelling the entire system is quite difficult, as the size of the model is limited by the finite element program. Only the interested cell is modelled and the mutual influence of neighbouring cells is neglected. The model is 7.1 m high, 2 m wide and 5.8 m deep. Figure 2 shows the conducting region in one cell.

CALCULATIONS VERSUS MEASUREMENTS

The interested cell is the main supply for the first feeder which is in the front part of the cell. The current was 910 A at the moment of the measurements. 464 A was running to the left and 446 A to the right. The second feeder, in the back part of the cell, carried a current of 237 A. The magnetic field was calculated and measured in front of the cell, in de middle between the cell and the wall and at a height of 1 m above the ground. Figures 3, 4 and 5 show the calculated values (solid line) and the measurements (dots).

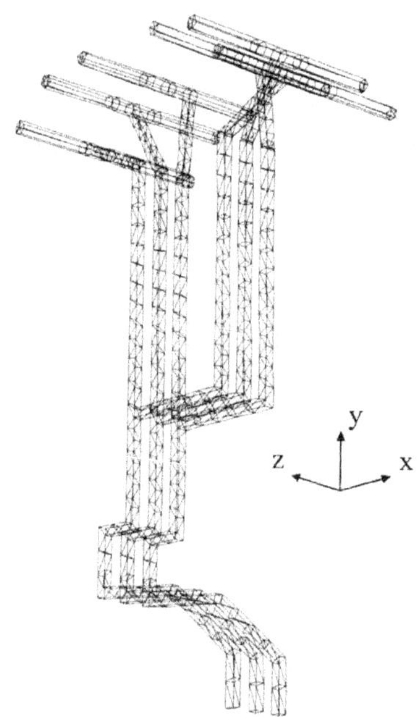

Figure 2. Conducting regions in one cell.

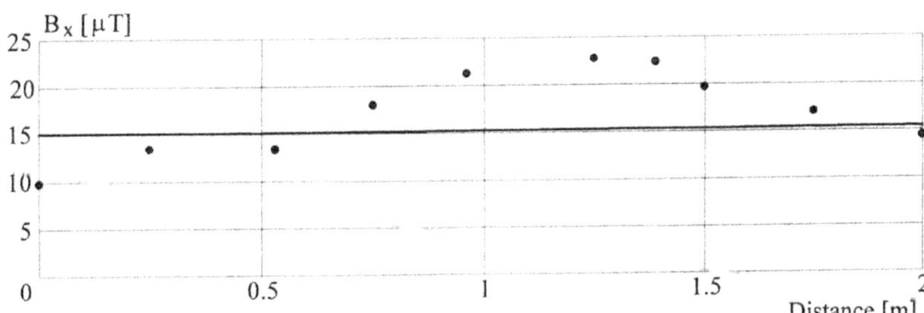

Figure 3. RMS-value of the x-component of the magnetic field.

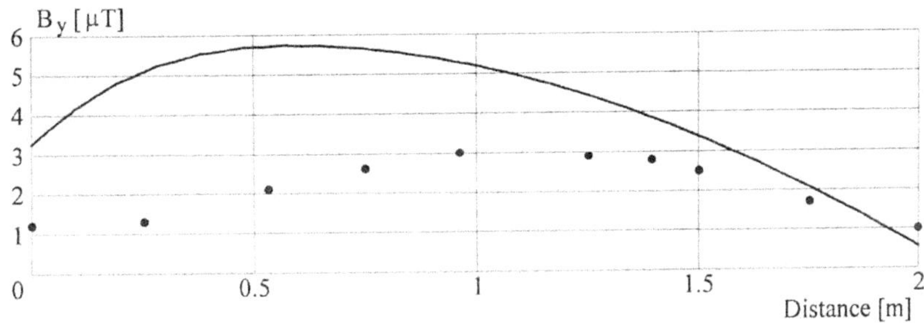

Figure 4. RMS-value of the y-component of the magnetic field.

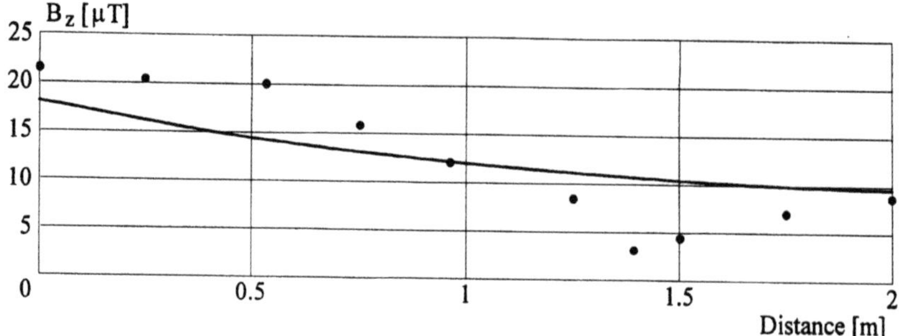

Figure 5. RMS-value of the z-component of the magnetic field.

CONCLUSIONS

The difference between calculations and measurements is unacceptable, although both values are far below the limiting values proposed by IRPA. Therefore further investigation is necessary, not only to improve the calculations for this particular problem but also to make it possible to generalize the results. The importance of this kind of problems is increasing due to the interest paid to it by the general public.

Earlier calculations of the electric and the magnetic field surrounding a high voltage overhead line showed good agreement between the calculations and the measurements. Therefore it is necessary the improve the finite element model. This can be done by making the model in better agreement with the real situation and by dealing with the mutual influence of neighbouring cells.

ACKNOWLEDGEMENTS

The authors are grateful to the Belgian Nationaal Fonds voor Wetenschappelijk Onderzoek for its financial support of this work and the Belgian Ministry of Scientific Research for granting the IUAP No. 51 on Magnetic Fields.

REFERENCES

Holaday Industries - 3604, "ELF Field Strength Measurement System: User Manual".

IEEE Standards Board / American National Standards Institute / Institute for Electrical and Electronics Engineers, 1987, "IEEE Standard Procedures for Measurement of Power Frequency Electric and Magnetic Fields from AC Power Lines", ANSI / IEEE Std 644 - 1987.

Infolytica Corparation, "MagNet 5 Reference Manual".

International Radiation Protection Association (IRPA), "Interim Guidclines on Limits of Exposure to 50 / 60 Hz Electric and Magnetic Filcds", Health Physics 58, p. 113 - 122.

Silvester, P.P., Ferrari, R.L., 1990, 'Finite Elements for Electrical Engineers", Cambridge University Press, 2nd Edition.

3-D MAGNETIC FIELD ANALYSIS OF EPSTEIN FRAME FOR MEASURING MAGNETIC CHARACTERISTICS

T. Nakata, N. Takahashi, K. Fujiwara, M. Nakano,
H. Ohashi and H.L. Zhu

Department of Electrical and Electronic Engineering
Okayama University
Okayama 700, Japan

INTRODUCTION

The effective magnetic path length L_m of the Epstein frame is specified as 94cm by the IEC (International Electrotechnical Commission) standard.[1] In order to measure accurately the magnetic characteristics, L_m which is a function of the flux density and quality of specimen should be determined exactly. As the magnetic circuit of the Epstein frame is three-dimensional, 3-D numerical analysis taking account of nonlinearity is necessary.

In this paper, the effects of flux density and quality of specimen on L_m are investigated using the 3-D finite element method with edge element.[2, 3]

ANALYZED MODEL FOR EPSTEIN FRAME

Fig.1(a) shows the model to be analyzed. The thickness of the specimen is 0.35mm. The interlaminar gap between specimens at the corner is assumed as 0.011mm (space factor : 97%). Fig.1(b) shows the analyzed region. It is assumed that the number of stacked specimens is infinite. The positions of the voltage winding (B coil) and the magnetizing winding are shown in Fig.1(c). The magnetizing winding is excited from a dc source.

ANALYSIS

The magnetic field is analyzed by using the 3-D finite element method with edge element. The unknown variable is the magnetic vector potential. As the average flux density in the specimen is specified, the flux Φ linking the voltage winding is given. The vector potential and the magnetizing current are unknown variables. Therefore, the fundamental equation (1) for magnetic field calculation and the relationship between the flux Φ and the magnetic vector

Electric and Magnetic Fields, Edited by A. Nicolet
and R. Belmans, Plenum Press, New York, 1995

potential $\boldsymbol{A}$ shown in Eq.(2) have to be solved simultaneously.

$$rot\,(\,\nu\, rot\,\boldsymbol{A}\,) = \boldsymbol{J}_0 \tag{1}$$

$$\Phi = \frac{n_B}{S_B}\iiint_{\Omega_B} \boldsymbol{A}\bullet \boldsymbol{t}_B dv \tag{2}$$

where ν and $\boldsymbol{J}_0$ are the reluctivity and the magnetizing current density in the magnetizing winding respectively. Ω_B is the region of the voltage winding. S_B and n_B are the cross-sectional area and the number of turns of the voltage winding respectively. $\boldsymbol{t}_B$ is the unit tangential vector parallel to the magnetizing current density vector.

Fig.2 shows the boundary condition. The discretization data is shown in Table 1. The specimen is discretized into 3 layers in the z-direction.

The effective magnetic path length L_m is obtained by

$$L_m = NI \,/\, H_m \tag{3}$$

where NI is the ampere turns. The average magnetic field strength H_m in the specimen is defined as

$$H_m = \iiint_{V_c} |\,H\,|\, dv \Big/ \iiint_{V_c} dv \tag{4}$$

where V_c is the region of the specimen within the winding.

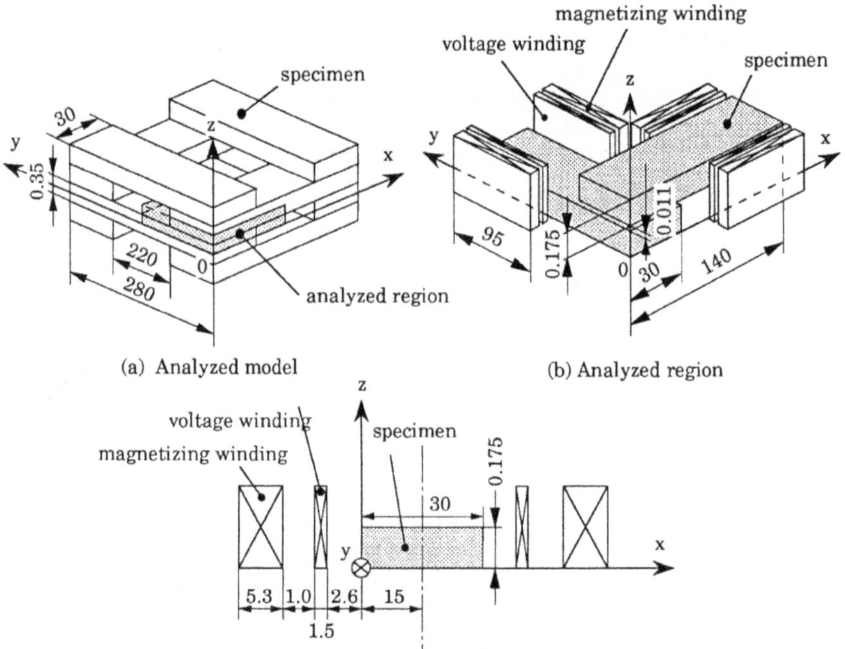

Fig. 1 Analyzed model.

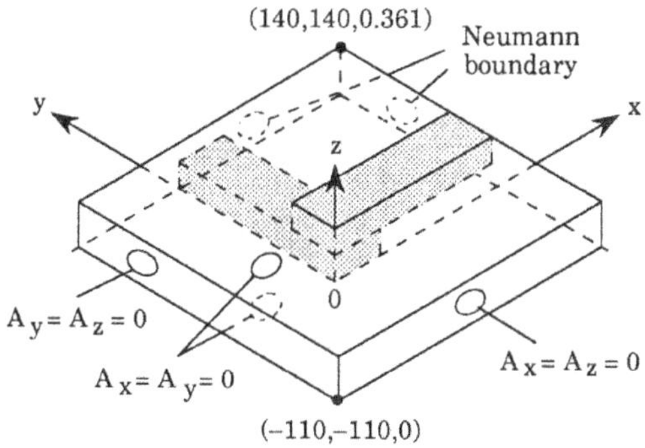

Fig. 2 Boundary conditions.

Table 1. Discretization data and CPU time

number of elements *1	11,552
number of nodes	13,689
number of unknowns	31,769
number of non-zeros	490,096
memory requirement (MB)	12.6
number of nonlinear iterations *2	8
number of iterations for ICCG method *3	2,181
CPU time for ICCG method (s) *4	881
total CPU time (s) *4	967

*1 element type : 1st-order brick edge element
*2 convergence criterion for Newton-Raphson method : 0.01(T)
*3 convergence criterion for ICCG method : 10^{-5}
*4 computer used : IBM 37T workstation (25 MFLOPS)

RESULTS AND DISCUSSION

Fig.3 shows the flux distributions near the corner in the two kinds of specimens. In the non-oriented silicon steel (AISI:M-15), the flux concentrates to the window side. In the highly grain-oriented silicon steel (AISI:M-1H), the flux flows in the rolling direction.

Fig.4 shows the effects of the flux density B and quality of specimen on L_m. The specified value of L_m in the IEC standard, which is equal to 94cm, is shown in the chain line. L_m of the non-oriented silicon steel under the dc excitation is lower than 94cm and that of grain-oriented silicon steel.

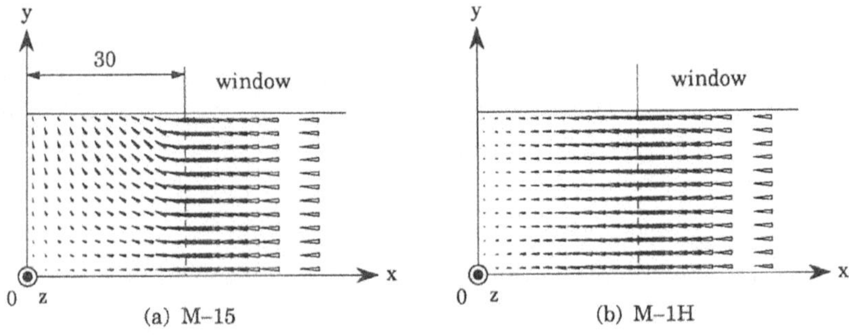

Fig. 3 Distributions of flux density vectors near corner (z=0.33mm).

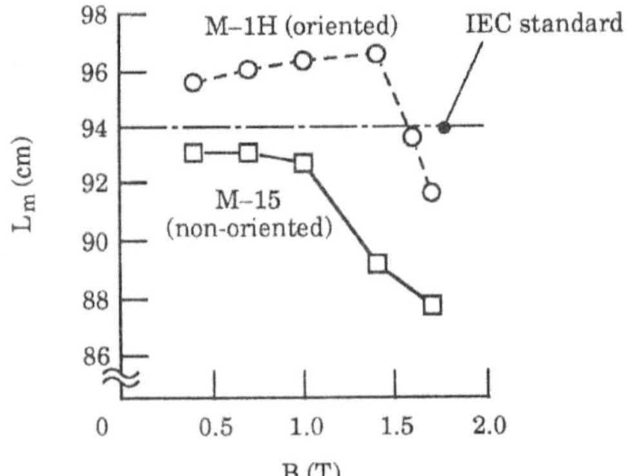

Fig. 4 Effect of flux density B on effective magnetic path length L_m .

CONCLUSIONS

It is shown that the effective magnetic path length L_m is affected by the flux density and quality of specimen. L_m for the non-oriented silicon steel is lower than that of grain-oriented silicon steel.

REFERENCES

1. IEC (International Electrotechnical Commission) 404-2 (1978).
2. T.Nakata, N.Takahashi, K.Fujiwara and Y.Shiraki, Comparison of different finite elements for 3-D eddy current analysis, *IEEE Trans. Magnetics*, vol.MAG-26, no. 2, pp. 434-437 (1990).
3. K.Fujiwara, 3-D magnetic field computation using edge element (invited), *Proceedings of the 5th IGTE Symposium on Numerical Field Calculation in Electrical Engineering*, pp. 185-212 (1992).

OPTIMAL DESIGN AND CONTROL OF AN INDUCTION HEATING-SYSTEM

M.Z. Liu, J. Fouladgar, Y.M. Li and A. Chentouf

Laboratoire de Recherches en Techniques Inductives
C.R.T.T. B.P. 406
44602 Saint Nazaire, France

Abstract A classical optimisation method is used to design the inductor form of an induction heating system. The inductor generates a desired magnetic field at the surface of a metallic load.

The finite element method is then used to solve the heat transfer equation and define the state equations binding the temperature distribution inside the load and the current in the inductor. From the state equations is calculated the inductor current which leads to a homogeneous temperature field in the load.

1 INTRODUCTION

For a heat treatment, the temperature variation in the material is often bounded by strict conditions such as the homogeneity of the temperature inside the load, the upper temperature boundary and the heating time.

Now the temperature distribution inside the load depends especially on the geometry of the system and the current variation in the inductor. For a given geometry it is not always possible to satisfy the temperature constraints only by making the inductor current vary. So one should solve the problem by determining both the inductor current and the inductor form. This requires the solution of the inverse electromagnetic and heat transfer problem.

The inverse problem has been addressed in some recent publications[1,2]. They use the classical optimisation technique for solution. The FEM is used to solve the electromagnetic and heat transfer equation and an optimisation technique such as gradient method is used in an iteration process to solve for the required optimised geometry.

In this paper we propose a classical optimisation method to determine the inductor form of an induction heating system for the heat treatment of the steel wheels.

Electric and Magnetic Fields, Edited by A. Nicolet
and R. Belmans, Plenum Press, New York, 1995

The inductor form is calculated so as to generate a desired magnetic field at the surface of the steel wheel. The FEM. is then used to solve the heat transfer equation and to define the state equations binding the temperature distribution inside the load and the current in the inductor. Having the state equations, one can determine the inductor current variation by applying an optimal control technique[3,4].

2 PROBLEM STATEMENT

Figure 1 demonstrates an induction heating system for the heat treatment of a steel wheel. It uses a planar circular inductor with practically the same dimension as that of the steel wheel.

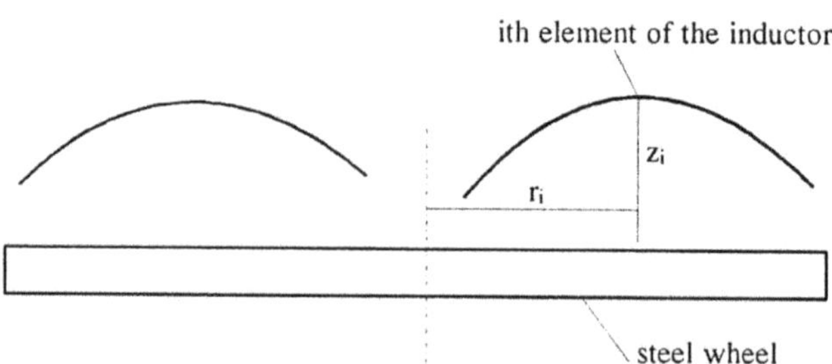

Figure 1 induction heating system studied

The heating of the steel wheel must be carried out under following conditions:

- at the end of heating, the temperature must be homogeneous and near to a desired value T_0 everywhere in the steel wheel, and ever be over T_{max} during the heating.
- the heating time of one steel wheel is limited to a few seconds.

If all the coils of the inductor are at the same distance from the steel wheel, the magnetic field at its surface is not uniform. This will lead to an inhomogeneous temperature in the steel wheel and increase the total heating time required to homogenise the temperature distribution. The idea is then to modify the distance between the individual coils and the steel wheel so as to generate an uniform magnetic field at the surface of the steel wheel.

3 SOLUTION PROPOSAL

3.1 Inductor Form

If one neglects the spiral form of the inductor, the geometry of the system is axisymmetric. So if r_i and z_i are the cylindrical co-ordinates of the coil n°i of inductor, the magnetic potential vector at each point (r, z) is given by

$$A_{i,k} = \frac{\mu_0}{2\pi} \iint_{S_i} J_s \frac{\sqrt{(r_i + r_k)^2 + z_i^2}}{r_k} \left[(1 - \tfrac{w^2}{2}) f_1(w) - f_2(w) \right] dS_i \tag{1}$$

where $A_{i,k}$: vector potential at the surface of kth element of the steel wheel generated by i^{th} inductor coil,

J_s : current density in the inductor,

$w = 2\sqrt{\frac{r_i . r_k}{(r_i + r_k)^2 + z_i^2}}$

r_i, r_k: rayon of the ith inductor coil and k^{th} element of the steel wheel,

$f_1(w)$, $f_2(w)$: complete elliptic functions.

At the working frequency (25KHz) the skin depth of the steel wheel is a few micrometers. This leads to an exponential variation of the vector potential inside the load in the z direction:

$$A = A_s e^{-(1+j)z/\delta} \tag{2}$$

If we divide the surface of the load in n elementary circular coils the vector potential at the point (r_k, z_k) of the surface is calculated by:

$$A_k = \sum_{i=1}^{n} A_{i,k} + \sum_{l=1}^{n} a_{k,l} . A_l \tag{3}$$

where $A_{i,k}$ is given by (1) and $\alpha_{k,l}$ is the complex coefficient related to the influence of the lth element to the kth element of the load.

Equation (3) can be written as

$$Q.A = F(r_i, z_i) \tag{4}$$

where Q is a constant matrix constituted of the coefficients $\alpha_{k,j}$, A is the vector of the potentials at the surface of the load and F a vector function of the coordinates r_i and z_i.

The inductor coordinates r_i and z_i are determined by minimising the objective function:

$$\min_{r_i, z_i} J = \frac{1}{2} \sum_{i=1}^{n} (A_i - A_0)^2 \tag{5}$$

where A_0 is the desired value of the vector potential.

3.2 Inductor Current

We intend to define the inductor current by an optimal control method. This needs to know the state equation:

$$T_{k+1} = M.T_k + N.(I_k)^2 \tag{6}$$

where T_k represents the state vector of the temperature at the kth time step, I_k the inductor current, and M and N the matrix to be defined. For a homogeneous temperature problem the best state vector is a two-dimensional maximal and minimal temperature vector. Solving the heat transfer equation we determine the vector (T_{max}, T_{min}) at the time step k and by the mean square identification method we calculate the matrix M and N.

The inductor current from first to N^{th} time step is calculated so as to minimise the objective function:

$$\min_{I_k} J_2 = \frac{1}{2}(T_N - T_0)^T (T_N - T_0) \quad (7)$$

with T_0 the desired temperature value in the steel wheel. This optimisation is carried by the dynamic programming method.

4 SIMULATION RESULTS

With the inductor form obtained above, we present, in figure 2, the simulation result.

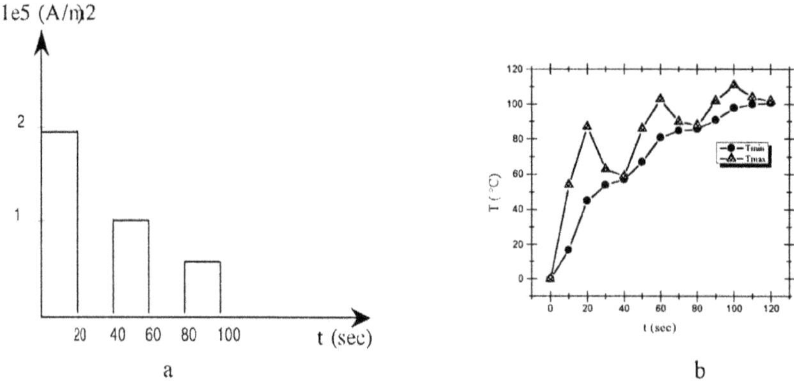

Figure 2 simulation result: a: current, b: temperature

According to these simulations, the temperature inside the steel wheel is homogeneous to 100°C (because: maximal temperature = minimal temperature = 100°C) after 120 seconds, and does not exceed 110°C. This is consistent with the imposed constraints.

5 CONCLUSION

We have presented a two step optimisation method to homogenise the temperature in an induction heating system. In the first step we determine the inductor form that gives an uniform potential vector at the surface at the load and in the second step we calculate the current variation which generates an uniform temperature inside the load in a given time.

Although the constraints are severe at the temperature variation, our method gives quite satisfying results in simulation.

References

1. S. Ratnajeevan, H. Hoole, et al. "Inverse problem methodology and finite elements in the identification of cracks, sources, materials and their geometry in inaccessible location," IEEE Trans. on Magn., Vol. 27, No. 3, pp. 3433-3443, 1991.
2. H. Hoole, K. Weeber, S. Subramaniam, "Fiction minima of object functions, finite elements meshes and edge elements in electromagnetic device synthesis," IEEE trans. on Magn., Vol. 27, No 5, Sept. 1991.
3. Marc A. Firestone and Charles E. Kessel, "Plasma kinetic control in a tokamak," IEEE Trans. on Plasma Science, Vol. 19, No. 1, Febr. 1991.
4. Y. M. Li, J. Fouladgar and G. Develey, "Inverse problem methodology in an induction heating system," Compumag, 1-4 Nov. 1993 Miami Florida.

MODELLING A MASSIVE IRON CORE ELECTROMAGNETIC BRAKE WITH EQUIVALENT CIRCUITS

Timo T. Vekara

Kone Cranes Int.
Box 6, FIN-05801 Hyvinkää
Finland

INTRODUCTION

When designing electromagnets for brakes, it is important to predict the response times of the electromagnet, i.e. the pickup and the pullout time. The models, based on the finite element and the finite difference methods, including motion, iron saturation and eddy currents, are accurate but slightly complicated for the design purposes[1-3]. Obviously using the finite difference method reasonable accuracy for common engineering purposes is achieved but a dedicated simulation software is required[4]. This paper describes a simple model for a DC electromagnet[5]. The model is simulated using PC-Simnon[6].

Modelling the transient axial anchor displacement of a cylindrical DC electromagnetic brake shown in Figure 1 is a complex problem at least for the following reasons:

1. The leakage fluxes in the air gaps and in the winding space depend on the anchor displacement and on the excitation.
2. The iron core area varies along the flux path in the anchor, in the yoke and in the corners. The inner and the outer iron ring areas are not equal. The area in both the inner and outer air gaps is not equal.
3. The iron core magnetization curve is nonlinear (hysteresis and saturation).
4. Eddy currents are significant because of the massive iron core in the magnet body and in the anchor.
5. The specific resistance of copper and iron as well as the permeability of the iron is highly temperature dependent.
6. Generally, the temperatures of iron and copper are not homogeneous.
7. The real construction is not ideal: The iron surfaces near the air gap are grooved. The springs generating the braking forces vary in tolerance. Because the air gap length varies, the anchor moves with one side first.
8. The electric, magnetic and mechanical transients are simultaneous.

Electric and Magnetic Fields, Edited by A. Nicolet
and R. Belmans, Plenum Press, New York, 1995

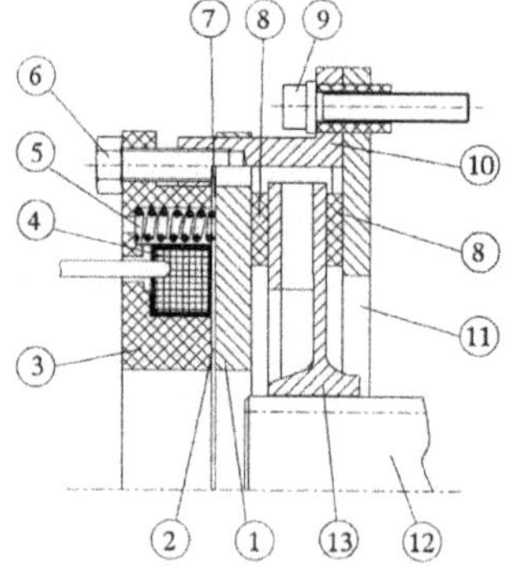

1. Anchor
2. Inner air gap
3. Magnet body
4. Winding
5. Spring
6. Bolt
7. Outer air gap
8. Friction material
9. Bolt
10. Brake body
11. End plate
12. Motor axis
13. Brake wheel

Figure 1. A cross section of a cylindrical disc type electromagnetic brake.

THE EQUIVALENT CIRCUIT MODEL

Developing the dynamic equivalent circuit model of the electromagnetic brake the following assumptions and simplifications were made:

1. The flux is homogeneous in the air gaps and inside the iron shells.
2. The direction of the flux path in iron is parallel to the core/coil boundary.
3. The leakage/main flux relation in the air gaps depends on the anchor displacement.
4. Air gap reluctances are functions of displacement, not of excitation.
5. Iron is fully saturated before demagnetization.
6. The force of the fringing leakage flux is insignificant.
7. The copper and iron temperatures are equal, uniform and constant

First in model building of an electromagnet, a lumped parameter magnetic circuit in a steady state is formed. In the inner and outer air gaps the flux is divided into a leakage flux and a force producing flux[7]. The magnetic force, f_m between two parallel plates of an area of A, through which a homogeneous flux links perpendicularly, is calculated simply: $f_m=B^2A/(2\mu_0)$. Beside the four air gap reluctances there exist in the steady state magnetic circuit leakage reluctances in the winding area and near the winding groove. The iron part is described with several reluctances because the flux linking area varies a lot along the path.

The magnetic circuit is solved using spreadsheet calculations with some excitations. As a result of these steady state calculations there appears the iron reluctance as a function of magnetic voltage across it, and four air gap reluctances as a function of the anchor displacement.

A dynamic model of a similar electromagnet is built[4] only in one dimension, i.e. the penetration depth. The magnetic properties of all the iron parts - anchor, yoke, inner and outer ring - in the magnetic circuit are described with an equivalent inner core, the length of which equals the length of the flux path in those iron parts. The lengthened inner core, the equivalent core describes the flux distribution in the massive iron core. It is divided into coaxial rings, called shells. The flux in each shell is considered homogeneous.

In former application the area of the inner and outer ring was equal. However, in brake magnet applications the outer area is often much larger than the inner one. For this reason when calculating the length of the equivalent inner core, the flux path length of the following parts was summed: inner ring, inner corner, winding groove in the yoke and the respective parts in the anchor.

The developed dynamic equivalent circuit model is shown in Figure 2. The model consists

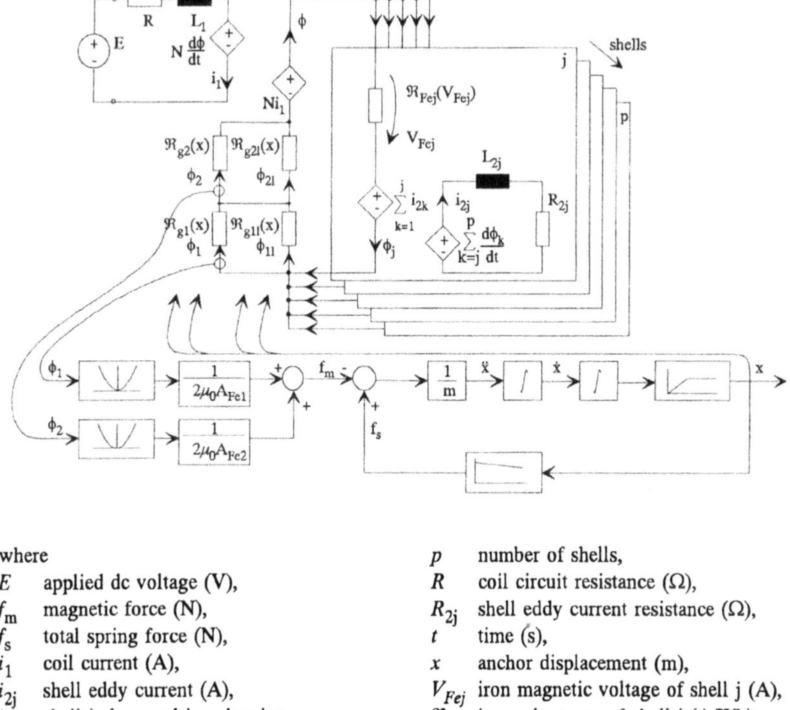

where

E applied dc voltage (V),
f_m magnetic force (N),
f_s total spring force (N),
i_1 coil current (A),
i_{2j} shell eddy current (A),
j shell index used in subscript,
L_1 coil leakage inductance (H),
L_{2j} shell leakage inductance (H),
m anchor mass (kg),
N number of turns in the coil,
p number of shells,
R coil circuit resistance (Ω),
R_{2j} shell eddy current resistance (Ω),
t time (s),
x anchor displacement (m),
V_{Fej} iron magnetic voltage of shell j (A),
$\Re_{Fej}$ iron reluctance of shell j (A/Wb),
$\Re_{g1}$ inner gap reluctance (A/Wb),
$\Re_{g11}$ inner gap leak. reluctance (A/Wb),
$\Re_{g2}$ inner air gap reluctance (A/Wb),
$\Re_{g21}$ inner gap leak. reluctance (A/Wb),
ϕ magnetic flux (Wb).

Figure 2. The dynamic equivalent circuit model of the electromagnetic brake.

of a coil circuit, magnetic circuits, eddy current circuits and a part describing the flux-motion-relationship. The coil circuit and eddy current circuits are modelled the same way, i.e. a resistor, a leakage inductance and a dependent voltage source.

The magnetic circuit has a common branch and several iron shell branches. Each shell branch has its own magnetic voltage dependent iron reluctance, and eddy current dependent magnetomotive voltage source. These iron shell branches are connected at the air gap boundary. So, it is possible for a flux to circulate between the shells. The coupling between the shells is presented via summations in the dependent voltage sources of the magnetic and eddy current circuits, c. f. Figure 2.

The magnetic force is calculated using the force producing flux in the two air gaps. The spring force is considered linear. General motion equation is used without damping. In the model there is also a block limiting anchor displacement.

ON THE SIMULATIONS AND MEASUREMENTS

The presented equivalent circuit model was simulated using PC-Simnon. The purpose of simulations was to calculate both the pickup and dropout times. An example magnet of inner and outer core area, 59 cm^2 and 144 cm^2 respectively, was simulated and measured.

The outer diameter of the magnet was 217 mm and the magnetic force was about 10 kN at 500 volts. The iron material was common carbon steel.

The number of iron shells was selected to equal five, as shown in Figure 2. The shell core area ratios were 1:1:2:2:2, indicating shell walls between 2.06...5.71 mm. The iron magnetization curve was modelled with two piecewise linear curves, one during magnetization and another during demagnetization.

Initially at turn on, the state variables were zero, except displacement when a step DC excitation voltage was switched on at the anchor displacement of 1.0 mm. At turn off, the state variables were initialized to the steady state values. At turn off a varistor, connected to the coil terminals, was modelled adding a constant resistance of 50 kΩ to the coil resistance of 280Ω. Somewhat greater varistor resistance values lead to numerical oscillations.

The model sensitivity on parameter variation was examined. As expected, the dropout time mostly depends on copper and iron resistance. The effects of iron and copper leakage inductances and the used magnetization curve are less significant.

The error in the pickup time calculation is partly due to the poor modelling of the leakage flux and the air gap flux distribution. The accuracy, when calculating the dropout time, can be increased by increasing the number of the shells and by modelling the varistor at the coil terminals more accurately.

CONCLUSIONS

When calculated pickup and dropout times were compared with the measured ones, it was shown that the transient anchor motion of the electromagnet can be described with the presented model. The accuracy of the model seems to be sufficient from the engineering point of view. Modelling accuracy depends mostly on the calculation accuracy of the varistor resistance, the iron resistance and the leakage inductance as well as the accuracy of the steady state calculations. Evaluating the values for iron resistances and iron leakage inductances requires further reseach. Naturally, the temperatures of the iron and copper via respective resistances have major effect on the response times of the simulated magnet.

ACKNOWLEDGEMENTS

The author would like to thank prof. Jarl-Thure Eriksson and associate prof. Juha T. Tanttu, both of Tampere University of Techonogy, for their valuable advices given during the model design.

REFERENCES

1. S. Yamada, Y. Kanamaru, K. Bessho, The transient magnetization process and operations in the plunger type electromagnet, *IEEE Trans. Magn.*,pp. 1056-1058, vol. 12, November (1976).
2. B. A. Aldefeld, Numerical solution of transient nonlinear eddy current problems including moving iron parts, *IEEE Trans. Magn.*,pp. 371-373, vol. 14, September (1978).
3. S.-Q. Zheng, D. Chen, Analysis of transient magnetic fields coupled to mechanical motion in solenoidal electromagnet exited y voltage source, *IEEE Trans. Magn.*,pp. 1315-1317, vol. 28, March (1992).
4. B. Lequesne, Dynamic model of solenoids under impact excitation, including motion and eddy currents, *IEEE Trans. Magn.*, pp. 1107-1116, vol. 26, March (1990).
5. T. Vekara, ''Nosturin jarrumagneetin dynamiikka,'' (The transient behaviour of an electromagnet used in crane brakes), Licentiate thesis in Finnish, Tampere University of Technology, Tampere (1993).
6. H. Elmqvist, K.J. Åström, T. Schöntal, B. Wittenmark, ''Simnon, User's Guide for MS-DOS Computers,'' SSPA Systems, Göteborg (1990).
7. H.C. Roters,'' Electromagnetic Devices,'' John Wiley & Sons Inc., New York (1941).

PLASMA MODELS FOR THE COMPUTATION OF 3D EDDY CURRENTS IN NEXT TOKAMAKS

Sophia Fantechi, Ioannis Sakellaris, Yves Crutzen

Commission of the European Union, Joint Research Centre
Institute for Systems Engineering & Informatics, Ispra site
I-21020 Ispra (Varese), Italy

ABSTRACT

This paper describes recent modelling procedures for 3D eddy current analysis in ITER (International Thermonuclear Experimental Reactor) blanket system during off normal operating conditions. Plasma disruption events characterised by current quenches and vertical displacements events (VDEs), as well as combined scenarios, have been investigated, to study the local and global effects of the eddy currents circulation and electromagnetic (EM) forces distribution.

INTRODUCTION

The analysis of the EM phenomena due to plasma - first wall interaction during off-normal operating conditions is of primary importance for tokamak technology[1]. The evaluation of the induced currents and related forces on plasma facing structures greatly affects the design of the fusion reactor and entails the verification of the behaviour of the internal reactor components during the fast transitions determined by current and magnetic field changes. Plasma disruptions are particularly dangerous events since they produce forces strong enough to cause mechanical deformations and missiles generation[2]. The assessment of the complex pattern of eddy currents and forces is based on 3D simulations with EM codes taking into account data from fusion experiments and using the finite element method (FEM) with edge elements[3]. Particularly crucial for the developpement of codes, as well as for the interpretation of the results, is the modelling of the plasma during the transient EM phenomena.

PLASMA MODELS AND EDDY CURRENT ANALYSIS FOR ITER

The numerical simulations for the ITER reactor have been carried out using the CARIDDI code[1,3]. "Centered" and "dynamic" disruptions have been investigated, i.e. plasma current decays without any changes in the plasma ring position, and disruptions

Electric and Magnetic Fields, Edited by A. Nicolet
and R. Belmans, Plenum Press, New York, 1995

consisting of a plasma displacement. We have considered a blanket system without internal shield; the same simulations carried out with an internal material have simply shown an increase of the blanket screening properties. The plasma is modeled by a toroidal axisymmetric conductor with specified position and time behaviour of the current (linear decay). The conductor has an elliptical or rectangular discretised cross-section and the plasma current is carried in each single element by a filament placed in its centre. The radial current density profile can be taken as uniform or picked. For the simulation of a current quench, one defines a conductor of suitable (fixed) section and gives a set of values of the plasma current at chosen instants. On the other hand, a plasma displacement is modeled by defining a set of conductors (of adjustable section) one above the other, such as the current switches on in one conductor while it decays in the preceeding one.

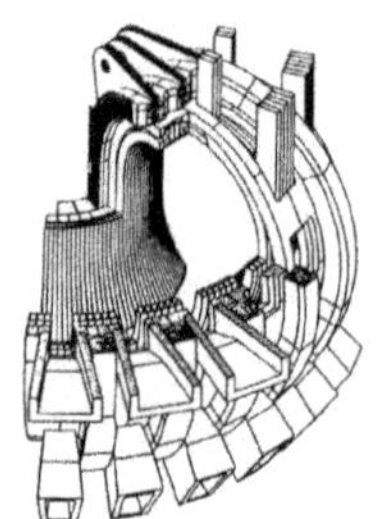

Fig. 1 - ITER blanket system.

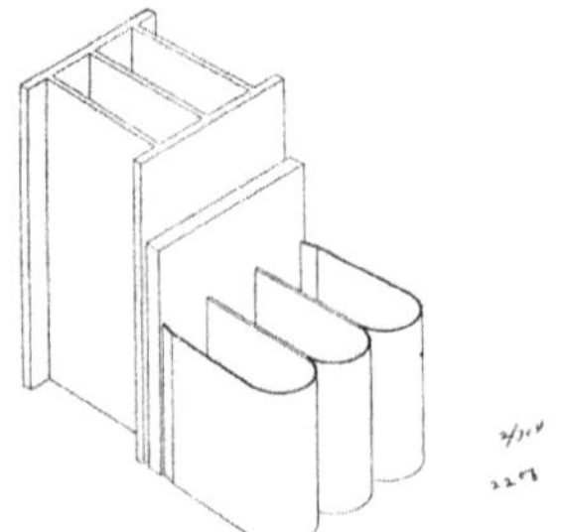

Fig. 2 - ITER blanket cells (FW, BP).

Preliminary Analysis in ITER Blanket System

In the first model, which considers a 5° sector of torus, the toroidally continuos first wall (FW) and back plate (BP) (fig.1) are made of stainless steel and the ondulated FW is represented by a smooth one of increased resistivity. The main results[4] are the eddy currents distribution in the blanket sector and the body forces, at various instants. Figures 3 and 4 give an example of the eddy currents and body forces distributions obtained, repectively, for a current quench of 25 MA in 25 ms and for a VDE over 2 m in 25 ms, at a plasma current of 25 MA.

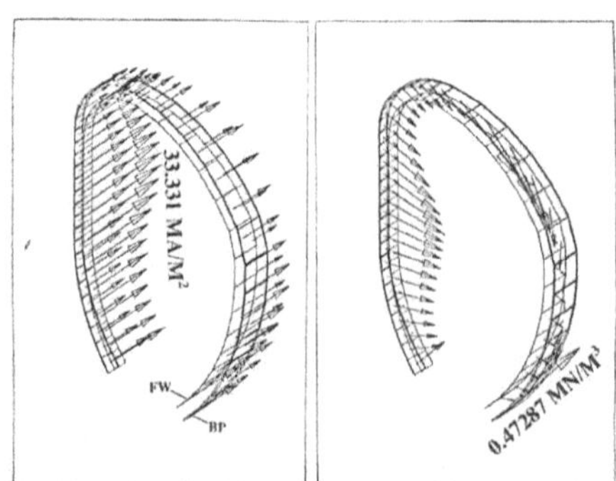

Fig. 3 - Eddy current density and body forces at 25 ms (VDE of 2 m in 25 ms).

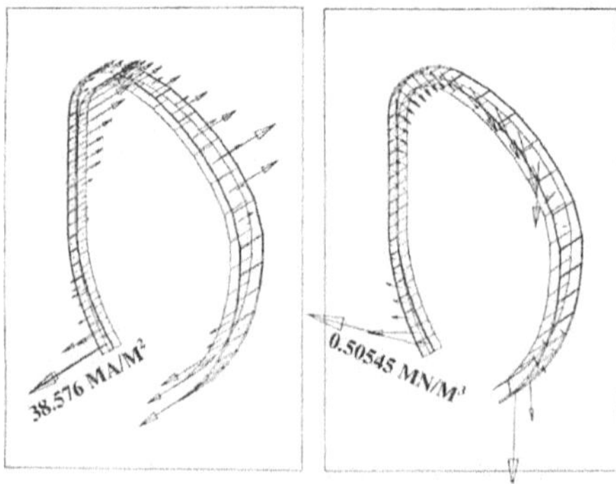

Fig. 4 - Eddy current density and body forces at 25 ms (current quench of 25 MA in 25 ms).

The inversion of sign of the toroidal currents is typical of a pure VDE, as is the vertical direction of the maximum forces, while a current quench always creates essentially radial forces and eddy currents circulating in one direction only. There is also a major edge effect, which, in the case of a VDE, is responsible for high vertical resultants localised at the bottom of the outboard blanket.

Analysis of a Curved Module of ITER First Wall

The FW is actually made up of cells (fig.2). This gives additional effects which have been studied modelling 1/4 of a cell slice: a thin shell of steel containing an homogeneous and isotropic conducting material, electrically insulated from it.

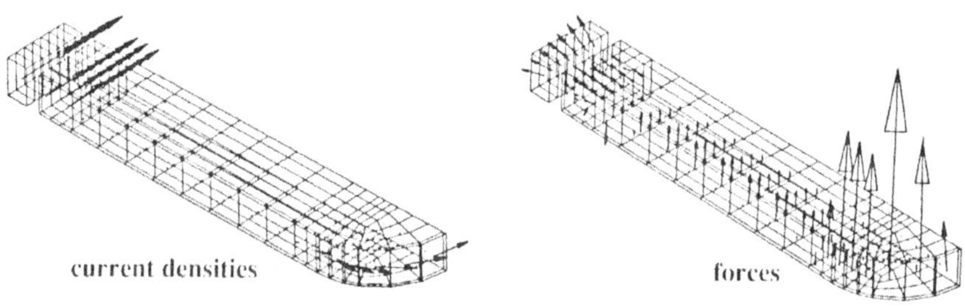

Fig. 5 - Eddy current density and body forces at 25 ms in a cell (current quench of 25 MA in 25 ms).

Due to a current quench, the structure becomes subject to high shear forces. This is due to the curved shape of the cells; it allows the circulation of non toroidal eddy currents, which are then able to interact with the large toroidal magnetic field (fig.5).

Upgraded Model of ITER Blanket System

A more sophisticated model taking into account the real FW geometry and including vacuum vessel (VV) and poloidal field coils has then been considered (fig.6) and, according to the latest outline design, a copper FW has been added. More significant disruptions for ITER have been investigated: rapid plasma current quenches and slow VDEs. Figures 7 and 8 show the eddy currents density patterns at 10 ms for a centered and a dynamic disruption respectively.

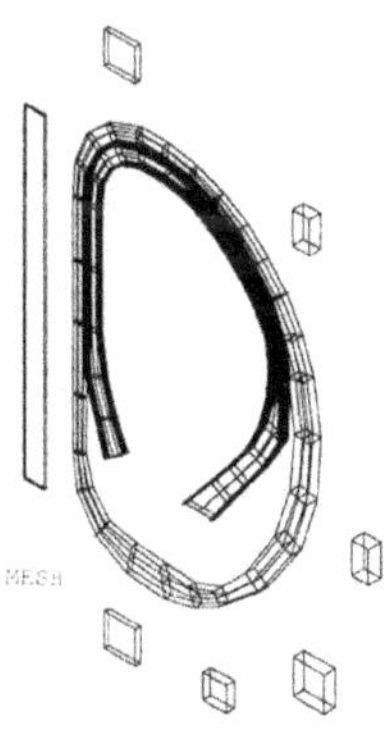

Fig. 6 - FEM model including vacuum vessel (VV) and poloidal field coils (1/2 of 1/96 of torus).

Due to the presence of the VV and copper FW, the eddy currents densities during a centered disruption are much higher than in the previous simulations, reaching values of 400 MA/m^2. Similarly, a VDE also induces higher currents, up to 140 MA/m^2.

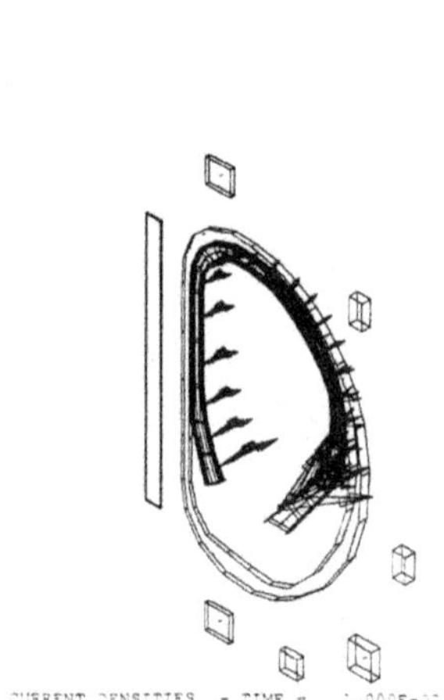

Fig. 7 - Eddy current density at 10 ms (current quench of 24 MA in 10 ms).

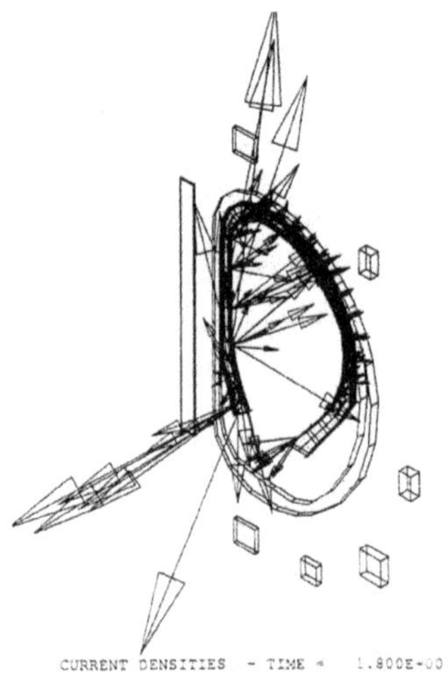

Fig. 8 - Eddy current density at 1.8 ms (downward VDE of 2.7 m in 1.8 s).

CONCLUSIONS

In terms of eddy currents patterns, centered disruptions and VDEs appear to have quite distinct effects on ITER in-vessel region. The strongest forces are produced by a rapid quench and act radially on the reactor blanket. In addition, the single blanket cells are subject to shear forces. If the (toroidally continuous) plasma facing FW is made of copper, a much more conducting material, the counteracting effect of the eddy currents will be much more efficient This is indeed used for passively stabilising the plasma and better controlling the disruptions. In any case, the observed edge effect should be carefully considered for the final design solution.

ACKNOWLEDGEMENTS

The authors wish to thank Messrs. R. Witty, G. Volta and F. Farfaletti-Casali for their interest in this research activity. They are also grateful to Messrs. M. Ferrari and V. Cristini of ENEA-Frascati for their contribution to the project and useful advices.

REFERENCES

1. Y. Crutzen, G. Molinari., G. Rubinacci, "Industrial Application of Electromagnetic Computer Codes", Computer & Information Science Series, Vol.1, Kluwer Academic Publishers, EUR 1313/ISBN 0-7923-0998-79 (1990).
2. Y. Crutzen, F. Farfaletti-Casali, A. Inzaghi, S. Papadopoulos, J. Schneider, N. Richard, Transient electromagnetomechanical analysis of the in-vessel region during off-normal conditions, *Fus. Tech. Special Issues/SOFT Rome* 22: 1027-1031 (1993).
3. R. Albanese, G. Rubinacci, Integral formulation for 3D eddy current computation using edge elements, *IEEE Proceedings* 135 (Pt.A., No.7): 457-462 (1988).
4. S. Fantechi, Modelling of tokamak plasmas in computational electromagnetic analysis, Technical Note No.I.94.11, ISEI/IE/2623/94, JRC-European Commission, Ispra site (1994).

NONLINEAR ANALYSIS OF THREE DIMENSIONAL MAGNETIC FIELD IN ELECTROMAGNETIC DEVICES

Vlatko Stoilkov, Milan Cundev, Lidija Petkovska and Krste Najdenkoski

Dept. of Electrical Machines and Transformers
Faculty of Electrical Engineering
91000 Skopje
Macedonia

INTRODUCTION

The most of contemporary principles of magnetic field computations appear as an interaction between numerical methods and fast regarding to time and powerful regarding to memory computers supported by graphical substations [1,2].

This paper shows an application of Finite Elements Method in order to analyse nonlinear magnetical and mechanical characteristics of low voltage actuator in 3D domain. Calculations are obtained by programme package FEM-3D developed by authors and adapted for PC use. Non linearity is involved by including different magnetic properties along x,y and z axes.

MATHEMATICAL MODEL

Three dimensional magnetic field distribution is computed on the base of Maxwell equations. An involving of magnetic vector potential **A** leads to Poisson equation:

$$\frac{\partial}{\partial x}\cdot\left[\nu(B)\cdot\frac{\partial A}{\partial x}\right]+\frac{\partial}{\partial y}\cdot\left[\nu(B)\cdot\frac{\partial A}{\partial y}\right]+\frac{\partial}{\partial z}\cdot\left[\nu(B)\cdot\frac{\partial A}{\partial z}\right]=-J(x,y,z) \tag{1}$$

Mathematical model for solution of the equation (1) is derived the on the basis of Weighted Residuals Method and Galerkin's procedure[4] of minimizing the error determined as:

$$\varepsilon=\nu_x\cdot\frac{\partial^2 A}{\partial x^2}+\nu_y\cdot\frac{\partial^2 A}{\partial y^2}+\nu_z\cdot\frac{\partial^2 A}{\partial z^2}+J(x,y,z) \tag{2}$$

Electric and Magnetic Fields, Edited by A. Nicolet
and R. Belmans, Plenum Press, New York, 1995

The values of magnetic reluctivities ν_x, ν_y and ν_z depend on magnetic field density because of nonlinear magnetic characteristic B=f(H), as well as an different magnetic properties of the ferromagnetic materials along x,y and z axes.

Different magnetic reluctivities along x, y and z axes due to core lamination are taken into account by following relations:

-for x and y axes:

$$\nu_x(B) = \nu_y(B) = \frac{\nu_{Fe}(B)\cdot \nu_o}{k_{Fe}\cdot \nu_o + (1-k_{Fe})\cdot \nu_{Fe}(B)} \tag{3}$$

-for z axis:

$$\nu_z(B) = k_{Fe}\cdot \nu_{Fe}(B) + (1-k_{Fe})\cdot \nu_o \tag{4}$$

where k_{Fe} is stacking factor and ν_{Fe} corresponds to magnetic characteristic.

Complex magnetic field computations are performed by the program package FEM-3D. The algorithm regarded to involvement of nonlinear magnetic properties is given by flowchart presented in Fig.1.

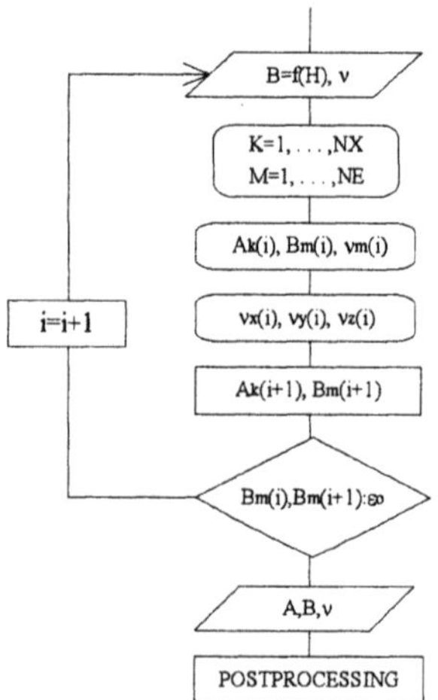

Figure 1. The flowchart regarding to nonlinear magnetic properties in 3D domain

NONLINEAR CHARACTERISTICS

Nonlinear analysis is applied on the three core low-voltage actuator CNM-31 with shading coil.

Threedimensional solution of magnetic vector potential is performed by discrete displacement from total opened position of the middle core (4.5 mm) through the total closed position of the middle core (0.5 mm) by displacing step of 0.5 mm. The distribution of magnetic field in open position is presented in Fig. 2.

On the basis of three dimensional field solution, several non linear characteristics are performed with the mechanical Force/distance characteristic as a target. As first, Flux linkage/Excitation current (Fig.3) with air gap as the parameter is obtained on the basis of magnetic vector potential. Inductivity/Excitation current (Fig.4) at constant air gap is obtained by numerical derivation:

$$L(I,\delta) = \frac{d\Psi}{dI} \tag{5}$$

An extreme non linearity in total closed position caused by influence of eddy currents in shading coil is obvious.

Numerical calculation of the force-distance characteristics (Figs.5,6) is obtained by the energy concept, as the numerical derivation of the magnetic coenergy at constant current and discrete dislacement as:

$$F = \frac{\partial W'(\delta,I)}{\partial \delta} \quad \text{at I=const. as parameter} \tag{6}$$

where the magnetic coenergy is obtained by numerical integration of the flux linkage:

$$W' = \int_0^I \Psi_\delta(\delta,I)\,dI \quad \text{at U=const.as parameter} \tag{7}$$

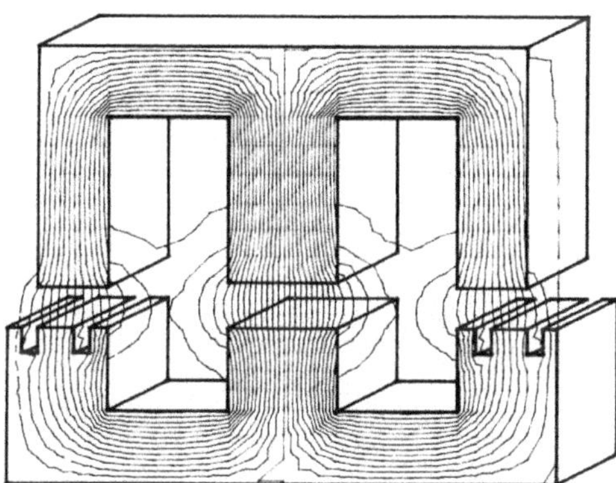

Figure 2. Magnetic field distribution of low-voltage contactor CNM-31

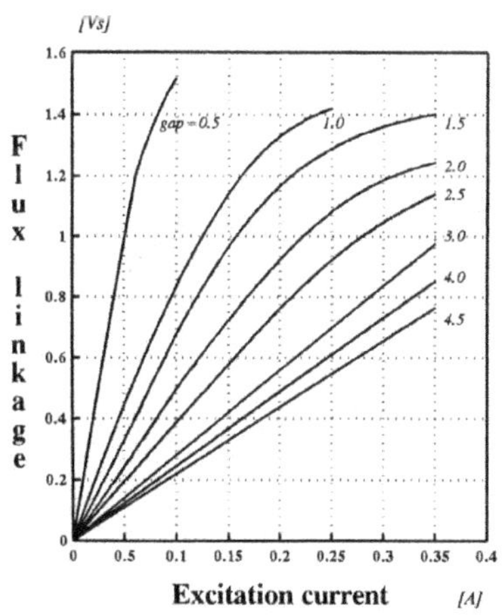

Figure 3. Flux linkage / Excitation current characteristics

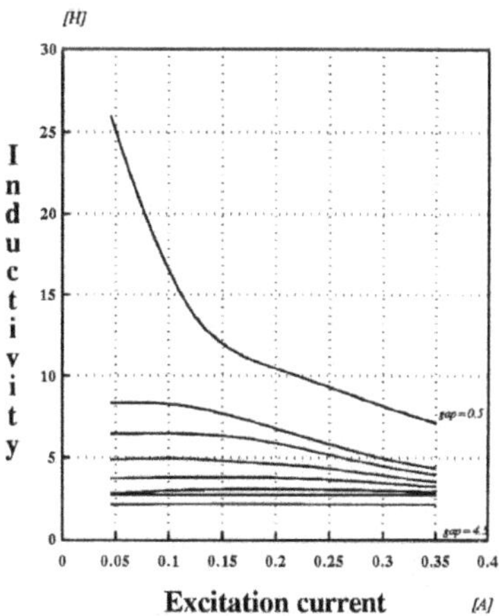

Figure 4. Inductivity / Excitation current characteristics

Finally, the numerically obtained results are compared with experimental ones obtained by measurement in laboratory surroundings. The comparison of the numeric and experiment characteristics shows an acceptable level of 10% error.

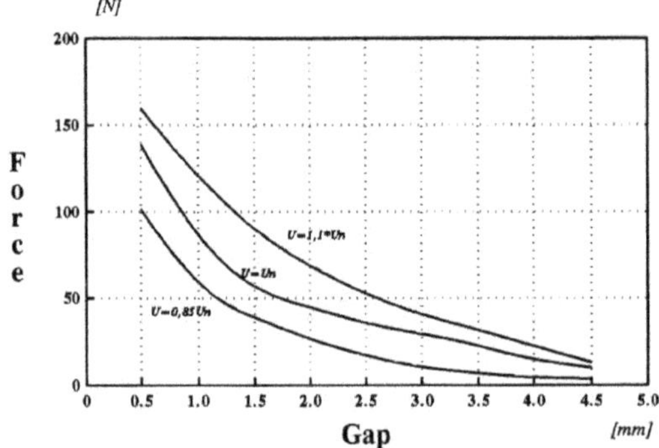

Figure 5. Force-distance characteristics

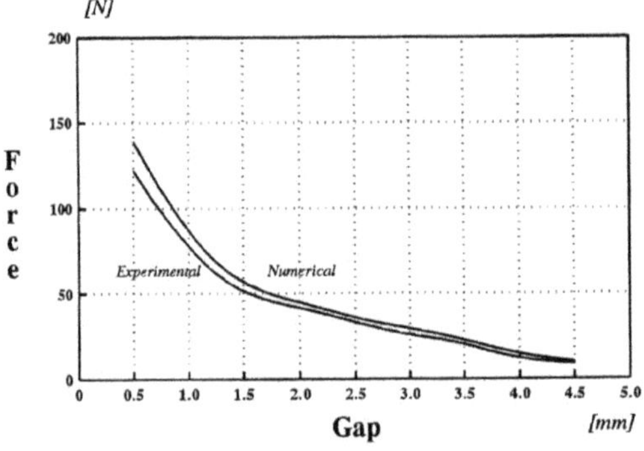

Figure 6. Numerical and experimental comparison

CONCLUSIONS

Having into consideration the necessity of optimal designing of electromagnetic devices, the paper offers an approach to electromagnetic and mechanic characteristics' computation during the period of designing, by simulating the working conditions.

The coincidence between numerical and experimental results within the 10% error express the proposed method as applicable one in the designing of new devices and improving the characteristics of present ones.

REFERENCES

1. M.Cundev and L.Petkovska. "Three-dimensional Magnetic Field Analysis of Electric Machinery by the Finite Element Method", 2as Jorn. Luso-Espanholas de Eng. Electrotec., Coimbra, Portugal (1991).
2. M.Cundev L.Petkovska and V.Stoilkov. "Three Dimensional magnetic Field Problems in Electromagnetic Devices", 9th COMPUMAG Conference, Miami, USA,1993
3. V.Stoilkov. "Contribution to the Analysis of Electromagnetic Characteristics of Low-voltage Actuators by Three dimensional Magnetic Field Solution", M.Sc. thesis, Skopje, Macedonia (1993).
4. P.P.Silvester and R.L.Ferrari. "Finite Elements for Electrical Engineers", Second edition., Cambridge University Press (1990).

IMPROVED LINEARITY LINEAR VARIABLE DIFFERENTIAL TRANSFORMERS (LVDTs) THROUGH THE USE OF ALTERNATIVE MAGNETIC MATERIALS

GW Midgley, D Howe, PH Mellor

Dept of Electronic & Electrical Engineering
University of Sheffield
PO Box 600, Sheffield, S1 4DU, UK

INTRODUCTION

The effectiveness of any controlled actuation system is, to a large extent, determined by the performance of its position feedback transducer. The demands on the transducer are often exacting, particularly concerning its linearity, repeatability and reliability. In this regard the Linear Variable Differential Transformer (LVDT) is capable of satisfying stringent performance requirements, whilst offering additional advantages such as environmental stability, ease of installation, relative simplicity and physical robustness.

In its simplest form the LVDT consists of an inner cylindrical plunger encircled by a uniformly wound a.c. excited primary winding, the position of the ferromagnetic plunger being deduced from the differential output of two outer secondary windings. A typical LVDT has a limited range of core displacements over which the linearity falls within a specified error band, leading to one of the major compromises of LVDT design, the ratio of the stroke to the overall device body length.The criteria used to assess the performance of LVDTs include linearity error and the linear range to which it applies in addition to an often stringent electrical specification including factors such as a high input impedance,to minimise input current and therefore ohmic loss, and a high output sensitivity, expressed in terms of output voltage for a given core displacement and a given input voltage.

Currently the design of LVDTs is based on the use of high permeability metals, such as nickel-iron alloys, and often does not consider the effects of eddy currents encountered at higher supply frequencies.The paper will examine potential design improvements based around increasing the LVDT supply frequency and to the advantages gained through the use of low loss powder composite materials.

THE INFLUENCE OF MAGNETIC MATERIALS ON LVDT PERFORMANCE

In order to address the low stroke-to-body ratio of most LVDTs the magnetic circuit of the device needs to be analysed under a variety of different operating conditions.One method of improving the stroke-to-body ratio involves reducing the effective airgap length, that is the distance between ferromagnetic plunger and shield, although the area available for windings will decrease as a result, leading to a lower input impedance and higher input current, factors which may be undesirable with low power consumption integrated circuit supply and decoding systems.

Electric and Magnetic Fields, Edited by A. Nicolet
and R. Belmans, Plenum Press, New York, 1995

Increasing the supply frequency would suggest itself as a solution to these problems although, in turn, additional eddy current losses are encountered in the ferromagnetic materials which can prove detrimental to transducer performance.

Typical LVDTs operate at frequencies between 400 Hz and 5kHz as significantly higher frequencies introduce eddy current losses detrimental to performance.For example Fig 1 shows the variation of sensitivity with excitation frequency for a typical device where the peak sensitivity is observed at a frequency of 2kHz .This effect can be explained with reference to the equivalent circuit, Fig 2, of the primary. At high frequencies the reactances X_l and X_m , representing the leakage and magnetising reactances, dominate the winding resistance,R_p and with no core loss would be expected to give a constant primary emf.At increased frequencies the core loss has the effect of reducing the magnetising impedance and therefore results in a poor output regulation.

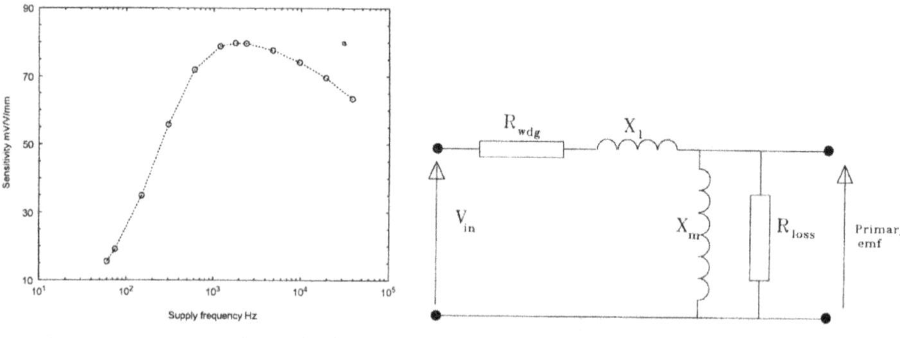

Fig 1 : Typical LVDT sensitivity characteristic

Fig 2 : Primary impedance circuit

In order to exploit the advantages of reducing airgap and increasing excitation frequency fully then alternative materials with lower loss properties need to be considered.For example the properties of the presently used nickel-iron are contrasted with two possible substitute materials, a sintered ferrite and an iron powder composite, in Table 1.The alternative materials have a much lower conductivity at the expense of significantly reduced permeability.However, the influence of material magnetic permeability is reduced in an LVDT due to the large reluctance represented by the device effective airgap.The effects of variation of permeability and conductivity in an LVDT were investigated using a steady-state linear a.c. finite element analysis coupled to values for primary and secondary winding resistances [1].

Table 1 : Comparison of magnetic materials for use in LVDT

Material	Relative permeability	Conductivity S/m	Skin depth at 1 Hz
Nickel-Iron	50 000	2.1×10^6	1.55 mm
Powder composite	70	100	6.02 m
Sintered ferrite	1500	0.25	26.00 m

The resulting characteristics shown in Fig 3 illustrate that increasing the relative permeability beyond 3000 gives only a marginal increase in input impedance. In contrast the use of materials with high conductivities leads to reduced input impedance, particularly at high excitation frequencies.

The high static hysteresis losses found in the magnetically harder materials is of lesser importance than eddy currents in an LVDT application where the core and screen are only lightly fluxed.Therefore the most appropriate material for use in an a.c. excited device such as an LVDT may not be simply the one with the highest permeability.

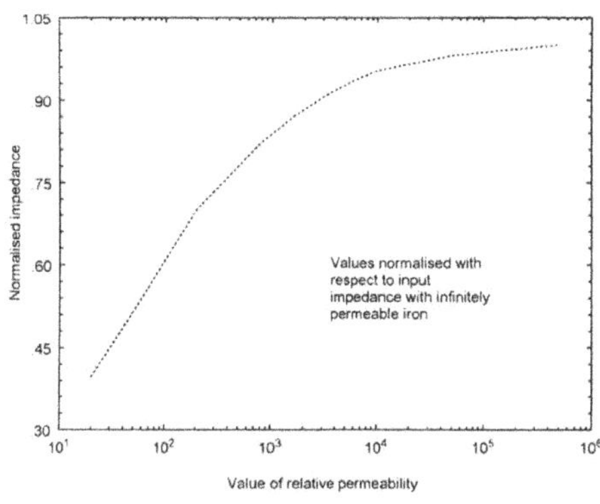

Fig 3a : Effect of permeability variation on input impedance

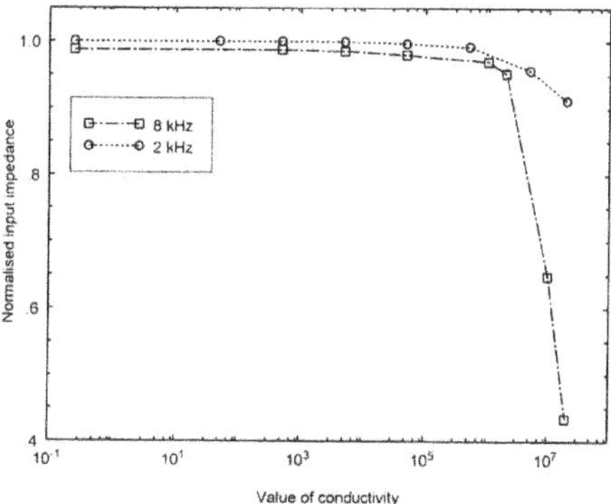

Fig 3b : Effect of conductivity variation on input impedance

COMPARISON OF LVDT PERFORMANCE

Candidate designs having the same body length but with differing airgap lengths were analysed for both nickel-iron and the powder composite Permite 55 [2]. Four prototype LVDTs were constructed from designs optimised using finite element analysis to the same body lengths and 2 airgap lengths, one being 15 % less than the other, using a CNC wire erosion process.The larger airgap length device constructed from nickel-iron was chosen as a reference, in terms of input impedance and sensitivity, against which the performance of the other devices were compared. In addition comparisons were made between the performance predicted from finite element analysis and the measured values.
The results presented in Table 2 show that substantial improvements in linearity, or linear range, can be achieved by utilising a reduced airgap in an LVDT constructed from either material.

Table 2 : Comparison of measured and predicted performance of LVDTs considered

Material and airgap length	Freq Hz	Linearity Error		Sensitivity mV/V/mm		Input current mA		Linear range for 0.56 % error
		Meas	Predicted	Meas	Predicted	Meas	Predicted	
NiFe large	1200	0.56%	0.73%	82.56	83.35	9.0	8.8	+/-4.0 mm
Permite large	2470	0.42%	0.30%	73.91	83.50	9.9	8.8	+/-4.2mm
NiFe reduced	4800	0.40%	0.42%	82.34	79.80	9.3	8.8	+/-4.2mm
Permite reduced	9830	0.29%	0.10%	77.73	82.10	10.0	8.8	+/-4.7 mm

The sensitivity variation with frequency of the four LVDTs was also measured and is shown in Fig 4.Again the predicted values are given and show good agreement, although a larger discrepancy exists between predicted and mesured values for the Permite devices which may arise from the material having a slightly lower permeability than specified.
Sensitivity values using Permite were, in general, lower than those using nickel-iron as the frequencies used to obtain the required input impedance were less than those at which optimum sensitivity is achieved. The flexibility of being able to scale an LVDT design linearly with frequency when using Permite enables an easy means of compensation.

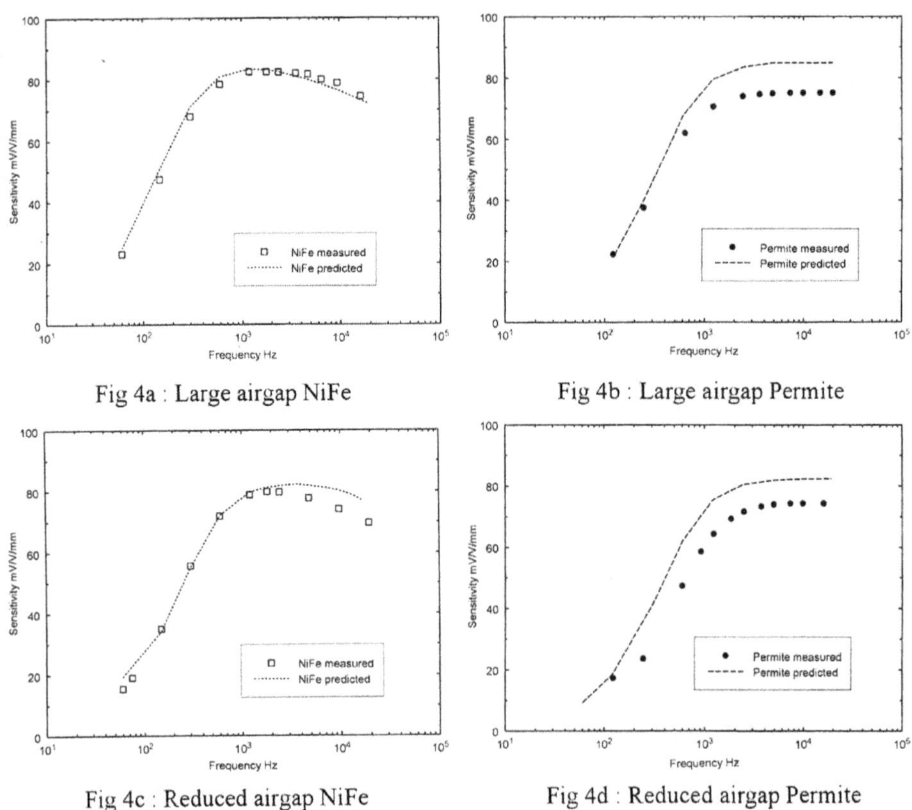

Fig 4a : Large airgap NiFe

Fig 4b : Large airgap Permite

Fig 4c : Reduced airgap NiFe

Fig 4d : Reduced airgap Permite

Fig 4 : Comparison of measured and finite element predicted sensitivities of four LVDT designs

CONCLUSIONS

The paper has demonstrated the substantial improvements in stroke-to-body length that can be achieved by utilising a reduced airgap in an LVDT operated at an increased frequency.

Applying the powder composite material enables additional flexibility in terms of scaling designs linearly with frequency where, through the absence of eddy currents, operation at greatly increased frequencies can now be considered.

REFERENCES

1. MEGA , Version 5 , Applied Electromagnetic Research Centre, University of Bath
2. Hoganas AB, S263 83, Hoganas Sweden

ACKNOWLEDGEMENTS

The authors would like to acknowledge EPSRC UK for the provision of a CASE studentship for Mr G.Midgley and the industrial collaborators Penny & Giles Ltd.

HARMONIC ANALYSIS OF THE FLUX DISTRIBUTION IN AN AXIALLY LAMINATED SYNCHRONOUS RELUCTANCE MOTOR

X. Feng and R. Belmans

Labo. EMA Dept. E. E. K. U. Leuven
Kard. Mercierlaan 94, B-3001 Leuven-Heverlee
Belgium

INTRODUCTION

The paper presents a numerical analysis of the radial flux density distribution in the airgap of the synchronous reluctance motor with axially laminated rotor. The stator has a standard three phase design. The rotor consists of axially interleaved magnetic iron laminations with non-magnetic spacer of different thicknesses. Due to the complex lay-out of the rotor, numerical methods such as finite elements (FE) and the Fast Fourier Transforms (FFT) prove to be powerful tools for investigating and evaluating the magnetic field in the airgap. Since any of the harmonic field in the airgap acts on the rotor, additional torques, losses and audible noise may arise. Harmonic analyses of the d- and q-axis radial flux density distribution in the airgap of the models have been performed. Comparing predictions of flux density distribution in the airgap by both the proposed analytical approach and numcrical methods are used and results are in good agreement.

ANALYTICAL CALCULATION FOR FLUX DENSITY DISTRIBUTIONS

The amplitudes of the respective fundamental components of the d- and q-axis rotating mmf resulting from a symmetrical three-phase current winding are F_{d1m} and F_{q1m}. The fundamental and harmonic components of the airgap flux density distribution along d-axis and q-axis may be calculated from analytical formulas:[1]

$$B_{d\nu m} = \frac{4\mu_0 F_{d1m}}{\tau(1+k_{sd})} \int_0^{\frac{\tau}{2}} \frac{1}{\delta_d(x)} \cos\left(\frac{\pi x}{\tau}\right) \cos\left(\frac{\nu \pi x}{\tau}\right) dx \tag{1}$$

$$B_{q\nu m} = \frac{4\mu_0 F_{q1m}}{\tau(1+k_{sq})} \int_0^{\frac{\tau}{2}} \frac{1}{\delta_q(x)} \sin\left(\frac{\pi x}{\tau}\right) \sin\left(\frac{\nu \pi x}{\tau}\right) dx \tag{2}$$

Electric and Magnetic Fields, Edited by A. Nicolet
and R. Belmans, Plenum Press, New York, 1995

NUMERICAL APPROACH FOR FLUX DENSITY DISTRIBUTION

An analytical approach to the magnetic flux density distribution in the airgap of the motor uses some equivalent coefficients. This may lead to large inconsistencies between analysis results and experiments. For the investigation of the behaviour of the models starting from the analysis of the magnetic field, FE and FFT are used to calculate the distribution of the magnetic flux density. For different stator currents, different airgap lengths and different non-magnetic insulation thicknesses of the rotor, the magnetic flux density distribution in the motor is computed separately, considering saturation of magnetic materials. Further harmonic analysis of the flux density distribution in the airgap is done using FFT. An example is shown in the figures 1 and 2. For the d-axis there are mainly fundamental and stator slotting components (35, 37, 71 and 73). For the q-axis field, there is mainly a leakage field and the corresponding amplitude of fundamental component is less than that of the stator slotting components (35 and 37).

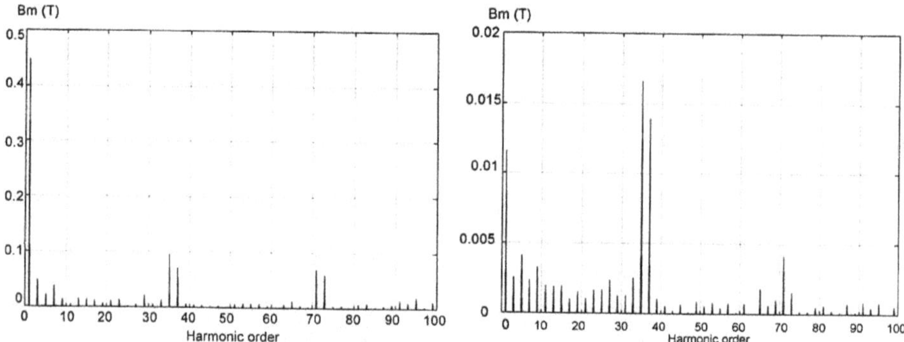

Figure 1: Harmonic analysis of the d-axis field

Figure 2: Harmonic analysis of the q-axis field

EFFECT OF ROTOR SALIENCY RATIO ON FLUX DENSITY COMPONENTS

The flux density distribution in the airgap depends on the saliency ratio of the rotor, the airgap length and the filling coefficient k_{fill} [1] of the rotor. With FE the d- and q-axis field distribution can be calculated and the corresponding radial airgap flux density distribution is plotted (figures 3 and 4). The analysis is concentrated on the effects of rotor saliency ratio on flux density components in the airgap.

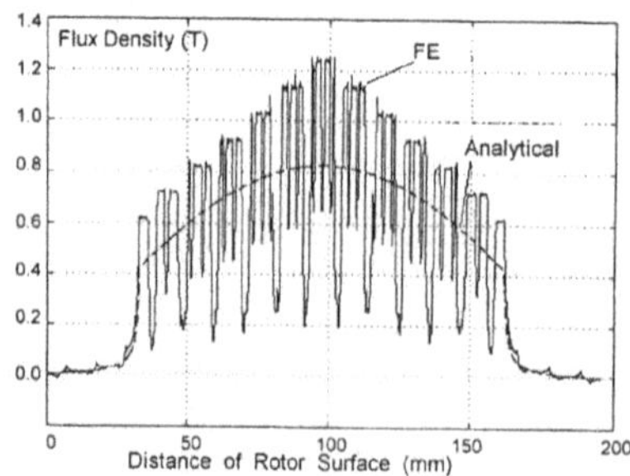

Figure 3: d-axis radial flux density distribution

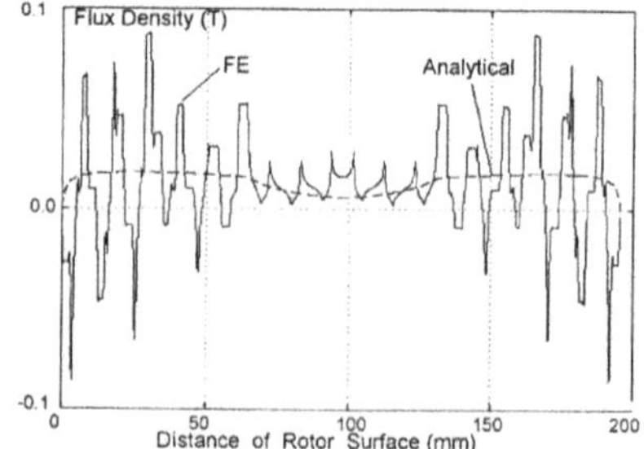

Figure 4:q-axis radial flux density distribution

Figures 5 and 6 indicate the ratio of the fundamental and harmonic amplitudes for d- and q-axis flux density distribution with the variation of rotor saliency ratio. About 0.66 of the saliency ratio of the rotor may be suggested to be adopted to reduce harmonic losses.

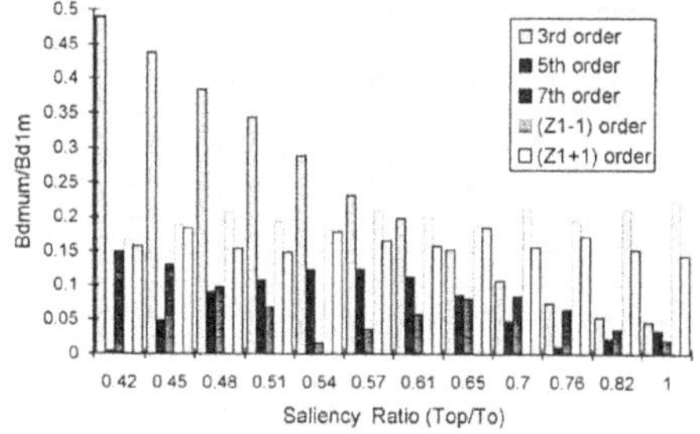

Figure 5: Ratio of d-axis harmonic to fundamental

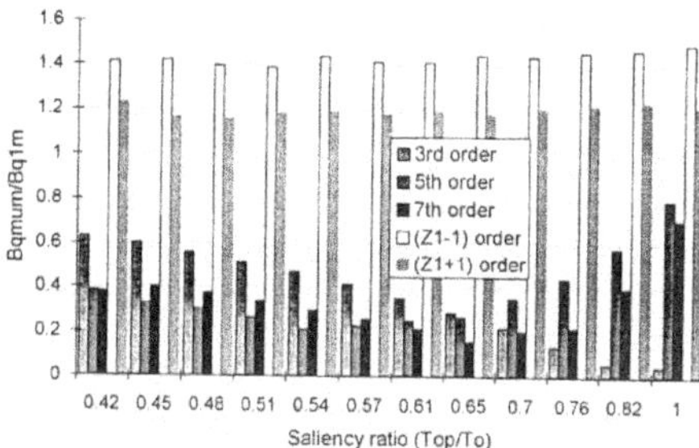

Figure 6: Ratio of q-axis harmonic to fundamental

COMPARISON BETWEEN RESULTS

To verify the agreement in the analysis and the FE, equations (1) to (2) and the numerical methods are used to calculate the d- and q-axis flux density distribution in the airgap of the linear models. Furthermore, fundamental and some harmonic components are evaluated.

Radial Flux Density Distribution in Airgap

The radial flux density distribution in the airgap along d- and q-axis is evaluated (figures 3 and 4). The d-axis results are in good agreement. However, the q-axis field waveforms are more different. The reason may be that some flux lines of the q-axis field cross the airgap a few times and specifically, some flux lines pass through the airgap twice under each stator tooth located at the pole span. This suggests that the q-axis airgap field is rich in stator slotting harmonic (figures 2 and 4).

Fundamental and Harmonic Components of Airgap Field

The waveforms of the radial flux density distributions resulting from FE are expanded in a Fourier series and the analytical results are also evaluated. Figures 7 and 8 show that d-axis results are in good agreement, but the amplitudes of q-axis airgap field show larger inconsistencies. This is clearly due to the construction of the radially laminated rotor.

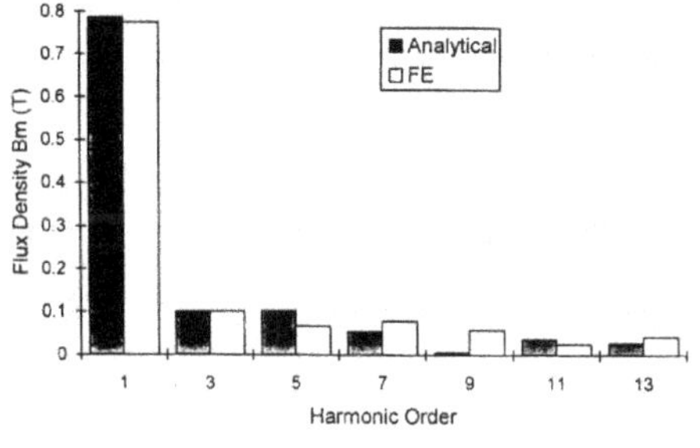

Figure 7: Component amplitudes of d-axis field

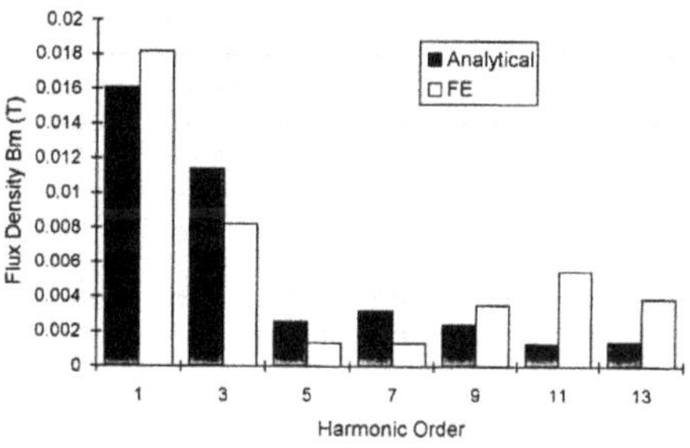

Figure 8: Component amplitudes of q-axis field

Comparison of the Different Components

The fundamental and 3rd order component variations of d- and q-axis fields with the saliency ratio resulted from analysis and FE are shown in figures 9 and 10. For d-axis and with the fluctuation of the saliency ratio, both results are coincided. For q-axis and the results of the fundamental and 3rd order, the related errors are larger. Especially, for q-axis fundamental component, the difference of both amplitudes produced from analysis and FE are decreased with the enlargement of the pole span. This may suggest that some modified coefficient has to be introduced to more exactly evaluate q-axis fundamental component of the field in the model with smaller saliency ratio.

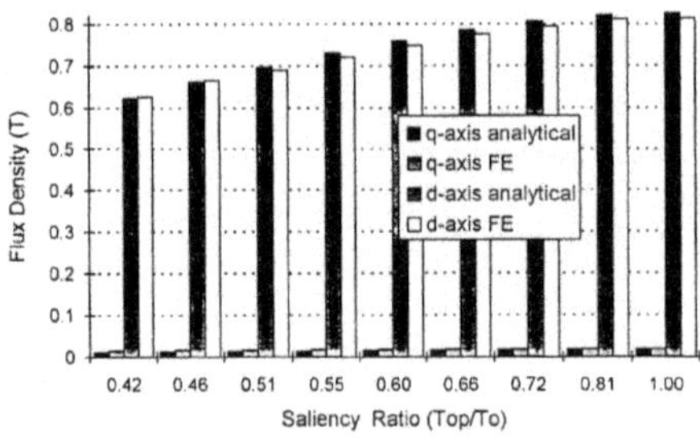

Figure 9: Comparison of fundamental components

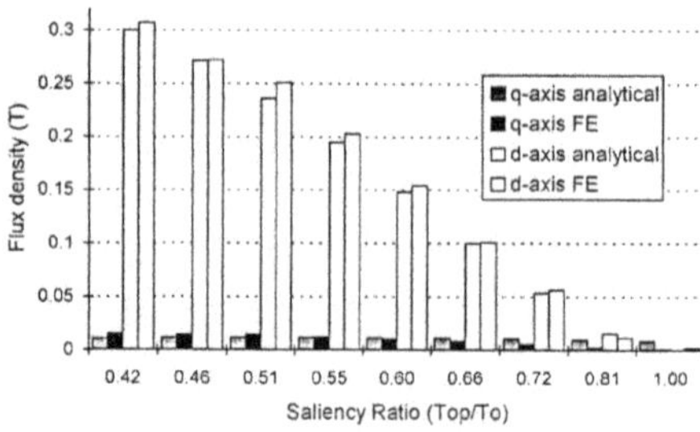

Figure 10: Comparison of 3rd-order components

CONCLUSIONS

An analytical approach and finite element method are used to evaluate the d- and q-axis field distribution in the airgap of the synchronous reluctance motor. Main conclusions are:

- For the d- and q-axis, the amplitudes of the fundamental and some harmonic components vary with the saliency ratio of pole span to pole pitch.
- The magnetic flux density distribution along d-axis contains mainly the fundamental component with relatively notable harmonics due to stator slotting, rotor lamination and insulation layers as well as the saliency ratio of the rotor.
- For the q-axis field, the amplitude of fundamental component is less than that of the harmonic components produced by the stator slotting. Therefore, there is mainly a leakage field and the corresponding q-axis magnetising inductance is very small.
- About 0.66 of the saliency ratio of the rotor, semiclosed stator slots and thinner rotor insulations are suggested to be adopted to reduce iron losses in the rotor laminations, caused by harmonics, especially in the q-axis field.
- In order to increase the ratio of d-axis inductance to q-axis inductance (L_d/L_q), some measure should be adopted to decrease stator and rotor slotting harmonics.

ACKNOWLEDGEMENTS

The financial supports for this work from the Belgian Nationaal Fonds voor Wetenschappelijk Onderzoek, Belgian Ministry of Scientific Research for granting the IUAP No. 51 on Magnetic Fields and K. U. Leuven are gratefully acknowledged.

REFERENCE

1. I. Boldea and S. A. Nasar, Emerging Electric Machines with Axially Laminated Anisotropic Rotor, *Electrical Machines and Power System* 19: 673 (1991).

THE SHAPING OF FLUX DENSITY AT THE AIR GAP OF SMALL DC MOTORS WITH DIFFERENT PERMANENT-MAGNET POLES

M. Rizzo,[1] A. Savini,[2] and J. Turowski[3]

[1] Department of Electrical Engineering, University of Palermo, Italy
[2] Department of Electrical Engineering, University of Pavia, Italy
[3] Technical University of Lodz, Poland

INTRODUCTION

Design, operation and reliability of permanent-magnet DC micromotors are highly dependent of the magnet employed, i.e. of its demagnetization curve, type of magnetization (either radial or tangential), size and position within the micromotor as well as of its shape.

In fact, it is known that, for a given configuration of rotor and stator, solutions with either radial or transverse permanent magnets are available[1]. Moreover, the shapes of polar shoes and air gap may bring about either extension or concentration of magnetic flux in the gap area, whereas pole shoes can be made up with solid mild steel or, in the case of slotted armatures, with laminated iron. In addition, demagnetization is an important problem, to prevent which gap widening can be employed[2]. Actually, knowing the distribution of flux density at air gap for different cases, it is possible to understand the influence of all these variables in a comparative way

The present work is just aimed at assessing the best configuration of flux density at the air gap of a permanent-magnet small dc motor in order to fulfil a given performance of the motor.

FINITE ELEMENT MODELLING

Four different models of 4-pole motors have been examined. All of them have the same geometrical dimensions: external rotor diameter D_r = 20.0 mm; external stator diameter D_s = 42.0 mm; width of air gap δ =0.6 mm. In all cases the so called copper rotor has been used with large slots and thin teeth. As to the number of slots, both possibilities have been considered, i.e. odd number, such as suggested by the classical theory of electrical machines, and even number[3]. As a result,it has been confirmed that an odd number of slots produces a smoother distribution of flux density at air gap because of a more uniform distribution of reluctance.

The permanent magnet employed for all four motors is Ferrite with B_r = 0.22 T and H_c = 120 000 At/m; in the various cases it has the same volume but different shape and position.

Electric and Magnetic Fields, Edited by A. Nicolet
and R. Belmans, Plenum Press, New York, 1995

The motor in Fig. 1 a) has a pole shoe made up with the same material as the yoke and an angular width equal to 66% of the pole pitch p (according to the classical theory the optimal width of pole shoe must vary between 60 and 70 % of p); the magnetization is of radial type.
The motor in Fig. 1 b) exhibits air-gap width increasing from the axis to the tips of pole shoes (the maximum value is 14 mm) and is given a radial magnetization.

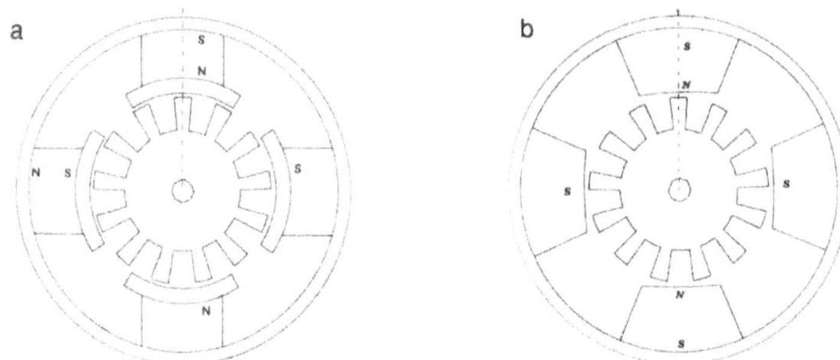

Figure 1. Permanent-magnet motors with radial magnetization: a) with pole shoe; b) with variable air gap.

The motor in Fig. 2 a) has a magnet with parallel magnetization. The motor in Fig. 2 b) has magnet with parallel magnetization as well but angular position. Both the latter motors exhibit large pole shoes made up with the same material as the stator yoke.

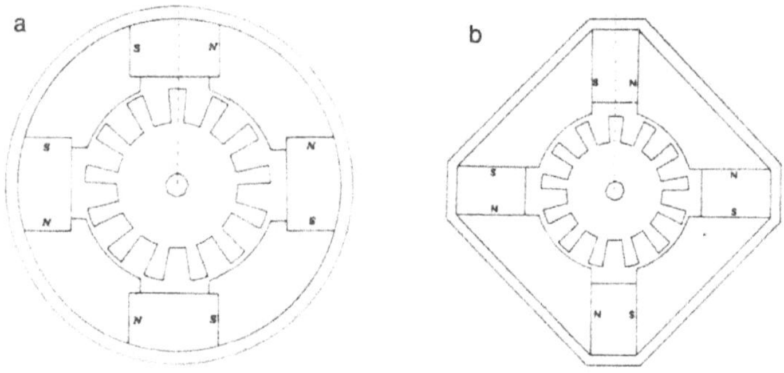

Figure 2. Permanent-magnet motors with parallel magnetization: a) with centered position; b) with angular position.

In order to investigate the influence of the different variables a numerical simulation of the magnetic field in the cross-section of the four motors has been carried out by the finite element method.

The demagnetization curve of the magnet has been approximated by a straight line even if the computer code employed is able to consider, if necessary, a recoil line.

The cross-section of the four motors has been subdivided into first-order triangular elements amounting to 1736 with 890 nodes.

The permanent magnet has been modelled in such a way that the magnetic field created by it, assuming a uniform magnetization, is equivalent to that produced by a current distributed along the surfaces of the magnet itself with density

$$\mathbf{J}_e = \mathbf{M}_o \times \mathbf{n} \tag{1}$$

where $\mathbf{n}$ is the unit vector normal to the surface and $\mathbf{M}_o = B_r/\mu o$.

The motors have been studied at no load in order to avoid the possible smoothing effect of load.

The results, in terms of radial component of flux density at air gap, for a double pole pitch, are reported in Fig.s 3 a), b), c), d).

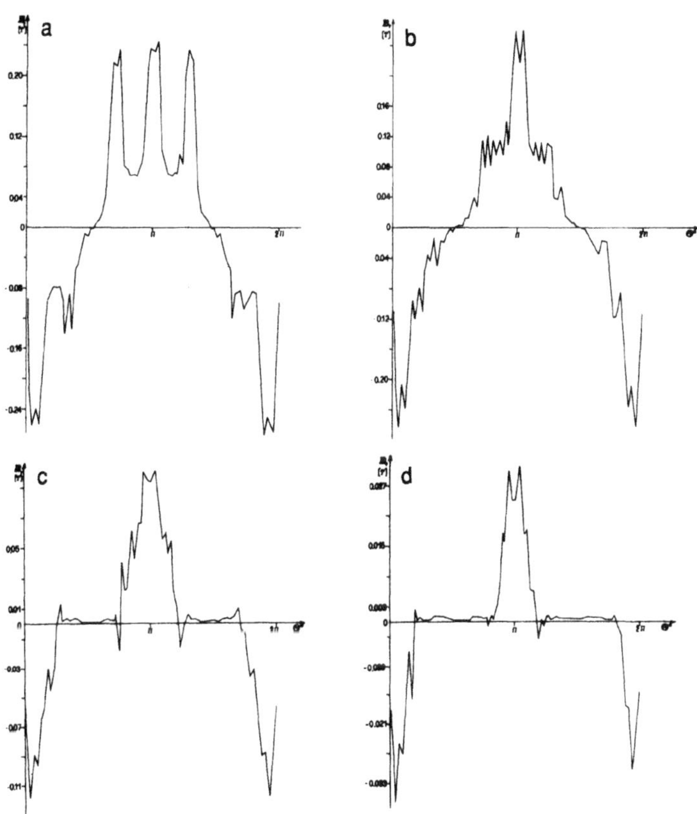

Figure 3. B_r as a function of Θ_e in a double pole pitch for the motors depicted : a) in Fig. 1 a); b) in Fig. 1 b); c) in Fig. 2 a); d) in Fig. 2 b).

As a reference position, that corresponding to the alignment of magnet axis and rotor tooth axis has been selected.

The two-dimensional magnetic field distribution has been investigated by solving the equations

$$\mathbf{curl}\ A = \mu H \tag{2}$$

$$\text{curl}\ \mathbf{H} = J \tag{3}$$

where A is the vector potential ($Wb\ m^{-1}$) with null divergence, J is the current density ($A\ m^{-2}$) and μ the permeability ($H\ m^{-1}$).

In a system of rectangular coordinates Maxwell's equations (2) and (3) take the form

$$(dA/dx_2, -dA/dx_1) = (\mu H_1, \mu H_2) \tag{4}$$

$$dH_2/dx_1 - dH_1/dx_2 = J \tag{5}$$

Instead of Eq.s (4) and (5) describing the field by means of magnetic field intensity $\mathbf{H} = (H_1, H_2)$, the following equations may be considered describing the field in terms of the rotated field $\mathbf{R} = (H_2, -H_1)$

$$-\mathbf{grad}\ A = \mu \mathbf{R} \tag{6}$$

$$\mathrm{div}\ \mathbf{R} = J \tag{7}$$

so that for the magnetic vector potential A the governing equation is

$$-\mathrm{div}\ (\nu\ \mathbf{grad}\ A) = J \tag{8}$$

where ν is the reciprocal of μ.

Homogeneous Neumann boundary conditions are assumed along all the boundary of the field region.

REMARKS AND CONCLUSION

Examining the diagrams shown in Fig. 3 it is possible to remark that the type of magnetization is, by far, more influential than the position of the magnet. In fact, the motor in Fig. 1b) and that in Fig. 2a) differ mainly because of the type of magnetization which in one case is radial and in the other one tangential; yet, the former has a peak value of flux density of 0.26 T while the latter a peak value of 0.10 T. Incidentally it must be noted that the fact that the flux at the air gap and the magnetization field of the magnet have the same direction brings about a reduction in losses. Moreover, for a given type of magnetization, the position of the magnet, either angular or centered, has remarkable influence as well. In fact the motor in Fig. 2b) has a waveform of flux density distribution at air gap quite comparable with that exhibited by the motor in Fig. 2a), but with a peak value of flux density notably different, i.e. 0.031 T. The motor in Fig. 1a), although having a peak value of flux density practically equal to that of the motor in Fig. 1b), has, compared to the latter, because of the presence of the pole shoe, an air-gap distribution less regular, which brings about a decrease of the magnetic energy corresponding to a given volume of magnet.

REFERENCES

1. M. Rizzo, A. Savini, J. Turowski, and S. Wiak, The influence of permanent magnets in brushless DC motors, in "Proc. CEFC 92 Conf", Los Angeles (1992)
2. M. Rizzo, A. Savini, and J. Turowski, FEM analysis and design of polar shoes in permanent-magnet DC motors, in "Simulation and design of applied electromagnetic systems", Elsevier (1994)
3. T. Kenjo and S. Nagamori, "Permanent Magnet and Brushless DC Motors", Clarendon Press, Oxford (1985)

FEM COMPUTATION OF L_d AND L_q IN AXIAL FLUX DISC MACHINES

Beatrice Mellara, Ezio Santini

Universita' di Roma "La Sapienza"
Dipartimento di Ingegneria Elettrica
Via Eudossiana 18
00184 Rome, Italy

INTRODUCTION

Axial flux disc machines (AFDMs) present interesting characteristics in terms of specific torque, efficiency and losses. Their development is mainly due to availability of innovative materials (permanent magnets, high permeability non-directional irons). They are considered one of the most interesting motors for individual electric traction. Their development is still in progress, especially from a theoretical viewpoint, and modelling problems are far to be completely solved [1, 2, 3, 4].

Numerical evaluation of L_d and L_q is a crucial point in the design of a rotating machine such as AFDM. Steady-state operation and transient performance of this machine are in fact described by these parameters. Their numerical evaluation is a great aid for the designer, and this paper shows a procedure for their accurate determination.

The algorithms used for the determination of the sequence inductances follows much the methods used for their experimental measurement. For sake of generality, both the rotor and stator structures will be hypothized as non-isotropic, with different space periods of geometric anisotropy. Materials will be assumed linear, and losses (eddy currents, hysteresis) will be neglected. The number of turns of each coil will be assumed as one, as well as the number of pole pairs. Windings will be three phase, each one having a number q of coils (in this case turns, because into each coil there is just one turn) series connected.

DEDUCTION OF THE GEOMETRICAL MODEL

Axial flux machines present in general a toroidal stator: the relevant windings are wound in a toroidal fashion around the stator, and their axis is orthogonal to the axis of the stator. One or two disc rotors face the stator and their axis is the same axis of the stator torus. The airgap is axial and the airgap surface is an annulus. The flux paths are mainly two straight segments in the airgap and two azimuthal arcs in the stator and rotor iron. Fields are therefore 3D. The number of poles for this kind of machines is high, and the reluctance in the stator iron is much less that one offered to flux paths in azimuthal direction in the air.

Electric and Magnetic Fields, Edited by A. Nicolet
and R. Belmans, Plenum Press, New York, 1995

Therefore the paths of the magnetic flux can be studied as 2D: the relevant geometry is the external surface of a cylinder having the same axis as the stator, and cutting this latter at a distance d such that $r < d < R$ (r and R denote the inner and outer stator radius). Moreover, optimisation of the design leads in general to a r / R ratio in the interval 0.6 - 0.8. This implies that no great difference exists between the inner and outer cylindrical sections; therefore a magnetic field analysis performed in the average stator radius describes the overall field behaviour of the machine with great accuracy.

ANALYTICAL CONSIDERATIONS

Inductances are geometric parameters, and they can be calculated in two ways: as the ratio between flux and current, or the ratio between the energy in the magnetic field and one half the current squared. Obviously, for linear systems (that will be studied in this paper) the two definitions leads to the same quantities [5]. The proposed algorithm (and the relevant computer program) uses both formulation in order to evaluate with high accuracy the desired numerical values.

In principle, numerical evaluation of L_d and L_q can be performed in a simple way. A time instant t must be fixed: the currents in the windings can be calculated, since

$$I_a = I \cos \omega t \quad ; \quad I_b = I \cos(\omega t - 2\pi/3) \quad ; \quad I_c = I \cos(\omega t + 2\pi/3)$$

The d-axis and q-axis currents, as well as the direction of the d-axis θ, are consequently known:

$$I_d = \sqrt{\frac{3}{2}}\, I \cos(\omega t - \theta) \qquad\qquad I_q = \sqrt{\frac{3}{2}}\, I \cos(\omega t - \theta - \pi/2)$$

It is enough therefore to place the rotor in the direction of the d-axis (or q-axis), to fix the currents in the windings (according with the sketch in Fig. 1), and to perform the FEM field analysis. Post processing allows to evaluate the energy in the field and the flux linked with the single coil or group of coils or phase [1], [2]. Projecting the fluxes linked with the phases along d-axis (or q-axis), the ratio between flux and current in the same axis allows to determine the relevant sequence inductances.

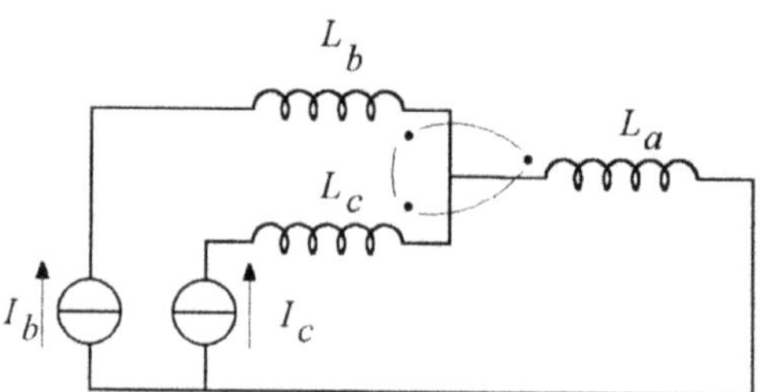

Fig. 1 Equivalent circuit and feeding of the windings for the numerical determination of the energy and of the fluxes linked with the phases.

The *a-priori* knowledge necessary for this task is only the geometrical direction of minimum reluctance of the rotor structure, which is evident in general by the view. No matter the angle θ (i.e. the relative stator-rotor position), the result of the numerical analysis leads to the same solution for L_d and L_q, apart from numerical approximations due to mesh irregularities, lack of symmetry, number of nodes in the mesh, constraints introduced by binary conditions, and so on. This holds, obviously, if no "magnetic interference" exists between stator and rotor: in other words if, moving the rotor, the reluctance of the stator flux paths does not change significantly. This is the case of reluctance machines with high rotor anisotropy (*equal pitch machines* [6]). In such machines there is a modulation of L_d and L_q along θ; moreover, there are relative stator-rotor positions in which a flux in the q-axis is evidentiated, even if there is no current along that axis (the same holds for the d-axis). In the following this case will be disregarded, even if the relevant calculations follow greatly the proposed method.

EVALUATION OF L_d AND L_q FROM PHASE INDUCTANCES

Sequence inductances can be calculated directly from the phase inductances. By definition, infact, the phase inductance $L_f(\theta)$ and the d- and q-axis inductances are [7]:

$$L_f = L_\sigma + L_1 + L_2 \cos 2\theta \qquad L_d = L_\sigma + 3/2\,(L_1 + L_2) \qquad L_q = L_\sigma + 3/2\,(L_1 - L_2)$$

If one phase is fed such that its magnetic axis is parallel to the rotor axis, the quantity $(L_\sigma + L_1 + L_2)$ can be evaluated. Feeding the phase in such a way that its magnetic axis is orthogonal to the direct axis, leads to the numerical determination of $(L_\sigma + L_1 - L_2)$.

The numerical value of L_σ must be previously determined in order to evaluate L_1 and L_2. Three possibilities exist:

1 - test at extracted rotor (rotor is taken off and windings are fed by a three-phase system);
2 - homopolar test (stator winding are series connected);
3 - numerical evaluation of the energy stored into the slots and in the end connections.

In test 1, flux paths into the slots are magnetically in parallel with respect to the paths into the region in air previously occupied by the rotor(s). Therefore, results of test 1 are highly approximated: paths in air, although longer than paths into the slots, present sections of great dimensions. In laboratory practice, the results of this test are commonly multiplied by an unknown factor 1.5 ÷ 2. Test 2, on the other hand, has a numerical meaning if the flux due to the currents into the slots is confined in the slots by concurrent flux (short-circuit test in induction machines). In the case that no excitation is present in the rotor, flux arising from stator windings is free to circulate into the rotor, and the only effect of homopolar feeding is to multiply by three the number of poles, with unpredictable numerical results, that are therefore are highly inaccurate. The possibility offered by test 3, that is characteristic of FEM and not feasible in laboratory practice, consists in the evaluation of the energy stored into the leakage regions. This task can be easily performed, if each slot and each tooth is defined as a region Ω, where the local contribution to the leakage energy

$$W_\Omega = \int_\Omega \int_0^B H(b)\,db\; d\Omega$$

can be calculated.

Since the leakage flux is practically independent on the relative stator-rotor position, its calculation can be performed in any stator-rotor position. Once L_σ has been determined, numerical evaluation of L_1 and L_2 is immediate, as well as L_d and L_q.

DIRECT EVALUATION OF L_d AND L_q

Direct evaluation of L_d (or L_q) is however possible, as described in the beginning of this paragraph. The windings must be fed with a given sequence of currents, the direction θ

of the d-axis (or the q-axis) must be determined, the rotor must be aligned in the same direction, and flux linkages in the phases must be determined by evaluating the average value of the vector potential $A(x,y)$ in the regions Ω occupied by the conductors (this holds if the current density into each conductor is constant). These quantities have to be projected in the direction of the d-axis (or q-axis) in order to evaluate the relevant flux. Numerical determination of L_d (or L_q) is readily accomplished. FEM analysis is greatly simplified if $\theta = 0$. In this case, the stator windings can be considered as a system of inductances equivalent to a single inductance, where the phase in which the current is maximum is series-connected to the remaining two phases parallel-connected. This fact is evident from Fig. (1) if $I_b = I_c$. In this case currents in the phases b and c are one half the current in phase a: the magnetic axis of this latter is superimposed to the minimum reluctance direction of the rotor, the remaining phases are magnetically in the same conditions, and therefore the whole machine can be considered as a single inductor.

This procedure requires a clever evaluation of the fluxes linked with the phases. The value of the d-axis (or q-axis) inductance can be computed starting from the vector potential A in the regions occupied by the windings. Standard numerical procedure for the evaluation of inductances must be modified as follows:

$$L = \frac{1}{i^2} \int_{\Omega} A\, J\, d\Omega = \frac{1}{i} \sum_{k=1}^{N} \frac{1}{N_{pk}} \frac{1}{\Omega_k} \int_{\Omega_k} A\, d\Omega_k$$

where i is the external instantaneous current value in the system of inductances, Ω_k is the region occupied by the k-th conductor (the total number of regions is in this case $N = 3q$), and N_{pk} is the number of regions parallel-connected when considering the current region k; in this case N_{pk} is one for the conductors of phase a carrying the total current, and N_{pk} is 2 for the conductors of the remaining two phases b and c.

The results of this procedure should coincide, apart numerical approximations characteristic of FEM, with the results obtained by means of analysis of phase inductances. Numerical tests show that a great accuracy is requested in FEM analysis, particularly when high precision is requested in flux linkages. Adequate choice of geometries allows to limit FEM analyses to just one pole: numerical tests show that use of binary conditions, even if not strictly needed, leads in general to greater symmetries in the evaluation of the fluxes. This leads to increase the time for the solution, but insures also a greater symmetry in the computation of fluxes when the geometry exhibits $q > 1$. Possibility of comparison between two different procedures, however, is characteristic in the computation of inductances [5], and this can be used as a criterion for the evaluation of the reached precision.

REFERENCES

[1] E. Santini et alii - **Axial-Flux Machine Having Counter-Rotating PM Motors for Propulsive Electrical Systems of Underwater Vehicles** - International Symposium on Ship and Shipping Research, Genova (Italy), 7 - 10 July 1992, pp. 6.8.1 - 6.8.11.

[2] E. Santini et alii - **Performance Evaluation of an Axial-Flux PM Generator** - International Conference on Electrical Machines (I.C.E.M.), Manchester (UK), 15 - 17 September 1992.

[3] E. Santini et alii - **Optimum CAD-CAE Design of Axial-Flux PM Motors** - International Conference on Electrical Machines (I.C.E.M.), Manchester (UK), 15 - 17 September 1992.

[4] E. Santini et alii - **Optimum Design of Iron-Less Stator Winding for Axial-Flux PM Machines** - 6th IEE International Conference on Electrical Machines and Drives, Oxford (UK), September 8-10, 1993.

[5] D.A. Lowther and P. P. Silvester. **Computer Aided Design in Magnetics.** Springer-Verlag 1985.

[6] A. Vagati, G. Franceschini, I. Marongiu, G. P. Troglia - **Design Criteria of High-Performance Synchronous Reluctance Motors** - IEEE IAS Annual Meeting, Houston (U.S.A.), October 1992.

[7] O. Honorati - **Armoniche di tensione, di corrente e di coppia nelle macchine sincrone** - Alta Frequenza, n. 3 Vol. XLII, 1973.

FEM DYNAMICS SIMULATION OF CONTROLLED-PM LSM MAGLEV VEHICLE

Kinjiro Yoshida, Hiroshi Takami, Shinichi Ogusa and Dai Yokota

Dept. of Electrical Engineering, Faculty of Engineering, Kyushu University
10-1 6-Chome Hakozaki Higashi-ku Fukuoka, 812 Japan

INTRODUCTION

The Maglev vehicle can be driven without wheels and with minimizing levitation power losses, by using a long-stator type linear synchronous motor (LSM) with controlled permanent-magnets (PM's), which has the integrated functions of LSM propulsion and attractive-mode levitation[1]. To reduce costs of the long-stator armature rails, it is practically important to simplify constructing the rails[2].

This paper presents FEM running simulations of 1/2 scale model Maglev vehicle which is propelled by a long-stator controlled-PM LSM[3] as shown in Figure 1. The Maglev vehicle is assumed to be levitated at a constant airgap length by the controlled-PM LSM. In a long-stator on the ground, semi-closed large slots are adapted and designed to install easily one-turn coils of a wave form. The large slots cause the controlled-PM LSM detent forces which give strong influences on dynamic operations of the running vehicle. A two-dimensional FEM is used for the dynamics simulation which is capable of precisely analyzing the detent forces produced between the stator teeth and the controlled-PM's.

Dynamics simulations show essential difference between the open-loop and feedback controls of the vehicle propulsion. The detent force problem in propulsion motion is successfully solved by adapting the feedback control. The dynamics simulations are verified from the experiments.

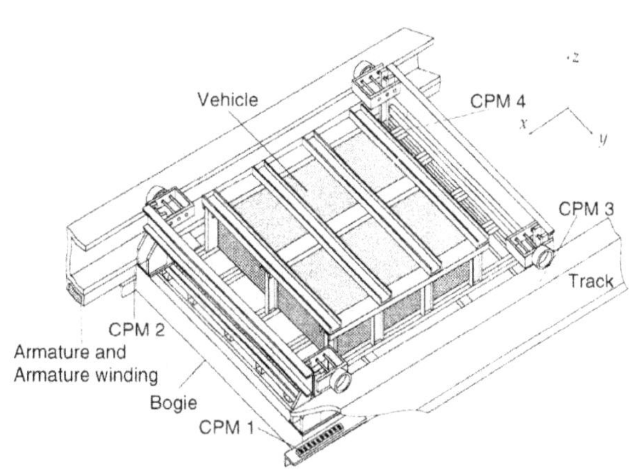

Figure 1. Configuration of controlled-PM LSM Maglev vehicle

Electric and Magnetic Fields, Edited by A. Nicolet
and R. Belmans, Plenum Press, New York, 1995

CONTROLLED-PM LSM MAGLEV VEHICLE

Figure 1 shows a configuration of the controlled-PM LSM Maglev vehicle which is a 1/2 scale model for experiments in our Laboratory. The vehicle is 1.70 m in length, 1.13 m in width and 180 kg in weight. The controlled-PM which consists of ten poles is mounted at the four corners connected through primary suspensions to the bogie. The travelling magnetic field is produced by supplying three phase currents for stator coils, and the vehicle is controlled to be propelled at the synchronous speed, with the controlled-PM's acting as the field exitation, while the vehicle is levitated stably by controlling the attractive force between the stator iron rail and the controlled-PM's.

THRUST FORCE ANALYSIS AND EQUATION OF MOTION

The equation of propulsion motion is simply described, neglecting the detent force, as follows:

$$M\dot{v}_x = F_{x0} - F_d \tag{1}$$

where

$$F_{x0} = F_x\left(\delta = \delta_0 \, , \, x_0 = \tau/2\right) = k_{F0}\, I_{10} \tag{2}$$

and F_x is the thrust force given in an analytical form[4], M the mass of the model vehicle, F_d aerodynamic drag force, v_x the vehicle speed, I_{10} an effective value of the stator current calculated in advance according to the demand speed pattern, k_{F0} the thrust force coefficient, δ_0 an air gap length where the weight of the model vehicle is balanced with the lift force, x_0 the mechanical load angle and τ is the pole pitch.

The discretization used for FEM thrust force analysis is shown in Figure 2. This shows discretization for one pole of the ten-pole controlled-PM, such as CPM1 as shown in Figure 1. Thrust force is calculated for one pole, considering its periodical structure. It includes the detent force depending on the relative position between the stator tooth and the controlled-PM. The vehicle has forty poles in all, so that the total thrust force is evaluated from forty times the one-pole thrust force. And the air gap length δ_o is assumed to be controlled at 10mm, where the PM-coil currents are assumed to be zero.

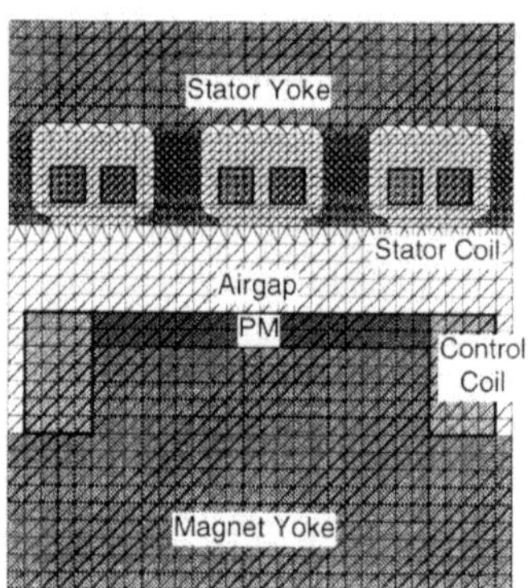

Figure 2. Discretization for one pole

DYNAMICS SIMULATION USING FEM FORCE ANALYSIS

F_{xo} needed for the vehicle to follow the demand speed pattern is calculated using equation (1) and then I_{10} is determined from equation (2). Instantaneous three phase armature currents, i_{ao}, i_{bo} and i_{co} are thus obtained in advance. In the open-loop control simulation, the equation of motion is solved for the thrust force analyzed by FEM subject to the specified armature currents.

In the feedback control simulation, using the demand speed v_{xo} and the vehicle speed v_x obtained from the dynamics simulation, the corresponding instantaneous armature currents are renewed each sampling time, according to the following control rule :

$$I_1 = G_1 \left(v_{x0} - v_x\right) + G_2 \int \left(v_{x0} - v_x\right) dt + I_{10} \tag{3}$$

where G_1, G_2 are feedback gains. In the discretization of the airgap region for the direction of linear motion, its width is 2mm for the tooth width of 10mm and the FEM thrust force analysis is carried out each 2mm as the vehicle moves. The time Δt_i, which it takes for the vehicle to pass through 2mm, is calculated from solving the equation of motion.

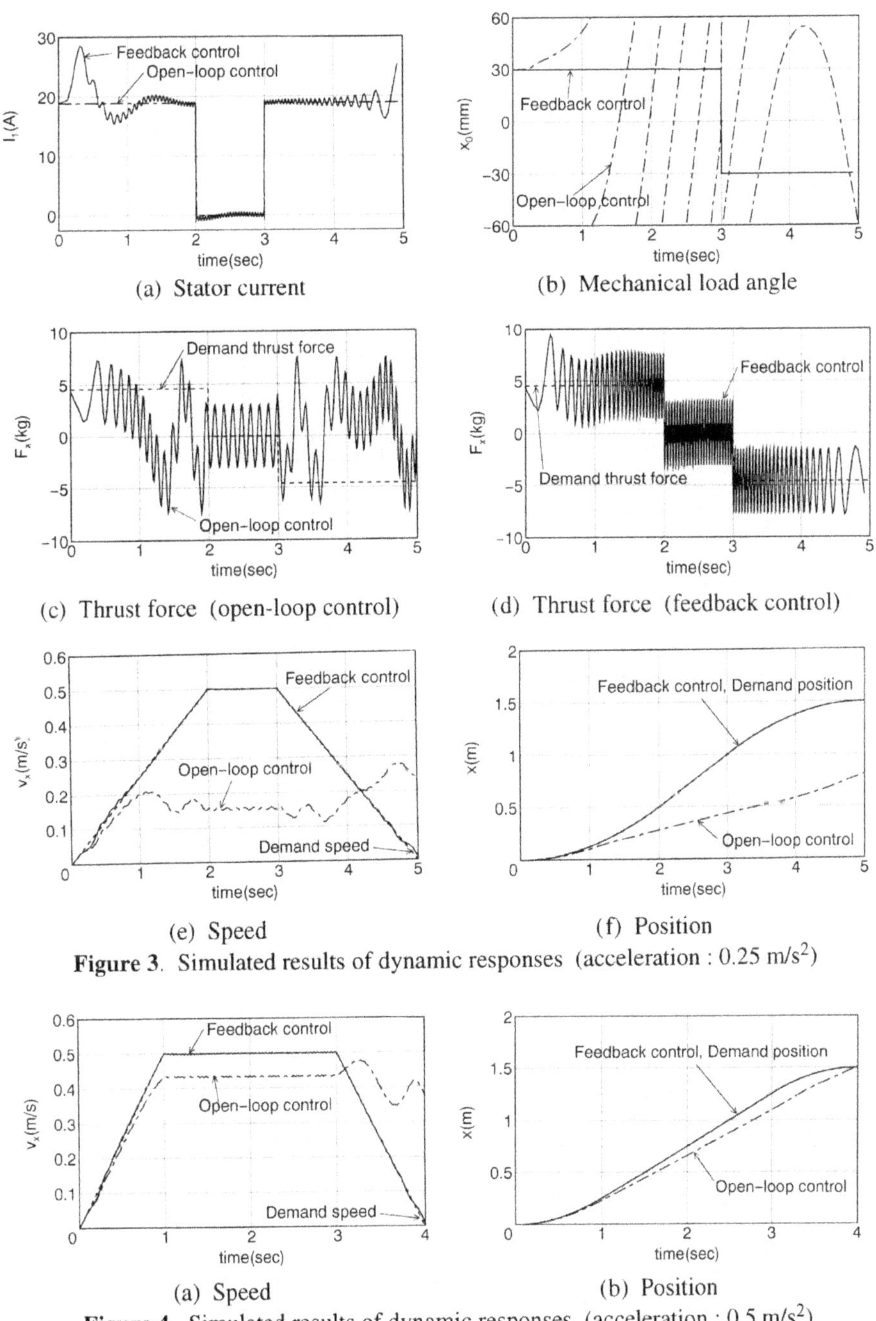

Figure 3. Simulated results of dynamic responses (acceleration : 0.25 m/s²)

Figure 4. Simulated results of dynamic responses (acceleration : 0.5 m/s²)

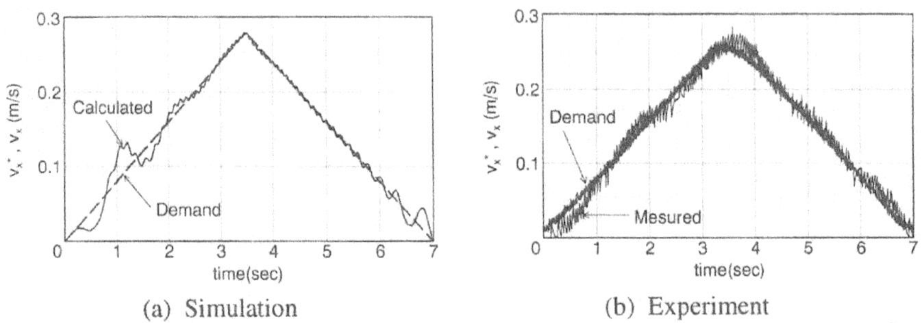

(a) Simulation (b) Experiment

Figure 5. Speed comparison of simulation with experiment (acceleration : 0.08 m/s^2)

SIMULATED RESULTS AND EXPERIMENTS

Figure 3 shows the simulated results for the vehicle to be controlled to follow the demand speed pattern with an acceleration of $0.25 m/s^2$ in Figure 3 (e), which are compared with respect to the open-loop and feedback controls. In Figure 3 (c) and (d), large ripples of the thrust force are produced in both control methods and they are detent force calculated by FEM. In the open loop control of propulsion motion, clearly from Figure 3 (b) and (e), after 0.5 seconds, the controlled-PM LSM pulls out synchronisum with the travelling magnetic field, due to the detent forces calculated by the FEM which cannot be compensated. In the feedback control, the effective value of the stator current is regulated quickly and the vehicle follows very well the demand speed pattern, as shown in Figure 3 (a) and (e).

Figure 4 shows simulated results for the speed and position responses in the case with twice acceleration as large as that in Figure 3. The feedback control also enables the vehicle to follow very well the more sharp demand-speed-pattern and to accomplish a good positioning.

Figure 5 shows very good agreement between the dynamics simulation and experiment.

CONCLUSIONS

For running simulation of the controlled-PM LSM Maglev vehicle, a two-dimensional FEM is applied for analyzing the thrust force including the detent force due to large slots which are adapted for reasonable design of constructing its long stator rail. It is found theoretically and experimentally that the proposed feedback control enables the Maglev vehicle to follow very well the required speed patterns with accelerations and decelerations of $0.08 m/s^2$, $0.25 m/s^2$, $0.5 m/s^2$ in spite of the large detent forces.

REFERENCES

1. H. Weh and H. May, "Fast Acting Magnets for Transportation Purposes", Proc. of Int. Conf. Maglev'85 in Tokyo, 155 (1985)
2. G. Heidelberg, K. Niemitz and H. Weinberger, "The M-Bahn system", Proc. of Int. Conf. Maglev Transport'84 in Solihul, 159 (1984)
3. K. Yoshida, S.Ogusa, D. Yokota, J. Hamamoto, I. Nakamura, "Running Simulation of 1/2Scale Controlled-PM LSM Maglev Vehicle", 5th Elect. Symp. Proc., 336 (1993)
4. K. Yoshida and H. Weh,: "A Method of Modelling Permanent Magnet for Analytical Approach to Electrical Machinery", Archiv für Elektrotechnik , 68, 229 (1985)

OPTIMIZATION OF CLAW-POLE ALTERNATORS USING 3D MAGNETIC FIELD CALCULATION

G. Henneberger, S. Küppers

Institut of Electrical Machines
University of Technology Aachen
Schinkelstraße 4, D-52056 Aachen

INTRODUCTION

Claw-pole alternators are used as automobile generators. The increasing amount of electronic components and small drives in cars help to make vehicles safer and more comfortable. So the alternator has to deliver more current, but at the same time its total volume must not change. Parallel the average speed of the combustion engine decreases because of increasing traffic density. So the alternator has to be improved at lower speeds especially.

The conventional optimization technique based on magnetic equivalent circuit and construction of a couple of samples is very time consuming, expensive and can be replaced by a more precise method. The tool that is used here bases on numerical field calculation with 3d finite elements. The electrical circuit with B6 rectifier bridge and battery is taken into account. The full calculation method was presented in an earlier paper [1]. A first modification to improve the output current by additional permanent magnets was investigated before [2].

An automatic parameter oriented meshing procedure was developed to investigate the influence of different sizes and to optimize the design of the alternator. In this paper the shape of the claws is modified to find an optimum of output current at standard stator.

THREE DIMENSIONAL MODELLING

The magnetic field of the claw-pole alternator is really three dimensional because of the small ratio of machine length to its diameter. The model for magnetic field calculation consists of machine shaft, the core with circumferrential exciting coil, two claw-pole wheels and the laminated stator. All parts of the rotor are made of solid iron. The full geometry of a typical alternator with 6 pole-pairs can be seen in figure 1.

Because of symmetry its enough to model only one pole-pitch. The periodicy of

Electric and Magnetic Fields, Edited by A. Nicolet
and R. Belmans, Plenum Press, New York, 1995

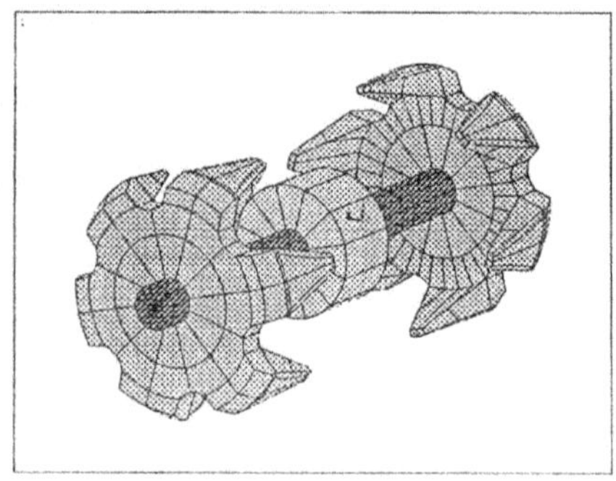

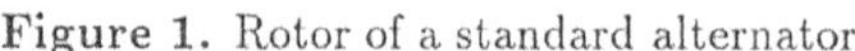

Figure 1. Rotor of a standard alternator

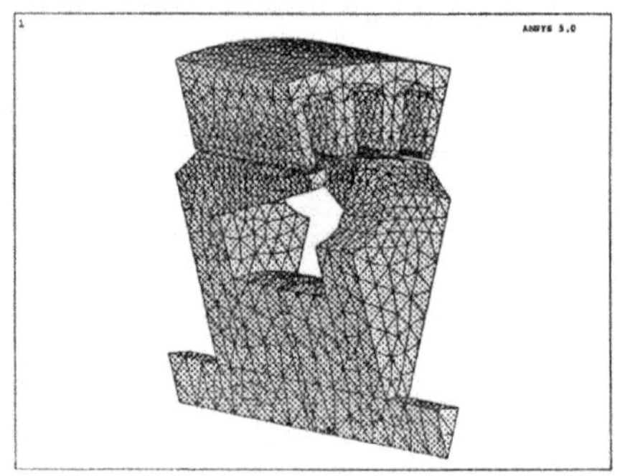

Figure 2. FE-mesh of the alternator

the magnetic field may be described at two points $P(x, y, z)$ and $P'(-x, y, -z)$:

$$\begin{aligned} B_x(P') &= -B_x(P) \\ B_y(P') &= -B_y(P) \\ B_z(P') &= B_z(P) \, . \end{aligned} \tag{1}$$

There are about 40 parameters to describe the alternator's geometry and some of them influence the distribution of field very strong. An automatic meshing procedure was developed to generate FE-meshes rapidly after changing one or more parameters. The mesh generation is done using the commercial FE-package ANSYS. As element type first order tetrahedra are used. Figure 2 shows a typical FE-mesh of the alternator. The generation of a mesh with 55.000 tetrahedra and 10000 nodes needs 20 minutes on a hp755 workstation. The mesh is converted to Patran-Neutralfile and is read into the FE-program MagNet.

CALCULATION METHOD

The nonlinear magnetostatic field problem is solved using the commercial 3d calculation program MagNet. A set of linear equations is solved for the magnetic scalar potential with the conjugate gradient technique [3]. The nonlinearity is taken into account in an additional Newton-Raphson procedure.

At no-load situation the induced voltage U_i can be calculated from the stator flux-linkage as function of the exciting current I_f. The position of stator coils is represented by a surface. At load situation an iterative solution method was developed [1]. The procedure starts with an initial guess of current I_1 and load-angle ψ_1 and is repeated, until the terminal voltage U_1, the output current I_1 and load-angle ψ_1 don't change any more. Usually there is convergence after 3 iteration steps.

DEFINITION OF PARAMETERS

The distribution of main and leakage flux as well as the output performance is dependent on a couple of geometric parameters. Here the effect of the claw's shape is investigated. Especially the pole-pitch factor α_i and the claw flank-angle φ_{cl} are varied to check the influence on the terminal voltage at no-load situation and the output current at load condition. Both sizes are defined at the surface of claws and are average values at middle of the claw.

$$\varphi_{cl} = atan\,\frac{0.5 \cdot (w_{max} - w_{min})}{l_{cl}} \tag{2}$$

$$\alpha_i = p \cdot \frac{w_{max} + w_{min}}{d_{ro} \cdot \pi} \quad (p = 6). \tag{3}$$

The shape of claws at the surface and the geometric parameters are shown in figure 3. For simplicity the shape is modelled without a chamfer that is usually put on one or both claws.

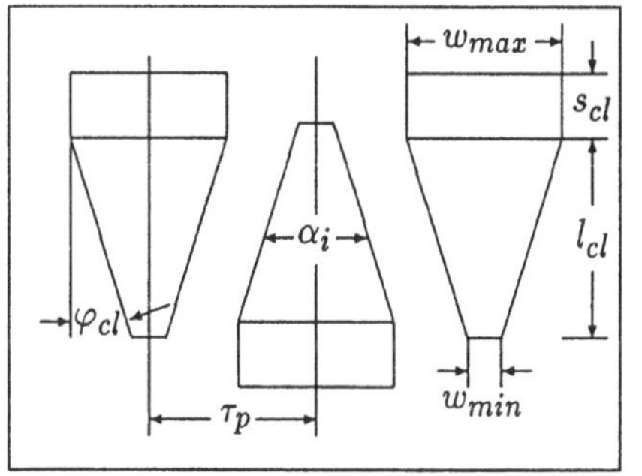

w_{min}: Min. width
w_{max}: Max. width
s_{cl}: Claw thickness
l_{cl}: Claw length
τ_p: Pole width
α_i: Pole-pitch factor
φ_{cl}: Claw angle
d_{ro}: Rotor diameter

Figure 3. Geometry of claw-poles

CALCULATION RESULTS

Variation of pole-pitch factor

The pole-pitch factor α_i is varied from 0.5 to 0.8 at no-load condition. The induced voltage U_i as a function of the exciting current I_f is presented in figure 4. It can be seen, that the optimum of induced voltage is dependent on saturation. At lower excitations the induced voltage is maximal at the highest pole-pitch factor. With increase of exciting current the machine becomes more and more saturated, and so the leakage flux between the claws increases, too. Consequently the optimum at higher currents decreases to the lowest pole-pitch factor of $\alpha_i = 0.5$.

At load-condition the situation is different because the optimum is dependent on the working point of excitation. All load calculations were done using a constant exciting current of $I_f = 4A$. The output current I_g is maximal at $\alpha_i = 0.5$ and decreases with increasing pole-pitch factor at total function of speed (Figure 5). The effect can be described looking at the leakage reactance $X_{1\sigma}$ and the reactance in quadrature axis X_q, which both increase with the pole-pitch factor.

Variation of claw-angle

As second size the claw-angle φ_{cl} is modified at constant pole-pitch factor of $\alpha_i = 0.62$. The angle is changed in five steps from 10 to 20 degrees. The results at no-load condition and load condition are presented below in figures 6 and 7.

There seems to be an optimum at an angle between 10 and 15 degrees. The maximum of induced voltage appears at $\varphi_{cl} = 12°$. At load condition and lower speeds

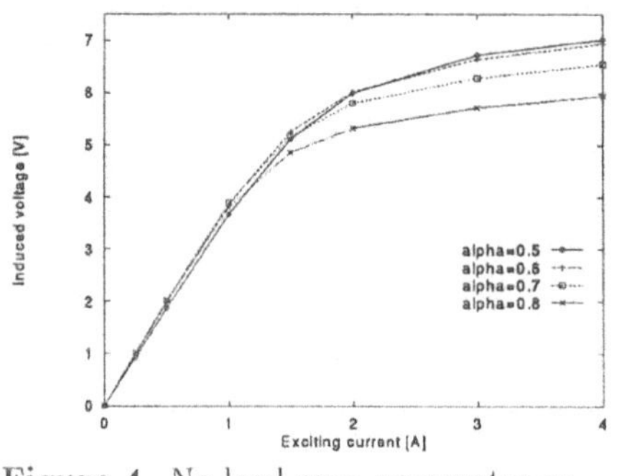

Figure 4. No-load case, parameter α_i

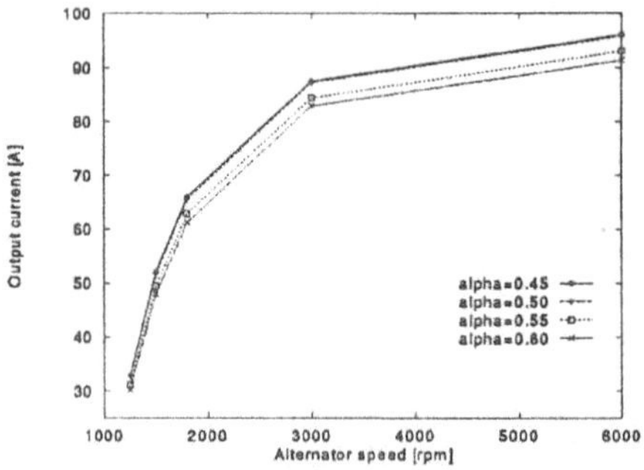

Figure 5. Load case, parameter α_i

the same maximum of output current can be found. At higher speeds the current decreases systematically with increasing claw angle. Usually the alternators have claw-angles ranging from 16 to 18 degrees.

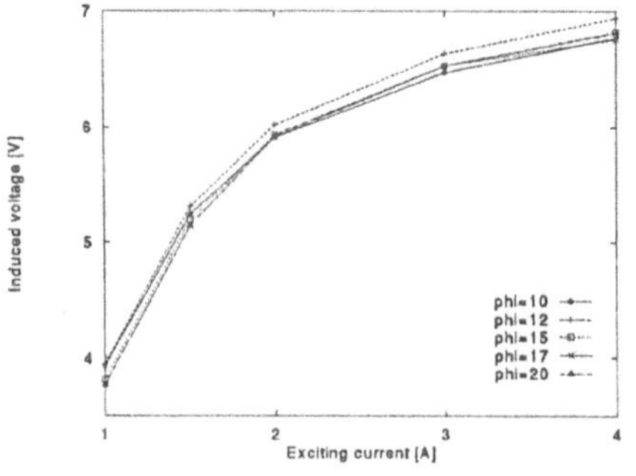

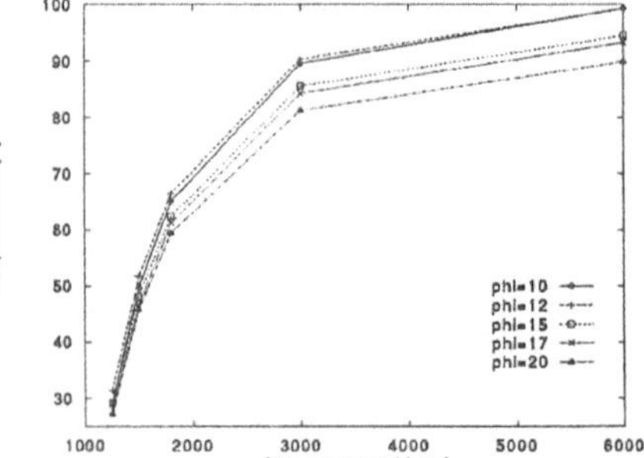

Figure 6. No-load case, parameter φ_{cl}

Figure 7. Load case, parameter φ_{cl}

CONCLUSION

An optimization technique has been presented to modify the shape of poles of a claw-pole alternator. As calculation tool a 3d FE method was used, based on an automatic parameter oriented meshing procedure. Induced voltage and output current were calculated changing the pole-pitch factor and the claw-angle. The optima of these geoemtric parameters were found.

REFERENCES

[1] R. Block and G. Henneberger, Simulation of the current and voltage waveforms of a claw-pole alternator, International Workshop on Electric and Magnetic Fields, Liege, Belgium, pp. 74.1 - 74.6 (1992).

[2] G. Henneberger and S. Küppers, Improvement of the output performence of claw-pole alternators by additional permanent magnets, ICEM Paris (1994).

[3] J.P. Webb and B. Forghani, A scalar potential method for 3d magnetostatics using edge elements, *IEEE Transactions on Magnetics Abbr.* 25:4126 (1989).

THREE-DIMENSIONAL FE ANALYSIS FOR EVALUATING SLIT EFFECTS IN LINEAR INDUCTION MOTOR

Takashi Onuki, Yushi Kamiya, and Tsugio Yamamura

Department of Electrical Engineering Waseda University
3-4-1 Ohkubo, Shinjuku-ku, Tokyo 169, Japan

INTRODUCTION

While a solid plate is usually adopted to the reaction plate of the linear induction motor (LIM), slits are often introduced to the plate in order to enforce the eddy currents effectively. In this paper, to find a suitable configuration of slits, we analyze the electromagnetic field of the LIM by 3-D FEM based on **A**-ϕ method. Adding to the ordinary approach, we propose a face-element method with double nodes for ϕ at slits to express the infinitesimal width of slits. In this paper, we adopt the periodical boundary conditions for simplicity of analysis, so it is virtually equal to analyze the usual rotating machine.

EXPERIMENTAL MODEL

Figure 1 shows the LIM used for our experiment. The specifications are Table 1.

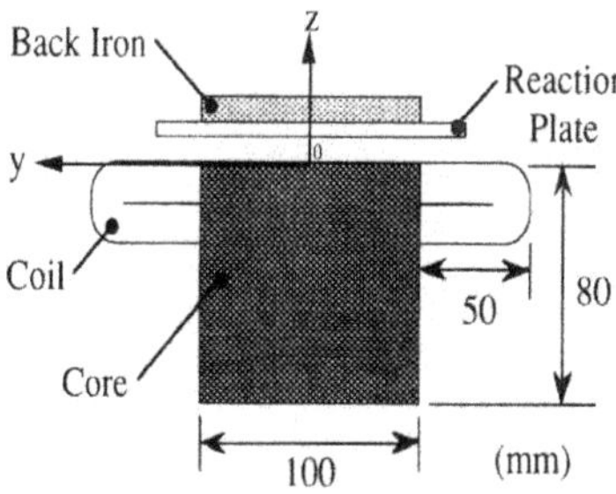

Figure 1. Basic configuration of the experimental LIM.

Table 1. Specifications of the experimetal LIM.

Pole pitch	225	[mm]
Coil pitch	225	[mm]
Winding factor	0.960	
Coil windings	180	[turns / pole]
Primary currents	2	[A]
Frequency	50	[Hz]
Length of rotor	450	[mm]
Thickness of reaction plate	1	[mm]
Thickness of back iron	2	[mm]
Air gap	4	[mm]

Electric and Magnetic Fields, Edited by A. Nicolet
and R. Belmans, Plenum Press, New York, 1995

THREE-DIMENSIONAL FE ANALYSIS

We analyze the electromagnetic field of the LIM by the three-dimensional finite element method adopting the magnetic vector potential **A** and the electric scalar potential ϕ with tetrahedron elements.

Analysis Model

Figure 2 shows the analysis model. Because of the symmetry, we can analyze only the left hand side of the LIM subjecting proper periodical boundary conditions.

Primary Winding MMF

For the primary winding mmf, we adopt a current sheet which consists of two regions (Fig. 3).

region 1: (coil side)

$$J_{1y} = J_s \cos\frac{\beta}{2}\, e^{j(s\omega t - \frac{\pi}{\tau}x)} \tag{1}$$

region 2: (coil end)

$$J_{2x} = jJ_s \cos\alpha \sin\left\{\frac{\pi-\beta}{2Y_0}(y-a)+\frac{\beta}{2}\right\} e^{j(s\omega t - \frac{\pi}{\tau}x)} \tag{2}$$

$$J_{2y} = J_s \sin\alpha \cos\left\{\frac{\pi-\beta}{2Y_0}(y-a)+\frac{\beta}{2}\right\} e^{j(s\omega t - \frac{\pi}{\tau}x)} \tag{3}$$

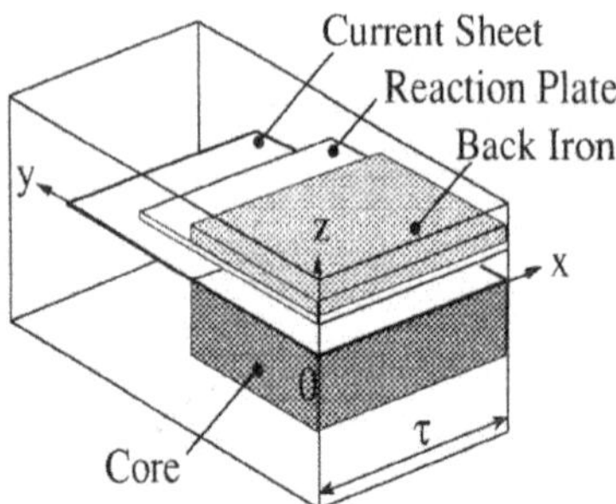

Figure 2. Analysis model.

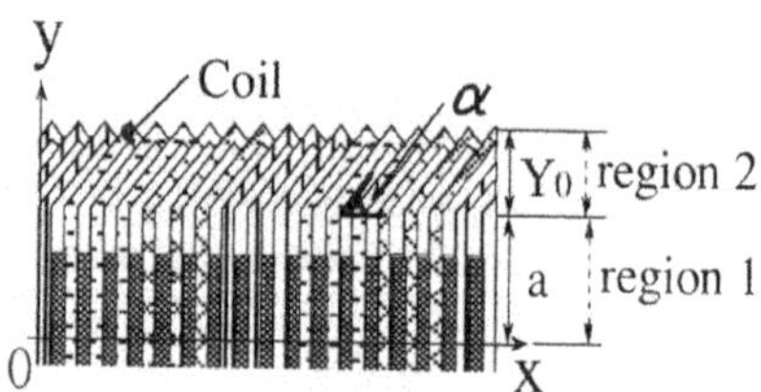

Figure 3. Diamond coils.

Face-Element with Double Nodes for ϕ

To express the infinitesimal width of slits, besides the discontinuity of the electric scalar potential ϕ [1] (Fig. 4), we propose a face element method with double nodes for ϕ (refer to method (a)). This method is compared with the usual approach that is $\sigma x=0$ at the slit region (refer to method (b)). Figure 5 shows the eddy currents in the reaction plate calculated by the new face element method.

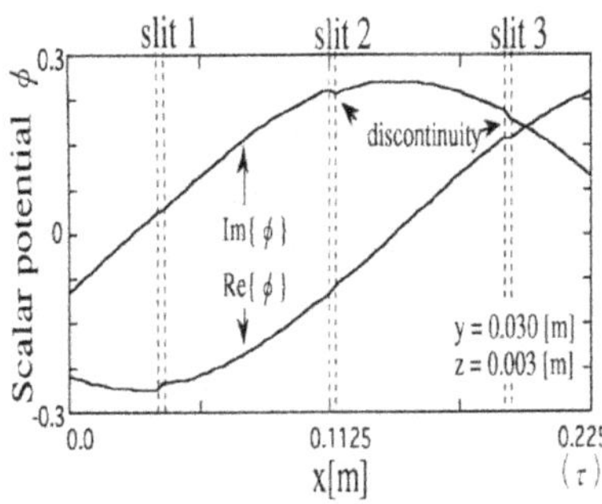

Figure 4. Distribution of the electric scalar potential ϕ in the reaction plate with three slits. (Calculated by the ordinary method that regard slits as air elements.)

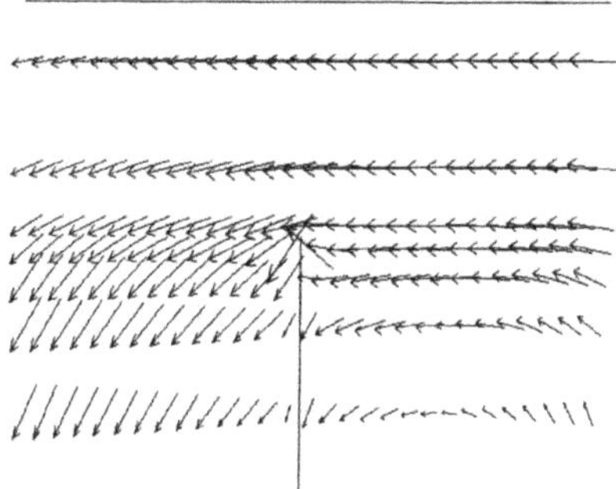

Figure 5. Eddy currents in the reaction plate with a slit. (Calculated by the face element method with double nodes for ϕ.)

RESULTS

Based on above approaches, we evaluate slit effects in the LIM, and compare it with the experimental data to conform the validity of the methods.

The Number of Slits

Figure 6 shows the flux density in the air gap with the various number of slits. The number of slits per slot-pitch is required more than 4 to obtain sufficient performances. These results are obtained by method (a), and (b).

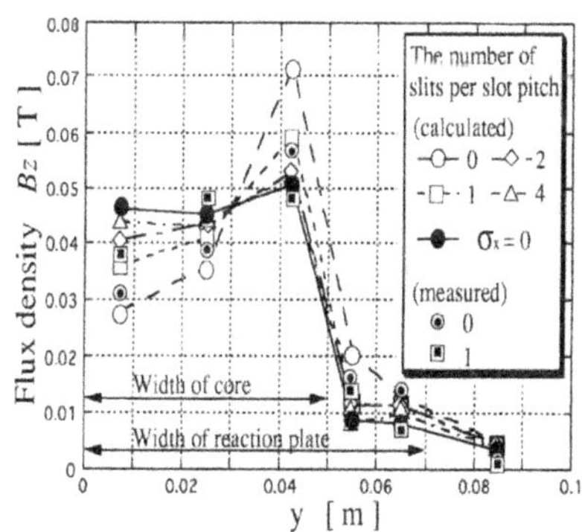

Figure 6. Distribution of the flux density in the air gap with various number of slits.

The Length of Slits

Figure 7, and 8 shows the eddy currents in the reaction plate. It indicates that the preferable slit length which enforce the eddy current effectively is a little bit longer than the stator width.

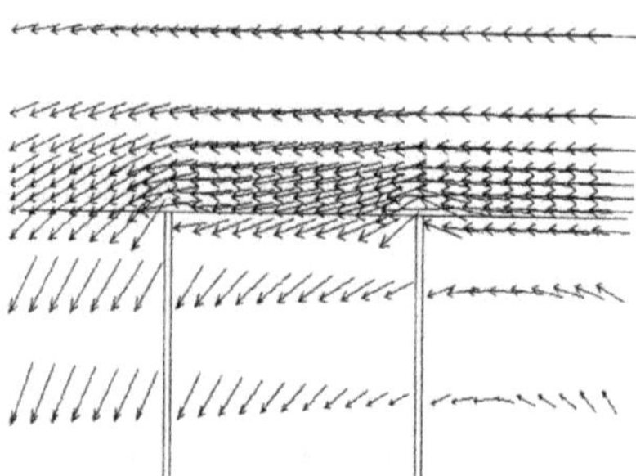

Figure 7. Eddy currents in the reaction plate with normal length of slit.

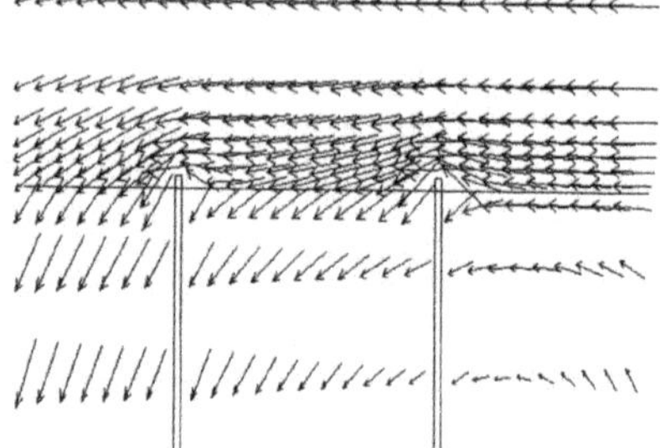

Figure 8. Eddy currents in the reaction plate with longer slit.

No Slit Region

A belt of no slit region in the middle part of the reaction plate is proposed from the structural point of view (Fig. 9), its width should be less than 40% of the stator width to obtain sufficient performances (Fig. 10). This results are obtained by method (b).

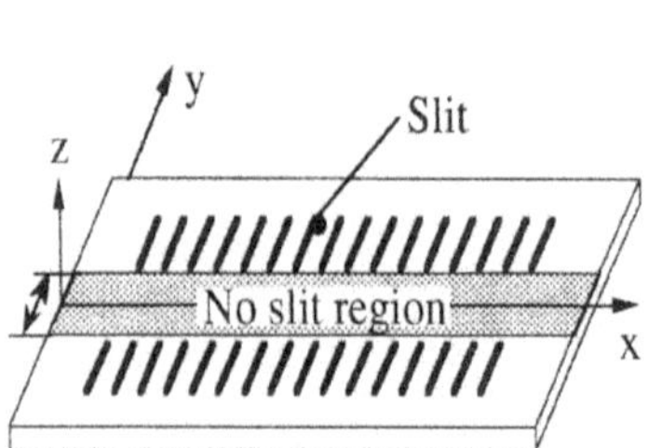

Figure 9. Reaction plate with no slit region.

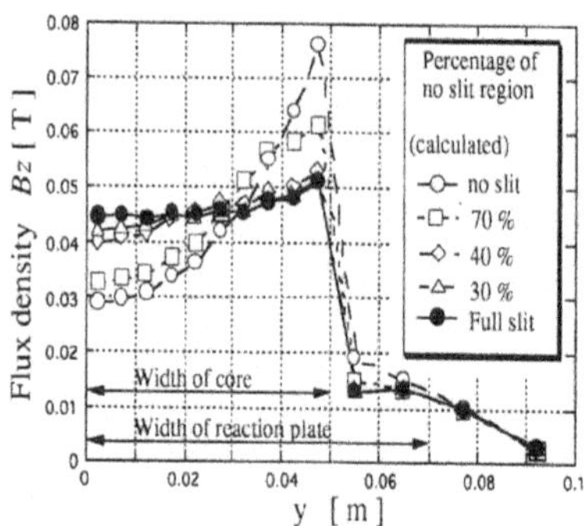

Figure 10. Distribution of the flux density in the air gap with no slit region.

CONCLUSION

In this paper, we investigate the slit effect in the LIM to find a suitable configuration of the slit by 3D FE analysis, and propose a face element procedure with double nodes for ϕ at slits to express its infinitesimal width. Results are following: (1) Number of slits per slot pitch is required more than 4 for obtaining sufficient performances. (2) Preferable slit length is a little bit longer than the stator width. (3) A belt of no slit region in the middle part of reaction plate is proposed from the structural point of view. However, the width of belt should be less than 40% of the stator width to enforce eddy currents effectively.

REFERENCE

1. T. Onuki, " Investigation of End Effect and grad ϕ on LIM with Short Secondary. ", IEEJ Trans. D, vol.108-11, (1988)

SMALL SIGNAL DYNAMIC MODEL OF SATURATED TURBOGENERATORS BY FINITE ELEMENTS METHOD

Muamba Idikayi, Jean C. Maun

Université Libre de Bruxelles
Génie électrique - c.p. 165
50 Av. Franklin Roosevelt
B - 1050 Bruxelles, Belgique

INTRODUCTION

Modern turbogenerators are characterized by high MVA rating per installed unit, this imply significantly higher levels of magnetic inductions throughout the magnetic paths of the machine. Consequently, to predict with accuracy the dynamic behaviour of turbogenerators, it is necessary to take more rigorous into account of magnetic saturation influence on values of the parameters which appear in the mathematical model of the synchronous machine.

Most of proposed methods improve the saturation of main magnetizing inductances, but saturation of leakage parameters like damper inductances and resistances is not well known.

The numerical solution of field equations of the machine by finite elements technique allows us to analyse the effect of magnetic saturation on all the parameters of a synchronous machine. However it is not practical to embed a finite elements routine in a general program of transient of a generator because of the cost in CPU time and storage needed by the finite elements computations.

Fortunately it is possible, from the results of finite elements calculations, to make simple but accurate saturation models for all the parameters that occur in the mathematical model of a synchronous machine. This paper shows how to built such saturation models and how to use them in the simulation of small signal behaviour of a synchronous machine.

FINITE ELEMENTS SIMULATIONS OF TURBOGENERATORS

Simulations of turbogenerators by finite elements technique is now well established[1]. With the usual assumptions, the steady-state magnetic field is described by the following general non linear equation in two dimensional form:

$$\frac{\partial}{\partial x}\left(\frac{1}{\mu}\frac{\partial A}{\partial x}\right)+\frac{\partial}{\partial y}\left(\frac{1}{\mu}\frac{\partial A}{\partial y}\right)=\sigma\frac{\partial A}{\partial t}-J \tag{1}$$

Electric and Magnetic Fields, Edited by A. Nicolet
and R. Belmans, Plenum Press, New York, 1995

with :
A : the magnetic vector potential, μ : the permeability, σ : the conductivity, J: the current density, t: the time.
The outputs of non linear finite elements calculations are then used to calculate the saturated synchronous inductances.
When small sinusoidal perturbations are superimposed on the state-steady, equation (1) is transformed in the following linear complex magnetic diffusion equation [1]:

$$\frac{\partial^2 \Delta A}{\partial x^2} + \frac{\partial^2 \Delta A}{\partial y^2} = j\omega \sigma \mu \Delta A - \Delta J \tag{2}$$

ΔJ : the magnitude of the small signal excitation current density, ω : the pulsation of the excitation, ΔA : the linear response of magnetic vector potential.

Equation (2) is solved using the finite elements method. The outputs are isochrone inductances from which the small signal dynamic parameters are derived by classical procedures [1]. Computed in this manner, the machine parameters reflect the actual magnetic state of the generator for the considered operating load point.

DEFINITION OF SATURATION FUNCTIONS FOR MACHINE PARAMETERS

The saturation function of a machine parameter is an analytical expression of the variation of this parameter in relation with a quantity which characterizes the level of the magnetic state of the machine[2]. The air-gap magnetizing flux or the total air-gap magnetomotrice force can be chosen as indicator of saturation level.

To construct saturation functions of the dynamic parameters of a synchronous machine, finite elements calculations are performed for several load operating conditions. When the evolution of any machine parameter versus the saturation indicator is graphed for a lot of operating load points , we observe that the set of the values of this parameter describe a shape which can be approached with an analytical function. This approximate analytical function can be expressed as a sum of exponential or polynomial simple functions like[3] :

$$f(x) = \sum_{l=0}^{n} c_l e^{d_l x} \tag{3}$$

$$f(x) = \sum_{l=0}^{n} c_l x^{d_l} \tag{4}$$

where: x is the saturation indicator, (x) is the saturation function of the considered machine parameter, c_l and d_l are the linear constants of the saturation function, n is the number of exponential or polynomial components needed to model more or less accurately the saturation function. By experience, we discover that $n = 4$ is sufficient to get a suitable saturation model. The constant coefficients of the saturation functions of machine parameters are calculated using a classical curve-fitting method[3].

The saturation functions of all the parameters of a synchronous machine can be included in a library of saturation functions. Thus, to include the effects of saturation on the dynamic behaviour of this machine, his library of saturation functions of dynamic parameters is linked with the main transient simulation code.

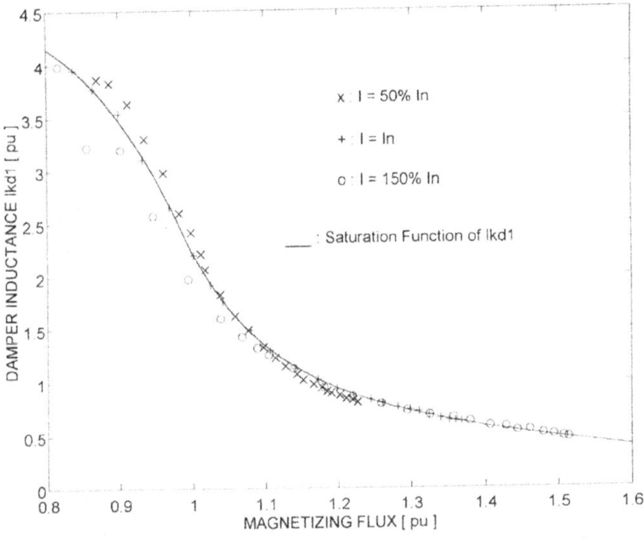

Figure 1. Saturation function of d-axis damper inductance l_{kd1} for CHINA machine.

APPLICATION TO SMALL PERTURBATION DYNAMIC MODEL

The saturation functions of small signal dynamic parameters have been made and functions libraries constituted for two large turbogenerators with the following ratings:

1. Turbogenerator CHINA: 361.4 MVA, 20 KV, 10.433 KA, cos φ =0.85, 2 poles.
2. Turbogenerator DOEL4: 1330 MVA, 24 KV, 31.995 KA, cos φ =0.85, 4 poles.

Figure 1 is an example that shows the curve of the saturation function of the d-axis damper inductance l_{kd1} of the CHINA machine and the finite elements points (n: nominal load current).

Now we use the saturation functions in the small perturbation dynamic model of these two generators to determine the saturation effects on the mechanical oscillations of the rotor.

The synchronous generator is connected to an Infinite Bus and is operated with a constant excitation voltage. The small signal dynamic model we have developed is like that described in[4,5], but it ihas been adapted to exploit the saturation functions of machine parameters.

The small perturbation equations is usually formulated in the standard state-variable form

$$\{\dot{X}\} = [A]\{X\} + [B]\{U\} \tag{5}$$

where $\{X\}$ is the state-variable vector which contains the deviations in currents, in rotor angle and in speed. $\{U\}$ is the forcing excitation vector which contains the incremental variations of the amplitude of the stator voltage and of the mechanical torque.

To include the saturation, all the machine parameters (inductances) which are used to form the elements of the system state matrix $[A]$, are evaluated from their respective saturation functions for the value of saturation indicator fixed by the quiescent conditions.

The effects of saturation are observed through the eigenvalues and the eigenvectors of $[A]$

Figure 2 shows the effects of saturation on the complex eigenvalues that characterize the mechanical oscillation of the rotor for DOEL4 turbogenerator. When the load grows the mechanical oscillations are less damped whereas their frequencies increase.

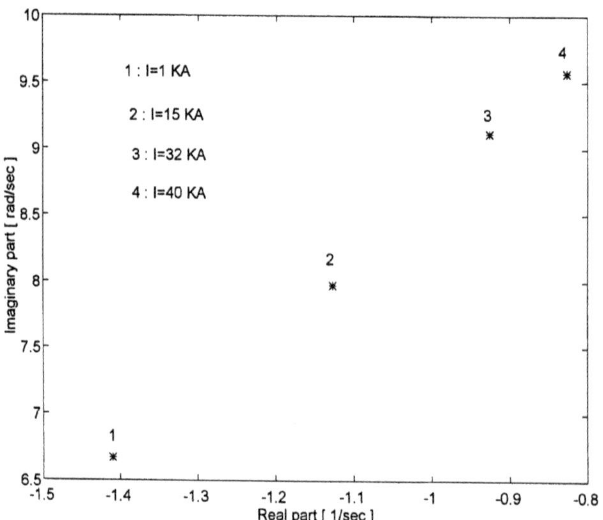

Figure 2: Saturation of mechanical eigenvalues for DOEL4 machine

CONCLUSION

A method has been proposed to construct simple but accurate saturation functions for all the parameters needed to formulate the mathematical model of a turbogenerator, from the finite elements magnetic fields computations. A library of saturation functions of machine parameters is built and linked to a program which simulate the small perturbation dynamic behaviour of a synchronous machine, to take in account the magnetic saturation.

REFERENCES

1. T. Kamabu, "Contribution au Calcul des Paramètres des Machines Synchrones par la Méthode des Eléments Finis", Thèse de Doctorat en sciences appliquées, Vrije universiteit Brussel (1987).
2. S.H. Minnich, R.P. Schulz, D.H. Baker, R.G. Farmer, D.K. Sharma and J.H. Fish, Saturation functions for synchronous gernerators from finite elements, *IEEE Transactions Vol. PAS-102,* 20:27 (1983)
3. W.H. Press, S.A. Teukolsky, W.A. Vetterling, B.P. Flannery, "Numerical Recipes in Fortran : The Art of Scientific Computing, Second Edition",: Cambridge University Press, New York (1992).
4. J. Lemay, T.H. Barton, Small Perturbation Linearization of The saturated Synchronous Machine Equations, *IEEE Transactions on Power Apparatus and Systems Vol. PAS-90,* 223:240 (1971).
5. P.C. Krause, "Analyse of Electric Machinery", McGraw-Hill Book Company, London (1986).

INFLUENCE OF LOAD AND AIR GAP DIMENSION ON MAGNETIC FIELD AND MAGNETIC CIRCUIT DESIGN OF A HIGHTORQUE STEPMOTOR

Wagner Juraj, Maga Dušan, Guba Roman,
Führichová Renáta, Opaterný Jozef

Slovak Technical University,
Faculty of Electrical Engineering,
Department of Electrical Machines and Devices,
Ilkovičova 3, 812 19 Bratislava, Slovak Republic

INTRODUCTION

In our previous papers we have presented our results on optimisation of dimensions and form of hightorque stepmotor magnetic circuit in dependence on maximal value of static torque. We have used the constant air gap. Only one phase has been excited by constant (rated) current.

In our paper we present our results by optimisation of magnetic circuit shape, shape of teeth and slots and depth of slots in dependence on maximal static torque.

MAGNETIC FIELD ANALYSIS OF MOTOR WITH CHANGING THE AIR GAP LENGTH

The solution has been carried out on simplified 2D model for tooth width (t_w) / slot width (s_w) ratio $t_w/s_w = 5/3$ resp. $3/5$; for a circle shaped slot and for δ/b ratio = 1/30 resp. 1/50 where δ = air gap width and b = 1/2 of slot pitch $t_w + s_w$. The results can be seen in tab. 1.

From our results is obvious that with decreasing air gap length the change of magnetic conductivity increases.

Table 1. The magnetic field analysis of motor with changing the air gap length.

b/δ	t_w/s_w	ΔB (T)
30	5/3	0.5000
50	5/3	0.5524
30	3/5	0.6077
50	3/5	0.6745

Electric and Magnetic Fields, Edited by A. Nicolet
and R. Belmans, Plenum Press, New York, 1995

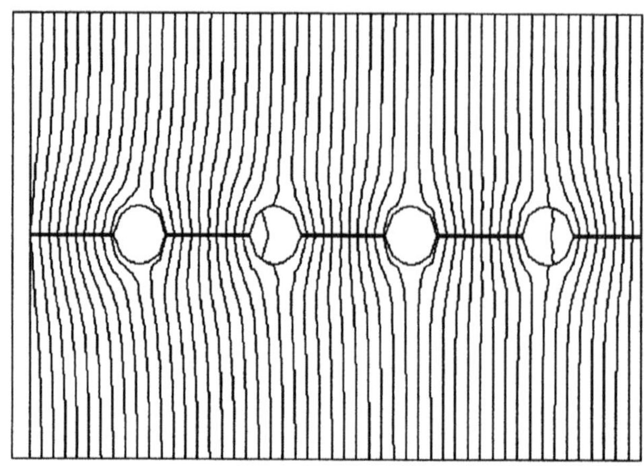

Figure 1. The simplified 2D model of hightorque step motor.

MAGNETIC FIELD ANALYSIS WITH CHANGE OF SLOT SHAPE AND DEPTH

The solution has been made for $t_w/s_w = 1/1.87$ for an oblong and a circle shaped slot as well; for slot depth $s_d = 0.25b$, resp. $s_d = 0.5b$. The results can be seen in tab. 2.

Table 2. The magnetic field analysis with change of slot shape and depth.

b/δ	t_w/s_w	slot shape	s_d	ΔB (T)
50	1/1.87	circle	0.25b	0.6397
50	1/1.87	oblong	0.25b	0.5535
50	1/1.87	oblong	0.50b	0.5197

It is obvious that the more suitable results have been obtained for the circle shaped slot face with slot depth $s_d = 0.25b$. our results did not confirm the conclusions of the most suitable t_w/s_w ratio published in [5].

THE INFLUENCE OF CHANGING CURRENT ON MOTOR MAGNETIC FIELD

We have again used the simplified model with $b/\delta = 12.5$; $s_d = 0.25b$; $I = (0.25 \div 1)I_n$; with one supplied phase. The results are in tab. 3. The torque increases properly to the increasing current.

Table 3. The influence of changing current on motor magnetic field.

I	$0.25I_n$	$0.50I_n$	$0.75I_n$	$1.00I_n$
ΔB (T)	0.0693	0.1385	0.2075	0.2754

THE MAGNETIC FIELD ANALYSIS WITH SIMULTANEOUS 2-PHASE SUPPLY

We have analysed the magnetic field of real motor. The Finite Elements Model (FEM) analysis was based on real sizes of motor cross-section with $s_w{:}t_w{:}s_d = 1{:}1{:}0.5$; an oblong slot shape.

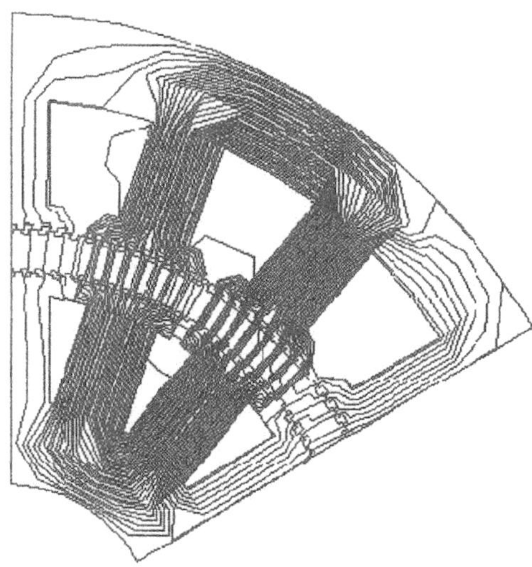

Figure 2. Cross-section of the motor.

for synchronising torque (from [1]) is valid:

$$T_S = \frac{1}{2} M_\delta^2 \frac{\Delta G}{\Delta \Omega} \tag{1}$$

where

ΔG is change of magnetic conductivity,

$\Delta\Omega$ is rotating through an angle,

M_δ is the magnetomotive force.

For torque from (1) follows:

$$T_S = 1.27 \frac{Q_P}{2m} \left[\left(B_{\delta_{1\,max}} . \delta \right)^2 + \left(B_{\delta_{2\,max}} . \delta \right)^2 \right] \frac{\Delta \Phi}{\Delta \Omega} . \frac{360}{2\pi} . \frac{\mu_0}{B_{\delta_{max}} . 4\delta} . l \tag{2}$$

where:

Q_P	is number of teeth of 1 pole,
$2p$	is number of poles,
$B_{\delta_{1\,max}}$	is the maximal value of induction in outer air gap,
$B_{\delta_{2\,max}}$	is the maximal value of induction in inner air gap,
$B_{\delta_{max}}$	is the mean value of induction in air gap with maximal change of conductivity,
l	is the magnetic circuit length.

We have realised the simultaneous supply of two phases for $\delta = 0.1$; 0.2 and 0.3 mm with:

a) currents of the same orientation;

b) currents of the opposite orientation.

The results can be seen in tab. 4.

Table 4. The magnetic field analysis with simultaneous 2-phase supply

The way of supply	δ (mm)	$\frac{\Delta G_{max}}{\Delta \Omega}$		T_s (Nm)
		outer δ	inner δ	
a	0.3	1.16	0.94	323.5
b	0.3	0.82	0.63	217.5
b	0.2	1.66	1.25	280.7
b	0.1	3.11	2.28	180.4

CONCLUSION

From simplified model we have obtained the following knowledge:

- the change of conductivity in air gap (and torque of motor as well) is proportional (linear) to current value (for single phase supply);
- the change of δ/b ratio from 1/30 to 1/50 leads to increasing of magnetic filed conductivity change in air gap. The rotor was revolved on one length unit (1/2 of tooth pitch).

In comparison with results presented in [2,3,4] and conclusions published in [5], it has been shown that the solution for $t_w/s_w = 1/1.87$ is (from viewpoint of maximal torque) not optimal.

From 2D model solution (which represents the periodically repeating cross-section of motor magnetic circuit with real dimensions) we have got the following conclusions:

- when choosing the optimal air gap length it is necessary to consider the magnetomotive force contribution too. As can be seen in tab. 4, despite of the maximal change of magnetic conductivity per unit of rotating angle for minimal $\delta = 0.1$ mm, the torque of motor is not the greatest;
- the motor has about 49% greater torque with 2 phase supply with currents of the same orientation.

REFERENCES

1. Wagner J., Dúbravčík P.: "High Torque Lowspeed Step Motor", Journal of Electrical Engineering No. 10, 45, 1992, p. 304 - 307.
2. Wagner J. et al.: "Ultra High Torque Motor - Principles of Design and Optimisation", Proceeding, International Conference, Electrical Drives and Power Electronics, Košice, Slovak Republic, 1992, Vol. 2, p. 359 - 363.
3. Wagner J., Guba R.: "Using the Finite Elements Method for Magnetic Field Analysis of a High-Torque Low-Speed Step Motor and for Design Optimisation", Proceeding, International Workshop on Electric and Magnetic Fields From Numerical Models to Industrial Applications, Liege, Belgium, 1992, p. 471 - 475.
4. Wagner J. et al.: "Determination of Step Motor Parameters with Magnetic Field Solution", Proceeding, ISEF '93, Warsaw, Poland, 1993, p. 105 - 108.
5. Stolařík M.: "Klidová brzda", Technika elektrických strojů, No. 4, 12, 1967.

3D FEM CALCULATION OF A LINEAR SYNCHRONOUS MOTOR

G. Henneberger and C. Reuber

Institut für Elektrische Maschinen der RWTH Aachen
Schinkelstraße 4
52056 Aachen, Germany

ABSTRACT

The present paper describes the 3D FEM calculation of a linear synchronous motor using Magnet 5.0. The position-dependent inductances and forces were calculated via the flux-linkage of the individual coils. The forces were also calculated using the Maxwell-Tensor.

CONVEYANCE VEHICLE

This motor was designed to drive a clean-room conveyance-system that is suspended und guided magnetically (fig. 1).

As the coils of ts + coils[1])

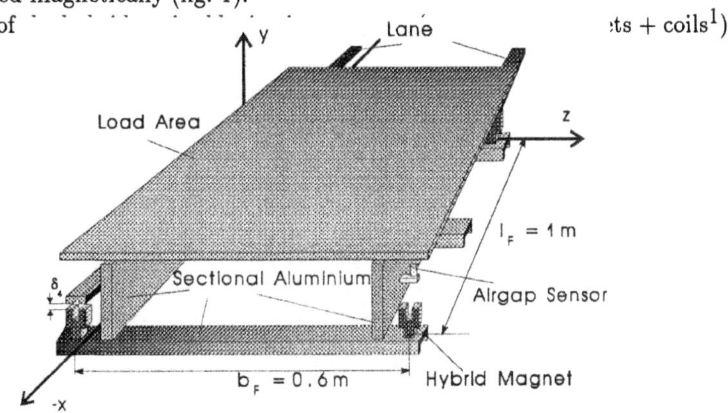

Figure 1. Conveyance vehicle

Electric and Magnetic Fields, Edited by A. Nicolet
and R. Belmans, Plenum Press, New York, 1995

are controlled to operate with minimal current, the levitation height, or y-position, of the vehicle may vary about ±2mm, depending on its total mass (50-100kg). Its z-position may also vary about ±4mm, as the lateral displacement is not controlled, but guidance is obtained by the stable equilibrium of the levitation magnets under the rails. So, the motor had to be designed with large air gaps and an ironless stator to avoid an unstable equilibrium in z-direction due to the normal forces of synchronous motors.

LINEAR MOTOR

Fig. 2 shows a view of the motor, while fig. 3 and 4 illustrate some cross-sections. Two coils of the two-phase stator winding form one pole pair. Fig. 4 shows two stator pole pairs consisting of 4 coils: phase R, coil 1 and 2, and phase S, coil 1 and 2. One pole has a x-extension of 6 cm. The whole motor consists of the shown two-pole moving part and a 16-pole stator (16 coils).

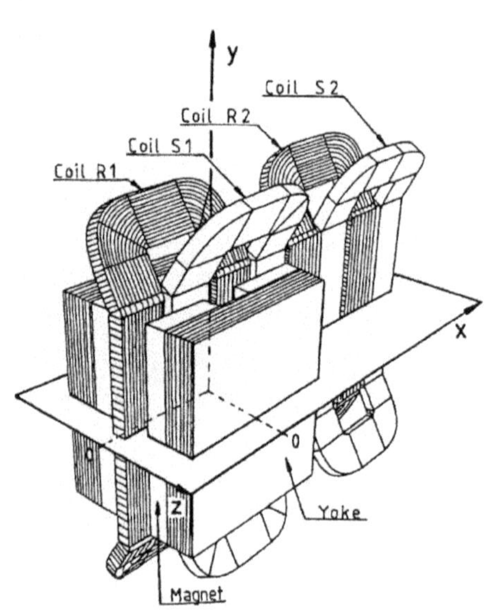

Figure 2. View of the motor

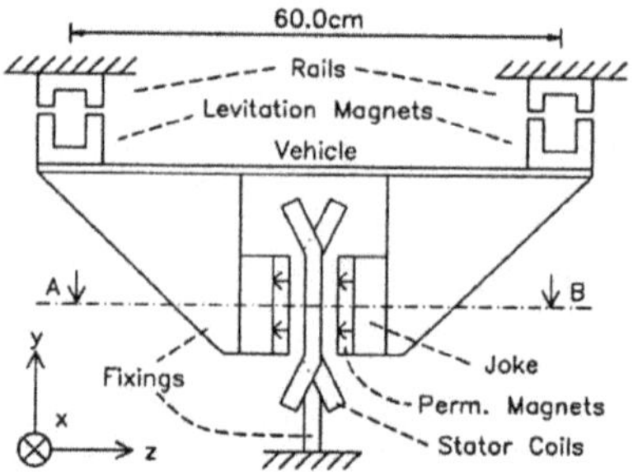

Figure 3. Sketch of the motor

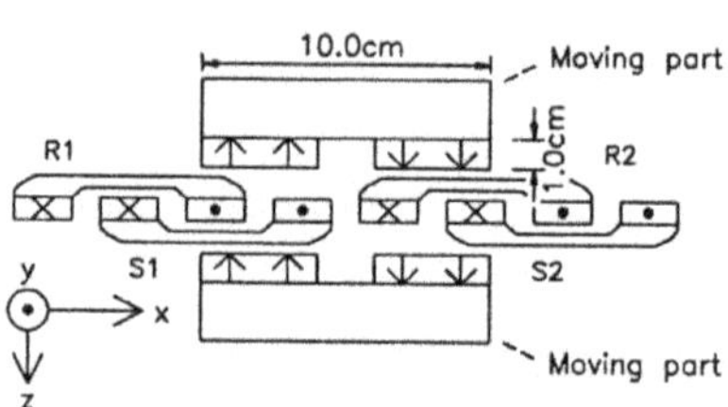

Figure 4. A-B cross section of fig. 3

To achieve the maximal propulsion force, the coils are driven field-oriented with sinusoidal currents. The coils outside the main-field of the moving part are not driven to minimize copper losses.

In a transport application the motor has to cover distances of some meters at low speeds (2-3m/s), so it is not necessary to provide stator coils along the whole way. A stator with a length of app. 1m will be adequate to accelerate the vehicle, then it travels with constant speed for some meters, and finally a second 1m long stator will decelerate and position it.

As each individual coil is driven only for a short time, the current may be very high without overheating the coils. For the same reason, the material for the permanenet magnets could be chosen with respect to a maximal B_r (1.32T).

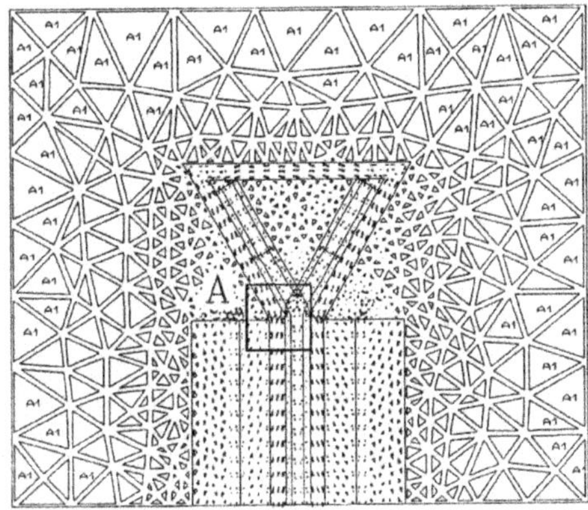

Figure 5. Sketch of mesh

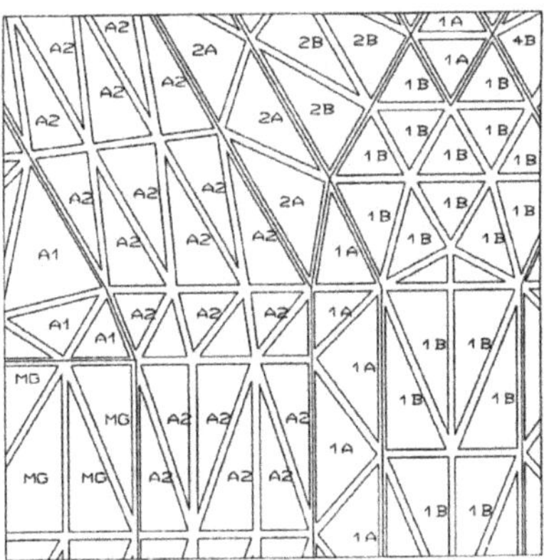

Figure 6. Detail 'A' of fig. 5

FIELD CALCULATION

A 3D FEM analysis by the software package MagNet 5.0 was applied to calculate the flux distribution of the motor exactly.

The mesh build for this problem consists of a y-z plane (fig. 5; see also fig. 3), that was extended into the x-direction (1799 elements, app. 80000 bricks). Due to the symmetry in y-direction, only the upper half of the motor (fig. 2) was modelled.

Three layers of elements surround the stator-coils in five sections to take up 15 slices for the maxwell-tensor. Here, local second-order computation was scheduled for improved accuracy (see label 'A2' in fig. 6). The same was done in the middle of the coils, where the magnetic induction passes through and shall be integrated, special layers of elements were provided for local second-order calculation (see label '1B' & '2B' in fig. 6)

POSTPROCESSING

Inductances

To determine the position-dependent self- and mutual inductances of the four coils, some extra problems were defined, where the permanent magnets were switched off and only one coil was driven. The inductances could be derived from the flux-linkages of all coils. These flux-linkages were calculated by integrating the induction perpendicular to the above mentioned slices that are situated in the centres of the coils. Fig. 7 shows, as an example, the self-inductance of the coil R1 (one turn; x=0 is the position shown in fig.2).

Forces using $\partial W_m / \partial x$

The force produced by an electrical motor can be expressed by the derivative of the electrical energy with respect to the position. Thereby, the currents have to remain unchanged. The resulting force consists of three different parts:[2] the electrical force, resulting from the interaction of excitation- and armature currents, the reluctance force,

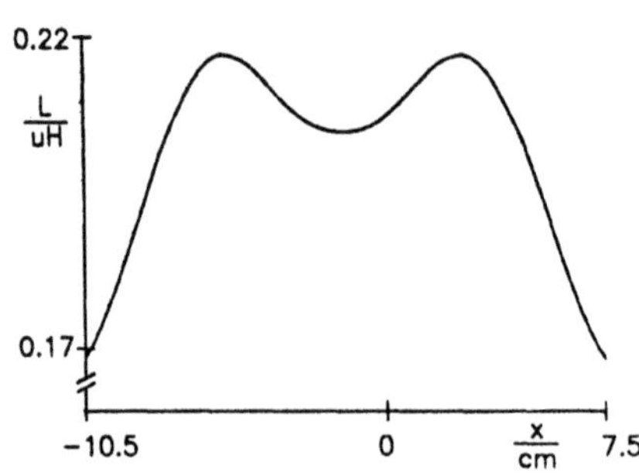

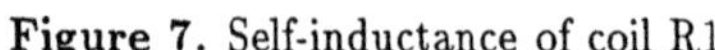

Figure 7. Self-inductance of coil R1

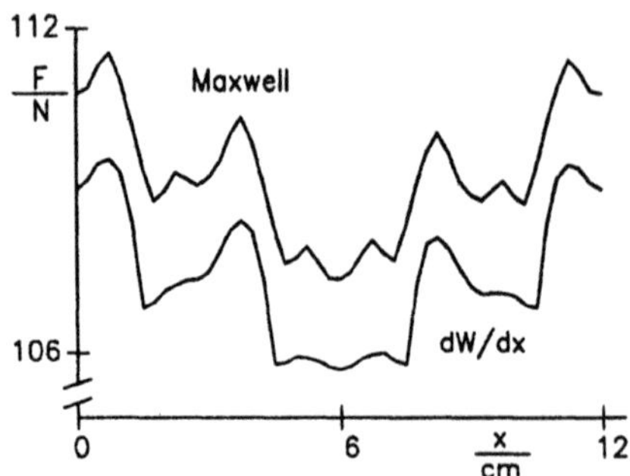

Figure 8. Position dependent forces

coming from the armature currents and the iron of the moving part, and the cogging force, that is produced by the excitation currents and the stator iron. As this linear motor has an ironless stator, it can not produce any cogging forces.

Fig. 8 shows the obtained force. That curve represents the sum of electrical and reluctance force. The latter is fairly small (some $0.3N_{pp}$), so that it is hardly visible; from x=0 to 6cm the forces are slightly higher than from x=6 to 12cm due to the reluctance force.

The minimum of the force (x=6cm) is due to the stray fields at the ends of the moving part. Here, the position of the coils is the same as in fig. 4. While the current on the right side of coil R1 is necessary for the propulsion force, the same current in the left side runs in the same (stray-)field, but with an opposite direction, so it brakes the vehicle. Fig. 2 shows the position x=0cm; here are no coils to drive in the stray fields, so the force is distinctly higher.

Forces using the Maxwell-Tensor

To carry out a calculation of the propulsion force with the Maxwell-Tensor $F_x = \frac{1}{\mu_0} \cdot \oint_S B_n \cdot B_t \, ds$, a surface S was formed by 5 slices that separate the moving part from the stator (fig. 5). These slices are situated in the centres of the provided element-layers to take all finite elements equally into account for the force-calculation. Due to the numerical drawback of the Maxwell-Tensor (the results are small in comparison with the values that are integrated), the set of 5 slices was extracted in 3 different, parallel element-layers (fig. 6), so that an average force could be calculated. Nevertheless, the result of the propulsion force in the no-load-case was not exactly zero. As there can not be any cogging-forces (ironless stator), this was interpreted as an offset due to the discretization. So, the difference of the rated-load-case and the no-load-case was evaluated as the rated force of the linear motor.

The calculated force is also represented in fig. 8. It shows the same shape as the $\partial W_m/\partial x$-force and differs only about 2%.

REFERENCES

1. G. Henneberger and D. Rödder, "Simulation und Realisierung einer energiesparenden Regelung für einen Tragmagneten mit Hybriderregung", Archiv für Elektrotechnik 76 (July 1993).
2. J. Kempkes, "Verbesserungsmöglichkeiten und Bewertung der Rundlaufeigenschaften von permanenterregten Synchronservomotoren", Diss. RWTH Aachen (1991).

CLASSICAL METHOD AND FIELD ANALYSIS

IN PERMANENT MAGNET STARTER DESIGN

Andrzej Sęk, Wojciech Białek

Warsaw Technical University
Institute of Electrical Machines
Plac Politechniki 1
Warsaw, Poland

PREFACE

Strong competition in automotive industry forces constant progress of all subassemblies of a modern car. It is relevant to necessity of development of new designers' tools making process of designing easier, more accurate and efficient. This paper is devoted to this problems, particularly to construction of permanent magnet car starters.

Popular coil excited starter type B76 (Poland) was a start point for this study. We tried to create permanent magnet motor of similar characteristics but smaller, lighter and more efficient.

We decided to use permanent magnet excitation as main quality change. Reduction of weight, size, current (which is related to battery size reduction) and simplification of construction were goals of optimisation. Leaving mechanical parameters unaltered makes comparisons possible.

Classical (grapho-analytical) method of design was assisted by computer simulation of magnetic field. Computation algorythm was designed using various references[1]. Field analysis based on finite element method and two dimensional model. Giving up 3-D simulation was caused by time consuming counting (PC 486 was used). This decision let us execute bigger number of iterations.

DESIGN COMPUTATION

In this design algorithm created by authors of this paper was used. This algorithm was translated into computer program to make calculations faster.

The basic assumptions were:

Electric and Magnetic Fields, Edited by A. Nicolet
and R. Belmans, Plenum Press, New York, 1995

- Number of poles - 4.
- Type of magnets - ferrite magnets Oxit 400 (Thyssen, Germany)[2]. Low cost and high resistance to demagnetisation are their main advantages.
- Straight wave winding - the most popular winding in low power starters.
- Power supplied from 12V battery.

From multidimensional space of construction parameters, stator outer diameter $D1$ (related to height of magnets in algorithm) and current density j were chosen as independent variables. Minimum $D1$ and maximum j securing proper values of mechanical parameters and magnets resistance to demagnetisation were defining the goal of optimisation.

After number of iterations we've obtained satisfactory results (given below):

I[A] current	M[N*m] torque	P[W] power	eta[%] efficiency	n[rpm] rpm
23	0.6	198	80	3.22E+03
45	1.1	356	73	2.86E+03
68	2	464	64	2.50E+03
90	3	521	55	2.15E+03
113	4	527	45	1.79E+03
135	4.5	482	35	1.43E+03
158	5	387	25	1.07E+03
181	6.1	242	14	7.16E+02

nominal power Pn[W]= 500
nominal voltage Un[V]= 12
nominal rpm nn[1/min]= 2500
outer diameter D1 [m]= 7.2E-02
efficiency = 0.6
air gap induction Bp[T]= 0.36
number of poles = 4
number of rods N= 152
number of slots = 19
section of wire[mm2]= 1.75
magnet induction Br[T]= 0.4
coercive force Hc [A/m]= 3.0E+05
rotor's diameter D[m]= 5.2E-02
rotor's lenght l=[m] 5.2E-02
internal power Pi[W]= 583
SEM En[V]= 8.4

nominal current In[A]= 69.4
flux[Wb]= 6.73E-04
rotor's lenght lm[m]= 5.72E-02
resistance of winding r[Ohm]= 4.16E-02
height of tooth hz[m]= 8.58E-03
width of tooth bz[m]= 2.12E-03
magnet's height hm[m]= 6.85E-03
stator's lenght ls[m]= 1.02E-01
stator's height [m]= 2.59E-03
margin of safety = 1.08
recomended(1.05..1.2)
flux error []=7.89E-03 (recomended<0.05)
real value of induction [T]= 3.63E-01
magnets' flux [Wb]= 6.78E-04
dissipation coefficient delta []= 1.10
short circuit current Isc[A]= 226

To obtain this results we had to overcome problems characteristic for permanent magnet motor design.. The main issue was big number of active rods necessary to obtain proper inducted voltage caused by low magnetic induction in air gap (0.36T in permanent magnet starter compared with 0.7T in coil excited one).

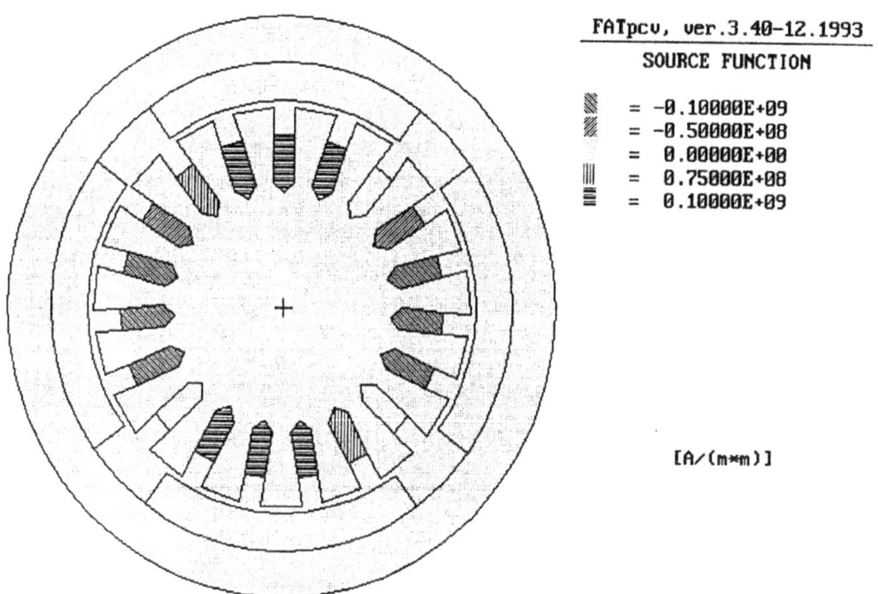

Fig. 1. Distribution of current sources

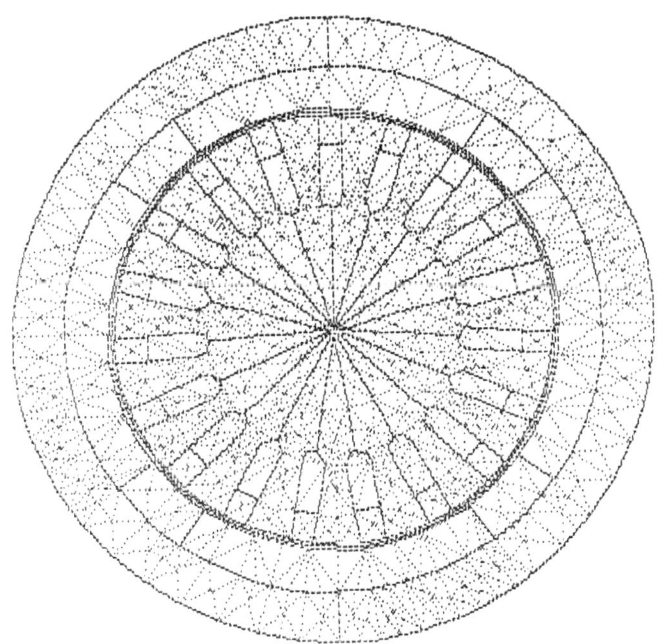

Fig. 2. Computation mesh

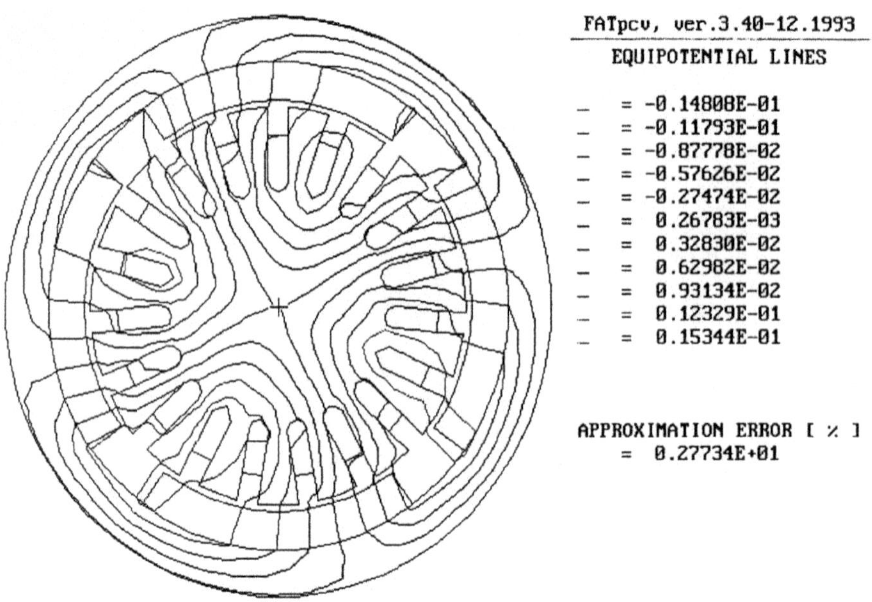

Fig. 3. Equipotential lines

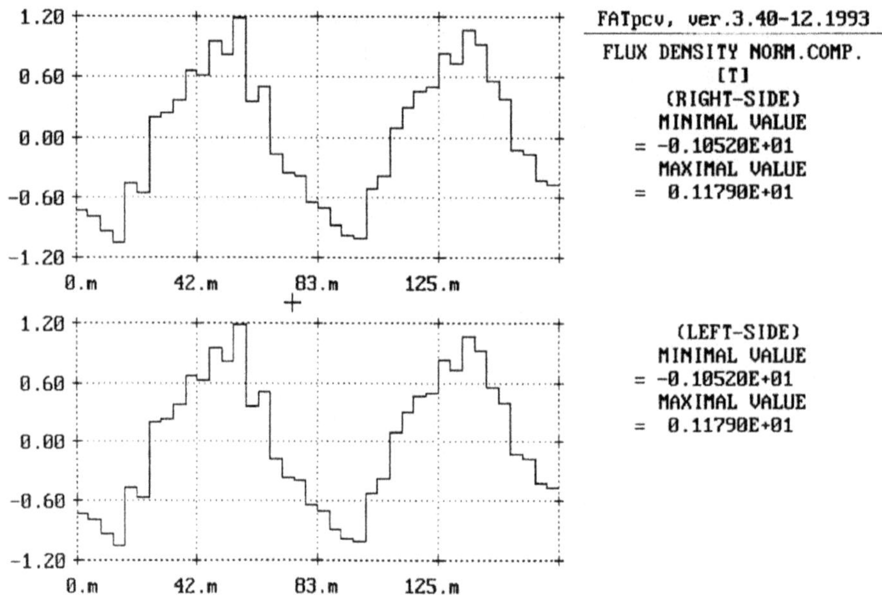

Fig. 4. Flux density normal component

MAGNETIC FIELD ANALYSIS

Use of computer field simulation let us examine more accurately following parameters:

- Magnetic induction distribution in particular elements of machine was necessary to state utilisation of materials and check results of "classical" computation.
- Working point of magnets (in discretised regions) - a very important in permanent magnet design factor, because of danger of demagnetisation.
- Starting torque - this value is significant because of character of starters work.

In simulation the latest version of FAT system was used. FAT is a specialised programming language developed in Warsaw Technical University. It employs finite element method and two-dimensional model of field. Our program in FAT describes geometry of motor and source function in short circuit state. It also performs automatic optimisation of discretisation network.

As a result of this program number of graphs and numerical values was obtained. Interpretation of these results makes possible valuation of design quality.

In permanent magnet car starter short circuit state is very important. Distribution of current sources is shown in Fig. 1. Note that in case of 19-slot machine position of slots under each pole differs.

Figure 2 shows discretization network generated by program. Multi step adaptive analysis caused compaction of network in regions of higher field gradient.

For qualitative evaluation of design distribution of equipotential lines (Fig. 3) is very important. More valuable information is given on graph of radial component of field in air gap (Fig. 4). It assures that remagnetisation of poles does not appear.

Calculation of starting torque was important element of magnetic field computation. Obtained value is 170Nm per meter of rotor length that equals 0,052m.

CONCLUSIONS

Because of exemplary character of design, the most valuable results of this research are conclusions gained during design and optimisation process:

- Joining classical design method with computer magnetic field simulation helps designer in more accurate computations and shows directions of optimisation.
- Possibility of reduction of outer diameter of starter is in case of small machines limited by increase of rotor's diameter caused by more complicated winding. However some reduction is possible even when low cost magnets are used. Decrease of weight is bigger because density of magnets is smaller then steel.
- Additional advantage of utilising permanent magnets is lower no-load running speed (3600 rpm against 6700 for B76). Higher nominal speed and intermediate gear can be used for higher utilisation factor.
- Lower value of short circuit current makes use of smaller battery possible.

This research reached its main goals. Starter that can by a replacement for B76 was designed. It is smaller (outer diameter 72mm against 76mm), simpler (no excitation coils) and more effective. Thanks to use of magnetic field computations verification of design algorithm was performed.

REFERENCES

1. I. Dudzikowski "Silniki komutatorowe o magnesach trwałych" Prace naukowe Instytutu Układów Elektromaszynowych Politechniki Wrocławskiej; Wrocław 1992

2. Thyssen "Product catalogue" 1141/26E 1987.

MODELLING AND DESIGN OF RECTANGULAR FED ELECTRONICALLY COMMUTATED PM AC MOTOR

Bogdan Kreča, Bojan Štumberger, Anton Hamler, Božidar Hribernik

University of Maribor, Faculty of Technical Sciences
Smetanova ul. 17, P.O. Box 224, 62000 Maribor, R. Slovenia

INTRODUCTION

Calculation of magnetic fields and torques presents the most important part for the motor calculation. Modern PM AC motor drives consist of the motor with the rotor magnets, inverter, feedback and controller. The discussion will be limited to the model for performance prediction of the motor. The paper describes a CAD package for calculating the magnetic and mechanical characteristics (torques) of the electronically commutated motor. The package comprises preprocessing of input data and preparation of data bases, mesh generation with automatic mesh generator, magnetic field calculation by Finite Element Method and postprocessing by use of the magnetic vector potential **A**.

TORQUE MODEL

The main focus is placed on the design of three - phase sinusoidally - fed motors and using surface - mounted rare earth of NdFeB magnet segments. The same principles are applied to rectangular - fed motors. The torque model applies to PM motors that are fed with rectangular pulses of the stator current [1,2]. Figure 1 shows a section of discretization of a four pole PM motor indicating some of the major materials. The automatic mesh generator is used [4]. The discussion will be limited to three phase motors which always operate with the current supplied to two phases in series, while the current of the third phase is equal to zero.

The stator is fitted with three phase windings that are fed with rectangular pulses of the stator current. The rotor consists of rare earth magnetic segments fixed to the rotor core.

The torque produced by the motor is given by

$$T = \sqrt{2}\,\frac{\pi\, r_1\, B_1\, l\, N_s}{2}\, I_1 \tag{1}$$

Electric and Magnetic Fields, Edited by A. Nicolet
and R. Belmans, Plenum Press, New York, 1995

where

B_1 -amplitude of the fundamental harmonic of magnetic induction in the air gap,

I_{efp} -the r.m.s. value of the stator current for rectangular fed, I_1=0.955I_{efp} , I_1 is the r.m.s. value of the fundamental harmonic of the stator current ,

0.955 -ratio between r.m.s. value of the fundamental harmonic of the stator current and r.m.s. value of the ractangular form of the stator current,

N_S -effective number of sine - destributed turns in series per phase,

l -the axial length of the motor,

r -inner stator radius.

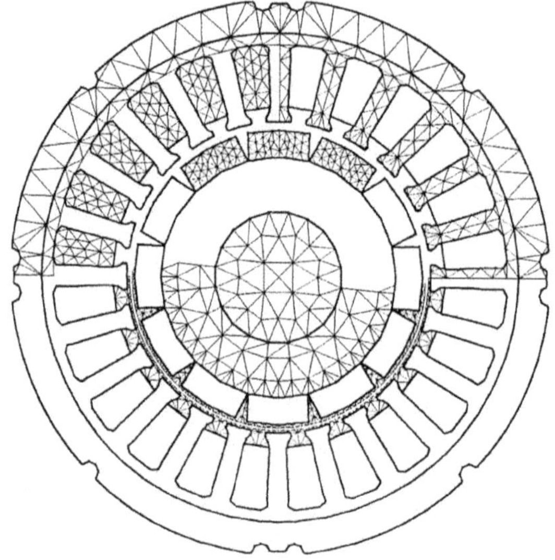

Figure 1. Section of discretization of PM motor by main materials.

THE MAXWELL STRESS METHOD

The Maxwlell Stress Method is based on the distribution of the magnetic field on the closed surface in the air gap arround the rotor [3].

$$T = \left[\frac{r}{\mu_o} \oint_c B_n B_t \, dc \right] L \tag{2}$$

where

r -distance from the integration path to the axis of rotation of the rotor,

c -closed integration path,

B_n,B_t -normal and tangential component of the magnetic flux density,

L -axial rotor length.

The accuracy of the torque calculation depends on the accuracy of B_n and B_t on the chosen integration path. For this purpose it is reasonable to refine the discretization arround the integration path. Figure 2 shows the discretization in the air gap. The integration path should

run through those elements in the air gap which are not deformed when the rotor position changes.

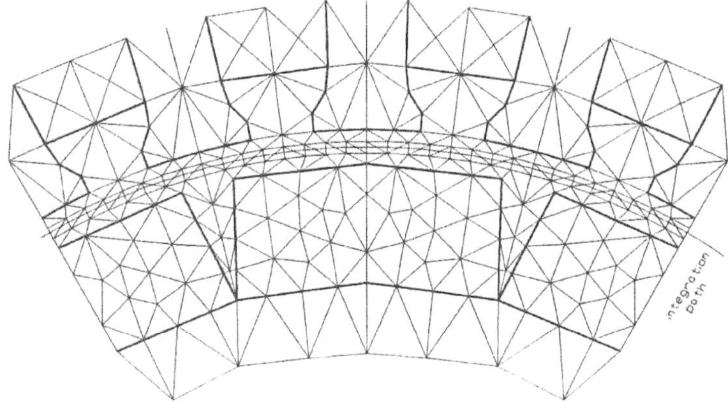

Figure 2. Discretization in the air gap and integration path for calculating of the torque by Maxwell Stress Method.

RESULTS

The results of the torque calculations obtained by Rectangular Fed Model and by Maxwell Stress Method compared with experimental results are shown in table 1. Figure 3 shows the magnetic field in the motor with equipotential lines sharing of 10% and flux distribution under normal load.

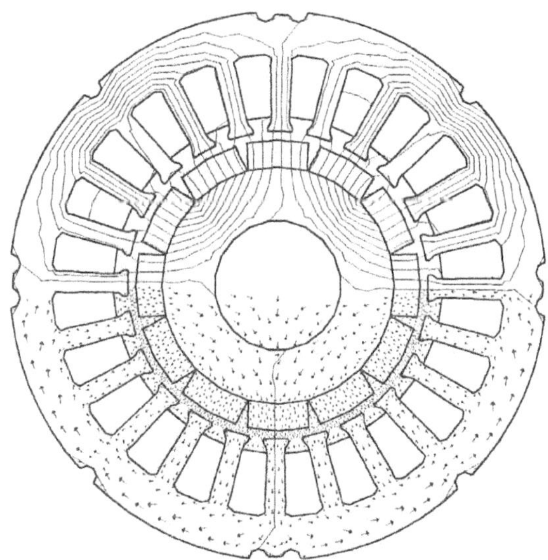

Figure 3. Magnetic field in the motor with equipotential lines sharing of 10% and flux distribution under normal load.

To obtain a complete form of the torque the calculation over one pole is required. The motor has 4 poles, the pole division is 90 degrees. Figure 4 shows the results of the torque calculation by the Maxwell Stress Method under normal load for different rotor positions and for axial length of the motor 52mm under normal load.

Table 1. The results of torque calculations compared by experimental results for different axial length of rotor.

axial length of motor	**Iefp**	Torque model	Maxwell Stress Method	Experimental
(mm)	(A)	(Nm)	(Nm)	(Nm)
26	1.42	0.605	0.552	0.595
52	3.22	1.342	1.377	1.443
75	5.48	2.223	2.045	2.101

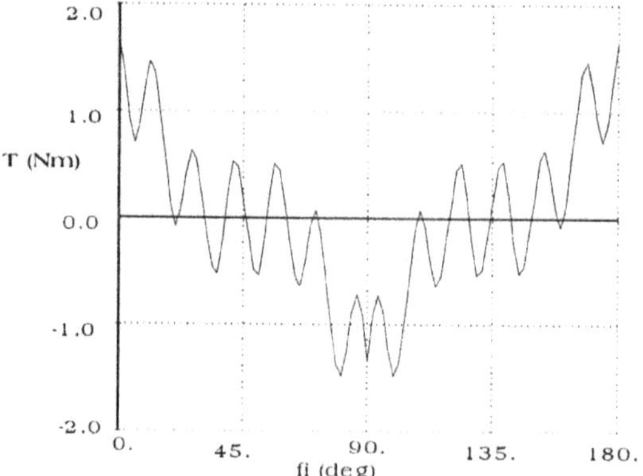

Figure 4. The results of the torque calculation by the Maxwell Stress Method under normal load for different rotor positions and for axial length of the motor 52mm.

REFERENCES

1. T.J.E. Miller, Brushless Permanent-Magnet and Reluctance Motor Drives, Oxford University Press, New York 1989, USA

2. G.R. Slemon, Design of Permanent Magnet AC Motors for Variable Speed Drives, IEEE Aplications Society Conference 26th IAS Annual Meeting, Michigan 1991, Publishing Services, 345E. 47th Street, New York 1991, USA

3. A.Hamler, B.Kreča, B.Hribernik, Investigation ot the Torque Calculation of DC PM Motor, IEEE Transaction on Magnetics, Vol.28, No.5, September 1992

4. M.Jesenik, M.Trlep, B.Hribernik, Automatic 2D Discretization for Numerical Methods, 5th International IGTE Symposium on Numerical Field Calculation in Electrical Engineering, Proceedings, Graz, September 1992, Austria

OPTIMIZATION OF THE SLOTTING OF HIGH-SPEED SOLID-ROTOR INDUCTION MOTOR

B. Laporte, G. Vinsard, J.P. Bock

Green, URA 1438, 2 Av. de la forêt de Haye
54516 Vandoeuvre-les-Nancy, France

INTRODUCTION

The solid-rotor induction motors can be used for applications that require high-speed because of their mechanical robustness. If the rotor is slotted then the value of the torque is improved. However it is a non obvious problem to find the optimal number of slots and their geometry.

The optimization needs the computation of the mean torque in the steady-state. The modelling of induction motors can be performed accurately by using time stepping methods[1,2]. However such methods are very expansive in CPU time because it comes to describe the entire transient from the start to the steady state. Then we prefer to use an approximated method that we have recently developed[2] and called the first harmonic method.

Even if the method that provides the torque is chosen, the optimization global problem itself remains difficult. It comes to maximize an objective function (the torque) in the space of all admissible shapes of the rotor. However the slots are identical and we assume that they are shaped as a rectangle and the total volume of the rotor remains constant. Then the set of admissible shape reduce to the product $R^+ \times N$ for a geometrical parameter (the depth of the slots) and the number of slots.

THE FIRST HARMONIC MODEL

This method is presented in this paper for steady-state induction motor running but it can be adjusted to the transient analysis[2].

The starting point is that only two time pulsations are in presence -the pulsation, in the stator, ω and, in the rotor, $s\omega$- because the motor is fed by balanced 3-phases currents and also because we consider that only the fundamental space harmonic can pass through the airgap.

The bidimensionnal geometry of the problem allows the introduction of the one component magnetic vector potential (MVP) fields a_r in the rotor D_r and a_s in the stator D_s as state variable. The assumption about their time form yields

Electric and Magnetic Fields, Edited by A. Nicolet
and R. Belmans, Plenum Press, New York, 1995

$$a_s(x_s,t) = \sqrt{2}Re(A_s(x_s)e^{j\omega t}) \ \ in \ D_s \ \ ; \ \ a_r(x_r,t) = \sqrt{2}Re(A_r(x_r)e^{js\omega t}) \ \ in \ D_r \tag{1}$$

where x_s and x_r are the positions in the stator and rotor in their own frame of reference. Introducing the rotor angular speed Ω, the relation between x_r and x_s is

$$x_s = \left| \begin{matrix} r_s \\ \theta_s \end{matrix} \right. ; \ x_r = \left| \begin{matrix} r_r = r_s \\ \theta_r = \theta_s - \Omega t \end{matrix} \right. \tag{2}$$

The mvp field A_r in D_r and the mvp field A_s in D_s verify

$$\nabla.1/\mu\nabla A_s + J_s = 0 \ \ in \ D_s \ \ ; \ \ \nabla.1/\mu\nabla A_r - js\sigma\omega A_r = 0 \ \ in \ D_r \tag{3}$$

where : μ, the magnetic permeability, is a function of the position x_s in D_s and x_r in D_r - σ, the electrical conductivity, is a function of x_r in D_r ; the conductivity of the stator region is neglected because it is assumed to be designed with sheet steel and the skin effect is neglected in the slots - J_s, the complex source current density, is a function of x_s in D_s that is closely related with the global source currents.

Both the rotor D_r and stator D_s region contains the air-gap (Figure 1.). Then the exterior boundary of D_r is Γ_s, the interior boundary of D_s is Γ_r. The first harmonic approximation is to consider that the problem can be divided into two parts :

a) The rotor problem - the mvp A_r is approximated as

$$A_r = \bar{A}_{s0}e^{jp\theta_r} \ \ on \ \Gamma_s \tag{4}$$

where $\bar{A}_{s0}$ is a provisionally unknown complex whose value is provided by the stator problem. Equations (3), (4) allow the computation of the field A_r parametrized by $\bar{A}_{s0}$. When this calculus is completed the complex

$$\bar{A}_r = 1/2\pi \int_{\Gamma_r} A_r e^{-jp\theta} d\theta \tag{5}$$

can be derived : it is the p order of the angular harmonic expansion of A_r on Γ_r.

b) The stator problem - In the same way, the mvp A_s is approximated as

$$A_s = \bar{A}_r e^{jp\theta_s} \ \ on \ \Gamma_r \tag{6}$$

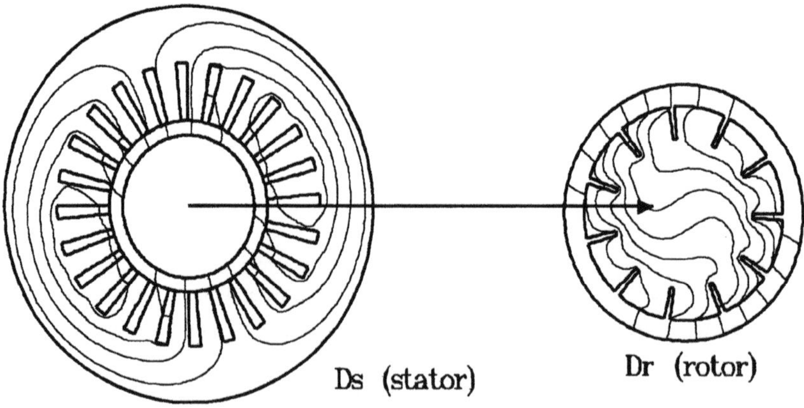

Figure 1. Geometry. The air-gap is zoomed. The stator and rotor fields are computed separately

where $\bar{A}_r$ is computed from (7). Equations (3), (6) allow the computation of the field A_s parametrized by $\bar{A}_{s0}$ via $\bar{A}_r$. Then $\bar{A}_s$ can be issued from

$$\bar{A}_s = 1/2\pi \int_{\Gamma_s} A_s e^{-jp\theta} d\theta \qquad (7)$$

This is the p order of the angular harmonic expansion of A_s on Γ_s.

The two steps a) and b) are the basic elements of the first harmonic method. Only the p order harmonic exists on the boundary Γ_r for A_s and Γ_s for A_r. The other harmonics exist on the boundary Γ_r for A_r and Γ_s for A_s but they are not used for the coupling between the two problems. From an operational point of view, the problem can be expressed as the algebraic equation only depending on one parameter

$$f(\bar{A}_{s0}) = \bar{A}_s - \bar{A}_{s0} = 0 \qquad (8)$$

The solution of (8) gives the final result.

THE LINEAR CASE

In the linear case, the effective algorithm becomes finite and can be expressed using only three steps:

a') same as a), $\bar{A}_{s0}$ to any value. The output is A_r and its p-harmonic $\bar{A}_r$ on Γ_r.

b') same as b), forcing J_s to zero in (3). The output is A_s and then $\bar{A}_s$.

c') same as b) forcing $\bar{A}_r$ to zero in (6). The solution field is called A_s^s, and the p order of its angular decomposition on Γ_s is $\bar{A}_s^s$

Then, introducing the scale parameter α, (8) becomes

$$\alpha\bar{A}_s + \bar{A}_S^S - \alpha\bar{A}_{s0} = 0 \Rightarrow \alpha = \bar{A}_S^S \,/\, (\bar{A}_{s0} - \bar{A}_s) \qquad (9)$$

Finally the actual mvp fields are: αA_r in the rotor and $\alpha A_r + A_s^s$ in the stator.

Formally, the first harmonic method is a classical computation of a virtual MVP field with the constraint that the MVP in the rotor vanishes on the stator boundary and the MVP in the stator vanishes on the rotor boundary, except the fundamental p order harmonic. As the other harmonics are quite negligible in a well designed induction motor, this assumption conforms with the reality of the motor.

THE OPTIMIZATION

Our objective is to realize a bipolar, high-speed, solid-rotor, induction motor with 24 slots in the stator. Then we optimize the mean torque with respect to the number and depth of slots with the constraint that the global volume of the rotor remains constant. The slip is a supplementary parameter that we remain constant (1%) for facility reasons ; but its rule is important and must be take into account in an actual optimization problem.

The model disconnects the computation of the magnetic vector potential in the rotor and stator. We compute once the field in the stator and vary the geometry of the rotor. For each configuration we perform a meshing and computation whose output is the mean torque. These points are plotted in a diagram (Figure 2): the map of the torque versus the number and the depth of the slots, each point correspond to a finite element computation and there is 2000 points in the map.

The interpolated values between two integer values of the number of slots have no sense but this representation is easier to read than a series of curves. The more surprising result is that the mean torque is almost a smooth function of the number of slots: the right figure is the direct result and the left figure a light smoothing (only one averaging with the neighbor for each points) of the results. This remark is important because it will allow, in a further work, to treat the dependance on the number of slots as a continuous function, by using a discrete derivative.

Another important result is that the problem has an optimum, approximately 19 slots whose depth is 20 % of the radius of the rotor. However we want to point out that such a diagram is dangerous because the reluctance torques are not computed and the optimum can be a number of slots that is incompatible with those of the stator. To overcome this drawback we can use semi-heuristic rules given by the electric constructor, or better, perform a full time computation at the optimum point.

CONCLUSION

The first harmonic method is very powerful to do a pre-optimization of a structure: we have made the computation of 2000 structures, which corresponds to 10 hour-CPU with a 4.2 Mflops computer.

The dependance of the mean torques on the number of slots is sufficiently smooth for that a sub-gradient method can be used in an automatic optimization.

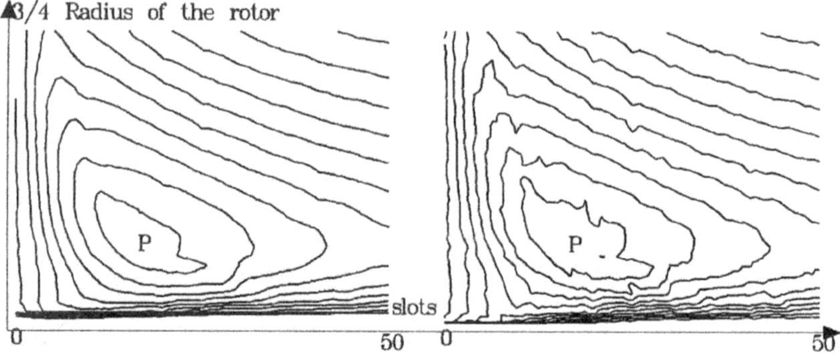

Figure 2. Map of the torque versus the number and depth. Left : smoothed ; Right : not smoothed.

REFERENCES

1. A.C. Smith, S. Williamson, J.R. Smith, "Transient currents and torques in wound-rotor induction motor using the finite-element method", IEE Proc., Vol. 137, Pt. B., No. 3, (May 1990).
2. J.P. Bock, B. Laporte, G. Vinsard, "A time stepping method using a space harmonic approximation for induction motors", ICEM, Paris 94.
3. B. Laporte, G. Vinsard, J.C. Mercier, "A computation method for induction motors in steady-state", Proc. IMACS'TC1'93, to appear in "Mathematics and Computers in Simulation" (1994).

MAGNETIC FIELDS CALCULATION OF INDUCTION MOTOR UNDER LOAD CONDITIONS

Stanislav Ferkolj, Rastko Fišer, and Hinko Šolinc

Faculty of Electrical and Computer Engineering
Laboratory for Electromotor Drives
Tržaška 25, 61000 Ljubljana, Slovenia

INTRODUCTION

The aim of the article is to present the mathematical model of current distribution in rotor squirrel cage bars of the uniform air-gap three-phase induction machine, taking into account skin effect. In the analysis of electric machines, transformations can often play a useful role. The space rotating frame method is used in the generalised theory of electrical machines on which exists a large number of references[1,2]. In conventional machine analysis where only the fundamental component of the flux wave is considered, the d-q-0 transformation contribute a major simplification by changing the time-varying differential equations to constant coefficient differential equations for constant rotor speed (steady state), which are much easier to solve than the original ones. Unfortunately, with m.m.f. harmonics included, finding a transformation that simplify the analysis or facilitate the solution of the machine equations is a difficult task.

The basic problem is to determine the nominal value of induced current in each rotor bar to create the input data for numeric calculation of magnetic flux distribution under steady state load conditions. The simplification have been made and only the fundamental component of the flux wave is considered, neglecting the influence of space harmonics on m.m.f. in the air-gap. Other assumptions in the mathematical model are: uniform air-gap, neglecting magnetic saturation, rotor winding is symmetrical cage placed in N_r generally skew slots.

MATHEMATICAL MODEL OF SQUIRREL-CAGE MOTOR

Figure 1a shows the starting point for mathematical model of induction motor with squirrel-cage. Stator excitations represents three-phase symmetrical winding and rotor cage consists of N_r identical and equally spaced rotor bars connected with end ring segments. The ordering of the stator phases and rotor bars (phases) are shown in Figure 1b, where

Electric and Magnetic Fields, Edited by A. Nicolet
and R. Belmans, Plenum Press, New York, 1995

each winding is represented by its axis and the numbering of the windings is ordered in the direction of rotation.

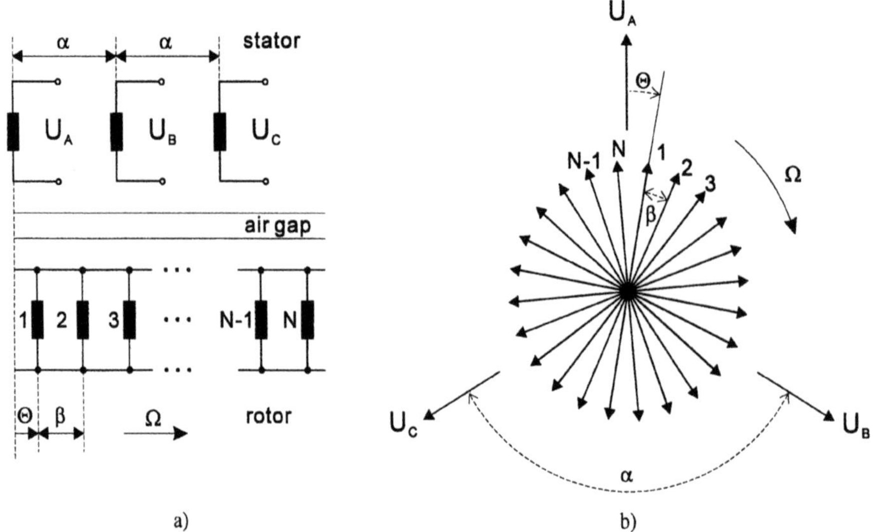

Figure 1. The stator and rotor system of axes

The voltage differential equations are formulated in "natural" reference frame. For symmetrical three-phase induction machine the following stator and rotor voltage equations can be obtained by physical considerations,

$$\bar{u}_s = R_s \bar{i}_s + \frac{d\bar{\psi}_s}{dt} \qquad \bar{u}_r = R_r \bar{i}_r + \frac{d\bar{\psi}_r}{dt} \tag{1}$$

where $\bar{u}_s, \bar{u}_r, \bar{i}_s, \bar{i}_r \bar{\psi}_s, \bar{\psi}_r$ are the stator and rotor voltage, current and flux linkage space vectors. If we express Equations (1) in the matrix form and split the total flux linkages into self and mutual flux linkages, the following equations can be written.

$$\begin{aligned} [U_s] &= ([R_s] + j\omega_s [L_{ss}]) \cdot [I_s] + j\omega_s [L_{sr}] \cdot [I_r] \\ [\,0\,] &= j\omega_r [L_{rs}] \cdot [I_s] + ([R_r] + j\omega_r [L_{rr}]) \cdot [I_r] \end{aligned} \tag{2}$$

The elements of resistance, self and mutual inductance matrix are calculated from construction parameters of stator and rotor lamella and stator three-phase winding[3]. Skin effect in deep rotor bar is taken into account as increasing value of resistance and slot leakage reactance as function of increasing slip value[4].

The complex form of Equations (2) and the dependence of the coefficients on rotor angular velocity are the main reasons for difficult calculations. Due to this reasons, transformation which simplify the equations must be made.

The mathematical model of current flow in each bar of squirrel cage rotor is based on space rotating frame transformation. Three phase stator variables are transformed to the d-q-0 frame fixed on the rotor. Such transformation is defined by Equations (3), (4) and (5), where f represents either stator voltage or current.

$$[C]=\begin{bmatrix}\cos\theta & \sin\theta & 1\\ \cos(\theta+\frac{2\pi}{3}) & \sin(\theta+\frac{2\pi}{3}) & 1\\ \cos(\theta+\frac{4\pi}{3}) & \sin(\theta+\frac{4\pi}{3}) & 1\end{bmatrix} \qquad [C]^{-1}=\frac{2}{3}\cdot\begin{bmatrix}\cos\theta & \cos(\theta+\frac{2\pi}{3}) & \cos(\theta+\frac{4\pi}{3})\\ \sin\theta & \sin(\theta+\frac{2\pi}{3}) & \sin(\theta+\frac{4\pi}{3})\\ \frac{1}{2} & \frac{1}{2} & \frac{1}{2}\end{bmatrix} \tag{3}$$

$$\left[f_{abc}\right]=[C]\cdot\left[f_{dq0}\right] \qquad \left[f_{dq0}\right]=[C]^{-1}\cdot\left[f_{abc}\right] \tag{4}$$

$$\theta=\Omega\cdot t=\left(\omega_s-\omega_r\right)\cdot t \tag{5}$$

CALCULATION OF MAGNETIC FIELD DISTRIBUTION

After solving the system of equations $[U]=[Z]\cdot[I]$ with Gaussian elimination method, the values of currents in each rotor bar (amplitude and phase) are known. This calculated values define the input data for excitations in further numeric computation. Magnetic field distribution was calculated as 2D non-linear magnetostatic by finite element software package MSC/EMAS[6].

Several cases from starting (blocked rotor, s=1, Figure 2), to breakdown torque ($s=s_{Tmax}$), rated torque ($s=s_n$, Figure 3) and no-load condition (s=0) of two pole-pairs induction motor operating conditions have been studied. The schematic distribution of stator and rotor currents is represented in Figures 2a, 3a. Figures 2b, 3b show magnetic field distribution in the cross-section of the motor and Figures 2c, 3c show the segment of the cross-section.

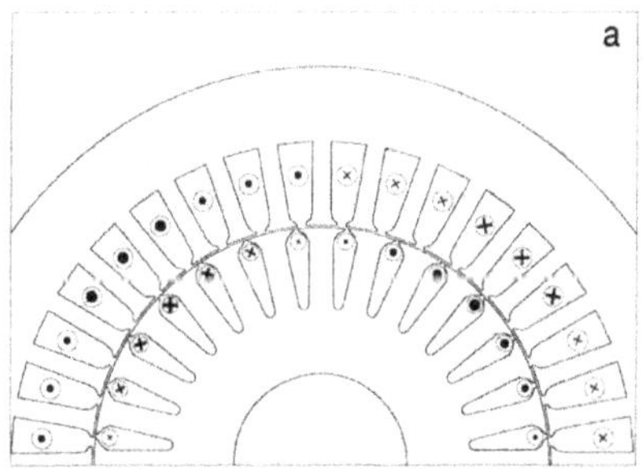

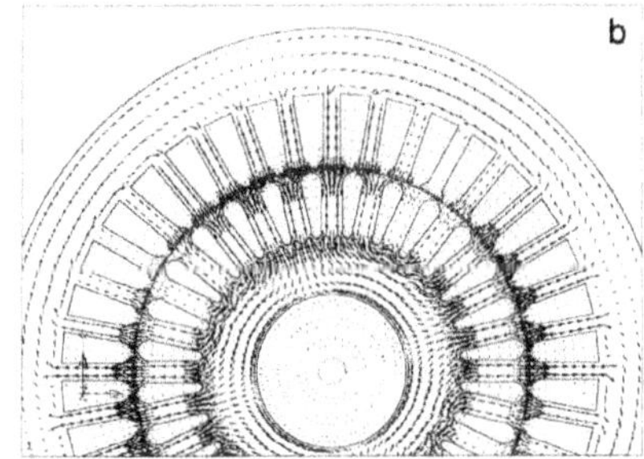

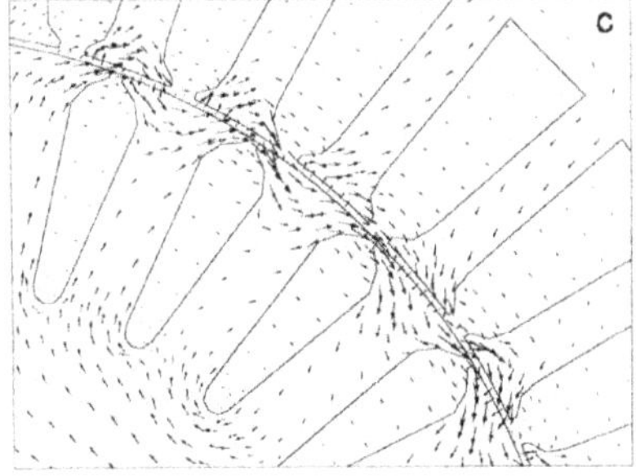

Figure 2.
a) Stator and rotor current distribution at s=1
b) Magnetic field distribution in the cross-section
c) Magnetic field distribution in the segment of the cross-section

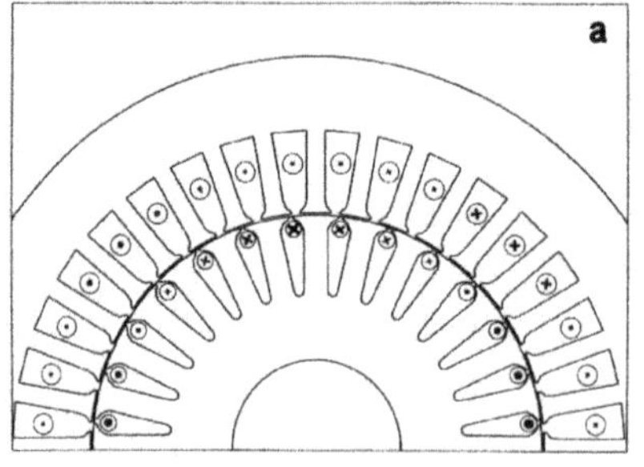

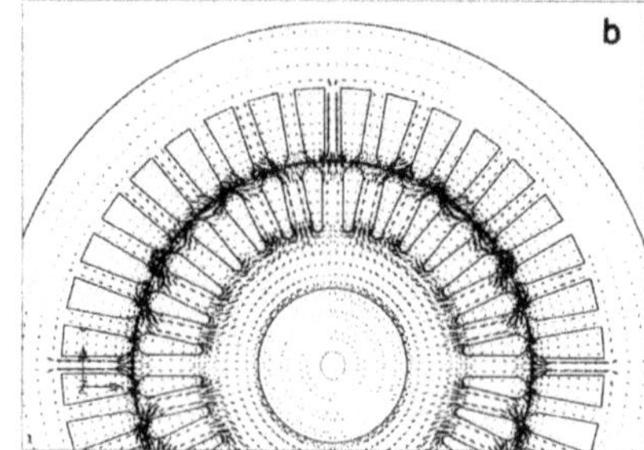

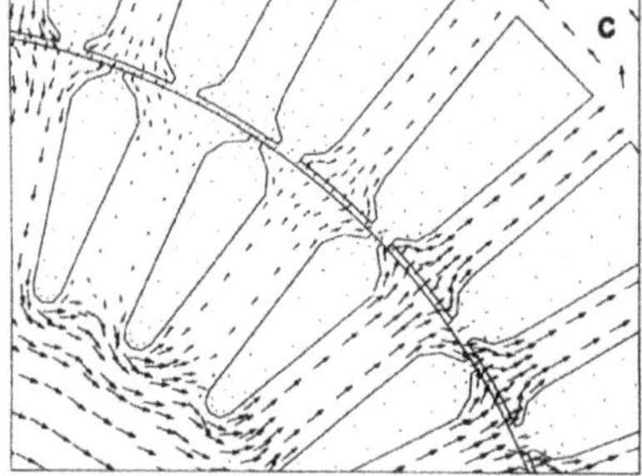

Figure 3.
a) Stator and rotor current distribution at $s=s_n$
b) Magnetic field distribution in the cross-section
c) Magnetic field distribution in the segment of the cross-section

CONCLUSIONS

The paper discusses method for the steady-state analysis of uniform air-gap symmetrically supplied three-phase induction machine. The equations have been established in natural and in rotor reference frame. The presented method of calculation of rotor bar currents allows us to determine current flow distribution for any slip value. An additional advantage of the method is the possibility of detecting local saturations in stator and rotor lamella, which can not be observed at no-load operating case, which is usually the reference for optimising the construction of stator winding. The method also makes possible to analyse the effects of rotor cage or stator winding damages on magnetic field distribution in the cross-section of the motor.

The presented model has been limited to the field of the basic harmonic. In further investigations the influence of other m.m.f. space harmonics will be incorporated in the equations and an algorithm for double squirrel-cage motor will be developed. Such motor construction provides the most complex variant which ensures full generality of the model. The calculation of stator and rotor currents as well as magnetic field distribution in the cross-section of the motor represents a practical tool for monitoring and investigations of different defects which can be obtained from harmonic analysis of frequency spectrum of stator current.

REFERENCES

1. P. C. Krause, "Analysis of Electric Machinery", McGraw-Hill Inc.,New York, 1986.
2. P. Jereb, Fundamental transformations at electric machines analysis, Electrotechnical Review vol.4, Ljubljana, Slovenia, 1979, 222-231.
3. R. Fišer, Influence of double squirrel-cage rotor construction parameters on induction motor torque characteristics, Master's thesis, Ljubljana, 1989.
4. H. R. Fudeh, C. M. Ong, Modeling and analysis of induction machines containing space harmonics, Part I, IEEE Transaction, Vol. PAS-102, No. 8, August 1983, 2608-2615.
5. A. Patyk, Application of forward/backward components to the analysis of failure states of the squirrel-cage motor, Archiv fuer Elektrotechnik 69 (1986), 429-437.
6. MSC/EMAS and MSC/XL Application Manuals, The MacNeal-Schwendler Corp., Los Angeles, 1993.

AUTOMATED OPTIMAL DESIGN OF A SMALL DC MOTOR WITH GLOBAL EVOLUTION STRATEGY AND FEM-MODEL

Kay Hameyer and Ronnie Belmans

Katholieke Universiteit Leuven, Dept. Elektrotechniek, Labo E.M.A.
B-3000 Leuven, Belgium

ABSTRACT

The design process of electromagnetic devices reflects an optimisation procedure. The construction and step by step optimisation of technical system in practice is a *trial and error*-process. In order to get an automated optimal design, numerical optimisation is necessary to achieve a well defined optimum. The application of automated optimal design of a non-linear magneto-static problem will be demonstrated in the paper. The characteristic features and behaviour of the numerical methods used are shown by minimising the overall material expenditure of a small dc-motor by optimising the rotor and stator shape.

INTRODUCTION

The design procedure of an electro-magnetic device may lead to sub optimal solutions because its success and effort strongly depends on the experience and individual parameters of the design engineer, as illustrated in fig. 1. To achieve faster design cycles, and to avoid the individual parameters, it is desirable to simulate the physical behaviour of the system by numerical methods. The increase of output power of an electromagnetic device by the raise of flux density is limited by the non-linear behaviour of the iron parts of the electromagnetic circuit. An improved output power can be achieved by optimising the non-linear magnetic circuit. A combination of evolution strategy and simulated annealing leads to a robust numerical optimisation method. Two dimensional field computation id done with standard finite element method using linear shape functions over triangular elements to approximate the vector potential.

STOCHASTIC OPTIMISATION

Search methods are known for their reliability, robustness and the easy treatment of the constraints. A complicated transformation into an unconstrained problem formulation is not necessary. Their main advantage is the insensibility to stochastic disturbance of the objective

Electric and Magnetic Fields, Edited by A. Nicolet
and R. Belmans, Plenum Press, New York, 1995

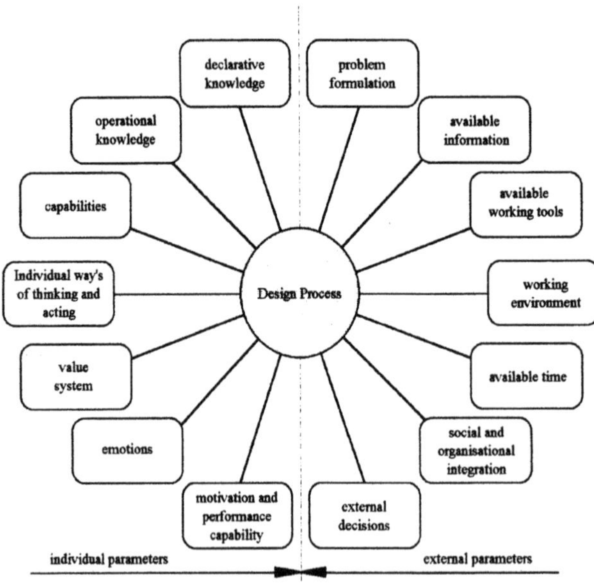

Figure 1. Parameters affecting the design process.

function caused by numerical evaluation. This insensibility results out of the non-deterministic search and disuse of derivatives of the objective function $Z(\mathbf{x}) = (x_1, x_2, \ldots, x_n)$. This function depends on all design parameters and represents the quality of the specific construction. Additional constraints limit the admissible parameter variation. Random based search methods only evaluate the objective function itself without use of derivatives. This problem is comparable to the search of a blind hiker for the mountain top only using his altimeter. The algorithm may get trapped in a local extremum and saddlepoints in the shape of the objective function are able to mislead the optimisation process to the wrong direction. To avoid the mentioned difficulties a combination of two search methods is used.

Here, random variations of the objective parameters x_i is done by the rules of evolution strategy. The introduction of the control parameter temperature T in the search process aims to avoid the system getting stuck in a local extremum. Barriers of height $\sim k_B T$, where k_B is the BOLZMANN constant, can be surmounted on the way to a better solution. The evolution strategy copies the natural principles mutation and selection (*survival of the fittest*) of biological evolution into the technical optimisation problem. The basic concept of the evolution strategy is found in the substitution of DARWIN's notation of fitness to the quality of a technical product. The driving force in the optimisation process is the repetition of mutation and selection in successive steps. In this optimisation example a so called (μ/ρ, λ) *comma* strategy has been chosen.

If μ is the number of parents, λ the number of offspring's or children and ρ a hereditary factor, the strategy can be described in the following way. ρ parent-vectors contribute to the creation of an offspring-vector. $1/\rho$ of the properties of one parent is transmitted to the child-vector. The mutation of the objective variables of an initial generation (parents) leads to a number of offspring's. The best offspring's are selected to form the next generation. In the optimisation process the repetition of mutation and selection of the objective variables leads from a temporary- to an improved solution. In order to transfer this procedure to an efficient optimisation method a self adaptive step length control is necessary. If step length is too small, the method will show a poor convergence rate. An improvement of quality will not be noticed. On the other hand, if mutation step length is too large, the algorithm results in pure random fluctuation. For more detailed information about different evolution strategies please refer to [1]

Mutation of an object is done with random additions to the object variables. For iteration step (k) the new variable $x_{C,i}^{(k)}$ is created with the parent variable $x_{P,i}^{(k)}$

$$x_{C,i}^{(k)} = x_{P,i}^{(k)} + \delta_{C,i}^{(k)}\ p_i^{(k)} \quad \text{with } i = 1(1)n \tag{1}$$

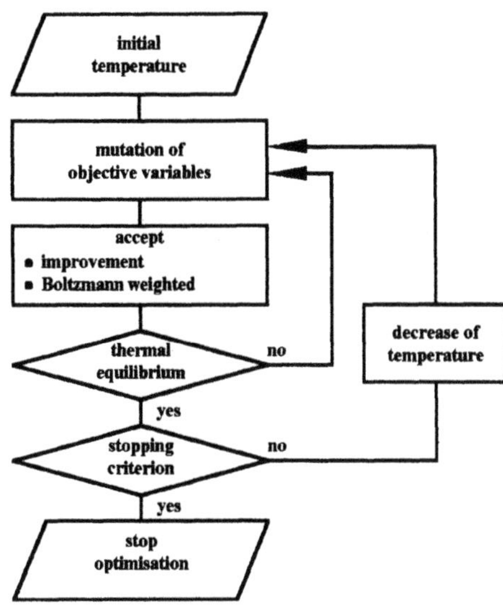

Figure 2. Block diagram of combined search algorithm.

To transmit the step width $\delta_{C,i}^{(k)}$ of the parent generation to the child generation, the step length of the randomly chosen parents is taken to build an average value. To obtain the mutation step width the before mentioned average value is multiplied by a step width factor α

$$\delta_{C,i}^{(k)} = \begin{cases} \delta_{C,i}^{(k)} \cdot \alpha & : i = 1(1)\lambda/2 \\ \delta_{C,i}^{(k)} \cdot 1/\alpha & : i = \lambda/2 + 1(1)\lambda \end{cases} . \quad (2)$$

The direction of the search process $p_i^{(k)}$ is created with a n-dimensional hyper sphere distributed random vector [2].

To get a global optimisation method the rules of simulated annealing [1,3] are combined with the evolution strategy. The substitution of the term energy in the idea of simulated annealing for the quality of the system leads to a robust optimisation technique. Figure 2 shows the block diagram of the optimisation algorithm used.

OPTIMISATION OF A DC MOTOR

The application of the mentioned stochastic search method is demonstrated by the optimisation of a permanent magnet excited dc motor. The objective is to minimise the overall material costs. Permanent magnet material, winding copper and iron lamination are taken into account. The quality function is formulated with the weighted sum of the different volumes of material multiplied with their cost / volume and a penalty term as:

$$Z(\mathrm{x}) = 10^{\left(\frac{cost(x) - cost_{max}}{cost_{max}}\right)} + penalty \quad (3)$$

The penalty term describes a weak form of a constraint to guaranty a minimum value of the torque T. The use of the term in the form:

$$penalty = \begin{cases} T < T_{min} & : \quad 10^{\left(\frac{T_{min} - T(x)}{T(x)}\right)} \\ T \geq T_{min} & : \quad 1 \end{cases} \quad (4)$$

allows the evaluation of the objective function even if the torque constraint is violated. Field computation is done with finite element method to obtain the field quantities. The torque is computed by integrating the MAXWELL stress tensor in the air gap region of the machine. Flux density depending rotor iron losses were taken into account at a

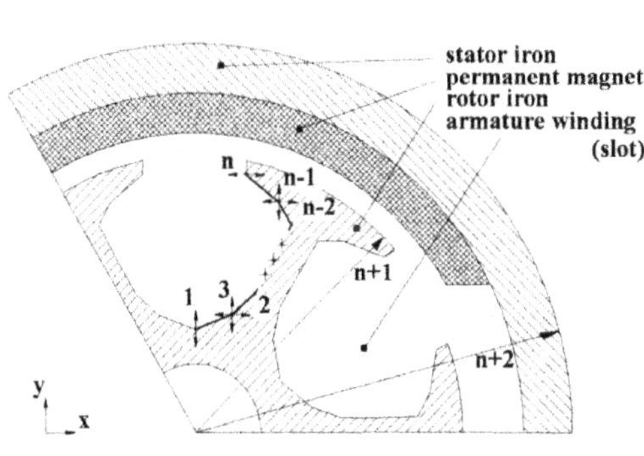

Figure 3. Motor geometry with objective variables.

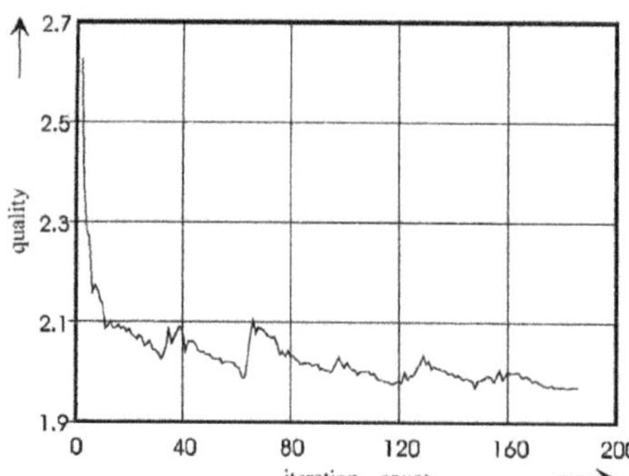

Figure 4. Quality versus iteration count.

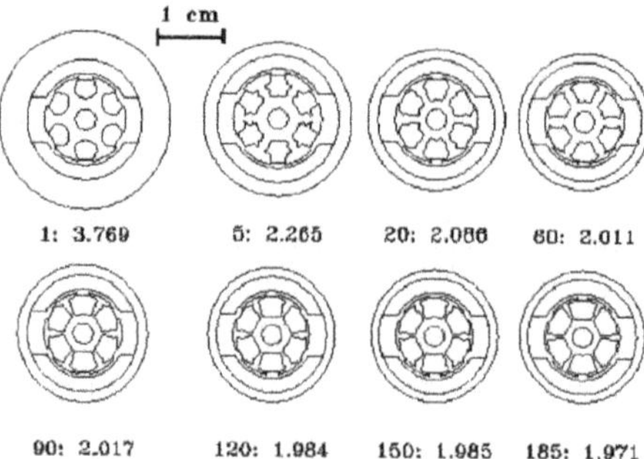

Figure 5. Initial, temporary and optimised shape.

rated speed of 200 RPM and subtracted from the air gap torque to form $T(x)$.

Figure 3 shows the shape of the dc motor to be optimised. The motor consists of two radial magnetised permanent magnet poles fixed to the inner radius of the stator lamination. The armature winding is placed into six rotor slots. The whole geometry is modelled by 15 design variables. The free parameters are the n/2 edges of the polygon describing the rotor slot contour and the outer dimensions of rotor and stator as indicated in fig. 3. The convergence rate of the optimisation process is plotted in fig. 4. The change in geometry, starting with a sub-optimal- via several temporary-solutions and finally the optimised shape of the dc motor is illustrated in fig. 5. It can be noticed, that the iron parts of the initial geometry are over dimensioned. The actual torque of this starting configuration was about 25% lower than the desired value of T_{min}. The final solution holds the torque recommended, which mainly is achieved by enlarging the winding copper volume about 20% compared to the initial geometry. Along the optimisation the overall volume of the motor was reduced about 38%.

CONCLUSION

A method of global minimisation of functions of continuous variables based on a combination of evolution strategy and simulated annealing was presented. The highly non-linear electromagnetic field of a small dc motor excited with permanent magnets has been optimised. Objective function evaluation was accomplished with finite element method using error estimation and adaptive mesh generation. This method guarantees the greatest possible facility in modelling and allows the optimisation without severe geometrical restrictions.

REFERENCES

1. Hameyer, K. and Kasper, M.: „Shape optimization of a fractional horsepower dc-motor by stochastic methods", ed. in: *Computer Aided Optimum Design of Structures III, Optimization of Structural Systems and Applications*, CMP, Elsevier Applied Science, London, New York, 1993, 15-30.
2. Schwefel, H.-P.: *Numerical optimization of computer models*, Wiley & Sons, Chichester 1981.
3. Kirkpatrick, S., Gelatt, C.D. and Vecchi, M.P.: „Optimization by Simulated Annealing", *Science 220:671-680, ed. in Neurocomputing*, Foundations of Research, (1988), 554-567.

Contributors

Aiello G.
Alfonzetti S.
Angeli M.
Anghel F.
Azzerboni B
Baland R.
Belmans R.
Benzarga D.
Besbes M.
Bialek W.
Bock J.P.
Bogdanov E.
Bossavit A.
Bottauscio O.
Bouissou S.
Bousbaine A.
Brandisky K.
Broche Ch.
Burais N.
Buret F.
Campolo M.
Cannistra G.
Cardelli E.
Chat-uthai C.
Chentouf A.
Chiampi M.
Chiarabaglio D.
Ciorba R.C.
Clark R.E.
Coco S.
Colamartino F.
Costa P.
Coulomb J.L.
Crappe M.
Crotti G.
Crutzen Y.
Cundev M.
Dableh J.H.
Dappen S.
De Weerdt R.
Domack S.
Dusan M.
Fantechi S.
Féliachi M.
Feng X.
Ferkolj S.
Findlay R.D.
Fiser R.
Foggia A.
Fouladgar J.
Freeman E.M.
Fujiwara K.
Gasmi N.
Gaspard J.Y.
Genon A.
Georgiev Zh.D.
Geri A.
Gorican B.
Gross P.W.
Guba R.
Hadji-Minaglou J.R.
Hadrys W.
Hameyer K.
Hamler A.
Hanitsch R.
Harmer K.
Hatziathanassiou V.
Hemead E.S.
Henneberger G.
Henninger P.
Henrotte F.
Howe D.
Hribernik B.
Idikayi M.
Ionesco B.
Ivanes M.
Jesenik M.
Johansson T.B.
Kamiya Y.
Kauffmann J.M.
Kedadra A.
Kladas A.G.
Kong L.
Kotiuga P.R.
Krähenbühl L.
Krawczyk A.
Kreca B.
Küppers S.
Labridis D.
Laporte B.
La Rosa M.
Le Dour O.
Legros W.
Leschi D.
Li Y.M.
Liu M.Z.
Lobry J.
Lombard P.
Louai F.Z.
Low W.F.
Lowther D.
MacKay D.K.
Mai W.
Mailfert A.
Marchand C.
Maun J.C.
McCormick M.
Mellara B.
Mellor P.H.
Mertens R.
Midgley G.W.
Minenna M.
Miraoui A.
Mitchell J.K.
Morabito F.C.
Moraru A.
Munteanu C.
Nabeta S.
Najdenkoski K.
Nakano M.
Nakata T.
Nava E.
Negro G.
Nicolas A.
Nicolas L.
Nicolet A.
Nilssen R.
Nysveen A.
Ogusa S.
Ohashi H.
Omekanda A.
Onuki T.
Opaterny J.
Pahner U.
Petkovska L.
Pichon L.
Piriou F.
Plejic M.
Ramirez J.A.
Razek A.
Remacle J.F.
Ren Z.
Renata F.
Renz W.
Reuber C.
Reyne G.
Riley C.D.
Rizzo M.
Sakellaris I.
Salerno N.
Sande G.
Santinelli M.
Santini E.
Savini A.
Savov V.
Schmitz M.
Sek A.
Sekkak A.
Simion E.
Skoczkowski T.
Solinc H.
Sonato P.
Stoilkov V.
Stranges N.
Stumberger B.
Sylos Labini M.
Takahashi N.
Takami H.
Tarczynski W.
Tegopoulos J.A.
Thollon F.
Topa V.
Trlep M.
Tuinman E.
Turowski J.
Umé M.
Van Dessel M.
Vasconcelos J.A.
Veca G.M.
Vekara T.
Vester M.
Vinsard G.
Wagner J.
Webb J.
Witczak P.
Yamamura T.
Yao-bi J.L.
Yokota D.
Yoshida K.
Zhu H.L.
Zollino G.

INDEX

GPSR Compliance
The European Union's (EU) General Product Safety Regulation (GPSR) is a set of rules that requires consumer products to be safe and our obligations to ensure this.

If you have any concerns about our products, you can contact us on

ProductSafety@springernature.com

In case Publisher is established outside the EU, the EU authorized representative is:

Springer Nature Customer Service Center GmbH
Europaplatz 3
69115 Heidelberg, Germany

www.ingramcontent.com/pod-product-compliance
Ingram Content Group UK Ltd.
Pitfield, Milton Keynes, MK11 3LW, UK
UKHW051523300726
14060UKWH00013B/652